Advances in Intelligent Systems and Computing

Volume 416

Series editor

Janusz Kacprzyk, Polish Academy of Sciences, Warsaw, Poland
e-mail: kacprzyk@ibspan.waw.pl

About this Series

The series "Advances in Intelligent Systems and Computing" contains publications on theory, applications, and design methods of Intelligent Systems and Intelligent Computing. Virtually all disciplines such as engineering, natural sciences, computer and information science, ICT, economics, business, e-commerce, environment, healthcare, life science are covered. The list of topics spans all the areas of modern intelligent systems and computing.

The publications within "Advances in Intelligent Systems and Computing" are primarily textbooks and proceedings of important conferences, symposia and congresses. They cover significant recent developments in the field, both of a foundational and applicable character. An important characteristic feature of the series is the short publication time and world-wide distribution. This permits a rapid and broad dissemination of research results.

More information about this series at http://www.springer.com/series/11156

Susumu Kunifuji · George Angelos Papadopoulos
Andrzej M.J. Skulimowski · Janusz Kacprzyk
Editors

Knowledge, Information and Creativity Support Systems

Selected Papers from KICSS'2014—9th International Conference, held in Limassol, Cyprus, on November 6–8, 2014

Editors
Susumu Kunifuji
Japan Advanced Institute for Science and Technology (JAIST)
Kanazawa
Japan

George Angelos Papadopoulos
Department of Computer Science
University of Cyprus
Nicosia
Cyprus

Andrzej M.J. Skulimowski
Department of Automatic Control and Biomedical Engineering
AGH University of Science and Technology
Kraków
Poland

Janusz Kacprzyk
Systems Research Institute
Polish Academy of Sciences
Warsaw
Poland

ISSN 2194-5357 ISSN 2194-5365 (electronic)
Advances in Intelligent Systems and Computing
ISBN 978-3-319-27477-5 ISBN 978-3-319-27478-2 (eBook)
DOI 10.1007/978-3-319-27478-2

Library of Congress Control Number: 2015957788

Printed on acid-free paper

This Springer imprint is published by SpringerNature
The registered company is Springer International Publishing AG Switzerland

Message from the Conference Chairs

This volume consists of a number of selected papers that were presented at the 9th International Conference on Knowledge, Information and Creativity Support Systems (KICSS 2014) in Limassol, Cyprus, after they were substantially revised and extended.

Since the first KICSS in Ayutthaya in 2006, the International Conference on Knowledge, Information and Creativity Support Systems aims to facilitate technology and knowledge exchange between international researchers and scholars in the field of knowledge science, information science, system science and creativity support systems. After Ayutthaya (2006) KICSS was successfully held in Nomi (2007), Hanoi (2008), Seoul (2009), Chiang Mai (2010), Beijing (2011), Melbourne (2012) and Krakow (2013).

Supporting creativity is becoming one of the most challenging areas in the twenty-first century, with research on creativity support systems and relevant fields being emerging. KICSS 2014 covered all aspects of knowledge management, knowledge engineering, intelligent information systems, and creativity in an information technology context, including computational creativity and its cognitive and collaborative aspects.

Putting together KICSS 2014 was a team effort. First of all, we would like to thank the authors of all submitted papers. Furthermore, we would like to express our gratitude to the program committee and to all external reviewers, who worked very hard on reviewing papers and providing suggestions for their improvements.

This volume consists of 26 regular papers and 19 short papers. Moreover, we are proud and thankful to have two renowned keynote speakers at the 2014 edition of KICSS:

- Boris Stilman on "From Fighting Creative Wars to Making Ordinary Discoveries"
- Andrzej M.J. Skulimowski on "The Art of Anticipatory Decision Making" as well as a tutorial by Boris Stilman on "Discovering the Discovery of the No-Search Approach".

Finally, we would like to thank our sponsors, Springer, JAIST, JCS, Austrian Airlines, the Cyprus Tourism Organisation and the University of Cyprus for their support of this conference.

We hope that you will find the contents of this volume interesting and thought-provoking and that it will provide you with a valuable opportunity to share ideas with other researchers and practitioners from institutions around the world.

George A. Papadopoulos
KICSS 2014 Conference Chair
University of Cyprus,
Cyprus

Susumu Kunifuji
KICSS 2014 Conference Chair
JAIST,
Japan

Andrzej M.J. Skulimowski
KICSS 2014 Conference Chair
P&BF and AGH,
Poland

Conference Organizer

The University of Cyprus

In Cooperation with

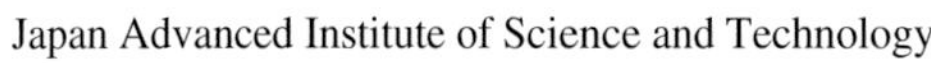
Japan Advanced Institute of Science and Technology

Japan Creativity Society

Financially Supported by

Cyprus Tourism Organisation

Patrons

Springer

Sponsors

Austrian Airlines

Keynote Speakers

Boris Stilman

Dr. Stilman is currently a Professor of Computer Science at the University of Colorado Denver (UC Denver), USA, and the Chairman and CEO at STILMAN Advanced Strategies (STILMAN), USA. Boris Stilman received MS in Mathematics from Moscow State University (MGU), USSR in 1972 and two Ph.Ds. in Electrical Engineering and Computer Science from National Research Institute for Electrical Engineering (VNIIE), Moscow, USSR in 1984. During 1972–1988, in Moscow, he was involved in the research project PIONEER led by a former World Chess Champion, Professor Mikhail Botvinnik. The goal of the project was to discover and formalize an approach utilized by the most advanced chess experts in solving chess problems almost without search. While program PIONEER had never played complete chess games, it solved a number of complex endgames and positions from the games of World Chess Champions. Based on these experiences over a number of years, in Moscow, Dr. Stilman developed experimental and mathematical foundations of the new approach to search problems in Artificial Intelligence. In 1990–1991, while at McGill University, Montreal, Canada, based on this approach, he originated Linguistic Geometry (LG), a new theory for solving abstract board games. LG allows us to overcome combinatorial explosion by changing the paradigm from search to construction (from analysis to synthesis). Since 1991, Dr. Stilman has been developing the theory and applications of LG at UC Denver. A leap in the development of LG was made in 1999, when he (with a group of scientists and engineers) founded STILMAN, LLC. A growing number of applications of LG developed at STILMAN have passed comprehensive testing and are currently being transitioned to the real-world command and control systems in the USA and abroad. Since 2010, Dr. Stilman is investigating the structure of the Primary Language of the human brain (following J. von Neumann). According to his hypothesis the Primary Language includes at least two major components critical for humanity, LG and the Algorithm of Discovery. Dr. Stilman has published several books (including "Linguistic Geometry: From Search to Construction"), contributions to books, and over 200 research papers. He is a recipient of numerous R&D awards, including the top research awards at University of Colorado, substantial grants from the US government agencies such as major multiple awards from DARPA, US Department of Energy, US Army, US Navy, US Air Force, etc.; Ministry of Defence of UK; from the world leading defence companies such as Boeing (USA), Rockwell (USA), BAE Systems (UK), SELEX/Finmeccanica (Italy-UK) and Fujitsu (Japan). More information about Dr. Stilman, history of LG and projects (including several narrated movies) can be found at www.stilman-strategies.com.

Andrzej M.J. Skulimowski

Prof. Skulimowski is a Professor and Director of the Decision Sciences Laboratory at the Department of Automatic Control and Biomedical Engineering, AGH University of Science and Technology. Since 1995, he has also been the President of the International Progress and Business Foundation, Kraków, where he has led over 80 research, consulting and policy support projects within the EU Framework Programs, ESTO, ETEPS, Interreg, LLP, ERDF, GTD and other programs. He graduated in Electronics at AGH University, and in Mathematics from the Jagiellonian University, Kraków, Poland. He was awarded a Ph.D. degree in Automatic Control with honors in 1985 and a D.Sc. degree in Operations Research in 1997, both from AGH University. He was a postdoctoral fellow at the Institutes for Automatic Control and for Communication Technology, both at the ETH, Zurich, Switzerland (1987–1989), and a visiting professor at the Institute of Information Management at the University of St. Gallen (1990–1995). His main field of expertise is multicriteria decision analysis, cognitive aspects of decision and creativity support systems, foresight, R&D policy, and forecasting. He is the author and editor of nine books and over 200 scholarly papers on these topics. He invented the anticipatory networks, coined the term mHealth and defined its major challenges, and is one of the pioneers of foresight support systems and advanced approaches to AI foresight. He was the General Chair of the 8th International Conference on Knowledge, Information and Creativity Support Systems (KICSS) held in Kraków, Poland, in November 2013.

Conference Topics

Anticipatory networks, systems, and decisions
Autonomous creative systems
Cognitive foundations of knowledge
Cognitive and psychological issues in creativity research
Cognitive foundations of knowledge
Collaborative activities in Living Labs
Collaborative idea generation and creativity
Collaborative knowledge creation
Collaborative working environments fostering creativity
Complex system modelling
Computer supported creativity
Creative approaches to model technological evolution
Creative business models
Creative conflict resolution
Creative coordination and cooperation mechanisms
Creative decision processes
Creative interaction techniques
Creative model building
Creative reasoning and decision making
Creative research environments and their performance
Creative social computing
Creative visualisation of data and knowledge
Creativity in finance
Creativity in augmented reality
Creativity in health care
Creativity in mobile applications
Creativity in social media
Creativity in the Cloud
Creativity measurement
Creativity support systems
Creativity transfer and stimulation
Creativity versus rationality
Creativity-enhancing interfaces
Creativity-oriented information system architectures
Decision sciences
Decision support systems (DSS)
Discovering opportunities, threats and challenges
Foresight support systems (FSS)
Future Internet and knowledge-based society
Future exploration and modelling
Future perspectives of knowledge, information, and creativity support
Game-theoretical aspects of knowledge

General creative systems (GCS)
Group recommendation, and advice
Heuristics and general problem solving
Identifying real options in complex business environments
Information fusion
Information quality
Intelligent analysis of Big Data
Knowledge extraction, creation and acquisition
Knowledge in multi-agent systems
Knowledge integration
Knowledge management in business, industry and administration
Knowledge representation and reasoning
Knowledge verification and validation
Living Lab support systems (LLSS)
Machine learning and creativity
Malicious creativity in the web, its discovery and remedy
Mathematical models of creative processes
Multi- and interdisciplinary approaches in creativity research
Multicriteria decision making
Natural language analysis
Non-monotonic reasoning
Ontology creation and management
Open innovation
Organizational learning
Preference modelling
Reasoning about knowledge
Recommender systems
Scientific information management
Search for a compromise in multicriteria decision making and collaborative games
Social Computing
Social factors of collaborative creativity
Software-based stimulation of creativity
Supervised and semi-supervised learning
Trust modelling
Uncertainty modelling
Virtual environment design
Visual Analytics and Intelligent User Interfaces
Web intelligence tools
World models

KICSS 2014 Statistics

- For KICSS Main track: of the 50 submitted papers, 29 were accepted as regular papers (acceptance rate 0.56 for regular papers), and an additional 10 were accepted as short papers.
- For KICSS Secondary track (short papers, research-in-progress reports): of the 11 submitted papers, 10 were accepted as short papers.
- This post-proceedings volume includes 26 regular papers and 19 short papers
- Number of countries in all submissions: 34

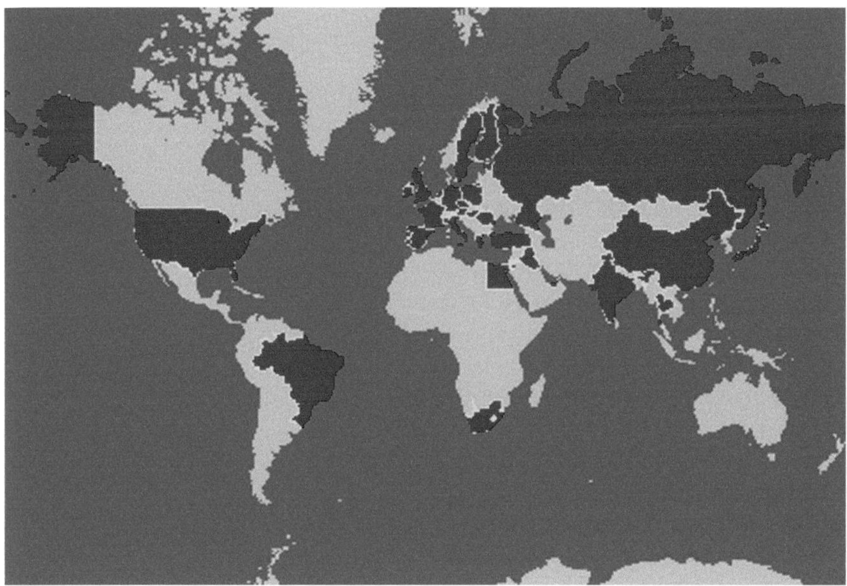

- Number of reviews for all submissions: 178
- External reviewers: 5
- External reviews: 8

Committees and Reviewers

Conference Chairs
Susumu Kunifuji, JAIST, Japan
George Angelos Papadopoulos, University of Cyprus, Cyprus
Andrzej M.J. Skulimowski, P&BF and AGH, Poland

Local Organizing Chair
George Angelos Papadopoulos, University of Cyprus, Cyprus

Award Chairs
Susumu Kunifuji, JAIST, Japan
Thanaruk Theeramunkong, Thammasat University, Thailand

Invited Session Chairs
Tomoko Kajiyama, JAIST, Japan
Vincent Cheng-Siong Lee, Monash University, Australia
Takaya Yuizono, JAIST, Japan

Program Committee
Ateet Bhalla, Oriental Institute of Science and Technology, India
Chi-Hung Chi, CSIRO, Australia
João Clímaco, University of Coimbra, Portugal
Simon Colton, University of London, UK
Eugene Dubossarsky, Presciient, Australia
Mark Embrechts, RPI, USA
Ali Eydgahi, Eastern Michigan University, USA
John Garofalakis, University of Patras, Greece
Tessai Hayama, Kanazawa Institute of Technology, Japan
Hidehi Hayashi, Naruto University of Education, Japan
Hidehiko Hayashi, Naruto University of Education, Japan
Christopher Hinde, Loughborough University, UK
Josef Jablonsky, University of Economics Prague, Czech Republic
Ivan Jelinek, Czech Technical University, Czech
Janusz Kacprzyk, Systems Research Institute—PAS, Poland
Hideaki Kanai, JAIST, Japan
Takashi Kawaji, Ohtsuki City College, Japan
Takahiro Kawaji, Ohtsuki City Colledge, Japan
Thomas Koehler, TU Dresden, Germany
Paul Kwan, University of New England, USA
Vincent C.S. Lee, Monash University, Australia
Antoni Ligeza, AGH, Poland
Ahmad Lotfi, Nottingham Trent University, UK
Akinori Minaduk, Kushiro Prefecture University, Japan
Motoki Miura, Kyushu Institute of Technology, Japan
Kazunori Miyata, JAIST, Japan
David C. Moffat, Glasgow Caledonian University, UK
Anna Mura, University Pompeu Fabra, Spain
Toyohisa Nakada, Niigata University of Interenational and Information Studies, Japan
Kazushi Nishimoto, JAIST, Japan
Maciej Nowak, University of Economics Prague, Czech Republic
Kok-Leong Ong, Deakin University, Australia
Francois Pachet, CSL Sony Paris, France
Robert Pearson
Przemyslaw Pukocz, AGH, Poland

Pawel Rotter, AGH, Poland
Jose L. Salmeron, University Pablo de Olavide, Spain
Jagannathan Sarangapani, Missouri University of Science and Technology, USA
Sheng Chai, Queen's University, Canada
Hsu-Shih Shih, Tamkang University, Taiwan
Mikifumi Shikida, JAIST, Japan
Marcin Skowron, Austrian Research Institute for Artificial Intelligence, Austria
Johan Suykens, K.U. Leuven, ESAT-SCD, Belgium
Ryszard Tadeusiewicz, AGH, Poland
I-Hsien Ting, National University of Kaohsiung, Taiwan
Brijesh Verma, Central Queensland University, Australia
Yongui Wang, Nanjiing University, China
Michal Wozniak, Wroclaw University of Technology, Poland
Fan Wu, National Chung-Cheng University, Taiwan
Takashi Yoshino, Wakayama University, Japan
Atsuo Yoshitaka, Japan Advanced Institute of Science and Technology, Japan
Takaya Yuizono, JAIST, Japan
John Zeleznikow, Victoria University, Australia
Songmao Zhang, Chinese Academy of Sciences, China
Constantin Zopounidis, Technical University of Crete, Greece

External Reviewers

Antonio Carmona, University Pablo de Olavide, Spain
Juan Manuel Berbel, University Pablo de Olavide, Spain
Jose Luis Barbero, University Pablo de Olavide, Spain

Contents

The Algorithm of Discovery: Making Discoveries on Demand 1
Boris Stilman

The Art of Anticipatory Decision Making 17
Andrzej M.J. Skulimowski

On Modeling Human-Computer Co-Creativity 37
Oliver Hoffmann

Comparison of the Two Companies' Staff by the Group Idea-Marathon Training on Creativity and Motivation 49
Takeo Higuchi

Using Appropriate Context Models for CARS Context Modelling 65
Christos Mettouris and George A. Papadopoulos

Computational Creativity for Intelligence Analysis 81
Robert Forsgren, Peter Hammar and Magnus Jändel

Bus Scheduling in Dynamical Urban Transport Networks with the use of Genetic Algorithms and High Performance Computing Technologies 97
V.A. Shmelev, A.V. Dukhanov, K.V. Knyazkov and S.V. Ivanov

MCDA and LCSA—A Note on the Aggregation of Preferences 105
João Carlos Namorado Clímaco and Rogerio Valle

A Model for Managing Organizational Knowledge in the Context of the Shared Services Supported by the E-Learning. 117
Agostinho Sousa Pinto and Luís Amaral

Online Communities of Practice: Social or Cognitive Arenas? 131
Azi Lev-On and Nili Steinfeld

Knowledge Extraction and Annotation Tools to Support Creativity at the Initial Stage of Product Design: Requirements and Assessment 145
Julia Kantorovitch, Ilkka Niskanen, Anastasios Zafeiropoulos, Aggelos Liapis, Jose Miguel Garrido Gonzalez, Alexandros Didaskalou and Enrico Motta

Web Board Question Answering System on Problem-Solving Through Problem Clusters 161
Chaveevan Pechsiri, Onuma Moolwat and Rapepun Piriyakul

A PHR Front-End System with the Facility of Data Migration from Printed Forms 177
Atsuo Yoshitaka, Shinobu Chujyou and Hiroshi Kato

Investigating Unit Weighting and Unit Selection Factors in Thai Multi-document Summarization 193
Nongnuch Ketui and Thanaruk Theeramunkong

On the Use of Character Affinities for Story Plot Generation 211
Gonzalo Méndez, Pablo Gervás and Carlos León

Vicarious: A Flexible Framework for the Creative Use of Sensed Biodata 227
Paul Tennent, Joe Marshall, Brendan Walker, Paul Harter, Steve Benford and Tony Glover

Application of Rough Sets in *k* Nearest Neighbours Algorithm for Classification of Incomplete Samples 243
Robert K. Nowicki, Bartosz A. Nowak and Marcin Woźniak

An Application of Firefly Algorithm to Position Traffic in NoSQL Database Systems 259
Marcin Woźniak, Marcin Gabryel, Robert K. Nowicki and Bartosz A. Nowak

Length of Hospital Stay and Quality of Care 273
José Neves, Vasco Abelha, Henrique Vicente, João Neves and José Machado

Bio-inspired Hybrid Intelligent Method for Detecting Android Malware 289
Konstantinos Demertzis and Lazaros Iliadis

Predicting the Impact of Advertisements on Web Pages Aesthetic Impressions 305
Gianni Fenu, Gianfranco Fadda and Lucio Davide Spano

OntOSN—An Integrated Ontology for the Business-Driven Analysis of Online Social Networks 317
Richard Braun and Werner Esswein

Educators as Game Developers—Model-Driven Visual Programming of Serious Games 335
Niroshan Thillainathan and Jan Marco Leimeister

First Year Students' Algorithmic Skills in Tertiary Computer Science Education 351
Piroska Biró, Mária Csenoch, Kálmán Abari and János Máth

Implementing a Social Networking Educational System for Teachers' Training 359
Stavros Pitsikalis and Ilona-Elefteryja Lasica

Virtual Environment for Creative and Collaborative Learning 371
Anna Bilyatdinova, Andrey Karsakov, Alexey Bezgodov and Alexey Dukhanov

A Semiotic and Cognitive Approach to the Semantic Indexation of Digital Learning Resources 383
Françoise Greffier and Federico Tajariol

The Significance of 'Ba' for the Successful Formation of Autonomous Personal Knowledge Management Systems 391
Ulrich Schmitt

The Significance of Memes for the Successful Formation of Autonomous Personal Knowledge Management Systems 409
Ulrich Schmitt

Computer Creativity in Games—How Much Knowledge Is Needed? 421
David C. Moffat and Paul Hanson

When Engineering and Design Students Collaborate: The Case of a Website Development Workshop 431
Meira Levy, Yaron Shlomi and Yuval Etzioni

Knowledge-Managing Organizational Change Effectors, Affectors and Modulators in an Oil and Gas Sector Environment 439
Anthony Ayoola

From Computational Creativity Metrics to the Principal Components of Human Creativity 447
Pythagoras Karampiperis, Antonis Koukourikos and George Panagopoulos

A Social Creativity Support Tool Enhanced by Recommendation Algorithms: The Case of Software Architecture Design 457
George A. Sielis, Aimilia Tzanavari and George A. Papadopoulos

The Approach to the Extension of the CLAVIRE Cloud Platform for Researchers' Collaboration 467
A.V. Dukhanov, E.V. Bolgova, A.A. Bezgodov, L.A. Bezborodov and A.V. Boukhanovsky

iDAF-drum: Supporting Practice of Drumstick Control by Exploiting Insignificantly Delayed Auditory Feedback 483
Kazushi Nishimoto, Akari Ikenoue and Masashi Unoki

Novel Methods for Analyzing Random Effects on ANOVA and Regression Techniques 499
Gergely Bencsik and Laszlo Bacsardi

nVidia CUDA Platform in Graph Visualization 511
Ondrej Klapka and Antonin Slaby

A Method for Opinion Mining of Coffee Service Quality and Customer Value by Mining Twitter 521
Shu Takahashi, Ayumu Sugiyama and Youji Kohda

Using Wiki as a Collaboration Platform for Software Requirements and Design 529
Irit Hadar, Meira Levy, Yochai Ben-Chaim and Eitan Farchi

Enhancing Software Architecture via a Knowledge Management and Collaboration Tool 537
Sofia Sherman, Irit Hadar, Meira Levy and Naomi Unkelos-Shpigel

Collaborative Requirement Prioritization for an E-Recruitment Platform for Qualified but Disadvantaged Individuals 547
Ahmet Suerdem and Basar Oztaysi

Text Comparison Visualization from Two Information Sources by Network Merging and Integration 557
Ryosuke Saga

Collective Knowledge and Creativity: The Future of Citizen Science in the Humanities 565
Milena Dobreva

Aligning Performance Assessments with Standards: A Practical Framework for Improving Student Achievement in Vocational Education 575
Metwaly Mabed and Thomas Köhler

Author Index 587

The Algorithm of Discovery: Making Discoveries on Demand

Boris Stilman

Abstract According to our hypothesis the Algorithm of Discovery should be an evolutionary component of the Primary Language of the human brain (as introduced by J. von Neumann in 1957). In our research we identified two such components, Linguistic Geometry (LG), and the Algorithm of Discovery. We suggested that both components are mental realities "hard-wired" in the human brain. LG is a formal model of human reasoning about armed conflict, an evolutionary product of millions of years of human warfare. In this paper we focus on discovering the Algorithm of Discovery, the foundation of all the discoveries throughout the history of humanity. This Algorithm is based on multiple thought experiments, which manifest themselves and are controlled by the mental visual streams. This paper reports results of our investigation of the major components of the Algorithm of Discovery with special emphasis on constructing a series of models and mosaic reasoning. Those approaches are demonstrated briefly on discoveries of the No-Search Approach in LG, the structure of DNA, and the theory of Special Relativity.

Keywords Linguistic geometry · Primary language · Artificial intelligence · Algorithm of discovery · Game theory · Mosaic reasoning

What if discoveries are produced routinely as an output of computer programs? What a leap this would mean for humanity? Approaching this time of making discoveries on demand is the purpose of our efforts.

More than 50 years passed since J. von Neumann hypothesized existence of the Primary Language [43]. Unfortunately, the nature of this language is still unknown. Our hypothesis is that the Primary Language is a collection of major algorithms crucial for survival and development of humanity, the underlying "invisible"

B. Stilman (✉)
University of Colorado Denver, Denver, CO, USA
e-mail: Boris.Stilman@UCDenver.edu; boris@stilman-strategies.com

B. Stilman
STILMAN Advanced Strategies, Denver, CO, USA

S. Kunifuji et al. (eds.), *Knowledge, Information and Creativity Support Systems*, Advances in Intelligent Systems and Computing 416,
DOI 10.1007/978-3-319-27478-2_1

foundation of all the modern languages and sciences. We suggested that one of the components of the Primary Language is Linguistic Geometry (LG), a type of game theory [1–4, 11, 12, 14–27, 36–41] that allows us to solve classes of adversarial games of practical scale and complexity. It is ideally suited for problems that can be represented as abstract board games, for example, military decision aids, intelligent control of unmanned vehicles, simulation-based acquisition, high-level sensor fusion, robotic manufacturing and more. The advantage of LG is that it provides extraordinarily fast and scalable algorithms to find the best strategies for concurrent multi-agent systems. Also, unlike other gaming approaches, the LG algorithms permit modeling a truly intelligent enemy. LG is applicable to the non-zero-sum games and to the games with incomplete information (i.e., imperfect sensors, weather, enemy deception, etc.).

We suggested [28] that every human brain "speaks" the LG language, though, only well trained commanders and, especially, advanced strategists are able to utilize it to full capacity. Most importantly, they are able to translate from the LG language, i.e., from the Primary Language, into the natural languages to describe strategies in the spoken language terms.

1 Towards Ordinary Discoveries

In our research on revealing other components of the Primary Language, besides LG, we assumed that they look like LG in some respects. Our contention is that the hypothetical Algorithm of Discovery must be one of such components. In a number of papers, we have been developing a hypothesis that there is a universal Algorithm of Discovery driving all the innovations and, certainly, the advances in all sciences [29–35]. All the human discoveries from mastering fire more than a million years ago to understanding the structure of our Solar System to inventing airplane to revealing the structure of DNA to mastering nuclear power utilized this algorithm. The Algorithm of Discovery should be a major ancient item "recorded" in the Primary Language due to its key role in the development of humanity. This line of research involved investigating past discoveries and experiences of construction of various new algorithms, beginning from those, which we were personally involved in [14–39, 42–44].

Thought experiments allow us, by pure reflection, to draw conclusions about the laws of nature [5]. For example, Galileo before even starting dropping stones from the Tower in Pisa used pure imaginative reasoning to conclude that two bodies of different masses fall at the same speed. The Albert Einstein's thought experiments that inspired his ideas of the special and general relativity are known even better [6, 9, 13]. The efficiency and the very possibility of thought experiments show that our mind incorporates animated models of the reality, e.g., laws of physics, mathematics, human activities, etc. Scientists managed to decode some of the human mental images by visualizing their traces on the cortex [5]. It was shown that when we imagine a shape "in the mind's eye", the activity in the visual areas of the brain sketches the contours of the imagined object; thus, mental images have the analogical nature. It appears that we simulate the laws of nature by physically reflecting the reality in our brain. The human species and even animals would have had difficulty to survive without even minimal "understanding" of the laws of environment. Over the course of evolution and during development of every organism, our nervous system learns to comprehend its environment, i.e., to "literally take it into ourselves" in the form of mental images, which is a small scale reproduction of the laws of nature. Neuropsychologists discovered that "we carry within ourselves a universe of mental objects whose laws imitate those of physics and geometry" [5]. In [28], we suggested that we also carry the laws of the major human relations including the laws of optimal warfighting. The laws of nature and human relations manifest themselves in many different ways. However, the clearest manifestation is in perception and in action. For example, we can say that the sensorimotor system of the human brain "understands kinematics" when it anticipates the trajectories of objects. It is really fascinating that these same "laws continue to be applicable in the absence of any action or perception when we merely imagine a moving object or a trajectory on a map" [5]. This observation, of course, covers actions of all kinds of objects, natural and artificial. Scientists have shown that the time needed to rotate or explore these mental images follows a linear function of the angle or distance traveled as if we really traveled with a constant speed. They concluded that "mental *trajectory* imitates that of a physical object" [5].

Our main hypothesis is that the Algorithm of Discovery is based not on formal logic but on the so called "visual streams", i.e., mental imaginary movies which run in our brain [10]. (By the way, LG is highly visual as well.) This is how it may work. Within the brain, the visual streams run consciously and subconsciously and may switch places from time to time (in relation to conscious/subconscious use). We may run several visual streams concurrently, morph them, and even use logic for such morphing, although this use is auxiliary. Then we mentally tag some of the objects shown in the movie and create the so-called symbolic shell around the main visual stream. This shell eventually becomes a standard symbolic algorithm that can be communicated to others employing familiar language, logic, mathematics, etc. I named this approach "visual reasoning". While the "visual" component (including pattern recognition) is, in general, pretty sophisticated, the reasoning component is relatively simple. Fortunately, the full scale mental visibility is rarely used in discoveries, and, in my opinion, the limited visibility can be simulated with a reasonable effort. The "reasoning" component is certainly within the scope of the modern software development.

Our approach to discovering the Algorithm of Discovery is analogous to an attempt to understand the algorithm of a program while watching its execution. Let us assume that this program's interface includes color movies on various subjects. In addition to this Algorithm, we are trying to discover the instruction set of the "computer" running this program, i.e., the means of the human brain to running it. With multiple published introspections of great scientists we can recreate clips from various movies, i.e., their imaginary thought experiments. What really helps is the assumption that all those movies were "demonstrated" by the programs running essentially the same algorithm. With our own past developments in LG, we have additional power of asking questions via morphing our own movies and getting answers by watching those morphed movies until the very end. Unfortunately, we do not have this power with the discoveries of other scientists.

2 The Algorithm of Discovery

In this section we briefly summarize the results introduced in [29–33]. The Algorithm of Discovery operates as a series of thought experiments, which interface with the rest of the brain and with external environment via imaginary animated movies (plays), which we named visual streams. These streams may or may not reflect the reality. This interface is constructive, i.e., visual streams could be morphed in the desired direction.

The input to the Algorithm is also a visual stream, which includes several visual instances of the object whose structure has to be understood or whose algorithm of construction has to be developed. Sometimes, the object is dynamic, i.e., its structure is changing in time. Then the input visual stream includes this visual dynamics. As a rule, neither the structure of the object nor the details of the dynamics arc present in the stream. It simply replicates (mimics) the natural or imaginary phenomenon. The task of the Algorithm of Discovery is to understand its structure including dynamics and/or develop an algorithm for reconstructing this object including its changes in time. This understanding happens in several stages. Importantly, it always ends up with the process of actual reconstruction of the object employing the construction set developed by the Algorithm on the previous stages. If the Algorithm investigates a natural real life object this imaginary reconstruction may be totally unrelated to the construction (replication) utilized by the nature. Usually, this reconstruction process is artificially developed by the Algorithm of Discovery with the only purpose to reveal the structure of the object. However, if the algorithm of natural replication is the goal of discovery than the Algorithm of Discovery will employ a set of different visual streams to reveal the relevant components utilized by the nature [35].

All the visual streams are divided into classes, Observation, Construction and Validation. They usually follow each other but may be nested hierarchically, with several levels of depth.

The visual streams operate in a very simple fashion similar to a child construction set. The Construction stream utilizes a construction set and a mental visual prototype, a model to be referenced during construction. This is similar to a list of models pictured in a manual (or a visual guide) enclosed to every commercial construction set. It appears that all the thought experiments in LG related to construction investigated so far, utilized those manuals. Imagine a child playing a construction set. He needs a manual to construct an object by looking constantly at its picture included in this manual. This model comes from the Observation stream as its output. It is not necessarily a real world model. It is not even a model from the problem statement. It is created by the Observation stream out of various multiple instances of the real world objects by abstraction, specifically, by "erasing the particulars". A final version of the object constructed by the Construction stream should be validated by the Validation stream.

The Algorithm of Discovery initiates the Observation stream, which must carefully examine the object. It has to morph the input visual stream and run it several times to observe (mentally) various instances of the object from several directions. Often, for understanding the object, it has to observe the whole class of objects considered analogous. If the object is dynamic (a process) it has to be observed in action. For this purpose, the Observation stream runs the process under different conditions to observe it in different situations. The purpose of all those observations is erasing the particulars to reveal the general relations behind them. A good example of multiple observations of processes is related to the thought experiments with various objects with respect to the inertial reference frames when discovering the theory of Special Relativity [9]. This includes experiments with uniformly moving ships, trains, experiments with ether as well as experiments for catching a beam of light (Sect. 5). Once the relations have been revealed, a construction set and a visual model have to be constructed by the Observation stream. Both are still visual, i.e., specific,—not abstract. However, they should visually represent an abstract concept, usually, a class of objects or processes, whose structure is being investigated. For construction, the Observation stream utilizes the Construction stream with auxiliary purpose (which differs from its prime purpose—see below). Note that the model construction is different from the subsequent reconstruction of the object intended to reveal its structure. This model may differ substantially from the real object or class of objects that are investigated. Its purpose is to serve as a manual to be used for references during reconstruction. Various discoveries may involve a series of models (Sect. 3).

When the model and the construction set are ready, the Algorithm of Discovery initiates the Construction stream with its prime purpose. This purpose is to construct the object (or stage the process) by selecting appropriate construction parts of the set and putting them together. If an object has a sequential nature the construction also takes place sequentially, by repetition of similar steps. If multiple models have been produced the final object construction can also be considered as a model construction. At some point of construction, the parts are tagged symbolically and, in the end, visual reasoning with symbolic tagging turns into a conventional symbolic algorithm to be verified by the subsequent Validation stream.

Models and construction sets may vary significantly for different problems. Construction of the model begins from creation of the construction set and the relations between its components. Both items should be visually convenient for construction. The Algorithm of Discovery may utilize a different model for the same object if the purpose of development is different. Such a different model is produced by a different visual stream.

In many cases the Algorithm of Discovery employs "a slave" to visually perform simple tasks for all types of visual streams. This slave may be employed by the Construction stream to "see" construction parts and put them together. More precisely, imagine a child playing a simplistic construction set. To avoid offending children, I had named this personality a Ghost. This Ghost has very limited skills, knowledge and, even, limited visibility. The Observation stream may utilize the Ghost to familiarize itself with the optional construction set, to investigate its properties. Next, the Construction stream may use the Ghost to perform the actual construction employing those properties. Eventually, the Validation stream may use the Ghost to verify visually, if properties of the constructed object match those revealed by the Observation stream. In all cases, the Ghost is guided by the Algorithm of Discovery or, more precisely, by the respective visual streams.

As was already discussed, the initial visual model is usually guided by a very specific prototype, where the Observation stream has actually erased the particulars. However, this specificity does not reduce generality in any way. This sounds like a paradox. Essentially, every component of this model carries an abstract class of components behind it. This way visual reasoning about the model drives reasoning about abstract classes, which is turned eventually into the standard formal reasoning. This happens as follows. A visual model drives construction of the formal symbolic model so that the key items in a visual model have tags representing the respective formal model. At first, the formal model is incomplete. At some stage, a running visual stream is accompanied by a comprehensive formal symbolic shell. Running a shell means doing formal derivation, proof, etc. synchronized with a respective visual stream. While the shell and the stream are synchronized, the visual stream drives execution of the shell, not the other way around. For example, a formal proof is driven by animated events within the respective visual stream. The visual streams, usually, run the creation of the visual model, the construction set and the final construction of the object several times. During those runs as a result of persistent tagging the symbolic shell appears. Multiple runs utilize the same visual components but during initial runs the synchronization of the stream and the shell is not tight. Further on, synchronization is tightened by morphing the visual model and/or adjusting symbolic derivation if they initially mismatch. Eventually, the stream and the shell switch their roles. In the end, it appears that the stream becomes the animated set of illustrations, a movie, driven by the running symbolic shell. For example, during the final runs (and only then), the visual streams, presented in [29–34], are driven by the constraints of the abstract board game, the abstract set theory and/or the productions of the controlled grammars. At this point the visual stream and the symbolic shell can be completely separated, and the visual stream can be dropped and even forgotten.

A stream may schedule other streams by creating almost a "program with procedure calls". Essentially, it may schedule a sequence of thought experiments to be executed in the future. These experiments will, in their turn, initiate new visual streams. In this case, the purpose, the nature, and the general outcome of those experiments should be known to the stream created this sequence. However, this sequence is different from the list of procedure calls in conventional procedural (or imperative) programming. The algorithms of those "procedures", i.e., the algorithms to be produced by the respective thought experiments are generally unknown. The experiments are not programmed—they are staged. The actual algorithm should be developed as a result of execution of such experiment. In a sense, this is similar to the notion of declarative programming when a function is invoked by a problem statement while the function's body does not include an algorithm for solving this problem.

The ability of a visual stream to schedule a sequence of thought experiments permits to create a nested top-down structure of visual streams with several levels of depth. Though, we suspect that the actual depth of the nested programmed experiments never exceeds two or three.

Proximity reasoning as a type of visual reasoning was introduced due to the need for approaching optimum for many discoveries. It is likely that all the technological inventions and discoveries of the laws of nature include "optimal construction" or, at least, have optimization components [13]. Thus, various construction steps performed by the Algorithm of Discovery require optimization, which, certainly, makes construction more difficult. As the appearance of this Algorithm is lost in millennia, for its main purpose, it could not certainly utilize any differential calculus even for the problems where it would be most convenient. For the same reason, it could not utilize any approximations based on the notion of a limit of function. Those components of differential calculus could certainly serve as auxiliary tools. In that sense, in order to reveal the main optimization components, the most interesting problems to be investigated should lack continuity compelling the Algorithm of Discovery to employ explicitly those components. Based on several case studies [34], we suggested that this optimization is performed by the imaginary movement via approaching a location (or area) in the appropriate imaginary space. Having such space and means, the Algorithm employs an agent to catch sight of this location, pave the way, and approach it. Contrary to the function based approach, which is static by its nature, the Algorithm operates with dynamic processes, the visual streams. Some of those streams approach optimum (in a small number of steps); other streams show dynamically wrong directions that do not lead to the optimum and prevent the Algorithm from pursuing those directions. Both types of streams represent proximity reasoning. We suggested that proximity reasoning plays a special role for the Algorithm of Discovery as the main means for optimization. Proximity reasoning is a type of visual reasoning. This implies that the Algorithm should reason about the space where distances are "analogous" to the 3D Euclidian distances. Roughly, when we approach something, the distance must be visually reduced, and this should happen gradually. The space for proximity reasoning should provide means to evaluate visually if the animated images representing various abstract objects approach each other or specific locations [34]. Construction of those spaces is the key component of the Algorithm of Discovery.

3 A Series of Visual Models for Discoveries

A discovery, i.e., a development of the final algorithm for the object construction is based usually on constructing a series of models. Each of those models may, in its turn, be based on multiple experiments and may result from multiple uses of the

Observation and Construction streams. Interestingly, those models may represent the same object, though, be totally different. The purpose of these models is to look at the object from different prospective to reveal different properties. The models do not appear at once. Experiments with one model demonstrate the need for the next one. The model construction is based on the wide use of the principle of erasing the particulars. For each model some of the particulars of an object under investigation are erased while other particulars are emphasized. A good example of such multiple models is the discovery of the No-Search Approach [24, 25, 27, 31, 33, 34]. This discovery is based on the four different models that represent the same abstract object, the State Space of the 2D/4A Abstract Board Game (ABG). This ABG is a reformulation of the well-known R. Reti chess endgame [4].

The first model is the so-called Pictorial LG that includes a network of zones, a pictorial representation of several types of local skirmishes. This representation is obtained by "projecting" optional variants of skirmishes from the State Space onto the Abstract Board. Moreover, to make those projections visible, an Abstract Board is mapped into the area of a 2D plain. This mapping permits to easily visualize Pictorial LG as a network of straight lines drawn on a sheet of paper or displayed on a screen. The straight lines represent trajectories of pieces, i.e., the planning routes of mobile entities. This type of representation permits to erase (abstract from) the particulars of movement of various entities such jumps, turns, promotions, etc. Small circles (representing stops) divide a trajectory into sections (the steps). This way, movement through the State Space of the ABG is visualized by the "physical" movement of pieces along the trajectories of the Pictorial LG. Moreover, the first model permits to conduct experiments that investigate visually if a piece moving along trajectory can approach a "dynamic" area on the Abstract Board while this area moves away, e.g., shrinks. However, a conclusion about approaching or non-approaching an area should be considered as local with respect to the ABG because both trajectories and areas are just "projections" of variants of movement and subspaces of the State Space on the Abstract Board. In order to expand from local to global conclusions the first model invokes the second one.

The second model is the so-called Mountain-triangle drawn on a sheet of paper or displayed on a screen. Essentially, it represents the same State Space of the ABG, though, with distinguished Start State at the upper vertex of the triangle. In addition, the Mountain-triangle represents the brute-force search tree of the ABG that grows top down from the upper vertex of the triangle and the terminal states located in the bottom side of it. Certainly, it is a rough representation of the State Space. However, it is convenient for visualizing a tree with top-down direction. It is called "Mountain" to reflect analogy with a climber's descends and ascends performed by the Ghost when visiting branches of the tree. The second model permits to elevate projection subspaces introduced in the first model into the full State Space. This elevation is based on the expansion of the ABG terminal states (introduced in the 2D/4A problem statement). The expansion experiments consist of the blowing inside the triangle various bubbles rooted in the bottom and directed to the top ("closer" to the Start State). Those bubbles represent various subspaces of the State Space. In order to establish link with their projections the second model invokes joint experiments with the first one. Those experiments reveal formal description of several bubbles, i.e., the subspaces of the ABG State Space. These descriptions are based on the zones, the components of the Pictorial LG. It appears that those bubbles and their complementary subspaces have complex structure and fill the full State Space. Further experiments demonstrate the need in decomposition of the State Space into multiple well-defined subspaces and their intersections to reveal the complete structure of the bubbles. Basically, we are talking about precise accounting for intersections of multiple bubbles which represent a clearly defined mosaic of tiles (Sect. 4). The second model invokes the third one.

The third model is the so-called State Space Chart. This model is again a representation of the ABG State Space. It is a square drawn on a plain and broken into four quadrants (by the vertical and horizontal lines) and a circle around the center of the square. This third model represents mosaic of eight tiles, four quadrants and four circular segments, the proper subsets of the respective quadrants. These tiles represent important subspaces of the ABG State Space which are described employing zones of the Pictorial LG. It appears that the Start State of the 2D/4A ABG (reflected by a small circle) belongs to the upper left quadrant. The subspaces represented by the circular segments have special value. All the states of those subspaces have a well-defined strategy leading to a specific result of the game, a white win, a black win, or a draw. Thus, the third model together with the first and second models provides means to investigate if there are strategies leading from the Start State to each of the circular segments. Those strategy-candidates are represented visually as lines linking the Start State to the appropriate segment. The investigation is based on four thought experiments that utilize effectively the visual dynamics of the first model. The experiments permit to eliminate two classes of strategy-candidates, the white winning strategies and the so-called Pure draw strategies. The rest of the candidates are preserved for the precise final testing on the fourth model. It should employ minimax search algorithm and choose the only real strategy existing in this problem.

The fourth model is the Solution Tree, the conventional search tree of the 2D/4A ABG. As usual, it grows top-down and employs minimax. However, the legal moves included on the Tree are those prescribed by the strategy-candidates preserved by the experiments with the third model. Specifically, these are classes of the black winning strategies and the so-called Mixed draw strategies. The final construction experiment yields the Tree with the optimal moves only. So, it is not a search tree in conventional sense. The two strategy-candidates being tested provide ultimate forward pruning that leads to constructing the final Solution Tree. Both candidates are described by the visual algorithms utilizing the first model of the Pictorial LG. Thus, every legal move to be included on the Solution Tree is selected employing the strategies generated as the outcome of the first three models. Moreover, it explicitly uses the first model to reflect visually the game state change in the Pictorial LG. The fourth model demonstrates that the only real strategy for this problem is the Mixed draw strategy. In terms of the mosaic reasoning (Sect. 4), application of the strategy-candidates to constructing the Solution Tree is the iterative application of the transformation matching rule leading to complete Solution Tree mosaic.

The construction of a series of visual models including the switch procedure from one model to another is a major component of the Algorithm of Discovery. Those series were identified in all the discoveries we investigated so far including discoveries in LG, in revealing the structure of DNA and in the theory of Special Relativity.

4 Mosaic Reasoning for Discovering Objects

Mosaic reasoning as a type of visual reasoning was introduced due to the analogy of the Construction stream operation with assembling a mosaic picture of small colorful tiles. Another, maybe, even more transparent analogy is known as a jigsaw puzzle when a picture is drawn on a sheet of paper and then this paper is cut into small pieces, mixed up, to be assembled later into the original picture. As Sir Thompson [42] pointed "… the progress of science is a little like making a jig-saw puzzle. One makes collections of pieces which certainly fit together, though at first it is not clear where each group should come in the picture as a whole, and if at first one makes a mistake in placing it, this can be corrected later without dismantling the whole group". Both analogies, the pictorial mosaic and the jigsaw puzzle, represent well the key feature of the Algorithm of Discovery construction set. However, we prefer the former because the jigsaw puzzle looks more like an assignment in reassembling a construct, a picture, which has already been created and, certainly, well known. In that sense, a tile mosaic is created from scratch, including choosing or even creating necessary tiles. In addition, a jigsaw puzzle is reassembled out of pieces based on random cuts. On the contrary, in pictorial mosaic, in many cases, every tile should have unique properties; it should be shaped and colored to match its neighbors precisely. A similar specificity is related to a group of adjacent tiles, the aggregate.

In the following sections we will utilize discoveries of the structure of DNA and Special Relativity to demonstrate mosaic reasoning for objects and processes, respectively.

For many discoveries, the components of the construction set should be developed with absolute precision, in the way that every part should be placed to its unique position matching its neighbors. We will use the same name, the tiles, for those construction parts. If precision is violated the final mosaic will be ruined and the discovery will not happen. Though a group of tiles, an aggregate, may be configured properly, its correct placement in the mosaic may be unclear and requires further investigation. Moreover, a tile itself may have complex structure which may require tailoring after placement in the mosaic. In some cases, a tile is a network of rigid nodes with soft, stretchable links.

Mosaic reasoning may stretch through the observation, construction, and validation steps of the Algorithm of Discovery operating with tiles and aggregates of tiles. Overall, mosaic reasoning requires tedious analysis of the proper tiles and their matching rules. Investigation of the matching rules is the essential task of the Observation stream. Multiplicity of those rules and their specificity with respect to the classes of construction tiles make the actual construction very complex. Selecting a wrong tile, wrong tailoring, choosing a wrong place, or incompatible neighbors may ruin the whole mosaic. The matching rules are the necessary constraints that control the right placement of the tiles. Missing one of them, usually, leads to the wrong outcome because the Algorithm of Discovery is pointed in the wrong direction.

Some of the matching rules impact mosaic locally while other rules provide global constraints. The global matching rules include the requirement of the top-down analysis and construction, the global complementarity rule, certain statistical rules, the transformation rules, etc. For many if not all natural objects and processes, their structure is not reducible to a combination of the components. Large groups of tiles, i.e., large aggregates, may obey the rules which are hardly reducible to the rules guiding placement of singular tiles. This matching rule must be understood globally first, implemented in the mosaic skeleton construction, and, only then, reduced to the placement of the specific tiles. An example of the global matching rule for the discovery of the structure of DNA is the choice of the helical structure of the DNA molecule including the number of strands [35, 44]. The rule of the global complementarity means that placement of one aggregate may determine precisely the adjacent aggregate. In case of DNA, one strand of the helix with the sequence of the base tiles attached to it determines the unique complementary second strand with the corresponding sequence of the base tiles. The global statistical rules related to the whole mosaic may reflect the relationship between major structural components, the large aggregates. If understood and taken into account by the Observation stream, they may focus the Construction stream and lead to a quick discovery. In the case of DNA, the so-called Chargaff rules reflect the structural relationship between the base tiles of the complementary strands of the double helix [35, 44]. Yet another class of global matching rules is called transformation rules. This is an algorithm for reconstructing an aggregate out of another aggregate and placing this aggregate in the proper location. Applied sequentially, such a rule permits to turn an aggregate, the so-called generator, into the set of adjacent aggregates. This way the whole mosaic could be constructed. For example, the whole mosaic of the DNA molecule could be constructed if the generator and the singular transformation are defined. Over the course of four experiments, the double helix generator was constructed. It includes a pair of nucleotides with sugar-phosphate backbone and purine-pyrimidine base. The transformation is a combination of translation and rotation. Interestingly, this type of construction may be utilized by the Algorithm of Discovery as a convenient procedure to reveal the structure of an object, e.g., the DNA molecule, while the nature may have used a totally different algorithm for producing the same object.

The local matching rules include the local complementarity rule, the interchangeability rule, etc. The local complementarity means, roughly, that a protrusion on one tile corresponds to the cavity on the complementary adjacent tile. For the

DNA molecule this is usually a hydrogen bond of a base tile (a protrusion) that corresponds to a negatively charged atom of the adjacent tile (a cavity). The local complementarity often expresses itself in the requirement of various kinds of symmetry within the pairs of matching construction tiles. The whole class of the local matching rules is based on interchangeability. In simple terms, if two aggregates that include several tiles are not identical but interchangeabe, their internal structure may be unimportant. There are several levels of interchangeability. Two aggregates could be essentially the same, i.e., their skeletons coincide. Importantly, those skeletons must include nodes which serve as the attaching points of the aggregates to the rest of the mosaic. The notion of an internal skeleton depends on the problem domain and is specific for different types of mosaic. For example, two different aggregates for the DNA mosaic may have identical ring structures but the atoms and respective bonds that do not belong to those structures may be different. Another lower level of interchangeability of the aggregates does not require their skeletons to coincide. The only requirement is that the attaching points of those aggregates are identical. In all cases interchangeability means that the stream can take one aggregate off the mosaic and replace it with another. This will certainly change the picture but the whole structure will stand. We named those aggregates plug-ins. It appears that plug-ins played crucial role in the discovery of the structure of DNA because such a plug-in was the key component of the helical generator, a purine-pyrimidine base [35, 44].

Besides mosaic structural components that include tiles, aggregates, global and local matching rules, there is an unstructured component that we named a mosaic environment. Such environment may impact the structure of tiles, aggregates, application of matching rules, and the whole mosaic while being relatively unstructured itself. In case of DNA, this was the water content whose lack or abundance could seriously impact the structure of the whole mosaic.

5 Mosaic Reasoning for Discovering Processes

A different type of mosaic, the mosaic of processes, was constructed by Einstein while discovering his theory of Special Relativity [6]. In reality, this was not a construction from scratch—it was a reconstruction of the Galileo-Newton mosaic into new one, the Einstein mosaic. Both mosaics consist of moving tiles, the inertial frames, i.e., those frames moving along straight lines with constant velocities with respect to each other. Contrary to the static mosaic of objects considered above, the inertial frames mosaics represent processes developing in time. Moreover, various entities like human beings or water waves could be moving within those frames. Essentially, these are processes of processes. Mathematically, all those frames should be considered as those in the 4D space with time as the fourth dimension. For the Galileo-Newton mosaic this is a 4D Euclidian space, while for the Einstein mosaic this is a Minkowski space. Note that none of those mathematical constructs were actually used by Einstein for his discovery [6]. Visualization of the 4D spaces is impossible; however, visualizing those mosaics as sets of the 3D processes developing in time supports fully various thought experiments and related visual streams. Typically, one of the frames is chosen as "static" like the one associated to the platform with the Ghost standing on it while the other frame is "moving" and is associated to a train passing by this platform and another Ghost walking inside a car of the moving train. There is no notion of an adjacent tile. However, there is still a notion of the transformation matching rules utilized by the Construction stream for transforming a generator into the whole mosaic of tiles. In those mosaics all the inertial frames are equivalent (or indistinguishable in terms of the laws of Physics), hence, any tile could serve as a generator and a plug-in simultaneously. For the Galileo-Newton mosaic the required transformation is just the Galileo transform while for the Einstein mosaic it is the Lorentz transform.

The Galileo-Newton mosaic has been around for several hundred years and was able to explain numerous experiments. There was no need for any reconstruction. Only during a couple of decades, before the Einstein's discovery in 1905, several questions were raised. Reconstruction of mosaics was preceded by two series of thought experiments. Some of them were pure thought experiments while many others were replayed in real world, though, initially, they were certainly conceived as the thought ones.

The first series has led to revealing the principle of invariance of the laws of physics, the foundation of both mosaics. These laws should be the same in all the inertial reference frames, i.e., for all the tiles. This series could be traced back to the Galileo experiments in the main cabin of a large swimming ship below its deck [8]. Uniform movement of entities inside the ship (also moving uniformly) is indistinguishable from those on the land. However, assuming that the ship is transparent their velocities would look different from the land due to addition of the ship's velocity (as a vector). While the Galileo's experiments dealt with mechanical movements the same principle should have covered the laws of electromagnetism by Maxwell-Lorentz. This meant, in particular, that there should not be any distinction in how the induction occurs in both cases, whether the magnet or the conducting coil is in motion. However, according to the classic theory this experiment was interpreted differently for those cases. This meant different laws for different tiles. It was noted by several scientists, including Föppl [7], and emphasized by Einstein [6].

The second series of thought experiments revealed special nature of light or, more precisely, electromagnetic waves. This series could be traced back to the experiments with water and sound waves as traveling disturbances in a medium. As is the case with movements of other entities, velocities of the waves inside the Galileo's ship are indistinguishable from those over the land (for the sound waves) or near the sea shore (for the water waves). Analogously to other moving entities on the transparent ship their velocities would look different from the land due to addition of the ship's velocity. According to the Maxwell's theory, light as well as all types of electromagnetic waves travels at a speed of approximately 186,000 miles per second. This includes AM and FM radio signals, microwaves, visible light, ultraviolet, X-rays, gamma rays, etc. Beginning from Maxwell himself, scientists believed that these waves propagate as disturbances of the invisible medium called the ether, and their velocity was registered relative to this ether. The ether should have had interesting properties. It should spread through the entire universe and should not affect big bodies like planets and stars as well as the smallest ones like specks of dust. In addition it has to be stiff for the light wave to vibrate at a great speed. Numerous thought experiments were intended to demonstrate that the ether waves are passing by the Ghost at a faster speed if he is moving through the ether towards the light source. Some of those experiments have actually been implemented in real world. These include experiments by Fizeau, Michelson and Morley as well as those contemplated by Einstein himself. The most influential was the thought experiment of the Ghost riding uniformly at the speed of light alongside a light beam and observing "frozen" light. The 13 real life experiments refuted all their thought prototypes by registering no difference to the speed of light. Multiple mental executions and morphing (over the period of ten years) of the riding light experiment led Einstein to conclusion that this was not a real effect—the light would not freeze but would run at the same speed according to the same Maxwell equations as for the Ghost standing on the land.

The visual streams utilized in the two series of thought experiments considered above led conclusively to adoption of the two principles [6]. The first was the rigorous spreading of the principle of relativity to all the physical systems that undergo change meaning that the laws governing this change are the same for all the inertial frames of reference. The second principle stated the constancy of the velocity of light c, weather the ray is emitted by a stationary or by a moving body.

Adoption of those new principles led in its turn to the construction of the Einstein mosaic of the processes. The first matching rule was the rule of simultaneity which was the algorithm based on the relation between time and signal velocity. The second matching rule was the length measuring rule which was the algorithm for applying the measuring-rod. Those rules utilized by the Construction stream for constructing processes involving rigid bodies moving at the speed close to the speed of light demonstrated visually (within visual stream) and mathematically the effects of time dilation and length contraction. The major matching rule derived from the above principles was the Lorentz transform. It could be visualized as a hyperbolic rotation. Applying it to the generator, an arbitrary inertial frame, permitted to construct the whole Einstein mosaic of processes, the universe of inertial frames.

Our research demonstrated that the Algorithm of Discovery does not search for a solution in the search space. Instead, it constructs the solution out of the construction set employing various tools and guides. The right choices of the construction tiles and the matching rules by the Observation stream permit focusing the Construction stream to produce a desired series of models with a proper mosaic and, eventually, to make a discovery. All the results on the Algorithm of Discovery are still hypothetical and have to be verified by software implementations. The very first implementations have been initiated at the University of Colorado Denver.

References

1. Botvinnik, M.: Chess, Computers, and Long-Range Planning. Springer, New York (1970)
2. Botvinnik, M.: Blok-skema algoritma igry v shahmaty (in Russian: A Flow-Chart of the Algorithm for Playing Chess). Sov. Radio (1972)
3. Botvinnik, M.: O Kiberneticheskoy Celi Igry (in Russian: On the Cybernetic Goal of Games). Sov. Radio, Moscow (1975)
4. Botvinnik, M.: Computers in Chess: Solving Inexact Search Problems. Springer, New York (1984)
5. Deheaene, S.: A few steps toward a science of mental life. Mind Brain Educ. **1**(1), 28–47 (2007)
6. Einstein, A.: On the Electrodynamics of moving bodies. Annalen der Physik **17**, 891 (1905) (transl. from German)
7. Föppl, A.: Introduction to Maxwell's theory of electricity. In: Teubner, B.G., Leipzig (1894) (transl. from German)
8. Galilei, G.: Dialogue Concerning the Two Chief World Systems. 1632 (tranl. by Stillman Drake)
9. Isaacson, W.: Einstein: His Life and Universe, p. 779. Simon and Shuster, New York (2007)
10. Kosslyn, S., Thompson, W., Kim, I., Alpert, N.: Representations of mental images in primary visual cortex. Nature **378**, 496–498 (1995)
11. Kott, A., McEneaney, W., (eds): Adversarial Reasoning: Computational Approaches to Reading the Opponent's Mind. Chapman & Hall/CRC, London (2007)
12. Linguistic Geometry Tools: LG-PACKAGE, with Demo DVD, p. 60. STILMAN Advanced Strategies, 2010. This brochure and 8 recorded demonstrations are available online: www.stilman-strategies.com
13. Miller, A.: Insights of Genius: Imagery and Creativity in Science and Art. Copernicus, an imprint of Springer-Verlag, New York (1996)
14. Stilman, B.: Formation of the Set of Trajectory Bundles, Appendix 1 to [3]. (1975)
15. Stilman, Ierarhia formalnikh grammatik dla reshenia perebornikh zadach (in Russian: Hierarchy of Formal Grammars for Solving Search Problems). Technical report, p. 105. VNIIE, Moscow (1976)
16. Stilman, B.: The computer learns. In: Levy, D., (ed.) 1976 US Computer Chess Championship, pp. 83–90. Computer Science Press, Woodland Hills, CA (1977)
17. Stilman, B.: Fields of Play, Appendix 1 of [4] (1979)
18. Stilman, B., Tsfasman, M.: Positional Value and Assignment of Priorities, Appendix 2 of [4] (1979)
19. Stilman, A.: Formal language for hierarchical systems control. Int. J. Lang. Des. **1**(4), 333–356 (1993)
20. Stilman, B.: A linguistic approach to geometric reasoning. Int. J. Comput. Math. Appl. **26**(7), 29–58 (1993)
21. Stilman, B.: Network languages for complex systems. Int. J. Comput. Math. Appl. **26**(8), 51–80 (1993)
22. Stilman, B.: Linguistic geometry for control systems design. Int. J. of Comput. Appl. **1**(2), 89–110 (1994)
23. Stilman, B.: Translations of network languages. Int. J. Comput. Math. Appl. **27**(2), 65–98 (1994)
24. Stilman, B.: Linguistic geometry tools generate optimal solutions. In: Proceeding of the 4th International Conference on Conceptual Structures—ICCS'96, pp. 75–99. Sydney, Australia, 19–22 Aug 1996
25. Stilman, B.: Managing search complexity in linguistic geometry. IEEE Trans. Syst. Man Cybern. **27**(6), 978–998 (1997)
26. Stilman, B.: Network languages for concurrent multi-agent systems. Int. J. Comput. Math. Appl. **34**(1), 103–136 (1997)

27. Stilman, B.: Linguistic Geometry: From Search to Construction, p. 416. Kluwer Academy Publishers (now Springer), New York (2000)
28. Stilman, B.: Linguistic geometry and evolution of intelligence. ISAST Trans. Comput. Intell. Syst. **3**(2), 23–37 (2011)
29. Stilman, B.: Thought experiments in linguistic geometry. In: Proceeding of the 3d International Conference on Advanced Cognitive Technologies and Applications—COGNITIVE'2011, pp. 77–83. Rome, Italy, 25–30 Sept 2011
30. Stilman, B.: Discovering the discovery of linguistic geometry. Int. J. Mach. Learn. Cybern. p. 20 (2012); Printed in 2013, **4**(6), 575–594. doi:10.1007/s13042-012-0114-8
31. Stilman, B.: Discovering the discovery of the no-search approach. Int. J. Mach. Learn. Cybern. p. 27 (2012). doi:10.1007/s13042-012-0127-3
32. Stilman, B.: Discovering the discovery of the hierarchy of formal languages. Int. J. Mach. Learn. Cybern. Springer p. 25 (2012). doi:10.1007/s13042-012-0146-0
33. Stilman, B.: Visual reasoning for discoveries. Int. J. Mach. Learn. Cybern. Springer p. 23 (2013). doi:10.1007/s13042-013-0189-x
34. Stilman, B.: Proximity reasoning for discoveries. Int. J. Mach. Learn. Cybern. Springer p. 31 (2014). doi:10.1007/s13042-014-0249-x
35. Stilman, B.: Mosaic reasoning for discoveries. Int. J. Artif. Intell. Soft Comput. p. 36 (2014)
36. Stilman, B., Yakhnis, V., Umanskiy, O.: Winning strategies for robotic wars: defense applications of linguistic geometry. Artif. Life Robot. **4**(3) (2000)
37. Stilman, B., Yakhnis, V., Umanskiy, O.: Knowledge acquisition and strategy generation with LG wargaming tools. Int. J. of Comput. Intell. Appl. **2**(4), 385–409 (2002)
38. Stilman, B., Yakhnis, V., Umanskiy, O.: Chapter 3.3: Strategies in Large Scale Problems, in [11], pp. 251–285 (2007)
39. Stilman, B., Yakhnis, V., Umanskiy, O.: Linguistic geometry: the age of maturity. J. Adv. Comput. Intell. Intell. Inform. **14**(6), 684–699 (2010)
40. Stilman, B., Yakhnis, V., Umanskiy, O.: Revisiting history with linguistic geometry. ISAST Trans. Comput. Intell. Syst. **2**(2), 22–38 (2010)
41. Stilman, B., Yakhnis, V., Umanskiy, O.: The primary language of ancient battles. Int. J. Mach. Learn. Cybern. **2**(3), 157–176 (2011)
42. Thomson, G.: The Inspiration of Science. Oxford University Press, London (1961)
43. Von Neumann, J.: The Computer and the Brain. Yale University Press, New Haven (1958)
44. Watson, J.D.: The Double Helix: A Personal Account of the Discovery of the Structure of DNA. Atheneum, New York (1968) [Scribner Classics Edition, New York, 1996.]

The Art of Anticipatory Decision Making

Andrzej M.J. Skulimowski

Abstract This paper presents the recent advances of the theory of anticipatory networks and its applications in future-oriented decision-making. Anticipatory networks generalize earlier models of consequence anticipation in multicriteria decision problem solving. This theory is based on the assumption that the decision maker takes into account the anticipated outcomes of future decision problems linked in a prescribed manner by the causal relations with the present problem. Thus arises a multigraph of decision problems linked causally (the first relation) and representing one or more additional anticipation relations. Such multigraphs will be termed anticipatory networks. We will also present the notion of a superanticipatory system, which is an anticipatory system that contains a future model of at least one anticipatory system besides itself. It will be shown that non-trivial anticipatory networks are superanticipatory systems. Finally, we will discuss several real-life applications of anticipatory networks, including an application to establish efficient collaboration of human and robot teams.

Keywords Anticipatory networks • Superanticipatory systems • Multicriteria decision making • Anticipatory collaboration • Preference modelling

---

A.M.J. Skulimowski (✉)
Decision Science Laboratory, Department of Automatic Control and Biomedical Engineering, AGH University of Science and Technology, al. Mickiewicza 30, 30-050 Kraków, Poland
e-mail: ams@agh.edu.pl

A.M.J. Skulimowski
International Centre for Decision Sciences and Forecasting, Progress and Business Foundation, ul. J. Lea 12B, 30-048 Kraków, Poland

S. Kunifuji et al. (eds.), *Knowledge, Information and Creativity Support Systems*, Advances in Intelligent Systems and Computing 416,
DOI 10.1007/978-3-319-27478-2_2

1 Introduction

This paper presents the theory of anticipatory networks, which generalizes the ideas related to anticipatory models of consequences in multicriteria optimization problems presented in [12, 13, 18]. It is assumed that when making a decision, the decision maker takes into account the anticipated outcomes of each future decision problem linked by the causal relations with the present problem. In a network of linked decision problems the causal relations are defined between time-ordered nodes. The future scenarios of the causal consequences of each decision are modelled by multiple edges starting from an appropriate node. The network is supplemented by one or more relations of anticipation, or anticipatory feedback, which describes a situation where decision makers take into account the anticipated results of some future optimization problems while making their choice. They then use the causal dependences of future constraints and preferences on the choice just made to influence future outcomes in such a way that they fulfill the conditions contained in the definition of the anticipatory feedback relations.

Both types of relations as well as forecasts and scenarios regarding the future model parameters form an information model called an anticipatory network [18]. In Sect. 2 we will show the basic properties of anticipatory networks as well as a method for computing them.

Following [12] and [18], in Sect. 3 we will present an application of anticipatory networks to select compromise solutions to multicriteria planning problems with the additional preferences provided in the form of anticipatory trees and general networks. We propose a more general notion of preference structure as compared to [16] and [18] that allows us to separate the preferences included in the anticipatory network from those used in present-time decision making. The study of properties of the anticipatory networks led us to introduce the notion of superanticipatory systems in Sect. 4. By definition, an *anticipatory system* in the Rosen sense [11] makes its decisions based on a future model of itself and of the outer environment. A *superanticipatory system S* is a system that is anticipatory and contains a future model of at least one other anticipatory system whose outcomes may influence the current decisions of *S* by a so called anticipatory feedback relation. This notion is idempotent, i.e. the inclusion of other superanticipatory systems into the model of the future does not yield an extended class of systems, but we can classify them according to a grade that counts the number of nested anticipations. We will observe that most anticipatory networks can be regarded as superanticipatory systems if we assume that future decisions can be based on similar anticipatory principles as the present decision. The class of superanticipatory systems has been introduced in [17] and [15].

The above mentioned theory arose from the need to create an alternative approach to selecting solutions to multicriteria optimization problems, where the estimation of an unknown utility function is replaced by a direct multi-stage model of future consequences of the decision made [12]. The anticipatory behavior of decision makers corresponds to the above definition of anticipatory systems proposed by Rosen [11] and developed further by other researchers [3, 8, 10].

A bibliographic survey of these ideas can be found in [8]. The ability to create a model of the future (of the outer environment and of itself), which characterizes an anticipatory system, is also the prerequisite for an anticipatory network, where nodes model anticipatory systems that can influence each other according to causal order. In this paper, anticipatory networks are restricted to model decisions made in so-called networks of optimizers, where each node models an optimization problem [18]. Similarly to anticipatory networks of optimizers, one can construct networks with nodes modelling Nash equilibria, set choice problems, rankings, random or irrational decision makers, or hybrid networks containing nodes of all types [16]. It should be pointed out that most anticipatory networks, and all those considered in this paper, are model-based, so that their nodes correspond to weak anticipatory systems in the Dubois sense [3].

The networks of anticipatory agents can be constructed by applying future decision problem forecasts and scenarios of anticipated consequences. The latter can be provided by foresight projects. In the final Sect. 5 we will discuss further extensions and applications of anticipatory networks that may be useful e.g. to build holistic future models or to establish efficient collaboration of heterogeneous teams consisting of robots and humans.

2 Anticipatory Networks as Generic Causal Models

The idea behind introducing anticipatory networks as models of consequences was formulated in [12, 18]. The basic principle is to use forecasts and foresight scenarios to estimate the parameters of future decision-making agents and to build a network of them. The anticipated future consequences of a decision made are modelled as changes in constraints and/or preference structures of future decision problems. The nature of these changes is assumed known to the present-time decision maker(s). It may result from model-based forecasts or foresight as well. Then, the anticipated outcomes of future decision-making problems that—of course—depend on constraints and preference structures, serve as a source of additional information that can be used to solve the current problem. In addition, future decision-making agents may use the same principle to make their decisions and this must be taken into account at the preceding decision stages.

Constructive algorithms for computing the solutions to the current multicriteria decision making problem taking into account the above anticipatory preference information feedback may be applied if we know that:

- All agents whose decisions are modelled in the network are rational, i.e. they make their decisions complying with their preference structures.
- An agent can assess whether the outcomes of some or all future decision problems causally dependent on the present one are more or less desired. This dependence is described as relations (usually multifunctions) between the

decisions to be made now and the constraints and/or preference structures of future problems.

- The above assessments are transformed into decision rules for the current solution choice problem, which affect the outcomes of future problems in such a way that they comply with the agent's assessments. The decision rules so derived form an additional preference structure for the decision problem just considered.
- There exists a relevance hierarchy in the network; usually the more distant in the future an agent is, the less relevant the choice of solution. However, this rule is not a paradigm.

Anticipatory networks which contain only decision-making agents solving optimization problems are termed *optimizer networks.* According to [18], an *optimizer O* is a multivalued function that assigns to a set of feasible decisions U and to the preference structure P a subset of the set of optimal decisions $O_F(U, P) \subset U$ that is selected according to P and to a fixed set of optimization criteria F with values in an ordered space E. Throughout this paper we will assume that the optimization problems solved by the optimizers have the form

$$(F: U \to E) \to min(\theta), \tag{1}$$

where E is a vector space with a partial order $\leq_\theta$ defined by a convex cone θ, i.e. iff

$$x \leq_\theta y \Leftrightarrow y - x \in \theta \text{ for each } x, y \in E.$$

The solution to (1) is the set of nondominated points defined as

$$\Pi(U, F, \theta) := \{u \in U: \; [\forall v \in U: \; F(v) \leq_\theta F(u) \Rightarrow v = u]\}.$$

Thus the criteria F and the ordering cone θ characterize a given optimizer uniquely. Most frequently, the decision maker's aim is to select and apply just one nondominated solution to (1). Thus the role of the preference structure P that occurs in the definition of an optimizer is to restrict the set of nondominated points in the solution process. Without a loss of generality we can assume that P is defined explicitly by pointing out for each $u \in U$ which elements of U dominate u. These are termed *dominating sets* and form a *domination structure* [2] which models the way the decision maker takes into account additional information about preferences when making the decision. Therefore P can be defined as a family of subsets of U in the following way

$$P := \{\pi(u) \subset U: \; u \in \pi(u) \; and \; [if \; v \in \pi(u) \; and \; w \in \pi(v) \; then \; w \in \pi(u)]\}_{u \in U},$$

i.e. for each $u \in U$ $\pi(u)$ is the set of elements preferred to u.

As in the case of orders defined by convex cones, the element $u \in U$ is *nondominated with respect to P* iff $\pi(u) \cap U = \{u\}$, which means that no other element of U is preferred to u. The set of nondominated points with respect to P will be denoted by $\Pi(U, F, P)$. If F and P are fixed or if F is an identity on U we will write just $\Pi(U)$.

In a common case, where the preference structure P is defined by a convex cone ζ,

$$\pi(u) := \pi(u, \zeta) = \{v \in U : F(v) \leq_{\zeta} F(u)\} \tag{2}$$

and $\Pi(U, F, P) = \Pi(U, F, \zeta)$. Conversely, in problem (1) $\Pi(U, F, \theta) = \Pi(U, F, P_\theta)$ with P_θ defined by (2). Now we can formulate the following:

Definition 1 The mapping $O(U, F, \theta, P)$, a multifunction of U, θ, P, as well as of F with values in the family of all subsets of U is termed a *free multicriteria optimizer* if for all U, F, θ, P $O(U, F, \theta, P) \subset \Pi(U, F, \theta)$ and the following implication holds

$$\Pi(U,F,\theta) \cap \Pi(U,F,P) \neq \varnothing \Rightarrow [O(U,F,\theta,P) \neq \varnothing \wedge O(U,F,\theta,P) \subset \Pi(U,F,\theta) \cap \Pi(U,F,P)].$$

If the latter condition is satisfied, but the Pareto optimality of $O(U, F, \theta, P)$ with respect to the problem (1) cannot be taken for granted, however either $O(U, F, \theta, P) \subset \Pi(U, F, \theta)$ or $O(U, F, \theta, P) \subset \Pi(U, F, P)$ then O will be termed simply a *free optimizer*.

If, beyond the criteria F, the ordering θ, and the preference structure P, an optimizer O takes into account an additional decision making rule R, such as a heuristics, numerical approximation, or a random choice rule from U then the set of solutions returned by this optimizer need not be contained in $\Pi(U, F, \theta)$. However, if $X := O(U, F, \theta, P)$ approximates in certain sense $\Pi(U, F, \theta)$, e.g. in terms of the Hausdorff distance, then O will be termed an *approximate multicriteria optimizer*. Since in most real-life optimization problems only approximate solutions are available, for the sake of brevity, whenever no ambiguity arises, approximate multicriteria optimizers will be referred to as multicriteria optimizers.

Observe that in a free multicriteria optimizer O with $P := P_\zeta$, where $\zeta \subset E$ is a convex cone, from the basic properties of domination structures it follows that for all $\theta \subset \zeta$,

$$\begin{aligned} O(U,F,\theta,P) &\subset \{u \in U:\ [\forall v \in U:\ F(v) \leq_\theta F(u) \Rightarrow v = u]\} \cap \{u \in U:\ [\forall v \in U:\ F(v) \leq_\zeta F(u) \Rightarrow v = u]\} \\ &= \Pi(U,F,\zeta). \end{aligned}$$

In the above case the preference structure represented by the cone ζ may result from an iterative process of gradually restricting the set of nondominated points to (1). This technique is referred to as the *contracting cone method* [4] since the dual cones to an increasing sequence of ordering cones $\theta \subset \zeta_1 \subset \zeta_2 \ldots \subset \zeta$ contract as do the sets $\Pi(U, F, \theta)$, $\Pi(U, F, \zeta_1)$, …, $\Pi(U, F, \zeta)$. Here, we refer to this methodology to show its similarity to the anticipatory network technique described in the

Anticipatory Decision-Making Problem (ADMP, cf. Sect. 3). Indeed, it can be seen [18] that the more anticipatory feedbacks taken into account in an anticipatory network with the starting node, the more opportunities exist to confine the choice in problem (1) to a smaller subset of the set $\Pi(U, F, \theta, P)$. If for all $u, v \in U$

$$F(v) \leq_{\theta} F(u) \Rightarrow v \in \pi(u)$$

then we will say that P conforms to the criteria F and order θ; in short P is *conforming*. Observe that this is the case if $P := P_{\zeta}$ and $\theta \subset \zeta$. If P is conforming then to select an $X \subset \Pi(U, F, \theta)$ the action of the optimizer can be stretched on the whole set U, without computing $\Pi(U, F, \theta)$, otherwise it must be restricted to $\Pi(U, F, \theta)$ yielding a bi-level optimization problem. Let us note that the computation or even an approximation of $\Pi(U, F, \theta)$ can be a hard task, so the conforming P are sought in the first order of importance when solving (1).

If in a free multicriteria optimizer we fix the preference structure P and the ordering cone θ then the resulting mapping $O_{P,\theta}$ will be termed a *multicriteria selection rule* if for all U, and a given class Φ of E-valued functions for each $F \in \Phi$ it selects a non-empty subset of the nondominated subset $\Pi(U, F, \theta)$ of U with respect to θ, i.e. if

$$O_{P,\theta}(U,F) := \Pi(U,F,\theta,P) \subset \Pi(U,F,\theta) := \{u \in U: [\forall v \in U:\ F(v) \leq_{\theta} F(u) \Rightarrow v = u]\} \quad (3)$$

It is easy to see that if P is conforming then $O_{P,\theta}$ is a multicriteria selection rule. A multicriteria selection rule is termed *proper* iff $\Pi(U, F, \theta, P)$ contains a single point. For instance, if P is defined by (2) and $\pi(u) := \{v \in U: vTx \leq 0\}$ for a certain $x \in \theta^*$, where θ^* is the dual cone to a convex, pointed θ (i.e. such that $\theta \cap (-\theta) = \{0\}$) with $int(\theta) \neq$, and Φ is such that $F(U)$ is closed, bounded and strictly convex then for all U and F $\Pi(U, F, \theta, P) = \{u_0\}$ and $u_0 \in \Pi(U, F, \theta)$. Thus the selection rule $O_{P,\theta}$ is proper.

As already mentioned, besides their optimizing capabilities, optimizers may form networks with several new properties compared to the theory of single or sequential decision problems. In particular, in feed-forward networks of optimizers constraints and preference structures in some optimizers are causally linked to the solutions of other problems and may depend on their preference structures. Thus, in a network of optimizers, the parameters of the actual instances of optimization problems to be solved vary as the results of solving other problems in the network.

Definition 2 If $O_1 := X_1(U_1, F_1, \theta_1, P_1)$ and $O_2 := X_2(U_2, F_2, \theta_2, P_2)$ are free multicriteria optimizers then a constraint influence relation r between O_1 and O_2 is defined as

$$O_1 r\, O_2 \Leftrightarrow \exists \varphi:\ X_1 \to 2^{U_2}:\ X_2 = \varphi(X_1). \quad (4)$$

Acyclic r are termed causal constraint influence relations—in short, causal relations.

Causal relations are represented by a (causal) *network of optimizers*. Definition 2 models the situation where the decision maker anticipating a decision output at a future optimizer can react by creating certain decision alternatives or forbidding them. This is described by influencing the constraints by multifunctions φ depending on the outputs from the preceding problems. As in [18] and [16], from this point on the term *causal network* will refer to the graph of a causal constraint influence relation. To complete the definition of anticipatory networks, we will define the anticipatory feedback relation.

Definition 3 Suppose that G is a causal network consisting of free optimizers and that an optimizer O_i in G precedes another optimizer, O_j, in the causal order r. Then the *anticipatory feedback* between O_j and O_i in G is information concerning the model-based anticipated output from O_j, which serves as an input influencing the choice of decision at optimizer O_i. Such a relation will be denoted by $f_{j,i}$.

By the above definition, the existence of an anticipatory information feedback between the optimizers O_n and O_m means that both condition below apply:

- the decision maker at O_m is able to anticipate the decisions to be made at O_n,
- the results of this anticipation are to be taken into account when selecting the decision at O_m.

The anticipatory feedback relation does not need to be transitive. As in the case of causal relations, there may also exist multiple types of anticipatory information feedback in a network, each related to the different way the anticipated future optimization results are considered at optimizer O_m. The multigraph of r (cf. (4)) and one or more anticipatory feedbacks define an anticipatory network of optimizers:

Definition 4 A causal network of optimizers Ω with the starting node O_0 and at least one anticipatory feedback relation linking O_0 or an optimizer O_n causally dependent on O_0 with another node in the network will be termed an *anticipatory network* (of optimizers). Ω is termed *proper* iff $n=0$.

In [18], the anticipatory information feedback in causal networks of optimizers was applied to selecting a solution to an optimization problem modelled by the starting element in an anticipatory optimizer network G. Specifically, while making the decision, the decision maker takes into account the following information contained in G:

- forecasts concerning the parameters of future decision problems represented by the decision sets U, criteria F, and the ordering structure of the criteria values θ,
- the anticipation concerning the behavior of future decision makers acting at optimizers, represented by the preference structures P,

- the forecasted causal dependence relations r linking the parameters of optimizers in the network,
- the anticipatory relations pointing out which future outcomes are relevant when making decisions at specified nodes and the anticipatory feedback conditions.

We will now present a few key definitions that refer to solving multicriteria decision problems using an anticipatory network of optimizers as a source of additional preference information.

Definition 5 An anticipatory network (of optimizers) is said to be *solvable* if the process of considering all anticipatory information feedbacks results in selecting a non-empty solution set at the starting problem.

Definition 6 A causal graph of optimizers G that can be embedded in a straight line will be called a *chain* of optimizers. If it contains at least one anticipatory feedback $f_{i,0}$ then G will be termed an *anticipatory chain (of optimizers)*

The causal constraint influence relations $\varphi(j)$ (cf. Definition 2) are defined as $\varphi(j) := Y_j^\circ F_i$, where the multifunctions $Y_j{:}F_i(U_i) \to\to U_j$ model the dependence of the scope of decisions available at O_j on the optimization outcomes of the problem O_i. Following [12], the total restriction of the decision scope at O_j generated by Y_j is denoted by R_j, i.e.

$$R_j{:} = Y_i(F_i(U_i)).$$

The resulting restriction of the set of nondominated outcomes at O_j is denoted by S_j. In an anticipatory chain, as exemplified in Fig. 1, there is only one predecessor of each optimizer O_i, for i > 0, so in the above formula i can be replaced by $j - 1$. By definition, the causal relation represented by $\varphi_{i,j}$ is *non restrictive* iff $S_j = \Pi(U_j, F_j, \theta_j)$. We will say that $\varphi_{i,j}$ *complies with* O_j iff $S_j \subset \Pi(U_j, F_j, \theta_j)$.

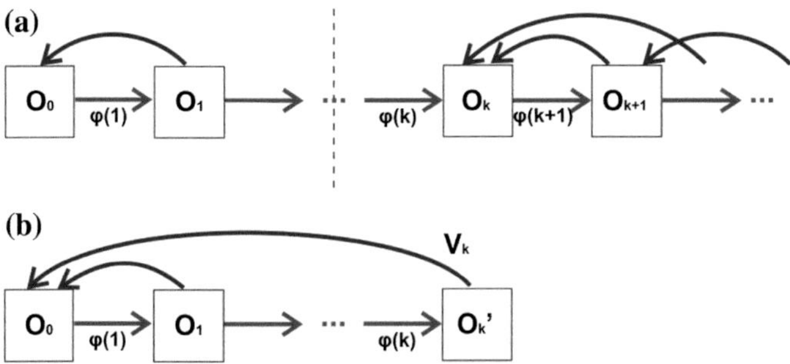

Fig. 1 Two examples of a chain of optimizers with anticipatory feedbacks: in figure **a** the anticipatory feedbacks linking O_k with O_{k+1}, and with further future optimizers have no influence on the decisions made at the starting node O_0. They are taking into account exclusively the decisions made at O_1, while the latter do not rely on any future considerations. The figure **b** depicts a case of anticipatively connected chain of optimizers. The temporal order complies in both cases with the causal relations defined by multifunctions $\varphi(j) := Y_j^\circ F_{j-1}$

Example 1 Figure 1 presents two examples of anticipatory chains of optimizers with different configurations of anticipatory feedbacks. In the case (a) above there are, in fact, two separate decision making problems, where a decision made at the optimizer $O_0 = (U_0, F_0, IR_+^{n(0)})$ is selected taking into account the anticipated outcomes at $O_1 = (U_1, F_1, IR_+^{n(1)})$ and $O_k = (U_k, F_k, IR_+^{n(k)})$ takes into account the outcomes of O_{k+1} and other future optimizers, but the decision made at O_0 is in no way related to the decisions made at the future decision nodes beyond O_1. Although O_1 influences future choices at $O_2, O_3, \ldots, O_k$, this optimizer does not select its decision in any way that may facilitate or hinder the achievement of any specific future goal. The solution of the problem $O_1 = (U_1, F_1, IR_+^{n(1)})$ is accomplished based on the local preference relation P_1 only. Thus the analysis of decisions in the anticipatory chain (a) may be decomposed into the analysis of the chains $O_0 \to O_1$ and $O_k \to O_{k+1} \to \cdots$. The case (b) presents an anticipatory chain equivalent to the decision situation considered in [12, 13], where the decision made at O_0 takes into account the outcomes of all subsequent problems, but there are no decision feedbacks between future optimizers.

Although the anticipatory chains constitute the simplest class of anticipatory networks, they are capable of describing a variety of sequential decision problems. A more advanced model is needed when an optimizer O_k influences two or more its immediate successors, say O_{k1} and O_{k2}, and the decision maker of O_k, or of any of its predecessors, is interested in the decision outcomes of both, O_{k1} and O_{k2}, or in the outcomes of two arbitrary optimizers that one following in the causal order O_{k1}, the other O_{k2}. If no optimizer is influenced by more than one immediate predecessor then the causal graph is a tree and we can formulate the following definition.

Definition 7 A causal graph of optimizers G that is a tree and contains at least two anticipatory feedbacks $f_{i,0}$ and $f_{j,0}$, each of them starting at optimizers that are not mutually causally conected, will be termed a *proper anticipatory tree* (*of optimizers*).

As a consequence, any anticipatory tree contains at least one optimizer that influences two or more its immediate successors. Such nodes in an anticipatory tree are termed *bifurcation optimizers*.

Along with chains, anticipatory trees are another special class of anticipatory networks that may be solved with dedicated algorithms. These are based on the decomposition of a tree into chains and on a subsequent analysis of them, starting from chains having a common bifurcation optimizer that is most distant in time [18]. Let us observe that if all anticipatory feedbacks in a tree were situated on one of its chains then the analysis of this tree could be reduced to just one chain. Anticipatory trees that possess this property are not proper.

Besides of bifurcation optimizers, the anticipatory trees contain a new phenomenon related to the anticipatory component of the multigraph, namely the *spurious anticipatory feedbacks*. These appear when the decision maker at an optimizer O_k would like to take into account the future outcomes of an optimizer O_m, but O_k and O_m are not causally connected, so there is no way to influence O_m to

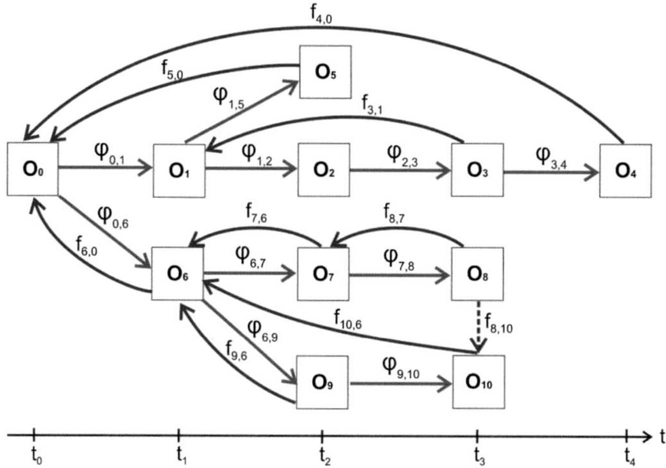

Fig. 2 An example of a tree of optimizers, with three bifurcation optimizers O_0, O_1 and O_6. Causal relations are defined by the multifunctions $\varphi_{i,j} := Y_i^\circ F_{i-1}$. Eight anticipatory feedback relations are denoted by $f_{k,m}$, where k is an index of a future node with the outcome taken into account by the optimizer O_m. There is also one spurious anticipatory feedback $f_{8,10}$

force or suggest a decision choice satisfying the condition defined in an anticipatory feedback.

An example of a tree of optimizers that contains a spurious anticipatory feedbacks is shown in Fig. 2.

3 Decision Making Problems in General Anticipatory Networks

In a non-trivial anticipatory network the following problem can be formulated:

Anticipatory Decision-Making Problem (ADMP, [17, 18]). For all chains of optimizers in an anticipatory network G with finite decision sets find the set of all admissible sequences of decisions $(u_0, \ldots, u_n)$ that minimize the function

$$g(u_0, \ldots, u_n) := \sum\nolimits_{i \in J(0)} h(u_i, q(0, i)) w_{0,i} \tag{5}$$

and such that for all i, $1 \leq i < n$, the truncated decision chain $(u_i, \ldots, u_n)$ minimizes

$$g(u_i, \ldots, u_n) := \sum\nolimits_{j \in J(i)} h(u_j, q(i, j)) w_{i,j}, \tag{6}$$

where $J(i)$, $i = 0, 1, \ldots, n$, denote the sets indices of decision units in G, which are in the anticipatory feedback relations with O_i and $w_{i,j}$ are positive coefficients corresponding to the relevance of each anticipatory feedback relation between the optimizers O_i and O_j. The function h may be defined as

$$h(u_j, q(i,j)) := \| F_j(u_j) - q(i,j) \|, \tag{7}$$

where $q(i, j)$ are user-defined reference levels of criteria F_i, for $i=0,1,...,n$.

From the formulation of the above decision-making problem it follows that the decision maker at O_0, while selecting the first element of an admissible decision sequence $u_0 \in U$ uses the anticipatory network G and the function g as an auxiliary preference structure to solve the problem (1). The key notion for the theory of anticipatory decision making can now be defined as follows:

Definition 8 A solution to the ADMP, a family of decision sequences $u_{0,m(0)}, \ldots, u_{N,m(N)}$ minimizing (5)–(7), will be called *anticipatory paths.*

Constructive solution algorithms for solving the ADMP take into account the information contained in an anticipatory network G. These have been proposed in [18] (Algorithms 1 and 2) for a class of anticipatory networks with discrete decision sets U_i, when the graph of causal relation r is either a chain or a tree. The anticipatory feedback conditions have been defined there as a requirement of O_i that the decisions at O_j, for j from a certain index set $J(i)$ such as O_i precedes O_j in causal order r are selected from the subsets $\{V_{ij}\}_{j \in J(i)}$, $V_{ij} \subset U_j$. Usually, this means that the values of criteria F_j admitted on V_{ij} are of special importance to the decision makers and can be defined as reference sets [14]. The general principles behind these algorithms are as follows:

- Decompose the anticipatory network into causal chains of optimizers linked by causal relations.
- Identify *elementary cycles* in each chain in the anticipatory network, i.e. cycles which do not contain other such cycles except themselves, consisting of causal relations along chains and anticipatory feedback relations.
- Solve the decision problem for each chain, by eliminating the elementary cycles.
- Use the logical conditions that defined the anticipatory requirements to bind the solution sequences to the common parts of the anticipatory chains.

Thus it is possible to reduce the analysis of anticipatory trees to a recursive analysis of anticipatory chains in the tree. Moreover, a general network can be decomposed into trees or chains, which makes it possible to apply solution rules for chains iteratively, gradually eliminating solved trees and chains. However, the solution procedures for anticipatory trees cannot be directly adopted for the solution of the problems where there may exist units that are influenced causally by two or more predecessors without taking into account synchronization problems.

Such networks can model problems where multiple resources, provided as outcomes of a number of different and independent decision processes, determine the scope of a later-stage decision. For example, to optimize the decisions in a potential future joint venture created to develop a new product (so-called NPD problem), the outputs provided by the potential future partners of this joint venture should be considered. It can be shown that taking into account the possibility of

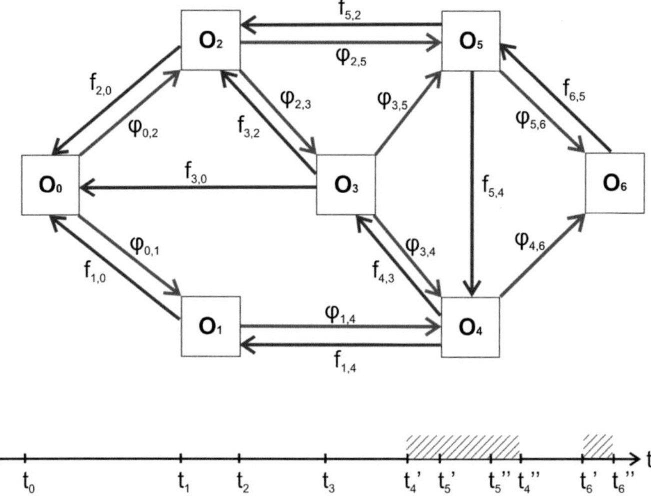

Fig. 3 A causal network of seven optimizers, where O_0, O_2, and O_3 are bifurcation optimizers, while O_4, O_5, and O_6 are each influenced by two predecessors. The shadowed areas between t_4' and t_4'' and between t_6' and t_6'' on the time axis denote the synchronization intervals for the simultaneous influence of O_1 and O_3 on the outcomes of O_4, and of O_4 and O_5 on O_6, respectively. The synchronisation interval for O_5, $[t_5', t_5'']$, is contained in $[t_4', t_4'']$. The anticipatory feedback $f_{5,4}$ between O_5 and O_4 is induced by the information flow from O_5 to O_4

creating future production alliances and representing such relations in an anticipatory network results in a competitive advantage over agents optimizing their own future outputs only. An example of a general anticipatory network is shown in Fig. 3.

To analyze general networked optimizers, it will need to be assumed that if an optimizer O_p is directly influenced by more than one predecessor then the aggregation rules are defined for each subset of influencing factors generated by the preceding optimizer (e.g. as an intersection or a union of the sets of feasible alternatives, each one imposed by a different preceding optimizer).

In addition, these rules must take into account the synchronization of influence that was not necessary in the case of anticipatory trees. Specifically, the simultaneous action of predecessors on O_p may be restricted to the prescribed time intervals. This is depicted above in Fig. 3, where t_i' and t_i'' denote the start and end of a synchronization time interval for the i-th optimizer.

In the most common situation, where the influence of preceding optimizers imposes a logical product of individual influences, the synchronization problem reduces to analysing the time intervals when the intersection of constraints resulting from multiple influencing multifunctions can still yield a feasible solution. However, in general, all combinations of logical conditions binding independent influences should be considered, including the situation where one agent's influence results in removing another agent's constraints. The analysis of such cases requires further studies, which, however, can be based on the solution scheme presented above and in [18].

The emergence of *induced anticipatory feedback* is a new phenomenon that could not occur in anticipatory chains of trees. In the example provided in Fig. 3 the anticipatory feedback $f_{5,4}$ between O_5 and O_4 is induced, but it is not spurious, as e.g. the spurious feedback $f_{8,10}$ in Fig. 2, although there is no causal relation between these optimizers. There are four necessary conditions for the existence of an induced anticipatory feedback. The first one is precedence in time of the decision made at the optimizer which is the target of the anticipatory feedback. The second is the existence of an information exchange between the optimizers that are source and target of induced anticipatory feedback. The third is the existence of at least one other optimizer that is influenced directly or indirectly by both, the source and the target optimizers of the induced anticipatory feedback, and the joint influence is restrictive, i.e. the influenced optimizer may select the decision from the intersection of sets provided as values of influencing multifunctions. The fourth condition is most specific and requires that the sets that define the anticipatory feedbacks at the source optimizer and commonly influenced optimizers were appropriately configured with respect to the values of influencing multifunctions. This is exemplified below.

Based on the example presented in Fig. 3, the phenomenon of induced feedback may be explained as follows: due to the assumed information exchange, the decision maker responsible for selecting the decision at O_4 knows both, the multifunction $\varphi_{5,6}$ that influences O_6 and the anticipatory feedback $f_{6,5}$. Similarly, the decision maker at O_5 knows the multifunction $\varphi_{4,6}$ and the actual choice made by O_4. Then O_4 can select such a decision $v \in \Pi(U_4)$ so that O_5 is forced to select a decision from the set $V_{5,4}$ that defines the anticipatory feedback $f_{5,4}$ to get a satisfactory solution of the problem solved by the commonly influenced optimizer O_6, specifically an element of $V_{6,5}$.

Of course, the more decision problems commonly influenced by the same pair of optimizers in a network, the more likely is the occurrence of induced anticipatory feedbacks. In addition, more complex configurations of induced feedbacks may arise if in an anticipatory network the same optimizer is causally influenced by three of more causally independent predecessors or when the information flow on the causal influences, anticipatory feedbacks, or preference structures is asymmetric.

The above presented phenomenon of induced anticipatory feedback, its extensions and relations to the topology and other properties of the network is a subject of an ongoing research that may potentially discover new applications of anticipatory networks as well as new problems to be investigated.

4 Anticipatory Networks as Superanticipatory Systems

Let us observe that in the above presented approach to solving anticipatory networks, we have assumed that anticipation is a universal principle governing the solution of optimization problems at all stages. In particular, future decision makers modelled at the starting decision node O_0 can in the same way take into account the network of their relative future optimizers when making their decisions. Thus, the future model of the decision maker at O_0 contains models of future agents including their respective future models. This has led us to introduce the notion of superanticipatory systems [15, 17] which directly generalize anticipatory systems in the Rosen sense [11] and weak anticipation in the Dubois sense [3]:

Definition 9 *A superanticipatory system* is an anticipatory system that contains at least one non-trivial model of another future anticipatory system.

Since a superanticipatory system is required to contain a model of *another* system, the above definition excludes the case where an anticipatory system models itself recursively. This is discussed later in this section.

By definition, this notion is idempotent, i.e. the inclusion of other superanticipatory systems in the model of the future of a superanticipatory system does not yield an extended class of systems since every superanticipatory system is also anticipatory.

Superanticipatory systems can be classified according to a grade that counts the number of nested superanticipations.

Definition 10 A superanticipatory system S *is of grade* n if it contains a model of a superanticipatory system of grade $n - 1$. An anticipatory system which does not contain any model of another anticipatory system is defined as superanticipatory of grade 0.

Let us note that the actual grade n of a superanticipatory system S depends on the accuracy of the model of other systems used by S. In addition, when constructing its model of the environment, S may underestimate the actual content of the other system models. Then, according to Definition 10, the grade of superanticipation of S should be regarded as a grade of the model, rather than the actual grade of the physical system.

It may be conjectured that if a superanticipatory system uses an empirical and rational modelling approach then it is more likely that the other systems will have models of a higher grade than S has estimated based on experiments. Thus the grade of the rational system S, when determined based on the information coming solely from the same system, can be regarded as the lower bound of an actual grade. Perfect knowledge of the grade can be attributed to a hypothetical ideal external observer only.

When referring to an anticipatory network, which is always a result of a certain modelling compromise, the following statement can be formulated

Theorem 1 Let $G = (O, r, f)$ be an anticipatory network, where O is the (finite) family of optimizers, r is the causal influence relation, and f is the anticipatory feedback relation. If G contains an anticipatory chain C such that there exist exactly

n optimizers in C, $\{O_{C,1}, \ldots, O_{C,n}\} \subset C = (O_1, \ldots, O_N, r, f)$, $N \geq n$, with the following property:

$$\forall i \in \{1, \ldots, n\} J_C(i) \neq \varnothing \text{ and } (\exists j \neq i\colon O_{C,j} r O_{C,i} \textit{ and } i \in J_C(j)), \quad (8)$$

where $J_C(i)$ is the set of indices of optimizers in G, which are in the anticipatory feedback relation with O_i. and no other chain in G has the property (8) with $m > n$ then G is a superanticipatory system of a grade of at least n.

The proof of the above Theorem 1 follows directly from the definitions of anticipatory networks (Definition 4) and superanticipatory systems (Definitions 9 and 10). Its first version appeared in [17].

It is easy to see that an anticipatory network containing a chain on n optimizers, each one linked with O_0 and with all its causal predecessors with an anticipatory feedback is an example of a superanticipatory system of grade n.

The notion of superanticipation is obviously related to the general recursive properties of anticipation. By definition, superanticipation makes sense only when the anticipation of the future is based on a predictive model. Problems to be solved that arise in a natural way are related to the accuracy of such models and to the grade of superanticipation. They are also related to the relation between *internal* (system) time, when the model is built and analyzed, and *external* real-life time, when the modelled objects evolve. A brief discussion of other recursive approaches related to anticipation such as recursion in Rosen's theory, Dubois' meta-anticipation, and information set models in multi-step games is given in [17].

A recursive anticipation can be applied in n-stage games, when one player anticipates the behavior of the others, cf. e.g. [7]. From the point of view of player G_1, anticipation is defined here for $k(G_1)$ steps forward and includes the anticipatory models for the other players $G_2, \ldots, G_N$. Each player can also possess a model of themselves (G_1) and of some or all the remaining participants with an anticipation horizon of $k(G_i)$ moves, $i = 2, \ldots, N$. Player G_1 thus fulfills the definition of a superanticipatory system and the game can be represented as an evolving anticipatory network. However, when the future moves of the other players result from a deterministic algorithm rather than from a decision-making model, anticipation may be based on the knowledge of the (deterministic) function identified with the operation of that algorithm. This may happen when a human player plays a deterministic game with a computer, or when a machine-machine interaction is modelled. Such games could be modelled by the master-slave (or driver-response) structure of the coupled system analogous to the leader-follower relation in multi stage Stackelberg games [5, 9].

5 Conclusions and a Discussion of Future Research Directions

The theory of anticipatory systems emerged originally in order to explain behavioral phenomena in systems biology, yet it turned out soon that it may also explain the collaboration and conflict patterns of human, artificial, as well as hybrid autonomous systems. Despite the efforts of its founder, Rosen, and Rosen's successors, in original formulation it contained a notion of 'internal model' of itself and other systems' future that has been regarded as vague and hard to implement constructively. The hitherto attempts to create a constructive theory of 'forward systems' symmetric to delayed control systems yielded a formal description that provided a correct computational framework in few cases only. A breakthrough was possible due to the introduction of the notion of anticipatory feedback that makes possible to describe the interaction between the models of the future systems and present-time decision makers, and of extending the anticipatory modelling to nested systems, whereas future generations of anticipatory systems have the same right as the present ones to define anticipatory feedbacks. The ability to define an anticipatory feedback is restricted to the case where the agent linked by an anticipatory feedback can causally influence the target agent of this feedback.

This paper examined the principle ideas concerning anticipatory networks as a new tool to model the multicriteria decision problems and the basic methods for solving them. Their extension, so-called superanticipatory systems, was also presented. We have also shown some potential further extensions of the theory of anticipation and presented research-in-progress on these topics.

Anticipatory networks may be applied to model and solve a broad range of problems. Apart from the above-mentioned potential uses in foresight, roadmapping, and socio-econometric modelling, there are further potential fields of application, such as:

- Anticipatory modelling of sustainable development: the underlying assumption of the anticipatory network theory, namely that the present decision maker wants to ensure that future decision makers have the best possible opportunities to make satisfactory decisions corresponds to the 'future generation' paradigm of sustainability theory. These 'future generations' are modelled by other network nodes.
- Anticipatory planning based on results of foresight studies, such as development trends, scenarios, and relevance rankings of key technologies, strategic goals, etc. Such planning can use deterministic as well as stochastic planning techniques and include multi-step game models.
- Anticipatory coordination of robotic swarms and human-robot systems, where anticipation is coped with multi-stage cooperative and leader-follower game models.

Anticipatory networks can also contribute to solving the stochastic optimization problems related to portfolio selection [1], road traffic management [6, 20], and to implementing the knowledge contained in foresight scenarios in a clear, formal way. Further real-life applications are discussed in [18] and [19].

Specifically, the development of the theory outlined above has been motivated by the problem of modelling the process of finding feasible foresight scenarios based on the identification of future decision-making processes and on anticipating their outcomes. Such an anticipatory network was applied in a recent information technology foresight project to build a strategy for a Regional Creativity Support Center [19]. Scenarios, such as those defined and used in foresight and strategic planning [4], may depend on the choice of a decision in one of the networked optimization problems and can be external-event driven. When included in a causal network of optimizers, the anticipation of future decisions and alternative external events would allow us to generate alternative structures of optimizers in the network.

Anticipatory networks, those that contain solely optimizers as well as hybrid ones [15], extend the plethora of modelling tools that can be used to formulate and solve decision making problems taking into account new future-dependent preference structures. When regarded as a class of world models for robotic systems, anticipatory networks provide a flexible representation of the outer environment, while superanticipation allows us to model collective decision phenomena in autonomous robot swarms. Further studies on this class of models may also contribute to the general theory of causality and lead to discovering surprising links to theoretical biology, quantum physics and the causal fields theory, as well as to the mirror neuron research in neurosciences. The theory of anticipatory networks links the ideas of anticipatory systems and models with multicriteria decision making, game theory, and the algorithmics. The formal methods that will be used to further develop the anticipatory networks theory include multigraphs and hypergraphs, dynamic programming in partially ordered spaces, controlled discrete-event systems, and general causality theory.

The above links and methods will make the research on networked anticipation truly interdisciplinary and may provide intriguing ties to the hard consciousness problem in cognitive sciences and the nature of time as qualia in the philosophy of mind. The relations to the theory of cooperative systems, specifically anticipatory robots, predictive and anticipatory control, foresight and backcasting, as well as to other areas of applicable basic research will assure the existence of a variety of potential real-life applications.

Finally, le us note that the above presented progress in the theory of anticipation and causality has appeared as a parsimonious effect of a foresight project devoted to modelling the ICT and AI futures, namely as a methodology to filter out the irrational technological and economic scenarios and to perform the technological and strategic planning.

Acknowledgments The author is grateful for the support of the research project No. WND-POIG.01.01.01-00-021/09:"Scenarios and development trends of selected information society technologies until 2025" funded by the ERDF within the Innovative Economy Operational Programme, 2006–2013.

References

1. Azevedo, C.R.B., Von Zuben, F.J.: Anticipatory stochastic multi-objective optimization for uncertainty handling in portfolio selection. In: Proceedings of the 2013 IEEE Congress on Evolutionary Computation (CEC), Cancun, Mexico, 20–23 June 2013. IEEE Computational Intelligence Society, pp. 157–164 (2013)
2. Bergstresser, K., Charnes, A., Yu, P.L.: Generalization of domination structures and nondominated solutions in multicriteria decision making. J. Optim. Theory Appl. **18**(1), 3–13 (1976)
3. Dubois D.M.: Mathematical Foundations of discrete and functional systems with strong and weak anticipations. In: Butz, M.V., et al. (eds.) Anticipatory Behavior in Adaptive Learning Systems, State-of-the-Art Survey, pp. 110–132. LNAI 2684, Springer (2003)
4. Godet, M.: Creating Futures—Scenario Planning as a Strategic Management Tool. Economica, London (2001)
5. Kaliszewski, I.: Quantitative Pareto Analysis by Cone Separation Technique. Kluwer Academic Publishers, Boston (1994)
6. Kanamori R., Takahashi J., Ito T.: Evaluation of anticipatory stigmergy strategies for traffic management. In: Proceedings of the 2012 IEEE Vehicular Networking Conference (VNC), Seoul, 14–16 Nov 2012, pp. 33–39 (2012)
7. Kijima, K., Xu, Ch.: Incentive strategies dealing with uncertainty about the followers MCDM behavior. Int. J. Systems Sci. **25**(9), 1427–1436 (1994)
8. Nadin, M.: Annotated bibliography. Anticipation. Int. J. General Syst. **39**(1), 35–133 (2010)
9. Nishizaki I., Sakawa M.: Cooperative and noncooperative multi-level programming. OR/CS Interfaces Series, vol. 48. Springer, Dordrecht, Heidelberg, London, New York (2009)
10. Poli, R.: An introduction to the ontology of anticipation. Futures **42**, 769–776 (2010)
11. Rosen, R.: Anticipatory Systems—Philosophical, Mathematical and Methodological Foundations. Pergamon Press, London (1985). 2nd edn, Springer, 2012
12. Skulimowski, A.M.J.: Solving vector optimization problems via multilevel analysis of foreseen consequences. Found. Control Eng. **10**(1), 25–38 (1985)
13. Skulimowski A.M.J.: Foreseen utility in multi-stage multicriteria optimization. In: Kacprzyk, J. (ed.) Seminar on Nonconventional Problems of Optimization, Warsaw, 9–11 May 1984. Proceedings, Part III, Prace IBS PAN 136, pp. 365–386 (1986)
14. Skulimowski, A.M.J.: Methods of multicriteria decision support based on reference sets. In: Caballero, R., Ruiz, F., Steuer, R.E. (eds.) Advances in Multiple Objective and Goal Programming, LNEMS 455, pp. 282–290. Springer, Berlin-Heidelberg-New York (1997)
15. Skulimowski A.M.J.: Hybrid anticipatory networks. In: Rutkowski, L., et al. (eds.) Proceedings of ICAISC 2012, LNAI 7268, Springer, Berlin-Heidelberg, pp. 706–715 (2012)
16. Skulimowski A.M.J.: Exploring the future with anticipatory networks. In: Garrido P.L., et al. (eds.) Physics, Computation, and the Mind—Advances and Challenges at Interfaces, Proceedings of the 12th Granada Seminar on Computational and Statistical Physics. La Herradura, Spain, 17–21 Sep 2012, AIP Conference Proceedings 1510, pp. 224–233 (2013)
17. Skulimowski A.M.J: Anticipatory networks and superanticipatory systems. Paper presented at the 10th International Conference CASYS, Liège, 5–9 Aug 2011. Int. J. Comput. Anticipatory Syst. **30**, 117–130 (2014)
18. Skulimowski, A.M.J.: Anticipatory network models of multicriteria decision-making processes. Int. J. Syst. Sci. **45**(1), 39–59 (2014). doi:10.1080/00207721.2012.670308

19. Skulimowski A.M.J.: Applying anticipatory networks to scenario planning and backcasting in technological foresight. In: 5th International Conference on Future-Oriented Technology Analysis (FTA)—Engage today to shape tomorrow, Brussels, 27–28 Nov 2014, 10 pp. https://ec.europa.eu/jrc/en/event/site/fta2014/programme/panel-session/cutting-edge-fta-approaches#28-Nov
20. Takahashi J., Kanamori R., Ito T.: Stability evaluation of route assignment strategy by a foresight-route under a decentralized processing environment. In: Raghavan, V., Hu, X., Liau, C.J., et al. (eds.) Proceedings of the 12th IEEE/WIC/ACM International Joint Conference on Intelligent Agent Technology (IAT 2013), Atlanta (GA), 17–20 Nov 2013, pp. 405–410. IEEE (2013)

On Modeling Human-Computer Co-Creativity

Oliver Hoffmann

Abstract Do we have a scientific model of creativity as emerging from contributions of computer users, computer systems and their interaction? Such a model would require describing the creative process in general, conditions for human creativity, the added value of human-computer cooperation as well as the role and power of computing. All of these topics have been the subject of research, but they have been addressed in different research communities. Potential obstacles for combining research results from research fields such as knowledge engineering and creativity research and properties of a general model of Human-Computer Co-Creativity are discussed.

Creativity Support Tools (CST) are a relatively recent topic in Human-Computer Interaction (HCI) [20]. But the goal of extending not just the limits of human productivity but also creativity has occupied a prominent place in the earlier history of computing. Computer-supported creativity had to be rediscovered after decades of scientific neglect. Is creativity a particularly hard challenge for computer science? On the one hand, creativity is a difficult topic in itself. On the other hand, we have developed a number of powerful computer technologies with the explicit goal of augmenting the human intellect, but we do not know whether these technologies have increased creativity. More than once, computer science has abandoned the topic of creativity as soon as a path towards extending computing power in itself was identified.

O. Hoffmann (✉)
Center for Medical Statistics, Informatics, and Intelligent Systems, Medical University of Vienna, Vienna, Austria
e-mail: oliver@hoffmann.org

S. Kunifuji et al. (eds.), *Knowledge, Information and Creativity Support Systems*, Advances in Intelligent Systems and Computing 416,
DOI 10.1007/978-3-319-27478-2_3

1 Augmenting Human Creativity

As early as 1945, aiding human creativity with computing machinery was proposed: "...creative thought and essentially repetitive thought are very different things. For the latter there are, and may be, powerful mechanical aids" [1]. Computers were expected to increase human creativity by alleviating the routine elements of scientific work. Did we reach this goal? With data-intensive science, we now have a research paradigm that relies completely on computational support [7]. But has research become more creative as a result? When Chaomei Chen developed CiteSpace and assessed scientific creativity, he still had to "identify principles that appear to be necessary for creative thinking from a diverse range of sources" [2]. Seven decades after Vannemar Bush, scientific creativity and its relationship to computer support are still poorly understood.

In 1960, J.C.R. Licklider extended the vision of supporting creative thought beyond the domain of scientific work as a generalized goal for achieving *man-computer symbiosis*: "The hope is that, in not too many years, human brains and computing machines will be coupled together very tightly, and that the resulting partnership will think as no human brain has ever thought and process data in a way not approached by the information-handling machines we know today" [8]. When Doug Engelbart followed up on the research into *augmenting human intellect*, he already emphasized the central role of *knowledge*: "The capabilities of prime interest are those associated with manipulating symbols and concepts in support of organizing and executing processes from which are ultimately derived human comprehension and problem solutions" [5].

This line of research was exceptionally successful: With the comparatively recent exception of touch devices, the combination of keyboard, graphical monitor, computer mouse and hypertext have been the standard computer interface over the last three decades. One might argue that our intellect was indeed augmented and our creativity supported by these technologies. If that is the case, we should in fact call them Creativity Support Tools. Why don't we? Because even though creativity support has been a recurrent research goal, a scientific method for identifying successful Human-Computer Co-Creativity (HC^3) was never developed. A recent systematic review of design theories for creativity support tools for instance states:

> Much work on creativity support systems provides design guidelines that inform about what set of features the class of information system should encompass in order to be successful [...] literature does not yet provide a wide range of design research efforts that test the effects of particular creativity support system design features [21].

We can evaluate computer system properties, but we do not have a reliable method for identifying a process of creative human-computer co-operation.

2 Research Fields Fragmentation

Human-Computer Co-Creativity is a process resulting in a creative outcome and involving one or more human individual(s) and one or more computer system(s). A model of HC^3 would therefore have to account for the role of each of these three elements as well as the relationships between them (Fig. 1). HCI has recently rediscovered Creativity Support Tools. But as a computing discipline, HCI is (again) focusing its efforts on the properties of the computer *tool*, rather than the creative process. Computational Creativity (CC) is investigating the entire creative process, but aims for developing creative computer systems as opposed to supporting human creativity. Human creativity itself is typically investigated independent of any computer interaction in Creativity Research (CR).

HC^3 research is fractured. While there are several research communities working on some of the relevant topics, there is no research community dedicated to understanding all of them in conjunction. The obvious task would then be combining research results from these communities. But trans-disciplinary research is facing fundamental differences in research culture. CR has grown out of psychology and its methods. HCI and CC are sub-disciplines of computer science. As engineering disciplines, the latter two are primarily concerned with working towards clearly defined outcomes, while the former places more emphasis on understanding a pre-existing phenomenon. When research is synthesized across these disciplines, key notions and their meaning have to be accommodated.

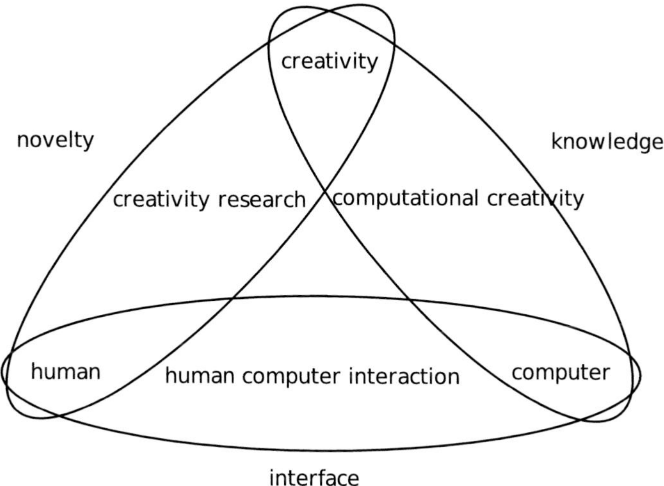

Fig. 1 Research topics for Human-Computer Co-Creativity

3 Knowledge in Computer Science

CC has its roots in Artificial Intelligence (AI). As in all AI-related disciplines, *knowledge* plays a key role in CC. In every day language, knowledge is understood as something an individual has or does not have. Whether or how this knowledge is or can be represented or transmitted through the use of symbolic representations is secondary. This is also the way *knowledge* is viewed in CC.

Computers do not develop and learn like human individuals, they are constructed by computer engineers. To give computers knowledge, AI researchers have invented various types of *knowledge representation*, leading to extended discussions on what constitutes the best type of knowledge representation.

Alan Newell has tried to settle this controversy within AI and his proposal is very illustrative of the content and method of AI research. He formulated a hypothesis claiming knowledge would reside in a virtual computer system level called the *knowledge level*. According to Newell [10] the levels in (Fig. 2) all share the following properties: Higher levels do not have to know about the specific implementation of lower levels. Each level can be reduced to the level below. And each level has its unique medium.

According to Newell's hypothesis, knowledge is a medium of computation just like electrons or software symbols and it consists of goals, actions and bodies, governed by the principle of rationality. By delegating knowledge to a new computer system level, Newell removed the need for deciding on the best form of knowledge representation: Any representation can implement knowledge in its own way.

But there is no knowledge without symbols, a view consistent with the earlier *physical symbol systems* hypothesis [11]. A discussion of competing views such as *subsymbolic AI* would be beyond the scope of this article, but as Newell correctly points out, his levels metaphor follows the practice of AI and computer science research in general.

The primary goal of computer science is to make computers more powerful and to ensure that computed results are correct at all times. Then how is correctness verified on the knowledge level? The highest directly observable system level contains symbols that might or might not encode some specific knowledge, but knowledge itself would be removed from direct observation. Newell proposed a mechanism of

Fig. 2 Computer levels according to Newell

knowledge level: knowledge

symbol level: software

register transfer sublevel: integrated circuits

logic circuit level: Boolean functions

circuit level: electric circuits

device level: individual transistors

indirect verification: If some (human or artificial) *agent* A can detect the impact of some specific knowledge in the actions of another agent B, agent A can verify the presence of such knowledge in agent B.

4 Objectivist/Realist World View

The world view underlying the computer science notion of knowledge can be described as *objectivist* or *realist*, which can be detected by discussing those aspects that authors like Newell do not discuss, and the questions that are not answered. A good example is: "Who is the agent that will verify the presence of knowledge in another agent?" More than one agent could take the role of the observing agent A in Newell's knowledge verification mechanism and the different As could come to different conclusions on whether knowledge is present in agent B (Fig. 3).

Newell does not explain how the A agents would be able to come to any consistent conclusion on agent B's knowledge, and since their symbol levels might be mutually incompatible, it is unclear how they could even communicate their different views on agent B. The only conceivable way Newell's knowledge verification might yield consistent results is by relying on a standardized observer agent A that serves as the absolute reference on verifying knowledge. Newell implicitly assumes *objective knowledge* that would be the basis, rather than the result of attempts at creating artificial intelligence.

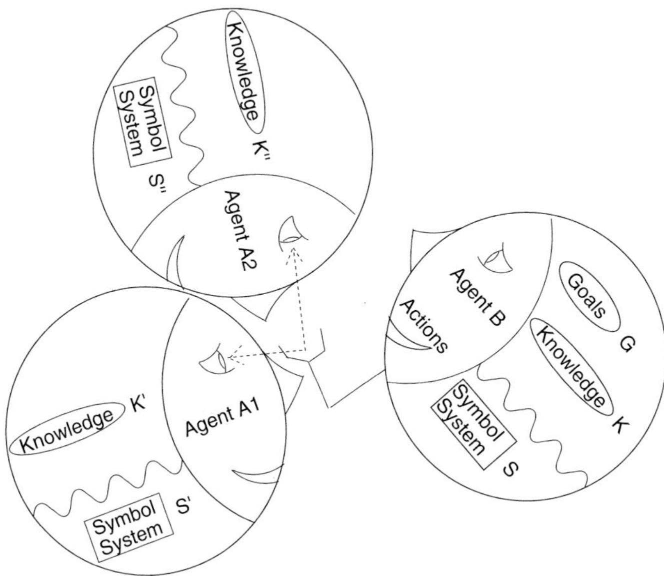

Fig. 3 How do different observing agents agree on the presence of knowledge?

Such assumptions are quite common in information technology. In fact, Shannon's definition of *information* [19] only works under the assumption of an objective observer for assessing the probability of the next symbol. Shannon actually used an object in the form of the English language dictionary as a means for replacing subjective meaning with an objective type of measurement.

For good reason, the objectivist/realist view has a long tradition in computer science. Computer systems are first of all objects which are required to work correctly and independent of the subjects that created them. The assumption (or implicit goal) of objective knowledge and objective correctness is therefore in general useful for this research field. But this view is in stark contrast to the more *subjectivist* or *constructivist* view in psychology and CR. Realist and constructivist world views each have their role in science and engineering [13], but the constructivist view is a better fit for topics such as learning or creativity. And some key notions from CR can only be fully appreciated within the constructivist view.

5 Divergent Thinking in Creativity Research

The standard definition of *creativity* in CR is bipartite: Creativity requires both *novelty* (also called *originality*) and *effectiveness* [17]. Effectiveness is quite compatible with the realist/objectivist world view adopted in CC, but novelty poses some challenges. Where does novelty come from?

In human creativity, two different modes of thinking have been observed: *Convergent thinking* and *divergent thinking* [6]. Problem-solving as search [9] is a typical objectivist approach in AI and a good example of convergent thinking: First all relevant constraints are determined and then possible solutions are checked for their consistency with these constraints. The end result will then be the solution with the least amount of divergence from what was required.

Divergent thinking does the exact opposite: exploring multiple paths of thought seemingly irrelevant and even inconsistent with the task at hand and only later merging some of the paths into a new, potentially surprising, and eventually also consistent result. Divergent thinking temporarily ignores all the criteria of correctness and objectivity that are at the heart of computer engineering. And divergent thinking has been shown to be an integral part of human creativity [16].

Convergent thinking is important for creativity as well. But the normal, convergent mode of thinking is regarded as a given in CR, and the divergent mode as a special addition required for true creativity. Divergent thinking is seen at the core, divergent thinking at the periphery of creativity. The four P's of creativity [15, 18] are a widely adopted framework of research consisting of Product, Person Process and Press. The Press portion stands for environmental and social constraints, which of course can only be accommodated with the appropriate amount of convergent thinking.

So the Press aspect of creativity can be seen as mandating a certain amount of realism/objectivism. But external constraints can also be interpreted from a radical constructivist point of view. In his highly influential study, Csikszentmihalyi [4] has for instance argued, that creativity can only be evaluated against its social frame of reference, denying any realist aspect of creativity.

Radical constructivism is of course not suited for describing computer systems, since computer systems are expected to work correctly and autonomously. The kind of radical realism adopted in AI on the other hand is ill suited for understanding human creativity which relies on the subject's ability of creating new and unexpected meaning. This does not imply that computer systems cannot be creative. It does however imply that the world view typically adopted in computing is inherently limited in its ability to describe and understand creativity. Divergent thinking is the one aspect of human creativity highlighting the shortcomings of the standard view of knowledge and meaning in computing: Any symbol, any meaning and the relevance of any information or assumption are up for re-interpretation in the divergent thinking stage of the creative process. HC^3 involves both human actors and computer systems, therefore a scientific model of HC^3 will have to accommodate both world views.

6 What a Model Could Look Like

In spite of the objectivist/realist notions for describing computer systems, AI researchers have repeatedly argued that their models do not have to conform to the objectivist requirement of falsifiability formulated by Popper [14]. Which assumptions from computing have been validated and should be incorporated in an integrated model?

Newells knowledge level hypothesis was for instance meant to be validated via the success of strong AI. Apart from rather technical implementation details on all the levels below the knowledge level, classical strong AI had only one falsifiable scientific hypothesis: That it would able to build intelligent robots and that it would be able to build them in a certain time frame. This one falsifiable hypothesis has turned out to be false and therefore the knowledge level hypothesis could be considered falsified.

But the research approach described by this hypothesis is still applied today, so his hypothesis should rather be regarded as a description of research culture. Rather than documenting all of their failures, computer engineers tend to modify their computer models until they arrive at a working prototype. Falsification has a very low priority in computing, to the degree that the criterion of falsifiability is argued to be irrelevant for instance in cognitive science [3].

While such a form of scientific discovery might be appropriate for the development of purely artificial systems, it certainly cannot be applied to a phenomenon involving human actors. A scientific model of HC^3 has to incorporate falsifiable hypotheses.

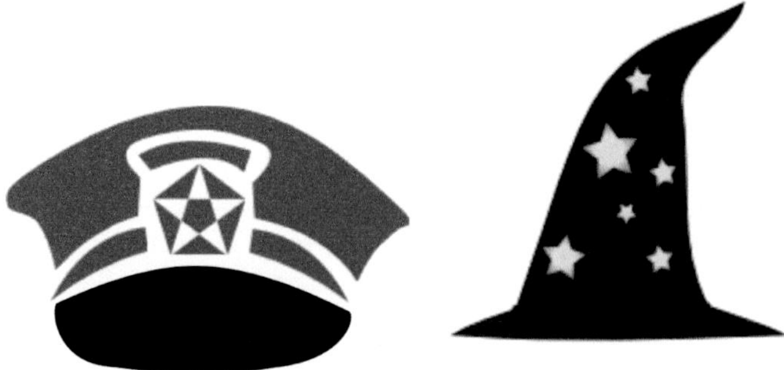

Fig. 4 Explicit convergent and divergent thinking roles

As opposed to the typical CC model, a model of HC^3 cannot be a computer model. The human and computer side of such an interaction are different and will have to be described in different ways. The objectivist/realist world view will be more suitable for describing the computer system, the subjectivist/constructivist view more suitable for describing the human partner. Their interaction will have to be described from both the human and computer perspective as well. The model should accommodate all the key notions from relevant research fields and describe compatible computer system properties. It should describe the interactions between humans and computers and the creative process. It should incorporate both an objectivist and a subjectivist understanding of knowledge and should explain the role of knowledge in the creative process.

I propose using an explicit description of human and computer *roles* as the starting point for such an integrated model. Human and computer partners might fulfill such a role in different ways, but in a creative co-operation, each partner should be able to perform identifiable roles at a given stage of the process (Fig. 4).

We know that humans cannot employ convergent and divergent thinking at the same time. If computers can in fact augment human creativity in the way described above, the computer partner should be able to augment the human partner by performing the matching, opposite role. The scenario described by Vannemar Bush can then be classified as the type of interaction with a human agent in the divergent and a computer agent in the convergent role (Fig. 5).

If both agents are to operate on a joint representation of the creative task, they will have to share some sort of knowledge representation. If divergent and convergent roles are identifiable, their impact on this knowledge representation should be identifiable as well. I propose the terms *diversion* and *consolidation* for the effect of applying divergent and convergent thinking on knowledge representation. Diversion of knowledge involves selectively disconnecting symbols, meaning and information and re-creating connections irrespective of external and internal constraints. Consolidation of knowledge involves merging previously inconsistent symbols, meaning and information into a new consistency. Concept blending [12] can for instance be seen as a type of knowledge consolidation.

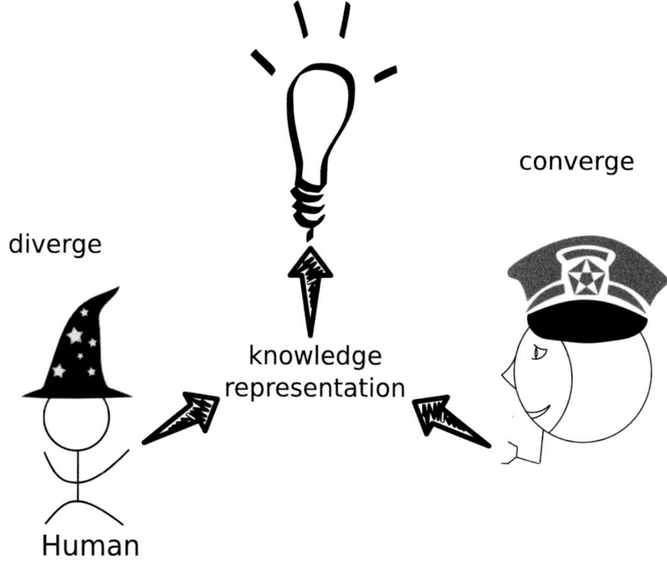

Fig. 5 Human agent in divergent and computer agent in convergent roles

In the general sense, this analysis is based on several hypotheses:

1. HC^3 exists as a unique phenomenon
2. Creativity as resulting from HC3 can be measured in the same way as human creativity
3. HC^3 is more than the sum of its parts: The creative co-operation of computer(s) and person(s) yield results that are either more creative or creative in a different way
4. The HC^3 process has observable properties and has general properties independent of application domain
5. the process of creative co-operation is different from other types of HCI
6. HC^3 requires both convergent and divergent thinking
7. divergent thinking involves the diversion of knowledge
8. convergent thinking depends on the previous consolidation of knowledge
9. human and computer partners can each perform the role of convergent or convergent agent, but in fundamentally different ways and never at the same time
10. Each step in the HC^3 process can be characterized by the flow of control and exchange of knowledge between human and computer partner

By such a deep integration of key research notions across disciplines, system properties such as openness or playfulness [21] can be analyzed in a systematic way, for instance as resulting from the availability of knowledge diversion tools. HC^3 is an iterative process, and in each of the process stages, human or computer can perform one of the following functions:

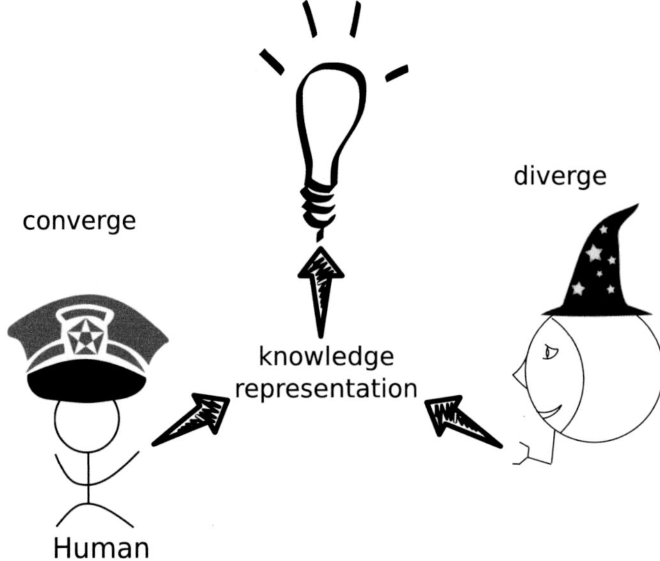

Fig. 6 Human agent in convergent and computer agent in divergent roles

1. diverge: multiple options are explored with little regard for consistency or correctness
2. compare: intermediate results are compared against original requirements
3. consolidate: previously inconsistent knowledge is blended into new consistent knowledge
4. converge: consistent solutions are generated

Among other things, these hypotheses entail that a scenario opposite to those described in the previous sections should be equally feasible: The computer as divergent agent and the human as convergent partner. One scenario would be the diverging computer partner generating seemingly inconsistent material with the human partner selecting and finally merging some of this material into something new (Fig. 6).

If this model is correct, human and computer can be creative in very different ways. The properties for the design of creativity support systems will be different for each of these scenarios, but if these hypotheses are correct, some properties will be relevant for all of them. Neither human nor computer would have to be creative on their own. But if we want to identify Human-Computer Co-Creativity, then we should be able to identify the stages of such a process as well as the overall creative result.

7 Conclusion

Human-Computer Co-Creativity is an old research goal which has seen renewed interest in the form of Creativity Support Tools in the last years. A general model of HC^3 would have to describe any kind of co-creativity and account for all relevant human and computer contributions as one system. Such a general model of the process of HC^3 is still missing.

Elements for such a model can be found in research fields for different aspects of HC^3. Therefore a synthesis of research results is an appropriate option for building a model. But the respective research communities are exhibiting fundamental differences in research culture and epistemology, which have to be acknowledged and considered. A unified model should include the key concept of divergent thinking and account for the processing and exchange of knowledge.

Explicit specification of roles played by human or computer partner can serve as a means for integrating these aspects. I am formulating the hypothesis that both human and computer can be the agents for either convergent or divergent thinking in the creative process. Verification or falsification of this model will depend on a mechanism for identifying these roles in an empirical setting.

References

1. Bush, V.: As we may think. Atlantic Monthly (1945)
2. Chen, C.: Turning Points : The Nature of Creativity. Springer, Berlin (2012)
3. Cooper, R.P.: The role of falsification in the development of cognitive architectures: Insights from a lakatosian analysis. Cogn. Sci. **31**(3), 509–533 (2007)
4. Csikszentmihalyi, M.: Creativity—Flow and the Psychology of Discovery and Invention. Harper Perennial, New York (1997)
5. Engelbart, D.C.: Augmenting human intellect—a conceptual framework, Technical Report AFOSR-3233, Stanford Research Institute, Menlo Park, CA, USA (1962)
6. Guilford, J.P.: Creativity. Am. Psychol. **5**, 444–454 (1950)
7. Hey, T., Tansley, S., Tolle, K.: The Fourth Paradigm: Data-Intensive Scientific Discovery. Microsoft Research, Redmond (2009)
8. Licklider, J.C.R.: Man-computer symbiosis. IRE Trans. Hum. Factors in Electron. HFE-1, 4–11 (1960)
9. Newell, A.: Human Problem Solving. Prentice-Hall Inc., New Jersey (1972)
10. Newell, A.: The knowledge level. Artif. Intell. **18**(1), 87–127 (1982)
11. Newell, A., Simon, H.A.: Computer science as empirical inquiry: symbols and search. Commun. ACM **19**(3), 113–126 (1976)
12. Pereira, F.C.: Creativity and Artificial Intelligence: A Conceptual Blending Approach. Walter de Gruyter and Co., Hawthorne (2007)
13. Peschl, M.F.: Constructivism, cognition, and science: an investigation of its links and possible shortcomings. Found. Sci. **6**(1–3) (2001)
14. Popper, K.: Logik der Forschung. Springer, Vienna (1935)
15. Rhodes, M.: An analysis of creativity. The Phi Delta Kappan **42**(7), 305–310 (1961)
16. Runco, M.A., Acara, S.: Divergent thinking as an indicator of creative potential. Creat. Res. J. (2012)

17. Runco, M.A., Jaegera, G.J.: The standard definition of creativity. Creat. Res. J. **24**(1), 92–96 (2012)
18. Runco, M.A., Pritzker, S.R.: Encyclopedia of creativity. Elsevier (1999)
19. Shannon, C.E.: A mathematical theory of communication. Bell Syst. Tech. J. **27**, 379–423 and 623–656 (1948)
20. Shneiderman, B., Fischer, G., Czerwinski, M., Resnick, M., Myers, B., Candy, L., Edmonds, E., Eisenberg, M., Giaccardi, E., Hewett, T., Jennings, P., Kules, B., Nakakoji, K., Nunamaker, J., Pausch, R., Selker, T., Sylvan, E., Terry, M.: Creativity support tools: report from a U.S. national science foundation sponsored workshop. Int. J. Hum. Comput. Interact. **20**(2) (2006)
21. Voigt, M., Niehaves, B., Becker, J.: Towards a unified design theory for creativity support systems. Design Science Research in Information Systems. Springer, Heidelberg (2012)

Comparison of the Two Companies' Staff by the Group Idea-Marathon Training on Creativity and Motivation

Takeo Higuchi

Abstract This paper presents the results of a comparative analysis of different levels of motivation for the R Company staff (R Group) and the P Company staff (P Group) through the 3 month Group Idea-Marathon training (G-IMS) on Creativity. Various observations during the training seminars and the support system show that the R Group was less serious in the course than the P Group. In this paper, the two companies' group training were analyzed by comparing creativity effects based on their respective motivation. The R Group participants were mainly forced by their department heads to participate in this Idea-Marathon training or just nominated to participate. Many staff of the R Group did not appear at the second or third training workshop within 3 months. They did not fulfill the recommended assignment of thinking and writing in their notebook every day. On the other hand, the P Group participants, however, mainly consisted of staff who voluntarily participated in IMS seminar. All the members of both groups were given the same creativity tests of the Torrance Tests of Creative Thinking (TTCT) as Pretest and Posttest with 3 months training of the G-IMS. The T-test analysis of both group scores for five Norms Reference Measures showed that the P Group improved more than the R Group in creativity while the Mann-Whitney U-test for 13 items Criterion Measures showed that the R Group had more variety of improvement in strength than the P Group. But the two-factor factoral ANOVA analysis between the R and P Groups, and within and Pretest and Posttest show that there were no difference statistically in TTCT main creativity Items. Reviewing the result of ANOVA and ETS score, IMS was found to be effective in improving creativity in the R & P Groups and we noticed that the R Group could be more creative if they had stronger motivation in their self-enlightenment.

Keywords Idea-Marathon system (IMS) · Torrance tests of creative thinking (TTCT) · Support systems · Motivation

T. Higuchi (✉)
Idea-Marathon Institute, Tokyo, Japan
e-mail: info@idea-marathon.net

S. Kunifuji et al. (eds.), *Knowledge, Information and Creativity Support Systems*,
Advances in Intelligent Systems and Computing 416,
DOI 10.1007/978-3-319-27478-2_4

1 Introduction

Today, any companies and laboratories in Japan and in the world are seeking higher creativity for their staff and researchers [1]. Many seminars with various kinds of newest popular creativity thinking methods are being offered to their staff, and the staff might voluntarily attend the seminar or attend with the feeling to be forced to.

Many major Japanese companies have been challenged by China, Korea and other globalized companies. Sony, Sharp and Panasonic have been in a painful situation in recent years. The only way to escape from this adverse situation is to create unique new products, for which before anything group creativity is required. But through Lost Decades in Japan, staff of many companies have been said to become less curious, non-confident and non-creative. Many staff might have been caught by a kind of learned helplessness, which was warned by Seligman [8].

When the author was giving IMS training in the two R & P Companies, it was observed that the enthusiasm in the P Group was stronger than that in the R Group. (1) The general attitude of the participants of both companies was almost the same and also both were fairly devoted during the training seminar period. (2) The rate of absentees of the R Group (66.7 %) at the third training seminar (3 months later) was much higher than that for the P Group (4.5 %). (3) One of the Support Systems of G-IMS training, ETS (See Sect. 2.2) had shown that the R Group recorded much less than the P Group (Table 1). (4) It was estimated that the P Group had stronger motivation than the R Group.

Almost all the original members of the P Group attended the third training seminar while many of the R Group were absent in their third training seminars.

Table 1 Comparison of R (*left*) and P (*right*) and no. of ideas after 3 months

Participants ID	5th ETS No.of days	5th ETS No. of Ideas	Balance of Ideas
1	87	30	-57
2	87	110	23
3	87	140	53
4	87	145	58
5	87	90	3
6	87	121	34
7	87	55	-32
8	87	142	55
9	87	27	-60
10	87	43	-44
11	87	87	0
12	87	175	88
13	87	88	1
14	87	54	-33
15	87	87	0
16	87	65	-22
17	87	36	-51
18	87	90	3
19	87	45	-42
20	87	30	-57
21	87	98	11
22	87	96	9
23	87	91	4
Idea no. Average per person		84.60	
Idea no. Average per person per day		0.97	

Participants ID	6th ETS No.of days	6th ETS No.of Ideas	Balance of Ideas
1	92	192	100
2	92	131	39
3	92	114	22
4	92	273	181
5	92	154	62
6	92	163	71
7	92	98	6
8	92	115	23
9	92	135	43
10	92	212	120
11	92	392	300
12	92	111	19
13	92	308	216
14	92	152	60
15	92	98	6
16	92	142	50
17	92	239	147
18	92	158	66
19	92	101	9
20	92	101	9
21	92	292	200
Ideano. Average per person		175.30	
Ideano. Average per person perday		1.91	

Therefore, we consider that the high rate of absentees in the R Group was a primary evidence of low motivation.

From these facts, we hypothesize that, if the participants had a high level of motivation, IMS support system would be well followed, and also the result of the TTCT score could be higher as well, and if the participants had a lower level of motivation, or a kind of helplessness, the support system would be less followed and the TTCT test result would be affected accordingly.

In this paper, two companies' group experimental results were studied.

In Sect. 1 Introduction, we explain the origin of our hypothesis of these two companies' staff. In Sect. 2, we explain IMS and the detailed effects of support system of ETS (e-Training System) and supply of thinking hints. We emphasize the difference of the effects between ETS and non-ETS in IMS training. In Sect. 3, we present the TTCT tests. In Sect. 4, we present the comparative analysis of two companies' staff. In Sect. 5, we discuss the results. We conclude in Sect. 6 and discuss future research, followed by references.

2 IMS and Its Support Systems

2.1 Basic Concept of IMS

In 1984, IMS System was developed and implemented by the author, who published the first book on IMS in Japanese [2]. Since 1984, the author has continued to practice IMS almost every day, thus for the last 30 years. IMS books were published also in English [3], Korean, Chinese, Thai, Hindi and Nepali.

The principles of IMS are defined as follows: (1) Use notebooks of the same kind. (2) Generate new ideas daily and write them in the notebook chronologically, numbering each idea regardless of any categories. (3) Draw pictures to illustrate your ideas as often as possible. (4) Talk to your neighbors. (5) Review your ideas. (6) Implement the best ideas [4].

IMS is a process which involves daily idea creation and immediate writing in notebooks individually following a chronological order. The ideas expressed are not limited to any specific area or topic but may encompass any subject area. As IMS becomes a daily habit, the practitioner will be encouraged by experiencing the power of continuity, instantaneity and self-confidence that stem from increased creativity. A synergistic effect can also be expected to emerge among those who practice IMS in groups if they discuss the ideas in their notebooks. Since IMS Institute was established in 2004, the institute administered the group training of many companies and laboratories. To make wider IMS effects in companies and laboratories in Japan, IMS has been studied quantitatively for its potential to enhance creativity. TTCT Figural tests were implemented for studies to prove the creative effect of IMS in details. Prior to this experiments at two companies, the creativity effect of IMS for researchers at one major food manufacturer's laboratory near Tokyo was already tested and analyzed by Higuchi et al. [5].

And the P Group, one of the two companies discussed in this paper, has been tested, analyzed and discussed in the author's dissertation [6]. After we made another group training of IMS for 3 months at R Group staff with Pretest and Posttest of TTCT, we made our comparative study between R & P staff using the TTCT figural test Pretest and Posttest.

2.2 ETS (e-Training System) and Its Weekly e-Hints

When the author in the past tried to persuade a group of company staff to start IMS by giving them several lectures over 3 months, many, many of them stopped their practice by twos and threes and eventually about 10 % remained practicing up to the end of 3 months. IMS participants might drop out of long term training easily before the final training schedule is completed if the trainees are left untouched except IMS lectures. Even though the Thinking Hints are supplied every week, some people do not utilize them, and, therefore, they still stop practicing IMS. This kind of drop-outs can happen, not only in IMS training, but also in any kind of company training. In usual company training, it is not rare or surprising to see low rates of participation unless management forces all members to participate without fail.

However, we do not want to force participants to continue IMS practices since the core of this training is self-discipline. In order to make "Thinking into Writing" a new habit, we have to avoid any compulsory pressure. To satisfy this condition, to prevent drop-outs from arising and to have all of the initial participants achieve the final goal of the training, we developed the e-Training System (ETS), through which all participants are requested to inform IMS training lecturer regularly by internet only the total number of ideas he or she has generated. If a participant does not do so on the due date, then we start attempting persuasion to that participant through the internet.

Upon receiving the total list of ideas sent by each participant, the training lecturer judges how each participant is doing by analyzing past progress, and sends to each participant a personal comment tailored to that person's particular situation. We have a rule that the feedback given to each participant has to be somehow different or original and suited to each participant in particular.

Even if this ETS support system is using the internet, it is not an automatic reply system that sends the same reply to everyone with just one part of the overall comment altered to fit each participant. This point is very important in eradicating the participants' drop-outs while he/she participates in the training. If the ETS were a fully automatic response system using computer software, we know that the participants would not listen to a machine's advice. Even if we use the internet, we believe that all the participants check if ETS comments are handled and answered by human. They might cross-check with other participants to compare the comments they received.

The ETS is a very human system that, though it takes time and pain, appeals to all participants to continue their habit-making effort. The problem is that in the case of a large number of participants, such as 250 participants simultaneously, it is quite daunting and time consuming to provide so much personal feedback to each of them.

Thus, we are planning to train in-house ETS Experts in companies so that in future, ETS can be carried out inside organizations. This highly personalized ETS is a major reason for the success of the continuation of IMS, that is, for accomplishing the goal of full participation in a 6 month program.

This highly personalized feedback to each participant, which is the core feature of the ETS, can be applied to many other internet programs and e-Learning systems.

Each comment makes the participants aware of their position and progress, as well as informs them on how to increase the number of their ideas and how to remember to practice IMS. This entire internet interaction is called ETS, and it includes the total numbers of ideas reported by participants along with the replies and comments from IMS lecturers (Fig. 1).

ETS is an essential aspect of IMS training course. In the event that someone stops engaging in IMS for an extended period during the training course, we can detect the stoppage within 2 or 3 week time and encourage the person to resume IMS by supplying extra emergency-hints for ideas. These direct comments, which are different for each participant, are appreciated by the participants since they

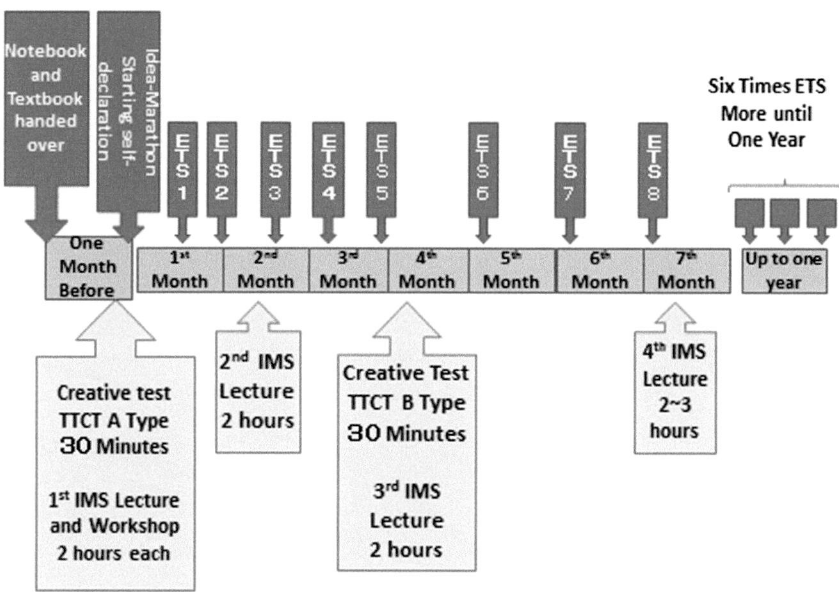

Fig. 1 Idea-Marathon 6 month training schedule

understand that these comments take time to create. We provide eight to nine ETS communication sessions within the first 3 months of IMS training. After the first IMS training session, the second training is provided one month later, and the third one after 3 months, and the fourth one after 6 months (Fig. 1).

2.3 Comparison of IMS with ETS and Without ETS

In the event that the ETS is applied for 3 months [2013]

In the event that the ETS was fully applied to IMS training for business staff, every 2 weeks checking took place in the number of ideas, returning comments. In Tokyo, in May 23, 2013, participants started IMS in a school for working students. After 3 months, all 23 participants were still continuing IMS since they had been backed up by the ETS and by a weekly supply of thinking hints (Fig. 2).

In the case where the ETS was not applied [2011]

In the case where the ETS was not applied and IMS participants were only asked to report the number of their ideas without returning comments, many participants stopped practicing IMS before the end of the first 3 months (Fig. 3).

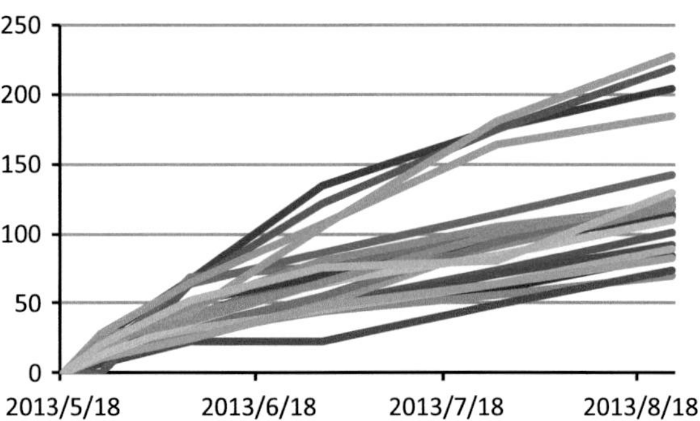

Fig. 2 In the case where the ETS was fully applied

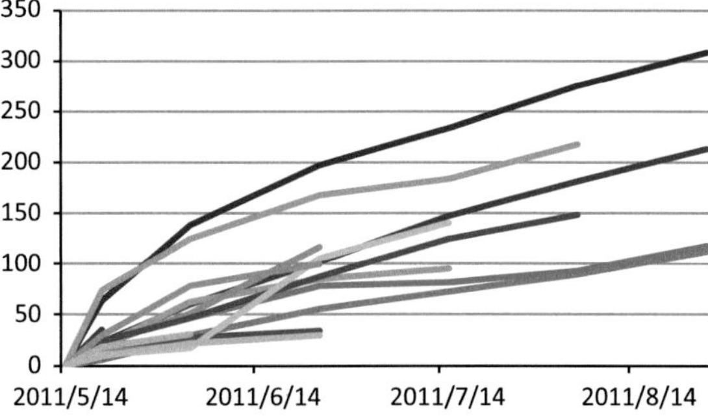

Fig. 3 In the case where the ETS was not applied and only the number of ideas was checked

2.4 Established Effects of ETS

After starting IMS, if IMS participants are not advised and encouraged regularly and properly before the next training seminar taking place a few months later, many participants will be naturally and gradually getting not steady, slowing down or stopping practicing IMS as their motivation decreases.

These comments from IMS Institute will keep positively encouraging participants to continue or accelerate IMS. This is named ETS (e-Training System). During 3 months of IMS training, 5 to 6 times ETS were implemented. In these experiments at two companies, we tried to study the correlation between the participants score of staff of both companies of creativity test and number of ideas recorded in ETS.

2.5 ETS Data of R & P Groups as Idea-Numbers, Average Per Person

We checked ETS data and found that the ETS result of both those companies greatly differed. Both the R & P Groups are large, busy electronic and software companies. However, in the case of the R Group, the rate of attendance to the third IMS lecture/workshop within a 3 month period was exceptionally low (Table 1).

Of course this rate of absence includes unavoidable business assignments for the staff to stop attending the third lecture/workshop. But still we understand many avoided the third IMS lecture/workshop intentionally and without any business reason. These absentees thought that there would be others attending in the training. Therefore, if they could be sleepers, the training seminar would go on without them.

This is a kind of irresponsible idleness that each of these absentees would like to be a sleeper as they knew that they will not be officially complained by the management of their absence in the training through their past experience.

In the case of the P Group, at the first lecture/workshop (with TTCT tests Type A), the number of the participants was 25 persons in total, but at the third lecture/workshop (with TTCT tests Type B) after 3 months, the participants decreased to 21 persons, which was an 84 % attendance rate of the total number of participants.

If we are supposed to create and write one idea per day, after almost 3 months, the balance of ideas can indicate how the participants are creating every day (Fig. 4).

In Fig. 1 above, it seems that both the R & P Groups remarked increasing ideas. However, it is clear that the number of ideas per person per day for the R Group

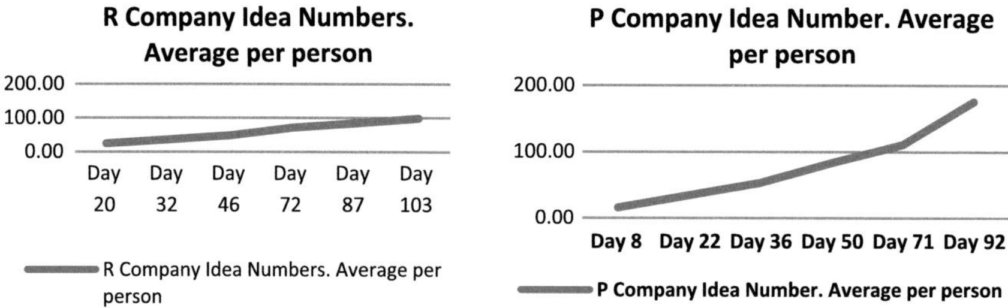

Fig. 4 ETS Data of R group (*Left*) and P group (*Right*) and numbers of ideas per person

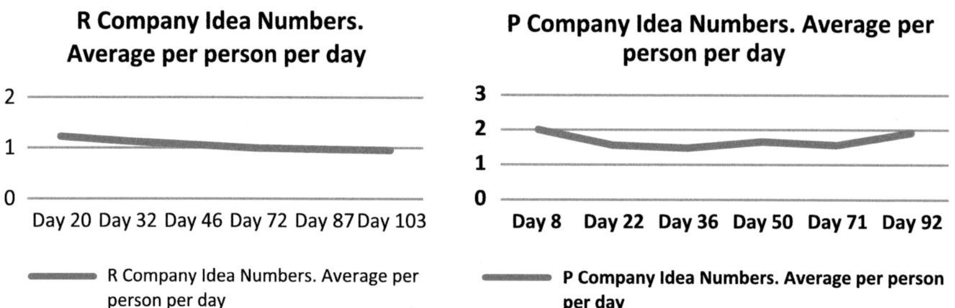

Fig. 5 ETS Data of R (*Left*) and P staff (*Right*) as numbers of ideas per person per day

(left) is slowly going down below "one idea per day per person", while it is going up to almost "two ideas per day per person" in the case of the P Group (right) (Fig. 2).

"Below one" means that the members of the R Group were not creating one idea per day per person but they were skipping some days, without any ideas. However, the P Group was creating nearly two ideas per day per person.

2.6 Weekly e-Hints

At the beginning stage of IMS training, it is important for beginners to find some thinking tips and hints for creating ideas in order to continue IMS every day.

To improve this situation, participants of IMS training can receive weekly hints from IMS Institute by e-mail (e-Hints), from which they may create new ideas until they get used to creating hints themselves. Some examples of e-Hints are: "Create a new Sushi recipe", "Think of a new type of vending machine" or "Think of new functions and uses for a calculator", and so on.

3 Creativity Measurement Using TTCT

To evaluate the creative effect of IMS training on both companies, we used the Torrance Tests of Creative Thinking (TTCT) Figural tests. We carried out the TTCT figural tests for groups of both companies by the Pretest in the first lecture and the Posttest in the third lecture to obtain quantitative creativity factors of IMS training. There are two types of TTCT Figural tests (Type A as the Pre-test and Type B as the Post-test) which were used to measure the creativity effects of continuous training of IMS for the 3 months period. TTCT Figural tests consist of the following five Norm-referenced measures [9]: **(1) Fluency**: The number of expressed ideas which meaningfully utilize the stimulus. Fluency is the gatekeeper of TTCT analysis. **(2) Originality**: The uniqueness of the ideas in terms of statistical infrequency. **(3) Elaborations**: The imagination and exposition of detail in the pictures. **(4) Abstractness of Titles**: The level of synthesis and organization evidenced in the titles in order to express the essence of the pictures. **(5) Resistance to Premature Closure (RPC)**: The ability to consider all relevant information and resist impulsive, premature conclusions.

In addition to the above-mentioned 5 Norms Reference Measures, there are 13 more Creative Strengths as Criterion–Referenced Measures in TTCT analysis to check the existence or nonexistence of: Emotional Expressiveness, Storytelling Articulateness, Movement and Action, Expressiveness of Titles, Synthesis of Incomplete Figures, Synthesis of Lines and Circles, Unusual Visualization, Internal Visualization, Extending or Breaking Boundaries, Humor, Richness of Imagery, Colorfulness of Imagery and Fantasy.

To make the inter-rater scoring, the first author obtained the official certificate for scoring TTCT tests Figural Type A test and Type B test at a scoring seminar held by the TTCT Center of the University of Georgia, from October 1 to 3, 2012.

4 Comparative Analysis of the Two Companies

4.1 t-Test Analysis of TTCT Score of Companies R and P

For the R Group, the *t*-Test for five Norm-Referenced Measures of TTCT Figural tests between Pretest and Posttest scores was studied. A statistically significant difference was found on Fluency $t(23) = 2.144$, $p < 0.05$. No other significant difference was found in the component of the Total Score, Originality, Elaborations, Abstractness of Titles and Resistance to Premature Closure. Figures in Table 2 depict average scores.

For the P Group, the *t*-test for TTCT Figural test Pretest and Posttest scores for five Norm-Referenced Measures showed statistically significant results in Score Total $t(20) = 2.187$, $p < 0.05$ and Originality $t(20) = 3.371$, $p < 0.01$. There was no significant result in Fluency, Elaborations, Abstractness of Titles and RPC.

Table 2 t-test results of TTCT Scores of R company (*Left*) and P company (*Right*)

R Company TTCT Pre-Post test t-Test

Measurement	Pre-test M(SD)	Post-test M(SD)	*t*-value	Sig
	Change of Scores between Pre-Post tests			
Total Score	562 (59)	574 (67)	0.79	*n.s.*
Fluency	89 (15)	99 (17)	2.144	*p*<.05
Originality	108 (22)	117 (20)	1.927	*n.s.*
Elaborations	150 (12)	150 (26)	0.219	*n.s.*
Abstractness of Titles	110 (26)	111 (25)	0.103	*n.s.*
RPC	105 (20)	98 (17)	1.196	*n.s.*

M=Score Means SD=Standard Deviation

df=23, *p*<.05 *p*<.01(Two sided t-test)

P Company TTCT Pre-Post Score t-Test

Measurement	Pre-test M(SD)	Post-test M(SD)	*t*-value	Sig
Score Total	557(97)	591(85)	2.187	*p*<.05
Fluency	99(19)	110(25)	1.99	*n.s.*
Originality	111(27)	126(23)	3.371	*p*<.01
Elaborations	145(18)	152(11)	1.878	*n.s.*
Abstractness of Titles	106(24)	106(34)	0.018	*n.s.*
RPC	95(22)	97(17)	0.576	*n.s.*

M=Score Means, SD=Standard Deviation

df=20, *p*<.05, *p*<.01 (Two sided t-test)

4.2 13 Items Criterion-Referenced Measures Scores of TTCT of R and P Company Staff

13 Items Criterion-Referenced Measures Scores of the TTCT test Pretest and Posttest in both R & P staff were analyzed with the Mann Whitney U-test: (Table 3).

Table 3 Comparison of both R & P staff on 13 items criterion-referenced measures score of TTCT Figural test between pretest and posttest with the Mann Whitney U-test

R Company 13 Items Criterion-Referenced Measures of TTCT Score

	Criterion Reference Measure	Pre-test Mean Average	Post-test Mean Average	*t*-Value	Sig
1	Emotional Expressiveness	23	26		*n.s.*
2	Story-telling Articulateness	21	28		*n.s.*
3	Movement and action	23	26		*n.s.*
4	Expressiveness of Titles	23	26		*n.s.*
5	Synthesis of Incomplete Figures	25	21		*n.s.*
6	Synthesis of Lines or Circles	18	31		*p*<.01
7	Unusual Visualization	18	31		*p*<.01
8	Internal Visualization	24	25		*n.s.*
9	Extending or Breaking Boundaries	22	27		*n.s.*
10	Humor	22	27		*n.s.*
11	Richness of Imagery	25	24		*n.s.*
12	Colorfulness of Imagery	23	26		*n.s.*
13	Fantasy	20	30		*p*<.01
14	Score Total of 13 Creative Strength M(SD)	3.4 (9)	3.5 (13)	5.03	*p*<.01

Item 1-13 Mann-Whitney U-test. Item 14 t-test

P Company 13 Items of Criterion-Rerenced Measures by TTCT Scores

	Criterion Reference Measure	Pre-test Mean Average	Post-test Mean Average	*t* Value	Sig
1	Emotional Expressiveness	21	22		*n.s.*
2	Story-telling Articulateness	20	23		*n.s.*
3	Movement and action	21	22		*n.s.*
4	Expressiveness of Titles	20	23		*n.s.*
5	Synthesis of Incomplete Figures	20	23		*n.s.*
6	Synthesis of Lines or Circles	17	26		*p*<.01
7	Unusual Visualization	21	22		*n.s.*
8	Internal Visualization	21	22		*n.s.*
9	Extending or Breaking Boundaries	21	23		*n.s.*
10	Humor	21	20		*n.s.*
11	Richness of Imagery	18	25		*p*<.05
12	Colorfulness of Imagery	22	21		*n.s.*
13	Fantasy	20	23		*n.s.*
14	Score Total of 13 Creative Strength M(SD)	10(6)	13(6)	2.568	*p*<.05

df=16, Item 1-13 Mann-Whitney U-test. Item 14 t-test

4.3 ANOVA Analysis Between Companies P & R and Pretest and Posttest

(1) Score Total: Then, the analysis of the Score Total by two factor factoral ANOVA within Pretest and Posttest and between R&P staff indicated a significant main effect for Pre-Posttest ($F(1, 43) = 5.222$, $p < 0.05$), the significance of which was confirmed by Bonferroni multiple comparison ($p < 0.05$). No interaction was found (Fig. 6).

(2) Fluency: The analysis of Fluency by two factor factoral ANOVA within Pretest and Posttest and between R & P staff indicated a significant main effect for Pre-Posttest ($F(1, 43) = 9.186$, $p < 0.01$), the significance of which was confirmed by Bonferroni multiple comparison ($p < 0.05$). No interaction was found (Fig. 7).

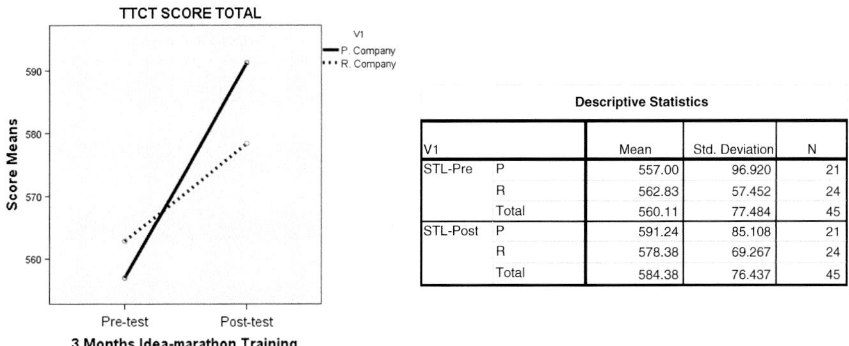

Descriptive Statistics

V1		Mean	Std. Deviation	N
STL-Pre	P	557.00	96.920	21
	R	562.83	57.452	24
	Total	560.11	77.484	45
STL-Post	P	591.24	85.108	21
	R	578.38	69.267	24
	Total	584.38	76.437	45

Fig. 6 ANOVA comparison between R & P staff of TTCT score total in pretest and posttest

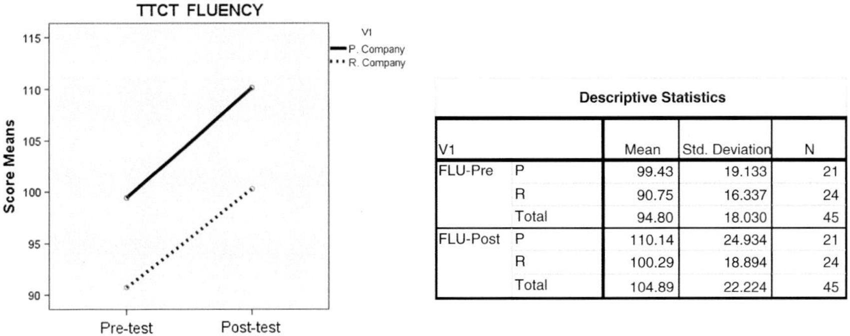

Descriptive Statistics

V1		Mean	Std. Deviation	N
FLU-Pre	P	99.43	19.133	21
	R	90.75	16.337	24
	Total	94.80	18.030	45
FLU-Post	P	110.14	24.934	21
	R	100.29	18.894	24
	Total	104.89	22.224	45

Fig. 7 ANOVA comparison between R & P staff of TTCT fluency in pretest and posttest

(3) Originality: The analysis of Originality by two factor factoral ANOVA within Pretest and Posttest and between R & P staff indicated a significant main effect for Pre-Posttest ($F(1, 43) = 15.089$, $p < 0.01$), the significance of which was confirmed by Bonferroni multiple comparison ($p < 0.05$). No interaction was found (Fig. 8).
(4) Elaborations: The analysis of Elaborations by two factor factoral ANOVA within Pretest and Posttest and between R & P staff did not indicate any significant main effect for Pre-Posttest. No interaction was found (Fig. 9).
(5) Abstractness of Titles: The analysis of Abstractness of Titles by two factor factoral ANOVA within Pretest and Posttest and between R & P staff did not indicate any significant main effect for Pre-Posttest. No interaction was found (Fig. 10).

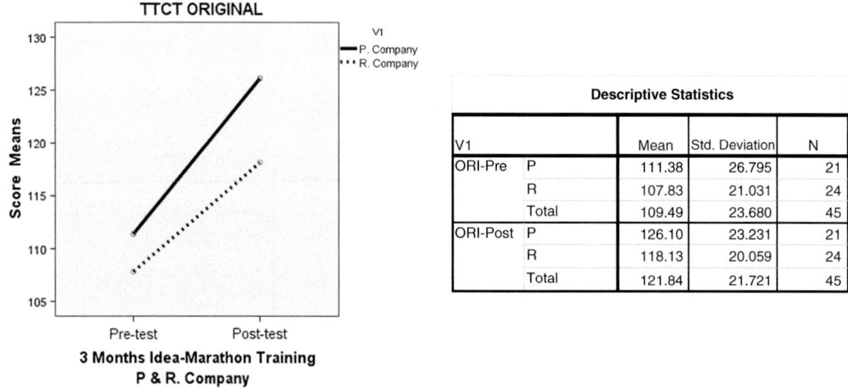

Descriptive Statistics

V1		Mean	Std. Deviation	N
ORI-Pre	P	111.38	26.795	21
	R	107.83	21.031	24
	Total	109.49	23.680	45
ORI-Post	P	126.10	23.231	21
	R	118.13	20.059	24
	Total	121.84	21.721	45

Fig. 8 ANOVA comparison between R & P staff of TTCT originality in pretest and posttest

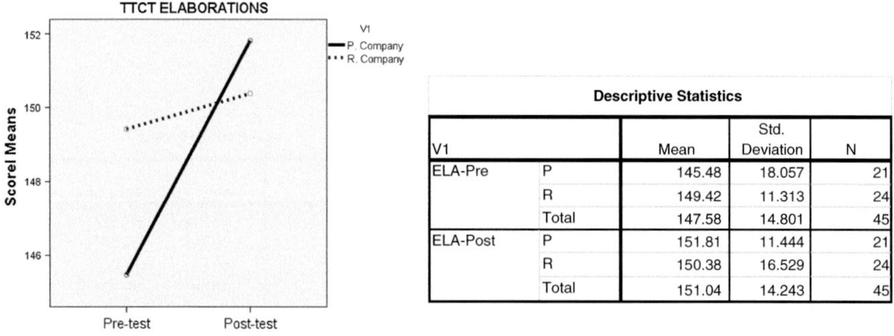

Descriptive Statistics

V1		Mean	Std. Deviation	N
ELA-Pre	P	145.48	18.057	21
	R	149.42	11.313	24
	Total	147.58	14.801	45
ELA-Post	P	151.81	11.444	21
	R	150.38	16.529	24
	Total	151.04	14.243	45

Fig. 9 ANOVA comparison between R & P staff of TTCT elaborations in pretest and posttest

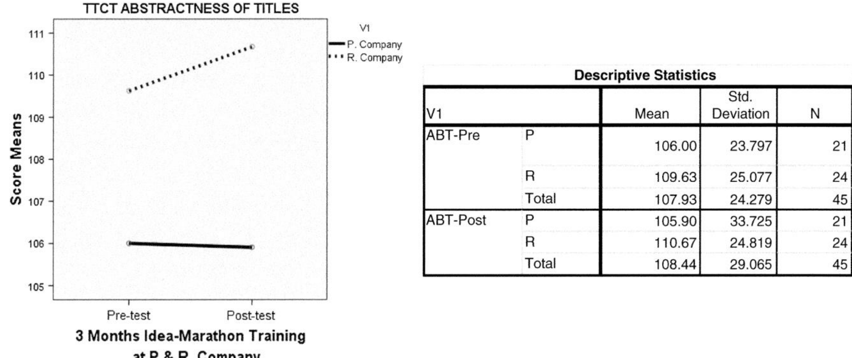

Descriptive Statistics

V1		Mean	Std. Deviation	N
ABT-Pre	P	106.00	23.797	21
	R	109.63	25.077	24
	Total	107.93	24.279	45
ABT-Post	P	105.90	33.725	21
	R	110.67	24.819	24
	Total	108.44	29.065	45

Fig. 10 ANOVA comparison between R & P of abstractness of titles in pretest and posttest

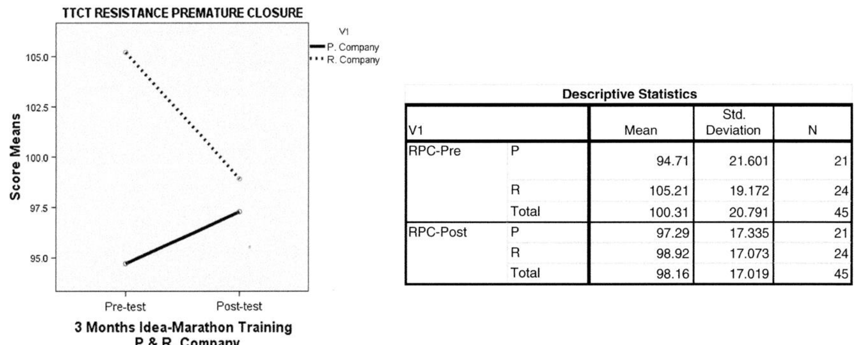

Descriptive Statistics

V1		Mean	Std. Deviation	N
RPC-Pre	P	94.71	21.601	21
	R	105.21	19.172	24
	Total	100.31	20.791	45
RPC-Post	P	97.29	17.335	21
	R	98.92	17.073	24
	Total	98.16	17.019	45

Fig. 11 ANOVA comparison between R & P staff of TTCT RPC in pretest and posttest

(6) Resistance to Premature Closure:The analysis of Resistance to Premature Closure by two factor factoral ANOVA within Pretest and Posttest and between R & P staff did not indicate any significant main effect for Pre-Posttest. No interaction was found (Fig. 11).

4.4 Analysis of Correlation Between TTCT Scores and ETC Number of Ideas in R & P Staff

In the case of the R staff, the correlation between the number of ideas recorded after 3 months in the ETS and each of the five Norm-Referenced Measures and Score Total was checked with the Pearson's product-moment correlation coefficient method. ETS's number of ideas correlated with the TTCT's Abstractness of Titles

($r = 0.534$, $p < 0.01$). Other Norm items of Fluency, Originality, Elaborations, RPC and Score Total were not found to be correlated.

In the case of the P staff, the correlation between the number of ideas recorded after 3 months in the ETS and each of the five Norm-Referenced Measures and Score Total was checked with the Pearson's product-moment correlation coefficient method. ETS's number of ideas correlated with the TTCT's Originality ($r = 0.516$, $p < 0.05$) and Elaborations ($r = 0.522$, $p < 0.05$) ETS's number of ideas correlated with the TTCT's. Other Norms of Fluency, Abstractness of Titles, RPC and Score Total were found not to be correlated.

5 Discussion

The discussion from the above-mentioned results is as follows:

(1) ETS is the essential part of a course of IMS training (Figs. 1 and 2) (2) Judging from the data of ETS (Figs. 5 and 6), the P Group was showing better results than the R Group. Actually, according to Fig. 2, the R Group was not creating ideas written in a notebook everyday while the P Group was creating very high daily rate of idea creation (Ref. Sect. 2.2 (2)). From this situation, prior to our Posttest of TTCT, we made one hypothesis that the R Group should have achieved lower scores of the TTCT Pretest and Posttest than the P Group. (3) The result of *t*-Tests to analyze the TTCT Pre-Posttest score of both the R & P staff showed a significant difference for the R Group for only Fluency with the P Group for Score Total and Originality. But this did not clearly show that the R Group was inferior in score to the P Group (Ref. Sect. 4.1). (4) From the viewpoint of the 13 items Criterion-Referenced Measures, the R Group had more significant items than the P Group (Ref. Sect. 4.2). (5) Then, we proceeded to two factors factoral ANOVA analysis between the R and P Groups with Pre-Posttest. The ANOVA test confirmed that there were no significant differences nor interactions between the R & P Groups, although there were significant results between Pretest and Posttest (Ref. Sect. 4.3). (6) Between the numbers of ETS ideas and the TTCT Five Norm-Referenced Measure Result, Abstractness of Titles was found to be significantly correlated in the R Group while Originality and Elaborations are found to be significantly correlated in the P Group (Ref. Sect. 4.4). (7) From all of the items (1) to (4), we may probably say that both the R & P Groups showed their creative improvement whatever the ETS results were. (8) The R Company is one of the major long-standing engineering companies in Japan and its business has been expanding steadily. The staff of the R Company is reputed to be very capable. However, reviewing the increasing number of absentees for the second and third lectures, the management was afraid that their R&D and other engineers were getting narrow-minded, interested in their own present assigned scope of work, especially lack of new business motivations. As the company is quite large, none of the staff imagines of going bankrupt. As a result, the creativity of these engineers has been getting stale.

6 Conclusion and Future Research

From our above-mentioned discussion, we can say that the training of Idea-Marathon was effective in improving the creativity of both the R & P Groups. Though the ETS of the P Group had significant correlations with TTCT in Score Total and Originality, the R Group had a significant correlation with TTCT in Abstractness of Titles only.

We tried to make our reasoning of what the reason for the large difference in the ETS result between the R & P Groups is. We estimated that the study of the R Group ETS and TTCT Scores was connected with the degree of their back-scratching, falling-in, taking-things-easy, living under the same roof with the same food, settling always amicably and self-estimating that someone else will solve problems. Ken-ichi Omae pointed out these natures [7]. We hypothetically found the reason for a poorer ETS result as the R Group was less motivated about self-enlightenment, and the R Group has much more real ability to improve in TTCT tests if its members get serious. They may have suffered from learned helplessness [8].

In future training seminars for any companies, we must make our survey to estimate the motivation and willpower level of the participant individual, and also we must make our investigation about the company business atmosphere so that we can improve our Idea-Marathon group training method for a staff like that of the R Group with more contents of encouraging motivation for self-enlightenment.

References

1. Drucker, P.: The Age of Discontinuity: Guidelines to Our Changing Society. Butterworth-Heinemann Ltd, Oxford (1969) (Reprinted in 1992, Transaction Pub.)
2. Higuchi, Takeo.: You can get Affluent Ideas from this Book". Diamond Inc, Tokyo (1992) (in Japanese)
3. Higuchi, Takeo: Ideas in Action. Adarsh Books, Digital Achievement of Idea Marathon System (2001)
4. Higuchi, Takeo: Idea-Marathon System, Encyclopedia of Creativity, Invention, Innovation, and Entrepreneurs. Springer, UK (2013)
5. Higuchi, Takeo., et al.: Creativity improvement by idea-marathon training, measured by torrance tests of creative thinking (TTCT) and its applications to laboratories. In: KICSS2012 (2012)
6. Higuchi, Takeo.: Dissertation "Enhancement Effects of IMS on Creativity, Chapter 6–2. Japan Advanced Institute of Science and Technology, Ishikawa, Japan (2014)
7. Omae, K.-i.: Cycle of Devil or Japanese Nature Leaning to Others. Shincho Bunko, Tokyo (1987) (in Japanese)
8. Seligman, M.E.P., Maier, S.F.: Failure to escape traumatic shock. J. Exp. Psychol. **74**(1). American Psychological Society, Washington, DC, USA (1967)
9. Torrance, E.P., Ball, O.E.: Torrance Tests of Creative Thinking, Streamlined Scoring Guide for Figural Forms A and B. Scholastic Testing Service, Inc, USA (1987)

Using Appropriate Context Models for CARS Context Modelling

Christos Mettouris and George A. Papadopoulos

Abstract Most context-aware recommender systems in the literature that use context modelling have the tendency to develop domain and application specific context models that limit, even eliminate any reuse and sharing capabilities. Developers and researchers in the field struggle to design their own context models without having a good understanding of context and without using any reference models for guidance, often resulting in overspecialized, inefficient or incomplete context models. In this work we build upon prior work to propose an enhanced online context modelling system for Context-Aware Recommender Systems. The system supports CARS developers in the process of building their own context models from scratch, while it supports at the same time sharing and reuse of the models among developers. The system was tested with a real dataset with positive results, as it was able to support context model development with instructions to the developer, model comparison, useful statistics, recommendations of similar models, as well as alternative views of context models to aid the developer's task.

Keywords Context modelling system · Context-Aware recommender systems · Application context model · Context instance model · Context variables · Context dimensions

---

C. Mettouris (✉) · G.A. Papadopoulos
Department of Computer Science, University of Cyprus, 1 University Avenue, 20537, CY-2109 Nicosia, Cyprus
e-mail: mettour@cs.ucy.ac.cy

G.A. Papadopoulos
e-mail: george@cs.ucy.ac.cy

S. Kunifuji et al. (eds.), *Knowledge, Information and Creativity Support Systems*, Advances in Intelligent Systems and Computing 416,
DOI 10.1007/978-3-319-27478-2_5

1 Introduction

A well-known and effective solution to the information overload modern life experiences at all fields is the usage of Recommender Systems (RS). Information overload refers to the vast amount of information users have to access nowadays: users can get lost, disappointed and frustrated for failing to retrieve the desired and needed information at a given time. RS use a variety of filtering techniques and recommendation methods to provide personalized recommendations to their users, mostly by using information retrieved from the user profile, from user's usage history, as well as item related information [5, 9]. However, traditional RS use limited or none contextual information to produce recommendations, as opposed to the Context-Aware Recommender Systems (CARS) that focus in using contextual information to enhance recommendations [2]. Context was first utilized into the recommendation process by Adomavicius by proposing three approaches: the Pre-filtering approach, the Post-filtering approach and the Multidimensional Contextual Modelling approach [1, 2]. Context modelling is important for modelling the contextual information to be used during the recommendation process.

An important contextual modelling issue in CARS is the development of domain specific and application specific context models that only represent information on the particular application domain (e.g. recommendation of movies). Our review on RS [11] had revealed that most CARS and semantic RS in the literature are domain and application specific, meaning that they cannot be applied in other domains. By designing domain and application specific context models, many different and very specific models are produced with no reuse and sharing capabilities.

Another problem is that developers and researchers attempt to design their own models based on their own knowledge and skills and more importantly without using any reference model, without any guidance and strictly focused on the application at hand, often resulting in overspecialized, inefficient or incomplete contextual models.

We had partially addressed the above contextual modelling problem in prior works, at first by proposing a generic, abstracted contextual modelling framework for CARS which developers and researchers can use theoretically to be guided through the process of properly defining the context for their application [11], and later by developing an online "Context Modelling System and Learning Tool" [10] based on [11], which is able to teach and guide developers towards a more efficient, effective and correct selection and usage of context attributes for building their own application model, allowing at the same time for sharing and reuse of context models among applications, regardless of the domain they belong to.

The modelling framework in [11] was essentially a model template in UML built in the Eclipse Modelling Framework (EMF) [7]. Although this framework was developed as a UML class diagram, it was mainly a theoretical tool rather than a modelling tool since: (i) it was not an easy and straight-forward procedure for developers to extend or instantiate a UML class diagram in order to build their own context models, (ii) it was time consuming, (iii) it required programming knowledge and skills and (iv) it did not offer guidance and learning of important concepts. The online "Context Modelling System and Learning Tool" [10] aimed at solving the aforementioned problems by specifying an easy to use UI for the developers and researchers to be able to effectively and efficiently build their models, share them with others, as well as reuse models of others. The above can be accomplished without any programming skills being required by the user. The system focuses also on learning, being able to introduce developers and new researchers with modern concepts from CARS research, as well as their role in a context model and a recommendation process [10].

In this work, we have extended and finalized the work conducted in [10] based on feedback received by experts, aiming to advance the functionality of the system towards a more effective context model development, sharing and reuse. The work was extended with important system functionality and the new *CARS Context Modelling System* [4] was tested by us in real settings by using a dataset released in the framework of the 2nd International Workshop on Information Heterogeneity and Fusion in Recommender Systems—HetRec 2011 [8]. More to the point, we have designed and implemented: (i) the validation of application contexts through context instances, (ii) the comparison of application context models regarding their common context variables, (iii) the recommendation of similar application context models to the user and (iv) the inclusion of the *context dimensions* concept [2] in the system and the enablement of a "context dimensions" view of the context models.

In Sects. 5–7 we discuss the above important additions, after presenting related work in Sect. 2, an introduction to the system concept in Sect. 3 and a presentation of its functionality in Sect. 4. Section 8 discusses system testing and Sect. 9 closes the paper with conclusions and future work.

2 Related Work

During our research [11] we have reviewed a number of CARS and semantic RS to opine whether the context models used were application/domain specific or generic. While domain specific models focus extensively on a particular domain, generic models do not and focus on being able to facilitate *any* application specific domain. This research revealed that most CARS and semantic recommenders in the literature use domain specific models [11], meaning that they cannot be applied for usage in other domains. A number of generic recommenders also exists that either apply to some generic application area, or can be applied to more than one domain by linking domain specific ontologies to their own data and knowledge pool in order to gain domain-aware knowledge and provide domain-aware functionality. Although some of the semantic and contextual models attempt to be more generic, the majority represent information that either concern a particular application domain (e.g. movies), or a more abstracted domain (such as products in general, web services, e-learning).

To the best of our knowledge no attempts have been made towards developing a context modelling tool that could facilitate the development of truly generic contextual models for CARS and the definition of their contextual entities, so that CARS developers be able to extend/update and reuse them to construct application specific models for their needs. Such a tool would simplify the process of contextual modelling in CARS and enable context uniformity, share and reuse; this is the motivation for this work.

3 System Concept

In our work we have followed the representational view of context [2, 6], meaning that, as in most CARS, the context of an application is defined through a predefined set of observable context attributes of static (not dynamic) structure which does not change significantly over time (as opposed to the interactional view of context where context is not necessarily an observable feature of an interaction [3]). Please note that by static structure we do not mean that the context itself is static, e.g. the context attribute *user location* has a well-defined static structure but is itself a dynamic context since it continuously changes values as the user changes locations. Therefore, we assume that there is a predefined, finite set of contextual attributes in a given CARS application and that each contextual attribute is defined in our system as a context variable. As an example of the interactional view of context the reader may refer to [3] in which the user is modelled based on human memory models proposed in psychology, where user preference models for previous interactions are stored within the user's long term memory, while the current user's model is stored in the user's short term memory. Then, the short term memory is used to retrieve information from the long term memory in order to be used to generate recommendations for the user.

The CARS Context Modelling System is presented online at [4]. On the right side of the main webpage, an image of the context modelling framework is presented for reference [11], while on the left side various options are presented. The red colour text throughout the system [4] is clickable and provides information on important system concepts that have to do with CARS research, as well as information on how to use the system. The information is provided in pop-up text boxes, as well as on the context modelling framework image on the right side for reference and easier comprehension of the information.

Many complicated concepts related to CARS research are used by the system in order to construct context models [4]. Concepts like "the multidimensional context-aware recommendations: *Users* × *Items* × *Context* *Ratings*", "context variables", "application contexts" and "context instances" need to be well understood by CARS developers and researchers in order to be able to use the modelling system to create their own application context models. To assist users on this difficult task the text provided within the text boxes is carefully selected from important published research papers, while references are provided as links wherever needed for further reading.

The fundamental concept of CARS research is to include the context in the recommendation process to result from the 2D un-contextual RS: Users × Items → Ratings to the multidimensional CARS: Users × Items × Context → Ratings [2]. The latter represents a single complete recommendation process and is the main idea behind this CARS Context Modelling System. For each recommendation attempt, a RS must examine whether each item is suitable for a user *in a certain context*. This can be depicted through the question: *what is the rating a particular user would assign to a particular item under a certain context?* This rating score is what a recommender must calculate. Therefore, in order to examine whether an item is suitable for a user, the recommendation scheme must have exactly one *user*, exactly one *item* but one or more *context* entities, for each of which a *rating* score can be assigned.

The *context variable* concept contains the actual contextual information to be inserted into a context model [4]. Each context variable has a name and a value to describe both the context parameter and its particular value, e.g. "Temperature" is a context parameter and "high" is its value. Therefore, to define the context: "temperature can be high, medium or low", a total of three context variables will be needed. Via a weight property developers may denote a particular importance for their context variable. The "static" property refers to whether the context variable is static (cannot change dynamically, e.g. user's date of birth) or dynamic (can change, e.g. weather).

An *Application Context* is a context model for a particular RS, e.g. a movies RS. It is built by a CARS developer in order to model the context for this particular RS. An application context model contains all context variables that the developer will select, along with their values. Since each context variable has a name and a specific value ("Temperature: high"), a developer must select all variables with a particular name for the model to be accurate and complete (e.g. all of the following: "Temperature: high", "Temperature: medium", "Temperature: low").

Although an application context model is built by a CARS developer to model the context for a particular recommender, the system enables sharing and reuse of such models by supporting other developers that want to build similar models in using the same context model and enhance/update it as needed. The idea is: since all RS of a specific type/field (e.g. online movie recommenders) interact in similar context settings, why having one (often incorrect or incomplete) context model for each such recommender built by each developer, when we can have just one (correct and complete) context model for all recommenders. We argue that application context models of recommenders of the same or similar fields should be identical, or in any case similar to a great extent. The system suggests to developers to use pre-existing application context models of similar applications (if any) and build upon them, instead of building a model of their own. Eventually, only one application context model for each type of recommender system will exist in the system that should be able to satisfy all developers.

A *Context Instance* is a "screenshot" of the context during an event of interaction between the user and the item that is involved in the recommendation process. For example, for a movie recommender, a context instance is the set of context variables that constitute the context during the event of a particular user (user = "Tom") watching a particular movie (item = "Rambo") at a particular time (we assume the first time that Tom watches Rambo). Such a context instance may have title: "Tom-Rambo C1" (C1 results from "Context1") and may be consisted of the context variables: the time of day, the day of the week, the IMDB ratings of the movie watched, whom did the user watch the movie with, etc. In a similar way, the context instance around the second time Tom watches Rambo will have a title "Tom-Rambo C2" and will again be consisted of a number of context variables. Another definition of the context instance is that it is the set of context variables with their corresponding values that constitute the context at the time a single recommendation is requested. In addition, a context variable may participate in a number of context instances, each of which is characterized by that context variable. For example, the context variable "time: morning" can participate in many context instances, all of which refer to morning time.

Based on the above, the context instances define all valid contextual information around a particular fact or event (in the example above around user Tom watching movie Rambo at a certain time). Ideally, a context instance should be automatically created using context sensing and retrieval during the occurrence of a fact/event and stored in the system. E.g. around the event of a user watching a movie the system should be automatically aware of: the time of day, the day of the week, the IMDB ratings of the movie watched, whom did the user watch the movie with, and any other context variables that could participate in the process. This is a very difficult process, in some way impossible to achieve with the current technology available (how is the system going to know whom did the user watch the movie with, unless the user states it?) and it is beyond the scope of this work. In this CARS Context Modelling System we provide the ability for CARS developers to create their own context instances for modelling purposes, in order to closely observe and study whether their application contexts are able to "catch" *any* context instance that may occur, and in that way *validate* their models. More on modelling validation follows in Sect. 5.

4 Building Context Models

In the main page in [4] an input form is provided in order for developers to add context variables. Context variables contain the actual contextual information that developers need to insert into their context model to populate it [11]. The set of all context variables currently available in the system constitute the *Generic Context Model*[1] (Fig. 1). The application context models, the context instance models and

[1]Due to limited space in the paper, not all context models are presented complete in figures; instead, we provide hyperlinks to the models on the online tool in footnotes for reference: http://www.cs.ucy.ac.cy/~mettour/phd/CARSContextModellingSystem/genericContextModel.php.

Generic Context Model: Context Variables

What is the Generic Context Model?
Options
- Display Application Contexts(info)
- Create an Application Context
- Display Context Instances (info)
- Create a Context Instance
- Display Context Dimensions(info)
- Back

Context Categories

ITEM CONTEXT (Weight: 5)	USER CONTEXT (Weight: 5)	SYSTEM CONTEXT (Weight: 3)	OTHER CONTEXT (Weight: 7)
movie title Static: TRUE	company: alone Static: TRUE	network: adequate Static: FALSE	temperature: cold Static: TRUE
movie genre: comedy Static: TRUE	company: with girlfriend Static: TRUE	network: excellent Static: FALSE	temperature: warm Static: TRUE
imdb ratings: 0-4 Static: TRUE	company: with friend(s) Static: TRUE	network: poor Static: FALSE	temperature: hot Static: TRUE
movie duration: <100mins Static: TRUE	user location (GPS): home Static: FALSE	batteryLevel:low Static: FALSE	time: weekday Static: TRUE
movie genre: drama Static: TRUE	user location (GPS): classroom Static: FALSE	batteryLevel:medium Static: FALSE	time: weekend Static: TRUE
movie genre: action Static: TRUE	research interest: Context-awareness Static: TRUE	batteryLevel:high Static: FALSE	Temperature: freezing Static: TRUE
movie genre: romantic Static: TRUE	research interest: Databases Static: TRUE		daytime: morning Static: FALSE
imdb ratings: 5-7 Static: TRUE	research interest: component based systems Static: TRUE		daytime: noon Static: FALSE

Fig. 1 The generic context model

the generic context model constitute the three types of context models supported by the system. The generic context model defines the basic contextual entities of RS, as well as their properties and associations in order for CARS developers to be able to extend it to construct application specific models for the needs of the application at hand. This generic context model simplifies the process of context model development and enables context uniformity, sharing and reuse.

All context models are based on the abstracted contextual modelling framework for CARS (refer to Fig. 2 in [11] and main page in [4]). These models constitute the "context" entity of this framework, as well as all the context related entities and relationships from that level downwards. In the generic context model[1] (Fig. 1), as in all context models, the four context categories are presented: "Item Context", "User Context", "System Context" and "Other Context" [4]. The rectangles depict the context variables (both name and value) and their parameters. Note that a context variable can belong to more than one context category; such decisions are made by the developer at the time of creation of the context variable or the context model. The "itemContext", "userContext", "systemContext" and "otherContext" constitute the four main context classes in the system and are meant to be perceived as the main context entities for any contextual model of CARS; any context information of any CARS should be able to be represented as a context property of one (or more) of the main context classes, as a context variable.

A simple and straightforward way is developed for CARS developers to build their own context models. They can either add a new context variable that is not

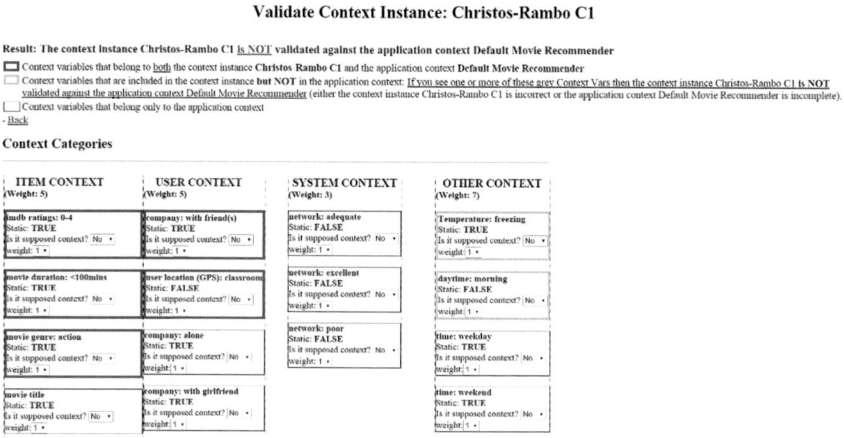

Fig. 2 The context instance "Christos-Rambo C1" is not validated against the application context default movie recommender

currently included in the system, or use the generic context model and simply select/unselect the context variables (already in the system) that are of interest to them by clicking on them. Before adding a new context variable in the system, developers are advised to first check whether the context variable they would like to add already exists in the system as part of the generic context model; if it does, developers are asked to use the existed variable in order to avoid redundant information in the system and confusion to other developers. In this way, the context variables created are universal among many applications and domains and hence they can be shared and reused in many context models of various CARS. For more information on the features and functionality offered by the developed online system, the reader is referred to the system [4].

5 Validating Application Contexts Through Context Instances

The system supports the validation of an application context model through one or more context instance models. As already stated in Sect. 3, an application context is a context model for a particular recommender application, e.g. a movies recommender system, while a context instance defines the context "screenshot" at the time a single recommendation is requested. It is evident that an application context model should be able to support any context instance related to the particular recommender; in other case, the application context model is incomplete. E.g. an application context model for a movie recommender should be able to support any movie recommender related context instance, such as "Tom-Rambo C1" (see Sect. 3) which can be consisted of a number of context variables such as the time of day, the day of the week, the IMDB ratings, etc. This means that the application context model must include all context variables of the context instance model.

The possibility to validate an application context model through context instance models can be a useful tool for the CARS developer who has just created her application context model in the system (let's suppose an application context model for a movie recommender) and wants to ensure that her application context is able to properly model the context of movie RS (i.e. model any context instance of these recommenders). The context instance can be created preferably by a RS user who will reflect her experiences regarding the activity of watching movies in the context instance model. If this is not feasible, the context instance model can be created by another developer. Following, the CARS developer will be able to access the context instance and by clicking a button, validate this model against her application context model. The system then provides a justified answer whether the context instance model was validated against the application context. If yes, then the application context is also validated. If not, the system provides information as to why the model was not validated. A context instance may not be validated against an application context either because the context instance is incorrect or because the application context is incomplete. Figure 2 provides an example of the context instance "Christos-Rambo C1" not being validated against the application context "Default Movie Recommender" (you may use the online system [4] to view the actual coloured page).

6 Recommendation of Application Context Models and Model Comparison

During the creation of a new application context model by the CARS developer, the system is able to recommend the top N (currently N = 5) most similar application context models regarding the percentage of common context variables (refer to Sect. 8 and Fig. 4). This is very important for CARS developers who are in the process of creating their application context model and would like to be informed about similar context models in the system, as well as the level of similarity. As soon as a new application context is created, the system automatically provides recommendations.

Moreover, the system provides an easy way to compare two specific application context models. Through the usage of colours, the system depicts the common context variables of the two application contexts, as well as the context variables that belong only to one of the two application contexts. Besides common context variables, the system also provides statistics regarding the percentage that each application context participates in the other application context. For an example on the above you may refer to Sect. 8.

If two application context models have many common variables, then the system proposes a merge. It is certainly a situation where the CARS developers must decide whether both application contexts are needed. This is especially interesting in the case where each application context concerns a different type of recommender system, e.g. a movie recommender and a book recommender, as context models of recommenders of different fields are very rarely similar.

7 Including Context Dimensions in the System

The basic concept of CARS research is to include the context in the recommendation process so that to result from the 2D un-contextual recommenders: *Users* × *Items* *Ratings* to the multidimensional context-aware recommenders: *Users* × *Items* × *Context* *Ratings* [2]. The term *Context* appears to be a single dimension itself, but in essence it represents the *Context Dimensions*, i.e. all the additional context-related dimensions that are being used in the recommendation process besides the user and the item (refer to Fig. 2 in [2] for an example of a three dimensional model for the recommendation space: User × Item × Time).

In our work, a context dimension can only be a context variable with all of its possible values. We use the following definition of the context dimension: *each individual context variable with a unique name and with all of its values can be perceived as a context dimension.* Therefore a context dimension name is a unique context variable name.

The system provides the "context dimensions view" option for each context model in the system: the generic model, an application context model or a context instance model. The "context dimensions view" is an alternative view of a model, besides the default "context variables view" of the system presented in Sect. 4. The "context dimensions view" is of particular importance for: i. observing the context dimensions of a model and ii. observing the context variables of a model with all of its values aggregated. Figure 3 presents the "context dimensions view" of the application context model Movie Recommender (refer to Sect. 8).

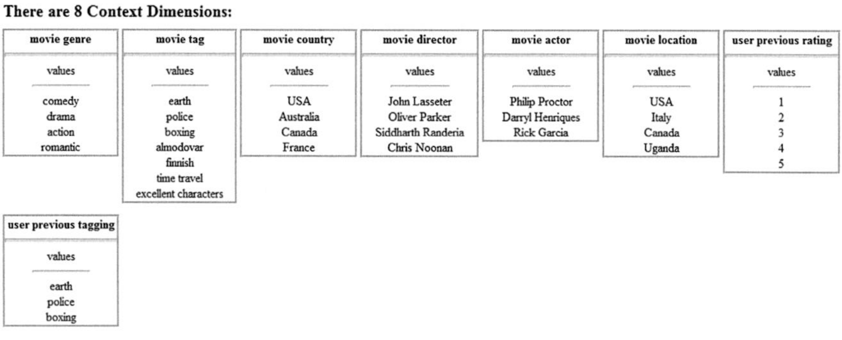

There are 8 Context Dimensions:

movie genre	movie tag	movie country	movie director	movie actor	movie location	user previous rating
values	values	values	values	values	values	values
comedy	earth	USA	John Lasseter	Philip Proctor	USA	1
drama	police	Australia	Oliver Parker	Darryl Henriques	Italy	2
action	boxing	Canada	Siddharth Randeria	Rick Garcia	Canada	3
romantic	almodovar	France	Chris Noonan		Uganda	4
	finnish					5
	time travel					
	excellent characters					

user previous tagging
values
earth
police
boxing

Fig. 3 Movie recommender application context model: "context dimensions view"

8 System Testing

The CARS Context Modelling System was tested by us in real settings by using a dataset aimed for usage with movie RS which was released in the framework of the 2nd International Workshop on Information Heterogeneity and Fusion in Recommender Systems—HetRec 2011 [8]. The dataset is an extension of MovieLens10 M dataset, which contains personal ratings and tags of users about movies. In the dataset, the movies are linked to Internet Movie Database (IMDb) and RottenTomatoes (RT) movie review systems. The dataset includes more than 2000 users, more than 10000 movies, more than 800000 ratings, as well as many related data such as movie genres, directors, actors, countries, locations and tags. It also includes information on user ratings on movies, as well as on user tagging of movies.

For system testing purposes we have played the role of a CARS developer who would like to model the context for building a movie recommender system based on this dataset. Similar datasets are frequently used by RS developers and researchers to develop their systems, both for commercial as well as research purposes. By using this dataset for building a context model in our system we can closely observe how a developer can be assisted in real settings.

We have created an application context model named Movie Recommender in the system based on this dataset and then, by using the available system functionality, we have compared this dataset with other application context models in the system to find similarities, as well as opine whether the context model built from the dataset was adequate to be used in a CARS for movies.

The first step was to (manually) extract the context attributes from the dataset. The context attributes we have identified are: "movie genre", "movie director", "movie actor", "movie country", "movie location", "movie tag", "user (previous) rating" and "user (previous) tagging". The first 6 context attributes were assigned under the context category "item context", while the final two under the context category "user context". Each context attribute is inserted into the system as a context variable in the form: "name: value". E.g. "movie country: France". Since the amount of raw data included in the dataset is huge (10000 movies) at this point we have decided to insert only a sample of the data in the system's database. This is adequate for our "proof of concept" type of testing. The context variables of the created application context model Movie Recommender can be seen via the online system,[2] while Fig. 3 shows the context dimensions of the same model (there are 34 context variables and 8 context dimensions in this model).

As we can observe from the online model[2] and Fig. 3, we were able to extract only 8 context dimensions from the dataset, populating the application context model with a sample of 34 context variables. If we had populated the model with all the information of the dataset, the context variables in the model would be hundreds

[2]http://www.cs.ucy.ac.cy/~mettour/phd/CARSContextModellingSystem/displayAppInstancesModel.php?appCont=Movie%20Recommender.

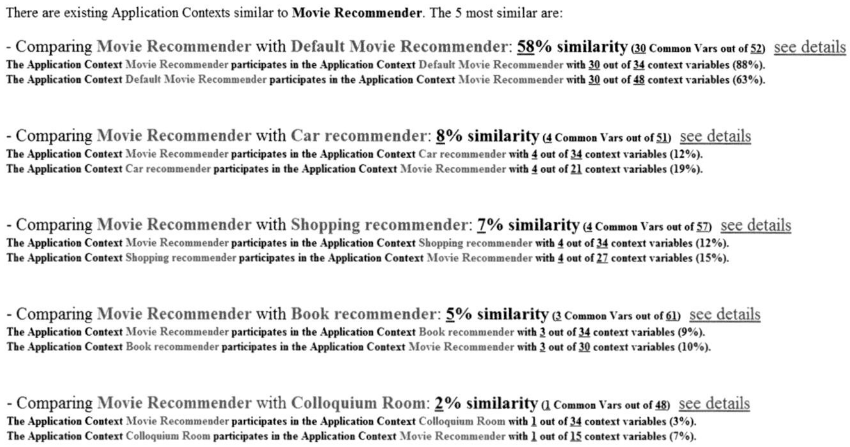

There are existing Application Contexts similar to **Movie Recommender**. The 5 most similar are:

- Comparing Movie Recommender with Default Movie Recommender: **58% similarity** (30 Common Vars out of 52) see details
The Application Context Movie Recommender participates in the Application Context Default Movie Recommender with 30 out of 34 context variables (88%).
The Application Context Default Movie Recommender participates in the Application Context Movie Recommender with 30 out of 48 context variables (63%).

- Comparing Movie Recommender with Car recommender: **8% similarity** (4 Common Vars out of 51) see details
The Application Context Movie Recommender participates in the Application Context Car recommender with 4 out of 34 context variables (12%).
The Application Context Car recommender participates in the Application Context Movie Recommender with 4 out of 21 context variables (19%).

- Comparing Movie Recommender with Shopping recommender: **7% similarity** (4 Common Vars out of 57) see details
The Application Context Movie Recommender participates in the Application Context Shopping recommender with 4 out of 34 context variables (12%).
The Application Context Shopping recommender participates in the Application Context Movie Recommender with 4 out of 27 context variables (15%).

- Comparing Movie Recommender with Book recommender: **5% similarity** (3 Common Vars out of 61) see details
The Application Context Movie Recommender participates in the Application Context Book recommender with 3 out of 34 context variables (9%).
The Application Context Book recommender participates in the Application Context Movie Recommender with 3 out of 30 context variables (10%).

- Comparing Movie Recommender with Colloquium Room: **2% similarity** (1 Common Vars out of 48) see details
The Application Context Movie Recommender participates in the Application Context Colloquium Room with 1 out of 34 context variables (3%).
The Application Context Colloquium Room participates in the Application Context Movie Recommender with 1 out of 15 context variables (7%).

Fig. 4 Top 5 recommendations of similar application context models

more; however, the distinct context variable names (and hence the context dimensions) would still be only 8. Another important observation is that the application context model[2] includes information in only two of the four context categories. This is certainly a limitation of the model, since it does not utilize the context entirely.

After creating the application context model, we have been provided with system recommendations of similar application context models already in the system (created earlier by us, by colleagues of ours and by experts on context-aware recommenders). The system provides the top 5 most similar models to our model. The most similar application context model is one of another CARS for movies (named Default Movie Recommender[3]) with 58 % similarity, following other models of other recommenders that are similar to ours in percentages between 8–2 % (Fig. 4). The similarities of the two application context models can be seen via the online system[4] (Fig. 5). In red colour the common context variables are depicted. In green are the context variables that belong only to the Movie Recommender context model, while in blue are shown the context variables that are included only in the Default Movie Recommender context model (you may use the online system[4] to view the actual coloured page).

It is quite easy to observe that the context model for the Movie Recommender is inferior to the Default Movie Recommender application context model, since it only includes a percentage of the context attributes of the Default recommender in all context categories. In Item Context category there is 65 % of common context, in

[3]http://www.cs.ucy.ac.cy/~mettour/phd/CARSContextModellingSystem/displayAppInstancesModel.php?appCont=Default%20Movie%20Recommender.

[4]http://www.cs.ucy.ac.cy/~mettour/phd/CARSContextModellingSystem/compareApplicationContexts2.php?sentData=$Default%20Movie%20Recommender$Movie%20Recommender.

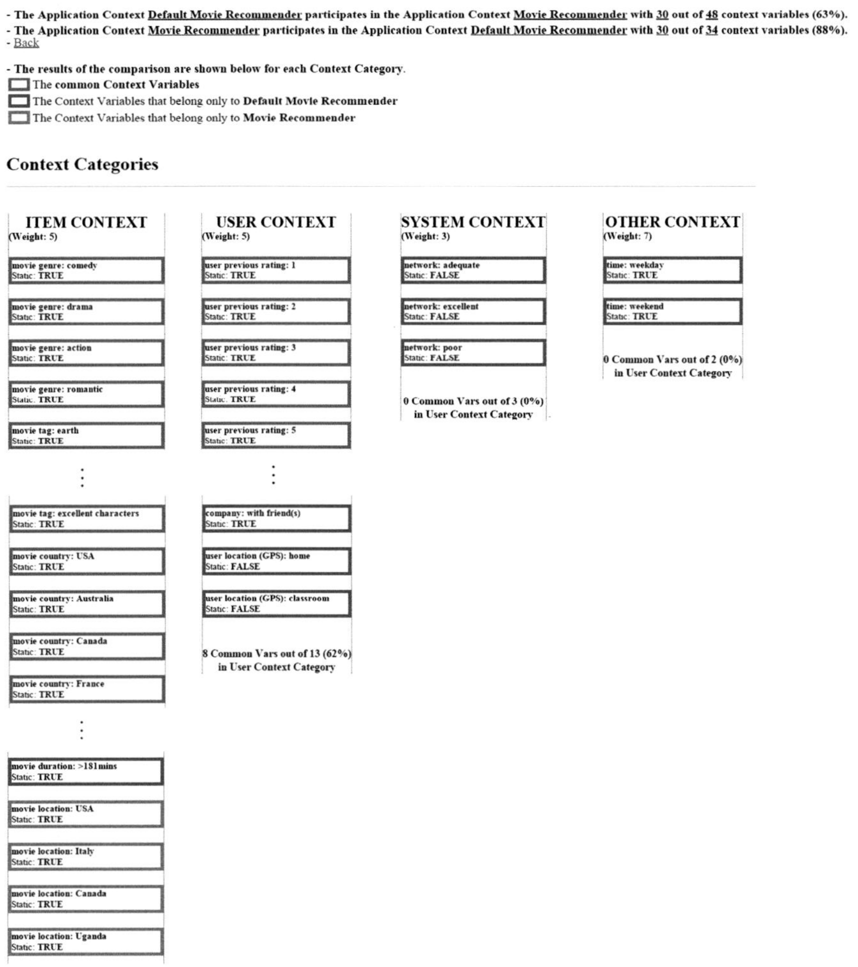

- **The Application Context Default Movie Recommender participates in the Application Context Movie Recommender with 30 out of 48 context variables (63%).**
- **The Application Context Movie Recommender participates in the Application Context Default Movie Recommender with 30 out of 34 context variables (88%).**
- Back

- **The results of the comparison are shown below for each Context Category.**

The **common Context Variables**

The Context Variables that belong only to **Default Movie Recommender**

The Context Variables that belong only to **Movie Recommender**

Context Categories

ITEM CONTEXT (Weight: 5)

movie genre: comedy Static: TRUE

movie genre: drama Static: TRUE

movie genre: action Static: TRUE

movie genre: romantic Static: TRUE

movie tag: earth Static: TRUE

⋮

movie tag: excellent characters Static: TRUE

movie country: USA Static: TRUE

movie country: Australia Static: TRUE

movie country: Canada Static: TRUE

movie country: France Static: TRUE

⋮

movie duration: >181mins Static: TRUE

movie location: USA Static: TRUE

movie location: Italy Static: TRUE

movie location: Canada Static: TRUE

movie location: Uganda Static: TRUE

USER CONTEXT (Weight: 5)

user previous rating: 1 Static: TRUE

user previous rating: 2 Static: TRUE

user previous rating: 3 Static: TRUE

user previous rating: 4 Static: TRUE

user previous rating: 5 Static: TRUE

⋮

company: with friend(s) Static: TRUE

user location (GPS): home Static: FALSE

user location (GPS): classroom Static: FALSE

8 Common Vars out of 13 (62%) in User Context Category

SYSTEM CONTEXT (Weight: 3)

network: adequate Static: FALSE

network: excellent Static: FALSE

network: poor Static: FALSE

0 Common Vars out of 3 (0%) in User Context Category

OTHER CONTEXT (Weight: 7)

time: weekday Static: TRUE

time: weekend Static: TRUE

0 Common Vars out of 2 (0%) in User Context Category

Fig. 5 Combined context model for the default movie recommender and the movie recommender

User Context category there is 62 % of commons context, while in the System Context and Other Context categories the common context is 0 %.

Switching to the "context dimensions view" (on the online system), we observe that the Movie Recommender context model includes the context dimension "movie location" consisted of 4 context variables which is not included in the Default Movie Recommender system. This is a deficiency of the Default model. However, there is a number of context dimensions that were included in the Default model but not in the Movie Recommender application context model: movie title, imdb ratings (the imdb ratings of a movie), movie duration, company (with whom the user watches the movie with), user location (GPS), network (describes the network connectivity/bandwidth) and time (time of day of watching the movie). We can observe that, while our context model includes only 8 context dimensions, the Default model includes 14.

9 Conclusions and Future Work

The purpose of the CARS Context Modelling System presented in this work is to serve as a tool for CARS developers (and researchers), enabling them to efficiently, effectively and correctly select and use context attributes for building their own application models, allowing at the same time for sharing and reuse of context models and information among applications, regardless of the domain they belong to. We have presented the system concept and its functionality, as well as discussed additional important features such as the validation of application contexts through context instances, the comparison of application context models, the recommendation of similar application context models to the user and the inclusion of the context dimensions concept in the system (via the "context dimensions view" of context models).

The system testing we have conducted in real settings by using a real dataset [8] showed that the system was able to depict important limitations of the application context model created based on the dataset, in comparison with other pre-existing CARS application context models in the system (built by us, colleagues and experts).

Based on the above, we conclude that if CARS developers were to define the context for a movie RS based on this dataset (which is a valid dataset for movie RS used by many developers) and without using the CARS Context Modelling System, the result would be an inefficient and incomplete context model. Instead, by using the CARS Context Modelling System we were able to build our application context model with the aid of system tools, compare our model with other models, as well as be recommended with similar models that could assist us in developing and enhancing our application context model.

Currently, the system provides all context models as visual images and will support the extraction of the models in xml and txt formats. As future work, we aim to support CARS developers in incorporating the context models within their RS by providing appropriate tools. An idea is to extent the system by providing automatic transformation of the context models to code. The idea is to support CARS developers in attaching their context models to their recommendation methods via an easy and straightforward way that will also support context model sharing and reuse.

References

1. Adomavicius, G., Sankaranarayanan, R., Sen, S., Tuzhilin, A.: Incorporating contextual information in recommender systems using a multidimensional approach. ACM Trans. Inf. Syst. (TOIS) **23**, 103–145 (2005)
2. Adomavicius, G., Tuzhilin, A.: Context-aware recommender systems. In: Ricci, F., Rokach, L., Shapira, B., Kantor, P.B.: Recommender Systems Handbook, pp. 217–253 (2011)
3. Anand, S.S., Mobasher, B.: Contextual recommendation. WebMine LNAI **4737**, 142–160 (2007)
4. CARS Context Modelling System. http://www.cs.ucy.ac.cy/~mettour/phd/CARSContextModellingSystem/
5. Deshpande, M., Karypis, G.: Item-based top-n recommendation algorithms. ACM Trans. Inf. Syst. **22**, 143–177 (2004)
6. Dourish, P.: What we talk about when we talk about context. Personal Ubiquitous Comput. **8** (1), 19–30 (2004)
7. Eclipse Modeling Framework Project (EMF). http://www.eclipse.org/modeling/emf/
8. hetrec2011-movielens-2k.: Dataset released in the framework of the 2nd International Workshop on Information Heterogeneity and Fusion in Recommender Systems (HetRec 2011) at the 5th ACM Conference on Recommender Systems (RecSys 2011). http://ir.ii.uam.es/hetrec2011/datasets.html (2011)
9. Karypis, G.: Evaluation of item-based top-n recommendation algorithms. In: Proceedings of the tenth international conference on Information and knowledge management, pp. 247–254 (2000)
10. Mettouris, C., Achilleos, A.P., Papadopoulos, G.A.: a context modelling system and learning tool for context-aware recommender systems. In: Hernandez-Leo, D., Ley, T., Klamma, R., Harrer, A., (eds.) Scaling up Learning for Sustained Impact, LNCS, vol. 8095, pp. 619–620. Springer Berlin Heidelberg (2013)
11. Mettouris, C., Papadopoulos, G.A.: Contextual modelling in context-aware recommender systems: a generic approach. In: Haller, A., Huang, G., Huang, Z., Paik, H.-Y., Sheng, Q.Z. (eds.) WISE 2011 and 2012 Combined Workshops. LNCS, vol. 7652, pp. 41–52. Springer, Heidelberg (2013)

Computational Creativity for Intelligence Analysis

Robert Forsgren, Peter Hammar and Magnus Jändel

Abstract We describe a decision support system for hypothesis assessment in which exploration is supported by computational creativity. A software tool for morphological hypothesis analysis and evidence handling is extended with a creative assistant that on demand suggests hypotheses that the analyst should consider. Suggested hypotheses are chosen so that they are far from hypotheses that the analyst previously has paid attention to but nevertheless are supported by evidence in an interesting way. For the purpose of providing thought-provoking suggestions, the creative assistant employs ensembles of novelty and value assessment methods and proposes hypotheses that stand out in this multi-ensemble analysis. Preliminary experiments investigate the system's potential for infusing novel and valid ideas into the decision making process.

Keywords Decision support system · Hypothesis assessment · Situation assessment · Computational creativity

1 Introduction

Situation assessment is the core of intelligence analysis. Once the analyst understands what really is going on it is comparatively easy to come up with reasonable actions and evaluate costs, risks and likely outcomes. Classical intelligence analysis means that the space of all relevant hypotheses is defined and that the analyst relates each piece of evidence to each hypothesis as for example in the *Analysis of*

R. Forsgren · P. Hammar (✉) · M. Jändel
Swedish Defence Research Agency, 164 90 Stockholm, Sweden
e-mail: peter.hammar@foi.se

R. Forsgren
e-mail: robert.forsgren@foi.se

M. Jändel
e-mail: magnus.jaendel@foi.se

S. Kunifuji et al. (eds.), *Knowledge, Information and Creativity Support Systems*,
Advances in Intelligent Systems and Computing 416,
DOI 10.1007/978-3-319-27478-2_6

Competing Hypotheses (ACH) method [6]. Realizing that such classical methods are impractical in face of the profusion of hypotheses and evidence provided by present day information handling systems, Gustavi et al. [4] pioneered that evidence should be connected to hypothesis attributes rather than to each hypothesis per se thus greatly reducing the number of evidence links and hence simplifying the analysis process.

The *Multi-Hypothesis Management and Analysis* (MHMA) method of Gustavi et al. [4] extends legacy *Morphological Analysis* tools in use by intelligence analysts. Morphological analysis is a three-phase qualitative analysis methodology in which a first phase maps out the hypothesis space, a second phase identifies consistency constraints and a third phase applies judgment to analyse and select preferred hypotheses. The method was pioneered by the astrophysicist and polymath Zwicky [11] and has been applied in diverse fields including future studies, policy analysis, law and technology. There is a vast literature on morphological analysis[1] for decision support that we will not attempt to review here (for a recent monograph see [8]).

Situation assessment is, however, not just about mechanically evaluating how the weight of pre-existing evidence is distributed over some hypothesis space. Based on what hypotheses that are found to be interesting, analysts will look for more evidence and re-evaluate existing evidence thereby following what essentially is a scientific research methodology. The creativity of this process is as important as the formal representations and information processing tools. Human creativity is, however, a fickle resource at best and succumbs easily to group-think and prejudice. In this paper we investigate how computational creativity can contribute to the analysis process.

Computational creativity is a new and burgeoning branch of computer science. For a recent review see [2]. In [7] we outline a research program for computational creativity in decision making where six different options for incorporating computational creativity in decision processes are described. One of these options pertains to using computational creativity for situation assessment i.e. coming up with the main hypothesis about what kind of situation that is as hand. The present paper describes a system and initial experiments that address this issue and provides hence a first step towards realizing the outlined research program. [7] provides also a brief review of the literature on computational creativity in decision support. As a harbinger of nascent interest in the decision analysis community, note the recent article [9].

In this paper we take a first step towards computational creativity for supporting hypothesis assessment in intelligence analysis by exploring the design space and design principles as well as experimentally comparing the simplest possible implementation with a few selected more complex designs. Section 2 introduces the intelligence analysis methods from the user's point of view and describes how

---

[1]Note that "morphological analysis" has different meanings in decision support and linguistics. In this paper we use the term only in the former meaning.

computational creativity is integrated in the user interface. Section 3 defines and motivates the computational creativity algorithms whereas Sect. 4 describes the implemented system that is explored experimentally in Sect. 5.

2 The User Perspective

We introduce evidence supported morphological analysis according to [4] by an example using the same scenario as used in our experiments in Sect. 4. The scenario is about the 2001 anthrax attacks [1] in which, shortly after the 9/11 events, letters containing lethal anthrax spores were mailed killing several infected victims. Both al-Qaeda and Iraq were suspected but B.E. Ivins, a U.S. biodefense scientist was declared to be the sole perpetrator although this conclusion is disputed [1]. Suppose now that we want to make a situation assessment in this scenario.

The hypothesis space is set up by morphological analysis. We define the *parameters* or conceptual dimensions of the problem and a domain of discrete *values* for each parameter. To facilitate presentation in readable figures and tables we use a small *morphological chart* with only three parameters and a handful of values per parameter. *Culprit*, *Motive* and *Source* (of the anthrax spores) are chosen as parameters. The set of values for each parameter is shown in Fig. 1. This morphological chart was developed in [4] where a detailed discussion of how it connects to the scenario is provided. A hypothesis is formed by picking a specific value (cell) from each of the parameters (columns), the hypothesis space being the set of all such combinations.

Using the MHMA tool, analysts enter evidence and connect evidence as having a positive, neutral or negative impact on each value cell. Hypotheses are scored based on the aggregated evidence weight on all the values that compose the hypothesis. The essential simplification of MHMA compared to ACH is that evidence is linked to cells rather than hypotheses thus avoiding the combinatorial explosion of evidence links in a large hypothesis space. To illustrate the method of coupling evidence, we have in Fig. 1 entered two pieces of evidence supporting each of the value cells *al-Qaeda*, *9/11 inspired* and *Simpler Afghan lab*. One piece of evidence supports each of the cells *B.E. Ivins*, *Target recipients* and *Gov. special lab*.

Culprit	Motive	Source
Al-Qaeda	9/11 inspired	Gov. special lab
Iraq	Domestic politics	Simpler lab U.S.
B.E. Ivins	Personal gain	Simpler lab Afg.
S.J. Hatfill	Target recipients	Iraqi special lab
People from U.S. gov.		

Fig. 1 Morphological chart with evidence weight represented by shade of colour. The cells with *darker colour* have twice as much supporting evidence as the *lightly coloured cells*. *White* cells are not related to any evidence

Assume now that a forensics team is strongly biased in favour of the al-Qaeda hypothesis maybe to the degree that it is considered disloyal to explore any alternatives to the current orthodoxy. A good computational creative assistant could help overcoming the ingrained bias of the group by proposing that analysts should look into the Ivins hypothesis. Given that the creative assistant knows that only the al-Qaeda hypothesis has been considered by the analysts, it would look for other hypotheses that both has evidential support and are the most different from the well-known hypotheses. The Ivins hypothesis fulfils both these criteria although some variations of the dominating hypothesis actually are better supported by the evidence.

The user interface of our tool consists of four fields as shown in Fig. 2. The *morphological chart* shows the hypothesis space. Analysts can enter evidence in *the evidence list* and connect each piece of evidence by positive or negative links to selected value cells of the morphological chart. *The hypotheses and evidence list* collates the hypotheses according to evidential support and allows the analyst to

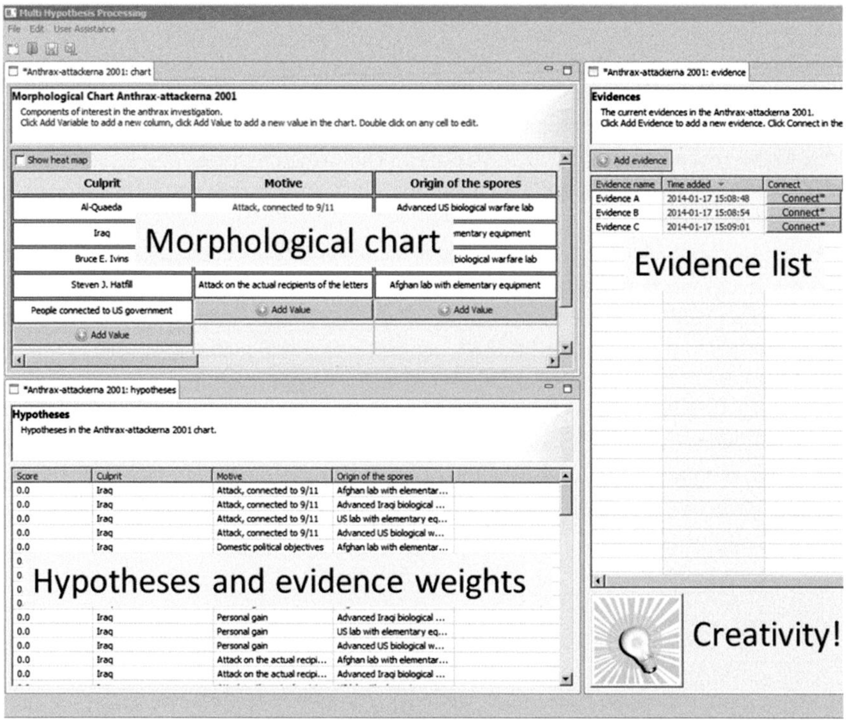

Fig. 2 Annotated user interface of the CC-MHMA tool consisting of the MHMA interface enhanced with the light bulb icon. Legacy MHMA features are the morphological chart, the hypothesis and evidence weight list and the evidence list

sort hypotheses according to evidence weight. By selecting a specific hypothesis the analyst can explore all evidence related to the hypothesis. The fourth field is the only new user interface component in our *Computational Creativity Supported Multi-Hypothesis Management and Analysis* tool (CC-MHMA). If the user clicks on the light bulb icon, the system suggests a creative hypothesis.

The purpose of the computational creativity function is to provide useful creative input to the analysts. Creative means in this context that the suggestions should be novel from the point of view of the analysts and germane to human goals and the situation at hand. The creativity of a suggestion can hence only be evaluated by human experts. One of the main problems is to provide the computational creativity agent with sufficient knowledge for producing novel and valuable suggestions. Novelty requires insight into what the analyst has considered before and might be aware of. Value calls for knowledge about the situation, the objectives of the analyst and general background information about the application domain. It is often much more difficult to compile the required knowledge in a computer readable form than to come up with suitable processing algorithms. The main hurdle for practical applications of computational creativity is that users cannot be bothered with extra work or cumbersome sensors in order to provide a computational creativity agent with the necessary knowledge. Therefore, we have in this work designed the system to harvest all inputs to the computational creativity from the pre-existing user interface and to include the light bulb as the single minimalistic and unintrusive extension of the user interface. The underlying computational creativity algorithm strives to suggest a hypothesis that has not yet been considered by the analyst and that is supported by the evidence in an interesting way. The next section will explain exactly what this means and how it is done. From the user perspective, the suggestion is shown by a colour-coded selection of values in the morphological chart, highlighting of the proposed hypothesis in the hypothesis list and highlighting of the evidence that impacts on the proposed hypothesis. The data provided by the tool to the computational creativity agent is the morphological chart, the evidence, the links between evidence and parameter values and the history of hypotheses selected for examination by the user. Note that any hypothesis that is suggested by the computational creativity agent is put into this list of known hypotheses to the effect that the computational creativity agent will not suggest something similar again.

3 The Algorithms

The first subsection describes the formal representation of the computational creativity agent while the following subsections flesh out the details. We discuss a broad range of possible algorithms leaving the definition of what has been explored in our preliminary tests to the next section.

3.1 Hypothesis Space and Formal Creativity Model

The hypothesis space H is the set of all hypotheses combined with a distance measure. The distance measure $d(h_1, h_2)$ is a real-valued function of the two hypotheses h_1 and h_2. The distance matrix D is formed by indexing the hypotheses according to an arbitrary order and defining matrix elements according to $D_{ij} = d(h_i, h_j)$.

A variable that is associated with a specific hypothesis and furthermore is defined for all hypotheses in the morphological chart forms a scalar field in hypothesis space. Any such scalar field will in the following be called a *charge field* and be denoted by a bold face symbol.

The next hypothesis to be recommended by the creative assistant is computed from by some multi-objective optimization algorithm Ω,

$$h_c = \Omega(\mathbf{n}, \mathbf{v}), \tag{1}$$

where **n** and **v** are the novelty and value[2] charge fields respectively. Multi-objective optimization algorithms (see [3] for a review) typically includes (1) finding the set of (Pareto) efficient hypotheses such that no other hypothesis have better novelty charge without having a worse value charge or vice versa and (2) selecting a solution among this set.

The *novelty charge* **n** of a hypothesis is a predictor for how novel the hypothesis appears to the user. It depends on how aware the analyst is of the hypothesis, the distance of the hypothesis to, and the level of awareness for, other hypotheses that the analyst is aware of. Formally we express this as,

$$\mathbf{n} = \Psi(\mathbf{a}, D), \tag{2}$$

in which **a** is the awareness charge field and D is the distance matrix. The function Ψ should be designed to allocate high novelty to hypotheses with low awareness charge that are far from any hypotheses with high awareness charge. The awareness charge represents how conscious the user is of the hypothesis.

The value charge **v** of a hypothesis is a predictor for how well the hypothesis describes the real-world situation that the analyst is interested in. Formally, the value charge field depends on the evidence E according to some function,

$$\mathbf{v} = \Xi(E). \tag{3}$$

[2]We suffer from a terminology collision regarding the term *value* between the fields of morphological analysis and computational creativity. Where the distinction is not obvious from the context *a cell in the morphological chart* is referred to as *value cell* whereas *usefulness* is called *value charge*.

Both the novelty charge and the value charge are real-valued variables in the domain [0, 1]. Ascending numerical value means rising novelty and value respectively.

3.2 Hypothesis Distance Measures

If the creative assistant has no insight in the semantics of the parameters and values of the morphological chart, the most obvious choice of distance measure $d(h_1, h_2)$ seems to be the Hamming distance [5] i.e. count the number of value cell substitutions that is required for transforming one of the hypotheses to the other and use the result as our distance measure.

Many possible enhancements of the distance measure depend on that the user can be coaxed to input more information. An example would be that some parameters in the table could be given more importance than others, creating a weighted Hamming distance where value substitutions would contribute to the distance in proportion to the weight of the parameter. In certain domains values could be ordinal, creating additional structure for a distance measure. Furthermore, we could ask the user to express the relations between the values in the domain of a given parameter as a graph. The contribution to the distance from each parameter could be counted as the number of edges that connect the value cells in each hypothesis. Also such a graph could be enhanced by weighing the edges. Other options for improving the distance measure with additional user input can also be envisaged.

3.3 Modelling Value Charge

We model the value charge of a hypothesis with a sigmoid form,

$$v_h = (1 + \exp(-\sum_{i=1}^{N_e} w_{ih}))^{-1}, \tag{4}$$

where the evidence weight w_{ih} describes how the piece of evidence with index i supports or contests the hypothesis h whereas N_e is the number of evidences. We take w_{ih} to be real-valued with positive values denoting degrees of support.

The sigmoid in (4) is symmetric, and saturates as the amount of (positive or negative) evidence increases, i.e. for a large collection of evidence the value v_h will change very little by the addition of one more piece of evidence. This captures the assumption that the belief (or lack thereof) in a hypothesis that is already well supported (or refuted) by a massive amount of evidence will not change much if another piece of evidence is added.

The MHMA software invites the analyst to define a negative, neutral or positive relation between each value cell in the morphological chart and each piece of evidence. By this device, MHMA implicitly connects each piece of evidence to many hypotheses in one fell swoop. To simplify, we will in the following only consider the resultant connections between evidence and hypotheses. The user input describing the relation between hypothesis h and the evidence with index i, is henceforth called the evidence impact factor and is denoted e_{ih}. In general, e_{ih} is real-valued with positive values representing levels of support.

The evidence weighting model relates the evidence weights to the evidence impact factors,

$$w_{ih} = \begin{cases} \gamma e_{ih} & \textit{if } e_{ih} \geq 0 \\ \gamma \alpha_{-} e_{ih} & \textit{if } e_{ih} < 0 \end{cases}, \tag{5}$$

in which γ is used for defining how much evidence that is needed for near certainty. The sigmoid function in (4) saturates when the absolute value of the sum of weights is about three which means that γ should be set so that this happens when the analyst would judge that the total weight of evidence is such that any further corroboration makes little difference for the conclusion. The parameter α_{-} is used for controlling the importance of negative evidence. The MHMA tool has primed the analyst to assume that $\alpha_{-} = 1$. The creative assistant could for example explore $\alpha_{-} = \infty$ according to which hypotheses that are disfavoured by any negative evidence have zero value or $\alpha_{-} = 0$ where the impact of negative evidence is disregarded. The creative assistant can furthermore use different perspectives on the relation between the evidence and the hypotheses by using several different evidence weighting models each represented by a choice of values of γ and α_{-}.

3.4 Capturing Awareness

To provide novelty, the computational creativity agent needs to know what hypotheses that the analyst already have considered. User interface actions, the timing of user interface actions and biometry are the three main means for learning about this. We represent the analyst's awareness of hypothesis h by a single real-valued variable a_h with values in the domain [0, 1] where $a_h = 0$ means no knowledge of h and $a_h = 1$ means maximum awareness of h.

Relevant user interface actions include selecting a hypothesis, selecting a group of hypotheses and examining evidence related to selected hypotheses. For each such action A and each hypothesis h that is associated with the action, the system should update the awareness charge according to some function $a_h = F(A, h, a_h)$ which operates on the present value of a_h and outputs an updated value of a_h. We can for example initialize all awareness charges to zero, allocate a baseline charge $a_h = 0.5$ the first time h is selected by the user and further increase a_h if the user scrutinizes related evidence or selects the hypothesis again.

The timing of user interface actions related to a hypothesis could also be taken into account which means that a_h is updated at regular time intervals $\{t_1, t_2, \ldots\}$ by an updating function that incorporates the history of user interactions according to,

$$a_h(t_n) = F(\mathbf{A}, h, a_h(t_{n-1}), t_n), \tag{6}$$

where t_n is the current time, $t_{n\text{-}1}$ is the time of the previous update and $\mathbf{A} = \{A(t_1), A(t_2), \ldots\}$ is the history of user actions. In (6) we could for example model the forgetfulness of analysts by letting a_h decay over time according to $a_h(t_n) = c(t_n - t_h)^{-\mu}$ where c, $t_h < t_n$ and μ are parameters. Psychological research indicates power laws for forgetting [10].

Timing analysis of user interactions is complicated by the lack of knowledge about what the user is doing while a given hypothesis is selected. The analyst may be vigorously ruminating over the hypothesis or alternatively be on a coffee break. Biometric methods, including for example video analysis or eye tracking, could provide crucial information about user behaviour to be encoded as special types of actions in (6).

3.5 Modelling Novelty Charge

Although there are many possible algorithms for computing the novelty charge field according to (2), we shall presently only define a simple baseline method and a somewhat more generic parameterized model.

According to the baseline method, the novelty charge of a hypothesis is zero if the awareness charge of the hypothesis is positive i.e. the user is considered to be aware of it. Otherwise, the novelty charge is proportional to the distance to the closest other hypothesis with positive awareness charge. Finally, we normalize so that novelty charges fall in the domain [0, 1].

The parameterized model is based on computing a *familiarity potential*,

$$\varphi(h) = \sum_{h' \in H} \frac{a_{h'}}{d^{\kappa}(h, h')}, \tag{7}$$

where the sum is taken over all hypotheses and κ is a positive integer. To handle the self-potential issue, we define $d(h, h) = \delta$ where δ is another parameter. In this equation, awareness charge is similar to electric charge and φ is analogous to electric potential. Consequently, the familiarity potential decreases as the distance to known hypotheses increases. The novelty charge is a sign-reversed normalized version of φ according to,

$$n_h = \frac{\varphi_{\max} - \varphi(h)}{\varphi_{\max} - \varphi_{\min}} \tag{8}$$

in which $\varphi_{\max}$ and $\varphi_{\min}$ are the maximum and minimum values of the familiarity potential respectively.

3.6 Putting It All Together: Creative Suggestions

The main assumption driving the design of our first prototype system is that users may be able to second-guess deterministic algorithms and will then feel that the creative clout wanes as they gain increasing experience of the tool. To avoid this, the creative assistant should choose randomly from an ensemble of different algorithms. In future experiments we intend to test if this assumption is valid. A generic creative suggestion method consists of, (A) a distance measure $d_i(h_1, h_2)$, (B) an algorithm for computing the novelty charge field Ψ_j (see (2)), (C) an algorithm for computing the value charge field Ξ_k (see (3)), (D) a multi-objective optimization algorithm $\Omega_m(\mathbf{n}, \mathbf{v})$ (see (1)). Each of these are selected from a corresponding ensemble of distance measures, novelty or value charge fields or optimization algorithms, respectively. The method for capturing the user awareness is considered to be fixed. At each new round of suggestion production, the creative assistant will select a creative suggestion method comprised of a randomly collection of components from these four aspects.

4 The Implementation

This section describes the experimental setups used in our initial explorative experiments. The computational creativity agent is implemented as an extension to the MHMA tool described in [4].

4.1 Baseline Implementation

The baseline implementation of the creative assistant is intended to investigate computational creativity in the simplest possible setting thereby providing a reference point for more complex implementations. The user awareness charge is initiated to zero for all hypotheses and is changed to one the first time that the user selects a hypothesis for examination. There is no time decay, and no other relations between awareness charge and user actions. Hence there are two distinct set of hypotheses: known hypotheses with $a_h = 1$ and unknown hypothesis with $a_h = 0$. This simplistic model of the user state was selected because it only uses information that is available in the MHMA tool.

The Hamming distance is the only distance measure employed by the baseline implementation. There is just one novelty charge algorithm according to which the novelty charge (n_h) of a known hypothesis is zero and the novelty charge of an unknown hypothesis is proportional to the distance to the closest known hypothesis. Furthermore, we employ one single value charge algorithm according to (4) and (5) with $\gamma = 1$ and $\alpha_- = 1$.

The multi-objective optimization algorithm applies the utility function,

$$f(v_h, n_h) = \beta \frac{n_h - \hat{n}}{\sigma_n} + (1 - \beta) \frac{v_h - \hat{v}}{\sigma_v}, \tag{9}$$

for selecting the hypothesis to suggest. In (9), $\hat{n}$ and σ_n are the average and standard deviation of the novelty charge with the corresponding notation for the value charge. Only unknown hypotheses are considered as candidates for selection and for calculating the averages and standard deviations. The parameter β balances the influences of novelty and value and is in our initial experiments ad hoc selected to 0.75. Since the optimum of this particular utility function always is Pareto efficient there is no need to explicitly compute the efficient set.

5 The Experiments

Experiments are performed on the morphological chart presented in the introduction. Three models are tested, representing a baseline implementation, an enhanced value model and an enhanced novelty model.

5.1 Testing the Baseline Implementation

The experiments use a collection of evidence comprising 35 pieces of information as further described in [4] and summarized in Fig. 3. Note that this set of evidence has been selected only for the purpose of testing the decision support tool and is not claimed to accurately represent the factual circumstances. A brief look at Fig. 3 suggest that there are much evidential support of Ivins being involved as well as

Culprit	Motive	Source
Al-Qaeda	9/11 inspired	Gov. special lab
Iraq	Domestic politics	Simpler lab U.S.
B.E. Ivins	Personal gain	Simpler lab Afg.
S.J. Hatfill	Target recipients	Iraqi special lab
People from U.S. gov.		

Fig. 3 Morphological chart showing the total weight of the evidence considered in our experiments. Positive evidence weight is indicated by *solid green* shading with *darker tone* representing more evidential support. Negative evidence weight is indicated by striped cells with the level of *red tone* indicating the absolute value of the evidence weight

domestic politics and the 9/11 events as motives. There is more evidence rejecting than supporting an Afghan lab as the source of the anthrax spores, whereas an Iraqi lab have an equal amount of positive and negative evidence.

We will first consider a situation in which the user just keeps pressing the light bulb icon and does not select any hypotheses other than those suggested by the creative assistant. This is not the normal mode of usage but serves to illustrate how the creative assistant works. The first ten hypotheses suggested by the creative assistant are shown in Table 1. We use abbreviated versions of the value cell names in the tables.

The first suggestion in Table 1 is the hypothesis with highest total value charge. For the next three suggestions, the assistant selects the hypothesis with the highest value charge that does not include any value cells from hypotheses already selected (i.e. they are maximally novel). The fifth hypothesis to be suggested is the one with highest evidence support but not sharing more than one value cell with any of the already known hypotheses.

Note that the wide scope of the hypothesis space combined with the simple MHMA evidence handling model occasionally give high value to intuitively rather unlikely hypotheses such as al-Qaeda targeting the individual mail recipients or Ivins using spores from an Iraqi lab. The user could, however, use even unrealistic suggestions as creative stepping stones rather than literally as candidates for the most likely solution. Although Ivins may not have had access to spores from Iraqi labs, investigators could be inspired to consider the possibility of a U.S. perpetrator producing a strain that appears to come from an Iraqi lab.

As we keep generating suggestions, the creative assistant eventually runs out of creativity at a point when all not known hypotheses have the same distance to the closest known hypothesis and thus the same novelty. The value charge is then the only decisive factor in (9) which makes any further suggestions trivial since the

Table 1 Hypotheses in the order suggested by the baseline creative assistant assuming that the user is aware only of previously suggested hypotheses

Culprit	Motive	Source	Evidence
Ivins	Domestic	Simple U.S.	+21, −7 = +14
U.S. gov.	9/11	Gov. special	+11, −2 = +9
Al-Qaeda	Recipients	Iraqi special	+9, −3 = +6
Hatfill	Personal	Simple Afg.	+5, −1 = +4
Ivins	Recipients	Gov. special	+18, −7 = +11
Hatfill	9/11	Simple U.S.	+12, −2 = +10
Hatfill	Domestic	Gov. special	+12, −2 = +10
Ivins	9/11	Iraqi special	+18, −8 = +10
U.S. gov.	Recipients	Simple U.S.	+11, −2 = +9
U.S. gov.	Domestic	Iraqi special	+11, −3 = +8

The evidence column shows the number of evidences supporting and refuting the hypothesis as well as the summed evidence weight

analysts have other tools for sorting hypotheses according to evidential support. Note that this effect is a consequence of using the Hamming distance as novelty measure and that analysts in practice will reach this state only in scenarios with quite small hypothesis spaces.

The fourth suggestion in Table 1 has a rather low value; only three out of 80 hypotheses have a lower value. Its high position in the list is caused by the combination of using the coarse-grained Hamming distance as novelty measure combined with the strong bias towards generating novel hypotheses engendered by the choice of $\beta = 0.75$ in (9). We tried setting $\beta = 0.25$ with the result that the fourth suggestion in Table 1 was removed from the top ten suggestions while all other hypotheses remained in the same order with a new hypothesis {*al-Qaeda*, *Personal gain*, *Simpler lab U.S.*} with evidence sum +8 appearing at the bottom of the list.

5.2 Enhanced Value Model

The sigmoid function in (4) saturates for most of the hypotheses in the baseline implementation. This means that differences in the creative utility function (9) is dominated by differences in novelty. Reducing the parameter γ should extend the range of value charge explored by (4). By running a series of experiments with different γ and β we found that the combination $\gamma = 0.3$ and $\beta = 0.3$ appears to give a reasonable balance of novelty and value. Table 2 provides an example of the output indicating a higher preference for value compared to Table 1.

As suggested in the discussion of modelling value charge, different value models can be obtained by varying, α_{-}. We have briefly investigated the case of $\alpha_{-} = 0$ which requires additional adjustment of γ in order to obtain results substantially differing from the baseline implementation. Setting $\gamma = 0.1$ and $\beta = 0.5$ produces a new and different set of creative suggestions.

Table 2 Hypotheses in the order suggested by creative assistant using the baseline implementation enhanced with the value model $\gamma = 0.3$, $\alpha_{-} = 1$ and using utility function parameter $\beta = 0.3$

Culprit	Motive	Source	Evidence
Ivins	Domestic	Simple U.S.	+21, −7 = +14
U.S. gov.	9/11	Gov. special	+11, −2 = +9
Ivins	Recipients	Gov. special	+18, −7 = +11
Hatfill	9/11	Simple U.S.	+12, −2 = +10
Hatfill	Domestic	Gov. special	+12, −2 = +10
Ivins	9/11	Iraqi special	+18, −8 = +10
Al-Qaeda	Recipients	Simple U.S.	+11, −2 = +9
Ivins	Domestic	Gov. special	+20, −7 = +13
Ivins	9/11	Simple U.S.	+20, −7 = +13
U.S. gov.	Domestic	Iraqi special	+11, −3 = +8

Table 3 Hypotheses in the order suggested by creative assistant using the baseline implementation enhanced with the novelty model of (7) and (8) in which both $\kappa = 1$ and $\kappa = 2$ give the same output

Culprit	Motive	Source	Evidence
Ivins	Domestic	Simple U.S.	+21, −7 = +14
U.S. gov.	9/11	Gov. special	+11, −2 = +9
Al-Qaeda	Recipients	Iraqi special	+9, −3 = +6
Hatfill	Personal	Simple Afg.	+5, −1 = +4
Iraq	Domestic	Gov. special	+10, −2 = +8
Hatfill	9/11	Simple U.S.	+12, −2 = +10
Ivins	Recipients	Simple Afg.	+14, −6 = +8
U.S. gov.	Personal	Iraqi special	+8, −3 = +5
Al-Qaeda	Personal	Simple U.S.	+10, −2 = +8
Iraq	9/11	Iraqi special	+8, −3 = +5

5.3 *Enhanced Novelty Model*

The enhanced novelty model according to (7) and (8) is controlled by parameters δ and κ. The self-potential (δ) has no effect in the present implementation since only hypotheses with zero awareness charge are candidates for creative suggestions. Using the baseline implementation with an enhanced novelty model with either $\kappa = 1$ or $\kappa = 2$ is, however found to make a significant difference in the output of the creative assistant. Table 3 shows the resulting hypotheses. Comparing this enhanced novelty model with the baseline implementation shows that the first four suggestions are the same but that the following suggestions are quite different reflecting that the potential model will differ more from the Hamming distance as the inventory of known hypotheses accumulates.

6 Discussion and Conclusions

Our initial experiments have focused on exploring key aspects of the generative algorithm and in particular the effect of varying selected components. The general impression from the experiments is that the creative assistant shows some promise and should be properly evaluated. This means that we must use much larger scenarios than in the present experiments. We cannot expect users to get a genuine eureka experience from creative suggestions unless the hypothesis space is so large that humans find it impossible to systematically consider all alternatives. Furthermore, we need a rich and complex evidence situation. In order to judge the creativity of the hypotheses suggested, and the value of the creative assistant, an experiment would use subject matter experts working on a realistic case for which they do not know the solution beforehand. The evaluation should be done in the process as well as post mortem.

We primarily regard the computational creativity as a tool for inspiring human analyst to explore a wider range of ideas and consider more alternatives rather than as generator of optional solutions per se. Furthermore, the infusion of creativity can stimulate users to expand the analysis model with new data, such as adding new value cells or looking for further evidence.

We recognize that morphological analysis may not be the ideal platform for building computational creativity. One could argue that a morphological chart spans a limited and static hypothesis space so that the only scope for creativity is to explore a predefined domain. However, we hold forth that the contribution of the computational creativity can only be evaluated by real users and that access to a user community of a perhaps less than ideal tool is a better starting point for researching computational creativity than to build a perhaps theoretically better tool with no practical opportunity for real-life evaluation. Likewise, we understand that a much better model of the user state can be built with state of the art behavioural research methods but we also know that more intrusive probes would deter professional analysts. By unobtrusively extending an existing, actively used baseline system for evidence supported morphological analysis we will in future more comprehensive experiments benefit from the crucial resource of the existing user community.

Although some of the suggestions produced by the creative assistant may seem to be obvious to a detached viewer, such ideas may still be useful in real-life decision making. Human free-thinkers may find it hard to get attention in a team that is locked into group-think. Originators of dissident ideas could be accused of disloyalty or having ulterior motives. That a divergent suggestion originates from a supposedly impartial and objective machine may help to make analysts consider it seriously and perhaps take it as a stepping stone for further exploration. Application of computational creativity to situation assessment in intelligence analysis, as described here, has not been explored before. Despite the limitations of the present implementation and the experimental scenario we feel that the novelty of the approach makes it interesting as a basis for further investigations and in particular full-fledged user trials.

Acknowledgments Christian Mårtensson suggested that we should consider how to apply computational creativity in morphological analysis. We thank Tove Gustavi, Maja Karasalo and Christian Mårtensson for making the MHMA tool available for our research. This research is commissioned by FMV, the Swedish Defence Materiel Administration.

References

1. Cole, L.A.: The Anthrax Letters. Skyhorse Publishing (2009)
2. Colton, S., Wiggins, G.A.: Computational Creativity: The Final Frontier? In: Proceedings 20th European Conference on Artificial Intelligence (2012)
3. Ehrgott, M.: Multiobjective Optimization. AI Mag. **29**(4), 47–57 (2009)

4. Gustavi, T., Karasalo, M., Mårtenson, C.: A tool for generating, structuring, and analyzing multiple hypotheses in intelligence work. In: Intelligence and Security Informatics Conference (EISIC), pp. 23–30 (2013)
5. Hamming, R.W.: Error detecting and error correcting codes. Bell Syst. Tech. J. **29**, 147–160 (1950)
6. Heuer, R.J.: Psychology of Intelligence Analysis. US Gov. Printing Office (1999)
7. Jändel, M.: Computational creativity in naturalistic decision-making. In: Proceedings of the Fourth International Conference on Computational Creativity, pp. 118–122 (2013)
8. Ritchey, T.: Wicked Problems—Social Messes: Decision Support Modelling with Morphological Analysis. Springer (2011)
9. Varshney, L.R.: Surprise in computational creativity and machine science. Decision Analysis Today **32**, 25–28 (2013)
10. Wixted, J.T., Ebbesen, E.B.: Genuine power curves in forgetting: a quantitative analysis of individual subject forgetting functions. Memory Cogn. **25**(5), 731–739 (1997)
11. Zwicky, F.: Discovery, Invention, Research Through the Morphological Approach. The Macmillan Company, Toronto (1969)

Bus Scheduling in Dynamical Urban Transport Networks with the use of Genetic Algorithms and High Performance Computing Technologies

V.A. Shmelev, A.V. Dukhanov, K.V. Knyazkov and S.V. Ivanov

Abstract Public transport is one of the main infrastructures in any city. It facilitates the smooth running of everyday life for ordinary people. Public transport services require constant improvement, and current methods of problem solving are not sufficient for dealing with high traffic congestion. In this paper, we present a genetic algorithm to optimize bus routes. We achieved a reduction of passengers' waiting times at bus stops.

Keywords Public transport scheduling · Genetic algorithm · Urban transport networks

1 Introduction

Public transport is a vital infrastructure for common citizens. Every day, many people use different kinds of public transport to get to work, to university, or to school. The use of public transport contributes to the reduction of air pollution, and prevents traffic jams. Public transport facilities could be improved. Our goal was to develop an effective bus schedule in order to increase public transport use.

V.A. Shmelev (✉) · A.V. Dukhanov · K.V. Knyazkov · S.V. Ivanov
ITMO University, Saint Petersburg, Russian Federation
e-mail: vad1611@yandex.ru

A.V. Dukhanov
e-mail: dukhanov@niuitmo.ru

K.V. Knyazkov
e-mail: constantinvk@gmail.com

S.V. Ivanov
e-mail: sivanov@mail.ifmo.ru

S. Kunifuji et al. (eds.), *Knowledge, Information and Creativity Support Systems*, Advances in Intelligent Systems and Computing 416,
DOI 10.1007/978-3-319-27478-2_7

The scientific and business communities have taken action regarding the optimization of public transport [1]. There is still a lot of work to be done in this area, as it is an pressing issue problem.

In the current paper, we describe an algorithm to reduce passengers' waiting time and driving time, taking into consideration the variations in passenger density and traffic conditions. We offer a solution that optimizes public transport dispatch, and the routes between bus stops.

2 Related Works

As mentioned previously, public transportation is a real problem. There are many approaches to optimize public transportation performance.

In the paper [2], traffic light span optimization was produced. The authors tried to implement a genetic algorithms approach to resolve the problem of controlling traffic lights.

Amoroso and Migliore [3] used a multi-objective function to design bus networks for small towns, such as Trapani. The applied approach produced a more efficient and effective bus network.

The bus network optimization problem was also was solved in work [4]. It proposed a heuristic approach to solving bus transportation network optimization problems. Their method involved genetic operators and a number of additional ingredients to aggregate a number of performance indicators to calculate a fitness function.

Work [5] solved the problem of the allocation of buses along routes. This problem was solved in two steps. Firstly, the frequency of buses required on each route was minimized by considering each route individually. Secondly, the fleet size in the first step was taken as an upper bound, and the fleet size was again minimized by considering all the routes together and using GAs.

3 Criterion of Efficiency and Genetic Algorithm

In the current paper, we tried to optimize the dispatch of buses to make public transport more convenient, which means decreased waiting time for buses, less driving time, and a lower cost of providing public transport activity. Therefore, we have a multiple-criteria optimization problem, in which each parameter should be minimized. We used a weighted sum model to compare solutions. Our criteria can be determined by the following:

$$F = w_0 \sum_{i=0}^{n} waiting\_time_i + w_1 \sum_{i=0}^{m} driving\_time_i + w_2 \sum_{i=0}^{routes\_number} \sum_{j=0}^{f(i)} bus\_driving\_time_j^i \quad (1)$$

w_i	relative weight of importance of the *i* criterion
n	number of came passengers
m	number of transported passengers
$waiting\_time_i$	waiting time by *i* passanger
$driving\_time_i$	driving time by *i*—passenger
$bus\_driving\_time_i^j$	time that needs *j*—bus on route *i*

Let us calculate the parameters described. Suppose we have *r* routes, and on each route there are *f*(*i*) flights. Our bus schedule can be represented as set of tuples, each of them containing bus dispatching times on separate routes:

$$\{(t_1^1, t_2^1, t_3^1, \ldots, t_{f(1)}^1), (t_1^2, t_2^2, t_3^2, \ldots, t_{f(2)}^2), \ldots, (t_1^r, t_2^r, t_3^r, \ldots, t_{f(r)}^r)\}$$

where *tij*—dispatch time of *i*th bus on route *j*. The passengers' arrival times and traveling times are described in the input data, which we obtained from a multi-agent model [6] of the city. Route information, fleet size, and bus information were given and were not changed during the optimization process.

For each route, there is given order of bus stops. This example is for route number 0:

$$(bus\_stop_0^0, bus\_stop_1^0, bus\_stop_2^0, .., bus\_stop_{bus\_stop\_count(0)}^0)$$

Where $bus\_stop\_count(i)$—the number of bus stops on route *i*, *bus_stopki*—number of bus stops *k* on *i*th route. *i*th bus arrival time on route *k* to bus stop *j*:

$$bus\_arriving\_time_{k_i}^j = bus\_arriving\_time_{k_i}^{j-1} + elapsed\_time_{j-1}^j(bus\_arriving\_time_{k_i}^{j-1}) \quad (2)$$

Where $elapsed\_time_p^q(time\_moment)$—time required to pass path $p \to q$ in the *time_moment* moment. But, $bus\_arriving\_time_{k_i}^0 = t_i^k$

Time required for *i*th bus to pass *j* route:

$$bus\_driving\_time_i^j = bus\_arriving\_time_{k_i}^{bus\_stop\_count(j)} - bus\_arriving\_time_{k_i}^0 \quad (3)$$

Define *i*th as passenger waiting time. Suppose that our passenger comes to the bus stop with the number *bus_stop* at $pass\_coming\_time(i)$ time moment, and takes the bus with bus number p and follows route k. Then, passenger waiting time can be represented as follows:

$$waiting\_time_i = bus\_arriving\_time_{k_p}^{bus\_stop} - pass\_coming\_time(i) \tag{4}$$

The time required by the passenger to drive to the bus stop *destination_bus_stop* on route k in bus p can be represented as follows:

$$\begin{aligned} driving\_time(i) = bus\_arriving\_time_{k_p}^{destination\_bus\_stop} \\ - (pass\_coming\_time(i) + waiting\_time_i) \end{aligned} \tag{5}$$

Genetic and evolutionary methods are usually used for the solution to the schedule's problems. In this paper, these methods are also used to find such schedules that allow for the reduction of passenger waiting and driving time. It is first necessary to represent our schedule as pertaining to an individual in the population. In our realization, each individual contains chromosomes and each of these chromosomes contributes to the buses' routes. A chromosome is an ordered set of the departure time t_j^i, where *i*—number of route, *j*—number of bus. Genes take their values from the routes' working intervals. There are r—number of routes, and *f*(*i*)—number of buses on *i*th route.

$$\begin{aligned} &[\{t_1^1, t_2^1, t_3^1, \ldots, t_{f(1)}^1), (t_1^2, t_2^2, t_3^2, \ldots, t_{f(2)}^2), \ldots, (t_1^r, t_2^r, t_3^r, \ldots, t_{f(r)}^r)\}, \\ &\{(t_1^1, t_2^1, t_3^1, \ldots, t_{f(1)}^1), (t_1^2, t_2^2, t_3^2, \ldots, t_{f(2)}^2), \ldots, (t_1^r, t_2^r, t_3^r, \ldots, t_{f(r)}^r)\}, \\ &\{(t_1^1, t_2^1, t_3^1, \ldots, t_{f(1)}^1), (t_1^2, t_2^2, t_3^2, \ldots, t_{f(2)}^2), \ldots, (t_1^r, t_2^r, t_3^r, \ldots, t_{f(r)}^r)\}] \end{aligned}$$

The current realization used a classic, genetic algorithm scheme. It used one-point crossing and tournament selection with the best individual transferring to a new population.

The fitness function was entered to estimate individuals (1). Discrete event simulation was used to measure passenger numbers in buses, stopping places, and times.

The fitness function calculation takes the most time, and individual estimation procedures are independent. Therefore, we can calculate that they are parallel. The parallel scheme of calculations is presented in Fig. 1. A master-slave [7] paradigm was used to make parallel calculations. Using such an approach, we obtain automatic, dynamic balancing, which is important when using heterogeneous networks

To analyze real data, we need large computing capacities. This problem can be solved by using cloud implementation. We used CLAVIRE [8], which allowed us to make data flow among the parts of the optimization package programs. The order of data flows was presented in [9].

The *Transport_genpairs* package creates pairs of all possible points that buses visit. The *Transport_findpath* package receives pairs and searches for optimal paths among them. The resulting data are sent to the schedule optimization program.

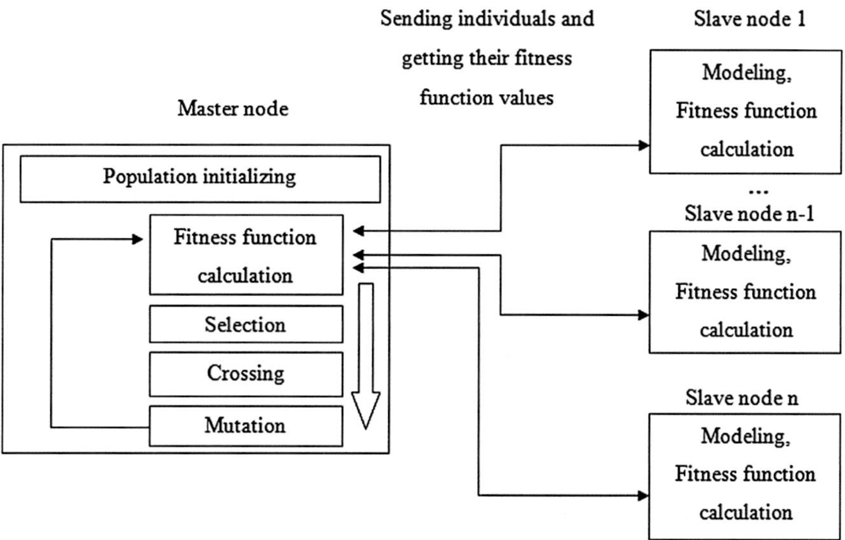

Fig. 1 Parallel calculation scheme

4 Scheduling Algorithm Examination

To check the algorithm, we performed an experiment on the data from the Petrograd district of Saint-Petersburg. Parameters for the genetic algorithm were found by a series of executions with parameter variations. The following parameters were used:

- Probability of mutation: 1 %
- Probability of crossing: 60 %
- Period of mutation: 15 iterations
- Population size: 3000 individuals
- Iteration number: 700

Figure 2 shows the mean waiting time, the passenger arrival density, and the number of passengers waiting at bus stops.

As shown in Fig. 2, the developed algorithm and the relevant software generate a more effective schedule compared to the traditional one, with equal headways. This difference is especially noticeable during rush hour periods (from 8:30 to 11:00), when the mean waiting time for passengers was reduced by three to five times, and the number of passengers waiting for a bus was reduced by two to four-and-a-half times. This advantage influenced the beginning and the end of the day, but this is normal if passengers know that public transport runs less frequently.

We also checked our algorithm results in more realistic case. Public transport is incapable of adhering strictly to the schedule, so we assumed that the time the buses needed to pass the current edge could be subjected to 20 %. The same holds true for people who come to bus stops. We set the arrival moment deviation to 5 min. Using

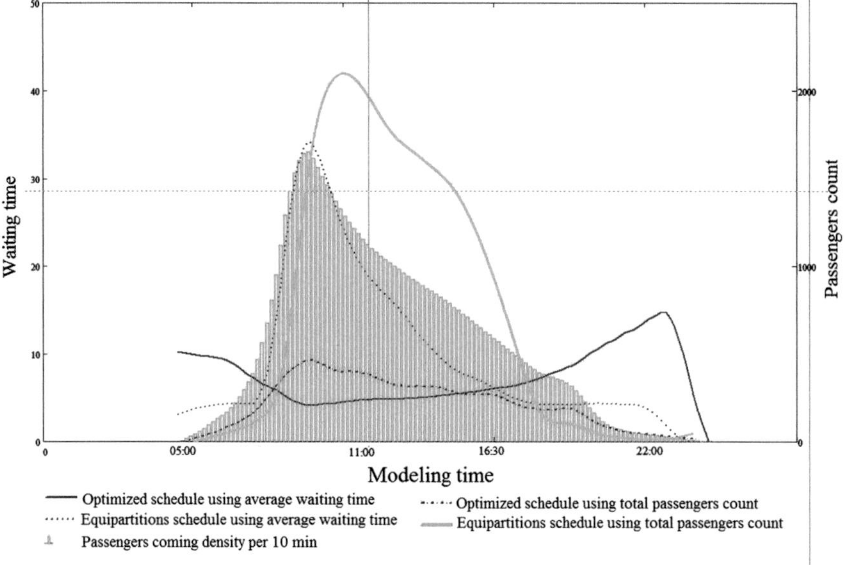

Fig. 2 Average waiting time of optimized schedule and evenly dispatching using

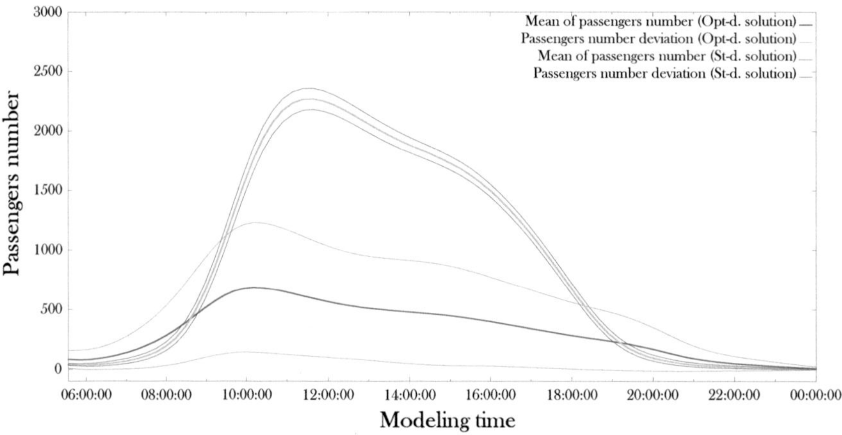

Fig. 3 Confidence intervals for the total number of passengers for optimized and uniform dispatch schedules

this assumption, we launched the algorithm 100 times and the results are presented in Figs. 3 and 4. A mean result of $\pm 3^*\sigma$ is shown.

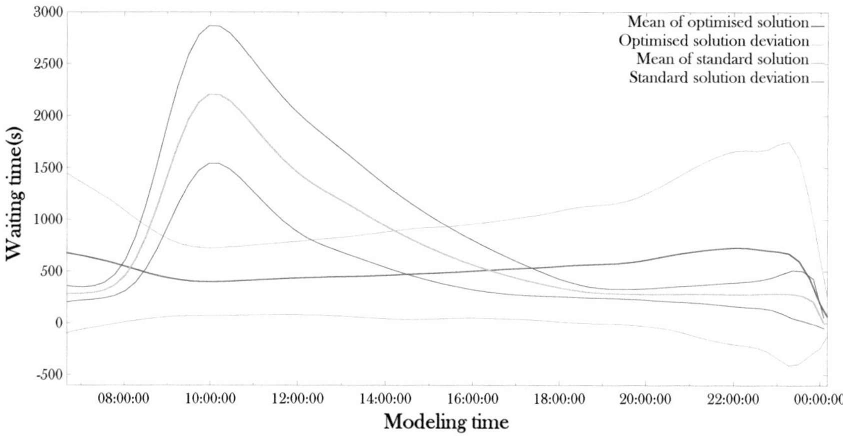

Fig. 4 Confidence intervals for the average waiting time for optimized and uniform dispatch schedules

5 Conclusion

In this paper, we described the method of bus scheduling in a dynamic transport network with the use of a genetic algorithm and high-performance technologies. The effectiveness criteria were based on waiting times and transportation times of all passengers, and the costs of bus movements. We examined the received method and relevant software of the Petrograd District of Saint Petersburg. The experiment showed that the components of the effectiveness criteria are two or more times better for an optimized schedule than they are for a traditional schedule with equal headway.

5.1 Future Work

During our algorithm examination, we took into account that buses have the ability to change their lines of march during the working period.

Each individual in the population contains a chromosome that, in the current case, is represented by a vector of pairs. Each pair consists of the bus dispatch moment and the route label that the bus will use during the trip. The vector is sorted by the moment of dispatch. The lengths of all chromosomes are equal, and this length is given as an application parameter. The number of pairs in the chromosome is more that the number of trips that a bus can perform. Each flight consists of the time spent on the flight performance, and the time spent on a technical break that occurs at the end of each flight. The technical break duration is also given as a parameter. If a bus is late at the starting point of the route, it is omitted and the next pair of chromosomes will be used.

The crossover, mutation and selection operation mechanisms are the same as they are in the algorithm mentioned above.

The main problem of such public transport optimization is that we rely to passengers arriving density; this could change because, if people were to spend less time commuting, they could go to the bus stop later, and our optimization approach would fail. This makes us wonder about implement in our model simple artificial intelligence to passengers. However, such an approach is beyond the scope of the present study and is therefore best reserved for future research.

Acknowledgement This paper is supported by the Russian Scientific Foundation, grant #14-21-00137 "Supercomputer simulation of critical phenomena in complex social systems".

References

1. Berlingerio, M., Calabrese, F., Di Lorenzo G., Nair, R., Sbodio, M.L.: AllAboard : a system for exploring urban mobility and optimizing public transport using cellphone data. Mach. Learn. Knowl. Discov. Databases, pp. 663–666 (2013)
2. Ceylan, H., Bell, M.G.: Traffic signal timing optimisation based on genetic algorithm approach, including drivers' routing. Transp. Res. Part B Methodol. **38**(4), 329–342 (2004)
3. Amoroso, S., Migliore, M., Catalano, M., Galatioto, F.: A demand-based methodology for planning the bus network of a small or medium town. European Transport Trasporti Europei n **44**, 41–56 (2010)
4. Bielli, M., Caramia, M., Carotenuto, P.: Genetic algorithms in bus network optimization. Transport. Res. Part C: Emerg. Technol. June 1998, **10**, pp. 19–34 (2002)
5. Kidwai, F.A.: A genetic algorithm based bus scheduling model for transit network. In: Proceedings of the Eastern Asia Society for Transportation Studies **5**, 477–489 (2005)
6. Ivanov, S.V., Knyazkov, K. V., Churov, T.N., Dukhanov, A.V., Boukhanovsky, A.V.: Modelng and optimization of city public transport in the CLAVIRE cloud computing environment. **3**(17), 1–11 (2013)
7. Huang, K.-C., Wang, F.-J., Tsai, J.-H.: Two design patterns for data-parallel computation based on master-slave model. Inf. Process. Lett. **70**(4), 197–204 (1999)
8. Knyazkov, K.V., Kovalchuk, S.V., Tchurov, T.N., Maryin, S.V., Boukhanovsky, A.V.: CLAVIRE: e-Science infrastructure for data-driven computing. J. Comput. Sci. **3**(6), 504–510 (2012)
9. Shmelev, V.A., Dukhanov, A.V., Knyazkov, K.V., Ivanov, S.V.: Bus scheduling in dynamical urban transport networks with the use of genetic algorithms and high performance computing technologies. In: 9th International Conference on Knowledge,Information and Creativity Support Systems."KICSS'2014. Proceedings", pp. 86– 92 (2014)

MCDA and LCSA—A Note on the Aggregation of Preferences

João Carlos Namorado Clímaco and Rogerio Valle

Abstract Cost and social dimensions are now being added to the existing environmental Life Cycle Assessment (LCA), leading to Life Cycle Sustainability Assessment (LCSA). LCSA is very complex with deep uncertainties and, generally, involves several stakeholders. Therefore, the analysis and interpretation of the outputs of LCSA is a difficult and complex task, requiring aggregation of preferences. The work in progress here presented deals with a study regarding the use of open exchange interactive software packages dedicated to Multi-criteria decision aiding in the context of LCSA output analysis and interpretation.

Keywords Life cycle sustainability assessment · Complexity · Multi-criteria analysis · Aggregation of preferences procedures

1 Introduction

In recent years, Multi-criteria models have undergone great development, especially, in the field we are particularly interested in this study, interactive methods based on a progressive and selective definition of preferences. Roughly speaking, the aggregation of preferences, in most of the cases, includes one or several of the following procedures: optimisation of weighted sums of the criteria (or other function of the criteria), pairwise comparison of alternatives and minimizations of a distance to the ideal point, or to other reference point. Interactive procedures, especially those rooted in constructivism, avoid a final aggregation of the preferences of decision agents based on a unique criterion, in some cases proposing the

J.C.N. Clímaco (✉)
INESCC, Universidade de Coimbra, Rua Antero de Quental, 3000 Coimbra, Portugal
e-mail: jclimaco@fe.uc.pt

R. Valle
SAGE/COPPE, Universidade Federal do Rio de Janeiro, Rio de Janeiro, Brazil
e-mail: rogerio.valle@sage.coppe.ufrj.br

S. Kunifuji et al. (eds.), *Knowledge, Information and Creativity Support Systems*,
Advances in Intelligent Systems and Computing 416,
DOI 10.1007/978-3-319-27478-2_8

combination of algorithmic protocols with the experience and intuition of decision agents in the process of preferences aggregation. We call them open exchange interactive procedures. As aggregation always implies loss of information, special care is needed. Ethical issues become very relevant. We believe that in the complex framework of LCSA, the most interesting Multi-criteria approaches have some kind of affinity with the Comparative Theory of Justice proposed by Amartya Sen [1]. So, interactive approaches should be favoured, and among them learning oriented tools, easier to accommodate to practical situations involving several stakeholders and not rarely also public opinion's points of view.

Preference aggregation is a widespread requirement in our societies. But perhaps the contemporary real life domain where they emerge in wider range and with greater intensity is in making decisions related to a sustainable way of life. Even though concerns with such decisions are not necessarily recent, sustainability remains a disputed concept. Yet, its association with development, as proposed by the Brundtland Report, has gathered an overwhelming acceptation, sustainable development being the one that "meets the needs of the present without compromising the ability of future generations to meet their own needs" [2]. Later, the notion has been incorporated into business and governmental decisions through a "triple bottom line" accounting framework that evaluates social, environmental and economic performance. Therefore, decisions about sustainability are inherently Multi-criteria, raising some theoretical and practical controversies. Moreover, it is consensual that sustainability assessment will be progressively more associated with the Life Cycle of the products. This means considering the whole set of cost, social and environmental impacts associated not only to the production, but also to the use and discard of goods, services and events, from the extraction of raw materials, through the several stages of transport and storage, till recycling, recovering or disposal in landfills. The tool to evaluate all those impacts is Life Cycle Sustainability Assessment (LCSA), which may help in public or private decisions on design options, transportation, and life use, etc. [3]. LCSA includes (environmental) Life Cycle Analysis (LCA), Social Life Cycle Analysis (SLCA) and Life Cycle Costing (LCC)—a complex procedure involving not only challenging uncertainty issues but also stakeholders with diverse backgrounds, interests, and points of view on the subject. The ISO LCA methodological framework is formalized since 2006. Now it is tentatively being extended to LCSA.

The major goal of this paper is discussing the potentialities and limitations of some Multi-criteria approaches when dealing with complex LCSA preference aggregation problems. Special emphasis is put in the application of a specific interactive package to a Social Life Cycle Assessment case.

2 Preference Aggregation in LCSA

The most relevant difficulties regarding Life Cycle Assessments are outlined and discussed in a paper by John Reap et al. [4]. The authors recognize that the methodology is still not very relevant in practice and give important hints on how to improve the available tools on the aggregation of preferences. Here we just outline the principal issues in short: defining impact indicators is not a simple task; impacts aggregation, usually based on the use of weights, tends to raise many controversies; the choice of adequate procedures of aggregation is conditioned by uncertainty associated with decision processes, and by the eventual existence of multiple actors; increasing time horizons leading to a bigger difficulty in measuring the socio-economical impacts; the interpretation of the results depending on the previously defined LCA/LCSA goals; the use of interfaces enabling a holistic evaluation is also a key issue.

3 MCDA Tools and Aggregation of Outputs in LCSA

This subject is methodologically discussed in [5, 6]. The phases of characterization and modelling of the process are crucial for the success of LCA and LCSA in practice: study goals and boundaries, data collection and management, definition of impacts, indicators/criteria and alternatives. In [5, 6] the potentialities of using problem structuring techniques in the LCA framework and the aggregation of outputs are discussed. It must be remarked that these issues are still more relevant in LCSA.

The aggregation of outputs in LCA is usually done using weighted sums of normalized outputs obtained by direct elicitation procedures. Many recent papers alert that the use of weights raises several questions/errors, still more relevant in LCSA. Compensatory procedures showed to be very problematic in these circumstances. Moreover, it must be remarked that the direct use of weights still involves normalization of the terms, which also may contribute to the distortion of the results in many situations. Multi-Attribute Utility Theory (MAUT) can help doing weights elicitation in a more scientific way, but it does not overcome the root of the problem. See, for instance, [4, 7, 8].

Taking into account the limitations of the additive model, some authors proposed to avoid total compensation approaches by using outranking methods such as those of the ELECTRE family, where the concepts of concordance and discordance are employed to build partial order outranking relations, followed by some kind of aggregation and exploitation of the results. Although avoiding the complete compensation, these approaches require fixing a large number of parameters, such as concordance and discordance thresholds, weights (even if here they are less problematic than in the additive model, because they just represent coefficients of importance of the criteria and not trade-offs, as in additive model), and in many cases indifference and preference thresholds to take into account uncertainty associated to the criteria scores. It must be remarked that, of course, changes in these parameters can influence drastically the results. See for instance [9]. Furthermore, simple non-compensatory procedures can also be useful in the aggregation of outputs in LCSA. Note that inter-criteria aggregation is particularly delicate in LCSA, considering that data can be either quantitative or qualitative. Moreover, uncertainty and lack of information can be dealt with by using different decision aiding tools that use several types of uncertainty representations. Sensitivity analysis and robustness analysis are very important approaches in LCSA framework and can be efficiently associated with Multi-criteria modelling. Finally, though the classical LCA is a "steady-state" assessment, real life cycles under study involve inter-connections and dynamic interactions and so dynamic system tools should be associated with MCDA [10].

4 On the Potentialities of Open Exchange Interactive Multi-criteria Procedures in the Aggregation of the Outputs of LCSA

The use of Multi-criteria analysis in the aggregation of outputs in LCSA is still very limited. Taking into account the difficulties associated with the use of many well-known methods in this context, the authors are testing flexible learning oriented Multi-criteria packages that seem to cope better with the problem. Reinforcing all previous remarks, we believe that the following issues should be considered carefully: the involvement of several stakeholders/actors (cooperating and/or negotiating); the desirable public participation in situations where public and private spheres and the evolution of their borders are important issues; the inevitability of coping with large uncertainties (see for instance, [11, 4]). In this section we outline the main characteristics of the open exchange interactive tools that we intend to test.

4.1 VIP-Analysis [12]

It is an interactive platform dedicated to the choice problematic regarding the evaluation of a discrete set of alternatives according to a Multi-Attribute additive value function. It does not require precise values for the weights. Rather, it can accept imprecise information (i.e. intervals and linear constraints) on these values. It enables the discovering of robust conclusions, i.e. those that hold for every feasible combination of the weights/scaling constants, and to identify what is the variability of the results resulting from the imprecision in the parameter values. This software is free.

As VIP Analysis does not requires accurate values for the weights, some of the drawbacks of using the additive model are really mitigated: the decision makers or stakeholders only need to identify linear constraints for the weights, normally by indirect ways (one example is by comparing equivalent swings); it is very appreciated, in the context of LCSA, the possibility of identifying which are the robust conclusions compatible with the use of incomplete information regarding the weights; this tool seems very adequate to support a group of stakeholders/actors meeting face to face around a computer, which is also very adequate in the LCSA context. However, a proficient use of VIP-Analysis requires the sharing of knowledge about the tool potentialities and limitations with the stakeholders, so a facilitator is required.

4.2 A Non-compensatory Software Package Integrating an Interactive Dashboard with an Extension of the Conjunctive Method [13]

The interactive package, based on an extension of the conjunctive method integrates an interactive multidimensional dashboard, in order to open options of analysis. However, here we do not follow the mainstream aggregation frameworks using weighted sums of normalized data regarding the considered dimensions, as in many situations when aggregating outputs of LCA/LCSA. We opted by a non-compensatory tool thence avoiding the most negative aspects of the additive model. The software of support to our proposal [13] is based on an interactive implementation of the conjunctive method, enabling the consideration of up to three performance thresholds, having in mind to classify the objects under evaluation. Quantitative and qualitative criteria are admitted. For details see [13]. The software is free.

For the applicability of this approach see the experiment reported below in paragraph 5. For details on the application of this tool in other type of applications see [14, 15].

4.3 ELECTRE Methods

The idea is using a non-conventional implementation of ELECTRE methods containing a control panel with slide bars that enables "continuous" variation of some parameters, providing real time sensitivity and robustness analysis. For instance, a simple visual inspection could allow a real time evaluation of the changes on the outranking relations graph according to parameter changes. This possibility is very suitable to our case because LCSA involves very high uncertainties. The choice of the adequate ELECTRE method depends on the problematic associated to the case under study. For details on ELECTRE methods see [16].

5 A First Experiment Based on a Brazilian Social Life Cycle Assessment Case Study

In this section we describe a first experiment regarding the integration of a Multi-criteria approach in a Social Life Cycle Assessment (SLCA) case. After outlining the case study (5.1), the software package referred to in 4.2 is introduced and applied to the case study (5.2). Although this experiment just deals with one of the three components of LCSA it is very relevant in the context of this paper because SLCA raises the most difficult problems regarding the aggregation of outputs in.

5.1 SLCA Case Study—on the Comparison of Wind and Thermo-Electric Power Stations

The case study refers to a wider on going research that analyses two energy production sites in Northern-East Brazil: a wind power plant (including a production equipment by company A and an installation, operation and maintenance of the equipment by a company B) and an oil thermal power plant (including plant installation by company C, oil supply by company D and operation and maintenance by company E) [17]. For illustration purposes, we restrict the social impacts here taken into account to those related to only one of the stakeholders, namely the local community of those companies. The considered SLCA subcategories are those of United Nations Environment Programme methodological guides, with exception of two of them that does not apply (respect of indigenous rights and delocalization and migration). For each one of them, three social indicators were taken from GRI (Global Reporting Initiative) guidelines, ISO 26000 and Ethos indicators were assessed. Data was collected in interviews, on-site observation and secondary sources (such as companies' sustainability reports).

Every Social Life Cycle impact Assessment must include a characterization procedure, i.e. inventory data attributed into a given subcategory must be modelled and expressed into a numeric indicator. The social impacts of each power plant correspond to an average value of the respective companies' scores (regarding the interviews and other sources), multiplied by a severity factor involving the social impact magnitude based on society consequences; the area of influence of the social impact in terms of space; the importance of each impact in the corresponding UNEP/SETAC category [18] (human rights, working conditions, healthy and safety, cultural heritage, governance and socio-economic repercussions); and the applicable legal requirements, or notifications from regulatory agencies. The calculation of those factors is beyond the scope of this paper. Details can be seen in [17]. The final outcome is the impact matrix presented in Table 1 (where an impact is better, when the corresponding value is smaller…).

Table 1 Impact matrix

Subcategories	Social impact indicators	Wind power plant	Thermal power plant
Access to material resources	When necessary and possible, the company helps improving its region public spaces (schools, health centers, green areas, etc.)	6	16
	Infrastructure and services provided primarily for public benefit through commercial activities, in-kind, or pro bono	24	18
	The organization has developed some project related to infrastructure with mutual community access and benefit	27	21
Access to intangible resources	Where necessary and /or possible, the company collaborates offering freedom of expression, access to information, community services, health, education, security	6	10
	The company organizes educational campaigns along with local companies in their community	6	10
	Presence/strength of community education initiatives	18	18
Cultural heritage	The organization is concerned about preserving its cultural heritage, mainly where the company activities have some impact	76	95
	The company promotes cultural activities such as appreciation of local cultures and cultural traditions	30	50
	Take action and support cultural activities enhancing minority, discriminated or vulnerable groups	63	78
Safe and healthy life conditions	Management effort to minimize use of hazardous substances	50	131
	The company promotes good health, contributing to access to medicines and vaccinations, encouraging healthy lifestyles (exercise, good nutrition, early diagnosis of diseases, etc.)	24	26
	The company seeks to eliminate negative impacts to health caused by any production process, any product or service supplied by the company	21	13
Community engagement	The company is concerned in maintaining contact with the surrounding community, seeking to minimize negative impacts their activities could cause	54	42
	The company invites local residents to attend meetings at which issues of collective interest are addressed	27	27
	Characteristics, scope and efficiency of any program or practice to assess and manage operation impacts in the community	84	56

(continued)

Table 1 (continued)

Subcategories	Social impact indicators	Wind power plant	Thermal power plant
Local employment	Hiring of employees residing in the neighborhood areas	72	36
	The company uses close NGO or cooperative services	18	22
	Proceedings for local hiring and proportion of high management positions from the local community in important operational units	81	66
Safe life conditions	Regarding the number of legal complaints per year against the organization with regard to security concerns	54	78
	Regarding the number of casualties and injuries per year ascribed to the organization	57	44
	The company has management policies related to private personal security	12	9

5.2 On the Application of a Non-compensatory Software Package to the SLCA Case Study

The software that supports our proposal [13] is based on an interactive implementation of the conjunctive method, enabling the consideration of up to three performance thresholds, having in mind to classify the objects under evaluation. Figures 1 and 2 present its interactive dashboard/control panel. In Fig. 1 a system of

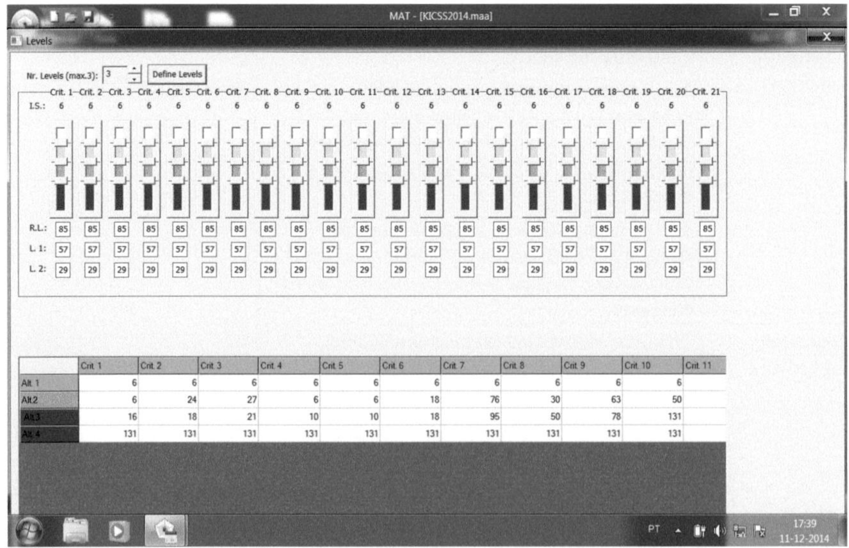

Fig. 1 Interactive dashboard: fixation of thresholds

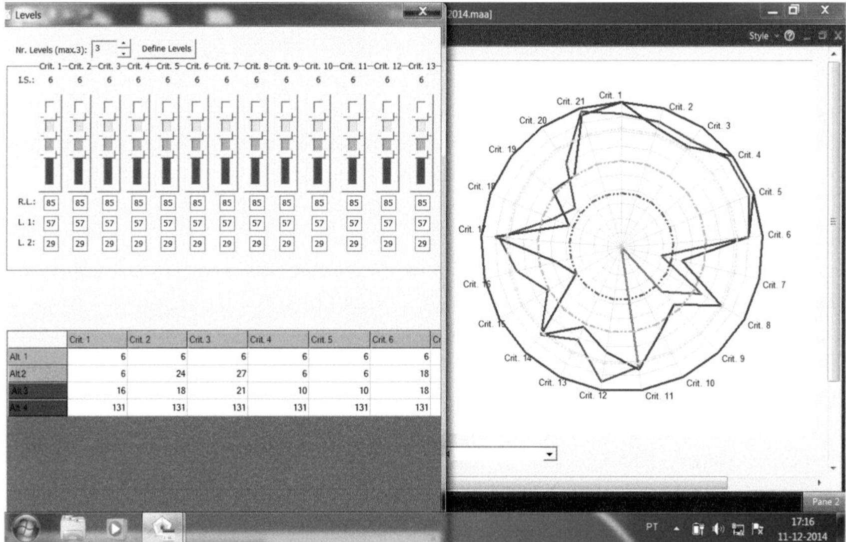

Fig. 2 Interactive dashboard: profile of each object

"elevator boxes" enables the fixation of thresholds for the various dimensions. In the bottom one find the matrix of the objects under evaluation with the corresponding performance on each dimension/attribute. In general, dimensions the evaluation of which is either quantitative or qualitative are admitted (note that Fig. 1 is a print screen showing just part of the criteria/indicators due to space limits; of course, an elevator is available enabling to see all of them). On the right of Fig. 2 the profile of each object concerning the evaluated dimensions is presented in a radar graphic. The fixation of each of the three thresholds which bound the four performance levels may be carried out through the "elevators" in Fig. 1. The representation of the thresholds in the radar graphic is made using coloured broken rings. The objects will appear with the following colours: red if at least one of the attributes of the object does not reach the red level; orange if every attribute satisfies the red level but at least one attribute does not reach the "good" threshold, yellow if every attribute satisfies the "good" threshold but at least one attribute does not reach the "very good" threshold and green if every attribute satisfies the "very good" threshold. Note that the described aggregation process is non-compensatory thence avoiding the problem that a weak performance in one dimension may always be compensated by a strong performance in another dimension, as in an additive model; there is no need to assume additive independence among the various attributes, an adequate property because that is a requirement too strong in the context of SLCA. Furthermore, it must be remarked that in the case of the proposed methodology there is no inter-criterion aggregation. This is very important to avoid the drawbacks related to the total compensation associated to the additive model.

The characteristics of this tool seem very appropriate to fulfill some requirements in SLCA, due to the intuitive aspect of the graphics, as well as to the simplicity of the user-oriented proposed approach, both conceptually and operationally.

It is also relevant that besides the classification of alternatives/scenarios, the system enables a holistic view of them. The combination of an analytic process with visual holistic views constitutes a remarkable added value, giving an important global feedback from the system to the stakeholders. Of course, this contributes to a global perception of the sustainability phenomena.

It is also relevant that besides the classification of alternatives/scenarios, the system enables a holistic view of them. The combination of an analytic process with visual holistic views constitutes a remarkable added value, giving an important global feedback from the system to the stakeholders. Of course, this contributes to a global perception of the sustainability phenomena.

In Figs. 1 and 2 are presented the data and some of the results provided by the non-compensatory package tool regarding the SLCA case above introduced. Note that alternatives 1 and 4 are virtual alternatives. Alternative 1 is the ideal solution (represented by the external contour limiting the radar graphic) and 4 is the anti-ideal solution (represented by the central point of the radar graphic), concerning the data of the two alternatives under evaluation (one regarding wind power generation and the other thermal power generation). So, the real alternatives under study are alternatives two and three of the figures. The thresholds that allow the classification of the alternatives were provided by specialists. They are the same for all the criteria. This is not very surprising because the scores generation procedure used in [17] led to normalized scores.

Regarding the spider web graphic in Fig. 2, we can conclude that alternative wind power (represented by the dark contour on the graphic) is classified as acceptable, because it passes the red threshold in all criteria but not all of them for the yellow one. So, it appears “orange” in the table on the right part of Fig. 2. On the other hand, the alternative 3 (represented by a pink contour on the sider web graphic) is unacceptable because it does not pass the red thresholds of criteria 7 and 10. So, it appears in “red” on the table on the right part of Fig. 2. Through an available very flexible and intuitive interactive manipulation of the performance thresholds the actors may acquire knowledge about possible variations in the classification of the objects under evaluation, taking into account the required levels for each attribute. This is particularly relevant for cases with a greater number of alternatives. Moreover, the holistic comparison of pairs of alternatives enables a visual inspection of the differences between them. Finally, a holistic view of the contours of the alternatives on the spider web graphic, together with the temporary exclusion of some criteria (which is very simple in operational terms) may help in the understanding of the process, forming global perceptions concerning the points of view of different actors involved in the study. We believe that schematic open and interactive systems can be useful for training people for thinking together and better, from the local and national political power agents to the corporation representatives, non-governmental organizations and other forms of public opinion expression, leading to a more creative society. This is particularly clear when we intend to join SLCA with environmental and economical LCA. Providing holistic views together with classical scientific analysis is crucial to promote a collective discussion in order to propitiate the evolution to a more creative way of living in our communities, not only from the technological point of view, but also considering the social and environmental dimensions.

6 Conclusion

In this paper we discussed the potentialities and limitations of some open exchange interactive Multi-criteria approaches dedicated to the aggregation of preferences, when applied to very complex problems, as the Life Cycle Sustainability Assessment problems. An illustrative example applying a non-compensatory software package to the analysis of the outputs of a SLCA study is presented.

References

1. Sen, A: Idea of Justice. Harvard University Press, Cambridge (2011)
2. Our Common Future–Report of the World Commission on Environment and Development: United Nations (1987)
3. Zamagni, A., Pesonen, H.L., Swarr, T.: From LCA to life cycle sustainability assessment: concept, practice and future directions. Int. Life cycle Assess. **18**, 1637–1641 (2013)
4. Reap, J., Roman, F., Duncan, S., Bras, B.: A survey of unresolved problems in life cycle assessment–part 2: impact assessment and interpretation. Int. J. Life Cycle Assess. **13**, 374–388 (2008)
5. Mazri, C., Ventura, A., Jullien, A., Bouyssou, D.: Life cycle analysis and decision aiding: an example of roads evaluation. http://sciencestage.com/d/1093635/life-cycle-analysis-and-decision-ainding-an-example-for-roads-evaluations-.html
6. Mietinen, P., Hamailainen, R.P.: How to benefit from decision analysis in environmental life cycle assessment (LCA). EJOR **102**, 279–294 (1997)
7. Benoit, V., Pousseaux, P.: Aid for aggregating the impacts in life cycle assessment. Int. J. Life Cycle Assess. **8**, 74–82 (2003)
8. Huppes, G., von Oers, L., Pretato, U., Pennington, D.: Weighting environmental effects: analytic survey with operational evaluation methods and a meta-method. Int. J. Life Cycle Assess. **17**, 876–891 (2012)
9. Cinelli, M., Coles, S., Kirwan, K.: Use of multicriteria decision analysis to support life cycle sustainability assessment: an analysis of the appropriateness of the available methods. In: 6th International Conference on Life Cycle, Gothenburg (2013)
10. Halog, A., Manik, Y.: Advancing integrated systems modeling framework for life cycle sustainability assessment. Sustainability **3**, 469–499 (2011)
11. Benetto, E., Dujet, C.: Uncertainty analysis and MCDA; A case study in life cycle assessment (LCA) practice. In: Proceedings of the 57th Meeting of the European Working Group on Multicriteria Decision Aiding. Viterbo (2003)
12. Dias, L., Clímaco, J.: Additive aggregation with interdependent parameters: the VIP analysis software. JORS **51**, 1070–1082 (2000)

13. Clímaco, J., Fernandes, S., Captivo, M.E.: Classificação MultiAtributo Suportada por uma Versão Interactiva do Método Conjuntivo (An Interactive version of the Conjunctive Method dedicated to Multiattribute Classification Problems). CIO—Working Paper 9 (2011)
14. Clímaco, J., Craveirinha, J.: Multi-actor multidimensional quality of life and sustainable impact assessment—discussion based on a new interactive tool. In: Proceedings of GDN 2013, Sweden (2013)
15. Valle, R., Clímaco, J.: Green economy in the state of rio de janeiro—a non-compensatory multidimensional interactive evaluation. In: XXII MCDM Conference, Malaga (2012)
16. Figueira, J., Mousseau, V.,Roy, B.: ELECTRE Methods, Multiple Criteria decision Analysis: State of the Art Surveys. In: Figueira, J., Erghot, M., Greco, S., Springer (2005)
17. Duarte, S.: Social impacts identification and characterization tool (SIICT): proposal and application for social LCA. SAGE/COPPE/UFRJ Internal Report (2014)
18. United Nations Environment Programme (UNEP). Guidelines for social life cycle assessment of products. Paris (2009)

A Model for Managing Organizational Knowledge in the Context of the Shared Services Supported by the E-Learning

Agostinho Sousa Pinto and Luís Amaral

Abstract Lifelong learning has become a fundamental element in any organization, as a tool for managing development, sustainability and innovation of knowledge capabilities. In this paper, we start by describing the combination of two phenomena—Shared Service Centre (SSC) and E-Learning—that expand knowledge capabilities. Also this intentional combination provides for the emergence of new knowledge. Then we produce a conceptual model for developing a continuous learning organization. The Delta model assumes—in essence—the characteristics of complex adaptive systems (CAS) and incorporates concepts of crowdsourcing. This model conceptualizes and integrates SSC for managing organizational knowledge and memory repository and E-Learning, as the enhancer for a push-pull communication process and the emergence of knowledge. Then, from the conceptual model, we mathematically deduct a Δ (Delta) factor. This factor presents characteristics that allow it to be used for determining the viability of continuous learning from the sharing of knowledge capabilities. In addition we present a set of recommendations designed Improvement Classes, which aim to guide a successful implementation of the Delta Model.

Keywords Knowledge management · Delta model · Delta factor · Lifelong learning · Shared services · E-learning · Complex adaptive systems

This work has been supported by FCT—Fundação para a Ciência e Tecnologia within the Project Scope: Pest OE/EEI/UI0319/2014.

A.S. Pinto
IPP/ISCAP/CEISE, Porto, Portugal
e-mail: apinto@iscap.ipp.pt

L. Amaral (✉)
Universidade do Minho/Centro Algoritmi, Guimarães, Portugal
e-mail: amaral@dsi.uminho.pt

S. Kunifuji et al. (eds.), *Knowledge, Information and Creativity Support Systems*,
Advances in Intelligent Systems and Computing 416,
DOI 10.1007/978-3-319-27478-2_9

1 Introduction

Learning organization is induced as a concept defining a structure, where knowledge is fully utilized to increase capacity, behaviour change and increased competition. As knowledge management evolved and developed, learning organization conceptualisation has become an interesting metaphor for the sustainability of contemporary organizations. It focuses on the importance and plausibility of the relationship between sustainability and knowledge, as a basis for innovation and business performance. This process of building a learning organization is founded primarily to be sustained by the availability of organizational memory. With the development of Information and Communication Technology (ICT), the conditions for increasing the amount of organizational memory, its availability and its relational power were created. The basic conditions were the interoperability of systems and knowledge of communication tools.

The Shared Services Centre (SSC) and E-learning characterize this evolution, respectively, by promoting interoperability and communication of knowledge. These two ICT developments are the focus for two reasons. The first reason is the adoption of the Shared Services Centre by the organization to build a common organizational memory. A Shared Services Centre in its concept is a repository of memories of different organizations, which integrates diversity. The second reason is the adoption of e-learning as a tool for the communication and management of knowledge. E-learning is in its concept and management tool for knowledge. These two developments of ICT overcome two of the facilitators of knowledge management—the existence of diversity and communication skills. These facilitators can be expanded, so that organizations can have access to other experiences that could develop unique knowledge. Moreover, if this knowledge is available to be communicated it may allow an organization what to choose to learn.

In this paper, we conceptually explore knowledge management organization and, consequently, its ability for continuous learning. This results in the development of lifelong learning, as a dependent from the Shared Services Centre and E-learning variables. We focus on the question "Does using a SSC and E-learning conceptual model positively influence organizational knowledge management?" Our work is based on the resources of the organization (RBV), where there is a live, independent and combined memory supported by resource stored in the systems of information technology and communication organizational memory. These features develop capabilities to support knowledge sharing. We are on the threshold of the definition of the Δ (Delta) factor for these knowledge sharing capabilities. This factor, initiated for SSC and E-Learning conceptual model for knowledge sharing can produce a positive effect on knowledge management and support the sustainability of the learning organization.

The sustainability of an organization relies mainly on the result of its responsiveness and adaptability to the environment that surrounds it. This environment is a competitive environment and each organization requires adaptability and the adoption of sustainability policies, which are supported by learning and innovation.

This work shows that this dimension integrates continuous learning from a perspective of knowledge management. As a "function", knowledge management is able to promote the goal of developing and maintaining an organization of continuous learning. Two elements of knowledge management are pointed out for promoting continuous learning: first, diversity management and, second, the communication of that diversity. These two elements are presented in any environment.

This work begins with an approach of the theoretical assumptions. Section 2 next discusses the conceptualization of the structure that can support the learning organization, while Sect. 3 presents the Research Design. Section 4 then describes the conceptual model, the Δ (Delta) factor and other results and, finally, Sect. 5 presents the conclusions and future work.

2 Conceptual Background

2.1 Organizational Knowledge Management

Organizations recognise the need for collaborative innovation, as it improves how they can create, accumulate and exploit knowledge that enhances their competitiveness and the sustainability of their organization. Additionally it not only revolutionises working ways and creation but it also promotes organizational learning. Moreover, it engenders the management of knowledge and interactions with the environment. Knowledge management should be used to stimulate organizational knowledge for the optimization of diversity and interoperable resources.

Through globalisation, the success of organizations depends on their ability to interact with the environment and their ability to operate globally [1]. There is a need to identify and define technological models that can effectively support this interaction for pursuing an innovative and entrepreneurial approach. In this pursuit knowledge management has a central role.

The creation of knowledge occurs in different forms. According to Davenport and Prusak [5, 6] there are five ways to generate knowledge; acquisition, dedicated resources, fusion, adaptation, and knowledge networks. Or as Nonaka and Toyama [10] said, in our Theory for Knowledge Creation, knowledge creation is featuring four stages of processing, creation and explicit knowledge: Socialization, Externalization, Combination and Internalization. Generated knowledge can then be analysed and this generates knowledge internalisation, if its analysis determines its usefulness for the organization. After verifying its usefulness, it is then systematised and filed, through a codification and co-ordination of knowledge. This usefulness aims to make knowledge accessible to those who need it. To determine how it should be encoded, the knowledge is defined into a tacit or explicit way.

Tacit knowledge is the knowledge that individuals or groups have but it is not consciously accessible, since it is not yet part of the reality [12, 13]. This knowledge acquiring requires an effort of understanding by processes that are not directly controlled by the learner. Explicit knowledge is the knowledge that is at the conscious level. Thus, not only the person or group recognise it, as they can convince others of this knowledge [14].

An important component of knowledge management is the existence of organizational memory. This memory enhances organizational knowledge gathering, organization, dissemination and re-use of the knowledge created inside the organization. Organizational memory is then a system capable of storing the resulting perceptions of experience or of keeping the abstract memory registers construction. This organization memory should be persistent in time and recoverable [8].

Organizational Knowledge Management is a field of multi-disciplinary research that cuts across areas such as information systems, computer science, human resource management and organizational sciences. It focuses on the sharing and re-use of knowledge. It includes individual and group skills, for improving quality, efficiency, increased customer and employee satisfaction and reduction of risks. Additionally, it improves knowledge development, through imagination, experience and experimentation.

The process of knowledge management can be structured into four fundamental areas [14]:

- Knowledge creation;
- Retention and retrieval of knowledge;
- Sharing and knowledge transfer; and
- Application of knowledge.

These four areas sustain the continuous learning, which is essential for keeping up to date human resources in relation to technological innovations and work practices. Additionally, information communication technology makes learning easier.

2.2 Shared Services

The promise of SSC comes from a hybrid conception of traditional models aimed at capturing the benefits with centralised and decentralised arrangements. By unbundling and centralising activities, the basic premise for SSC seems to be that services provided by one local department can be replicated in others with relatively few difficulties. This is a conceptualisation that is only possible through the support of information and communication technologies.

Centralised governance structures are characterised by substantial economies of scale and scope, because procurement of assets and services is done on the broadest scale possible within the organization. A centralised staff can eliminate redundant functions and improve the clarity of strategic alignment. Moreover, the organization evolves a simpler communicating structure. However, centralised governance makes response time slower and often there is a higher distance from customers.

In the decentralised governance structure business units respond faster and with more flexibly to necessary changes, since they have the knowledge and choice about the usable resources to support business priorities and the costs allocated to business unit initiatives. However, the company, as a whole, will have higher costs due to natural inefficiencies related to the duplication of services.

A Shared Services Centre should combine ideally the advantages of the structures of the two worlds, the centralised and the decentralised. On the one side, this should result in economies of scale, scope and standardisation. On the other side, this should result in a flexible and effective alignment of IT with the needs of business. Additionally, synergy and mutual learning will increase, while SSC will provide a clear management focus [16, 17].

2.3 E-Learning

E-Learning is one way for enterprises to improve their process of information flow for knowledge improvement and achievement. Its web-based system nature removes the users or learners time restrictions or geographic limitations. Moreover, availability and flexibility are often presented as advantages, when compared with traditional face-to-face systems. However, too many projects have high failure costs or users find them difficult to adopt.

E-Learning development followed two pathways. One pathway was as a distance learning tool and the other was as assisted computer learning. Common to the two pathways is the need for technological support with bottlenecks in the communication networks (availability) and in the experience of use (usability). These two pathways are subsumed under E-Learning, as the Internet becomes the integrating technology.

This adds pressure in the development of the learning materials—a task, which is most of the time unplanned but which is fundamental for the learning experience that reflects in the usability. The Internet allows for the instant widespread dissemination of the learning content simultaneously and at anytime and anywhere.

An additional strength of E-Learning is its use as a tool for standardisation of content, when compared with site formation of the different sectors of one organization. E-Learning can have control points that make it possible to have assessment of user interactions and outcomes.

In these scenarios E-Learning has properties of potential usefulness that goes far beyond the delivery of learning. The technology of E-Learning has characteristics for evolving into a structure for supporting organization knowledge management. It allows learning by relating new knowledge with past experiences, through the linking of learning to needs, and then by practically applying the learning. This can potentially develop a more user-oriented and effective deployment of knowledge availability and the development experiences. The adoption of E-Learning for knowledge management develops an environment of interactivity and it promotes efficiency, motivation, cognitive effectiveness and flexibility of the learning style [4, 7].

2.4 Complex Adaptive System (CAS)

The paradigm that has been used to model organization development has changed. This change can be found in a fundamental core of articles and books that deal with enterprise dynamics [10, 11].

Now, organizations are centred on developing and maintaining dynamic core structures in order to support the unpredictability of change on the edge of chaos. However, for supporting this in a viable manner, that is, with efficiency and effectiveness, information systems become a core structure.

In this context, information systems, as a fundamental element supported within an attached reductionist vision, will never have relevance. A paradigm shift should be achieved and developed. Prosperity in this new world demands dynamics for complexity and CAS is a great way to provide those [15].

Complex adaptive systems are systems with a great number of components, sometimes called agents, that interact, adapt and learn and many contemporary problems are under the theory of a complex adaptive system [18].

The dependency of complex systems on initial conditions can be expressed in a very exponential level by the butterfly metaphor, whereby a butterfly flapping its wings in South America can affect the weather in Central Park [9]. Complex systems exist on the edge of chaos presenting a regular and predictable behaviour but, suddenly, they can start a mass change response to what can be seen as a minor change [2].

3 Research Design

3.1 Aims and Objectives

The primary aim of this study is the evaluation of what can be obtained by using a SSC with E-learning for developing shared knowledge capabilities.

We develop this aim through a conceptual modelling of SSC and E-Learning sustained in the evaluation of the initial motives and expectations for adopting a SSC and E-Learning for knowledge management. This conceptual model evaluation results in the development of a mathematical deduction. This mathematical deduction validates whether knowledge increases or not, through the use of the conceptual model.

This study results into a conceptual model for sharing knowledge capabilities. Moreover, it has developed the identification of its knowledge sharing capabilities, through the measuring of its Δ (Delta) factor. This conceptual model is sustained under the concepts of SSC and E-Learning.

This analysis can be used by the organization to support a decision-making process related to the introduction of such a shared service centre. This research should contribute to the limited body of research on SSCs available and the use of E-Learning as a knowledge management tool.

This research focuses on the question "Is it possible to add value, through sharing knowledge capabilities by using a SSC and E-learning conceptual model for knowledge management?" We start by trying to define the Δ (Delta) factor for having a SSC for sharing knowledge. This Δ (Delta) factor will positively affect the management decisions regarding the use of Shared Services Centres with E-Learning for developing the sharing of knowledge capabilities.

3.2 Research Methodology

Qualitative research can take a positivist, interpretive or critical approach. The present work will take an interpretive approach supported by a critical approach. Absence from the outset is that there is no hypothesis to confirm or refute that the positivist approach is not appropriate.

The methods used tend to be qualitative, interpretive and constructive but do not exclude any quantitative analysis. For triangulation and the ability to be used on occasion, a quantitative approach is considered.

For convenience our contacts and references are organizations at national and international levels. Four organizations were contacted—one in the area of postal services, another service in the area of banking and two in the national government and regional administration. It was considered important for the sample that interviews were conducted in organizations that acted in different branches of activity.

The collection of materials analysed for this study used qualitative methods, which were carried out using the techniques of documentary analysis, participant observation and interviews (unstructured or semi-structured).

From the point of view of information processing techniques, all materials collected in the interviews were recorded initially using manuscripts and then recorded in a second step using audio recording. For the application of Grounded Theory several empirical materials and data were collected by using interviews in large Portuguese organizations. After the data was systematically collected it was coded according to the three encoding types: open, axial and selective.

From the selective coding emerged a set of concepts or classes that led to the theory presented here in Fig. 1.

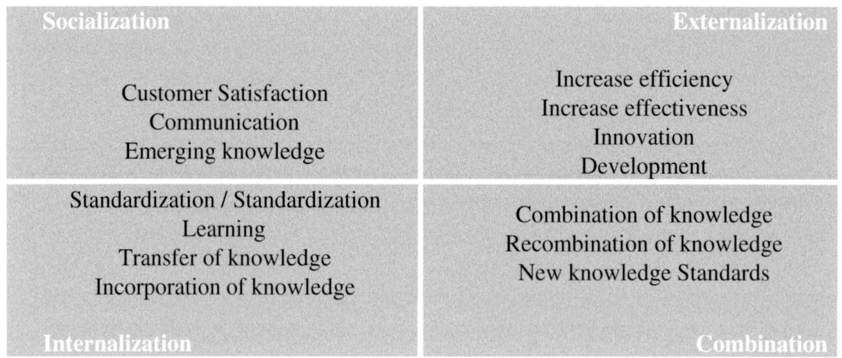

Fig. 1 Relationship between the classes and the *socialization*, *externalization*, *combination* and *internalization*

4 Conceptual Model

In a conceptual model for SSC use for knowledge management it is necessary to go beyond the process systematisation, standardisation and cost reduction that are the fundamental philosophical assumptions for SSC management. In this case SSC should developed unique and new combined knowledge emerging from the combination of sharable organizations. On one side, by adopting a SSC an organization has shareable knowledge available with potential for being integrated into other organizations. When this process occurs we have a process of sharing knowledge. On the other side, the SSC can combine different organizational memories and potentially develops new patterns. These two sides are conceptualized into the model of Fig. 2. This conceptual model describes how the SSC is a live, combined organizational memory and develops unique sharable combinations of knowledge using E-Learning communication flow.

Theoretical support and guarantees of scientific rigor research work was supported in Nonaka Theory for Knowledge Creation, featuring four stages of processing, creation and explicit knowledge: Socialization, Externalization, Combination and Internalization.

In this conceptual model A, B and C represent organizations that use and integrate the shared services centre and where the shared services centre holds the larger part of the individual organizational knowledge. The resulting Organizational Knowledge of the Shared Service (OK SS) comes from the addition of the Organization Knowledge (OK) that is sharable in each organization. The organizational knowledge of shared services exists by itself and also in one of the independent organizational memories (OK). The total amount of available knowledge will always be different from that existing individually [OK(A+B+C) <> OK SS]. The E-Learning is the artefact of technological nature that allows the communication of the organizational knowledge and also part of the organizational memory. The property of internalisation of Nonaka [10, 11], which is fundamental for learning,

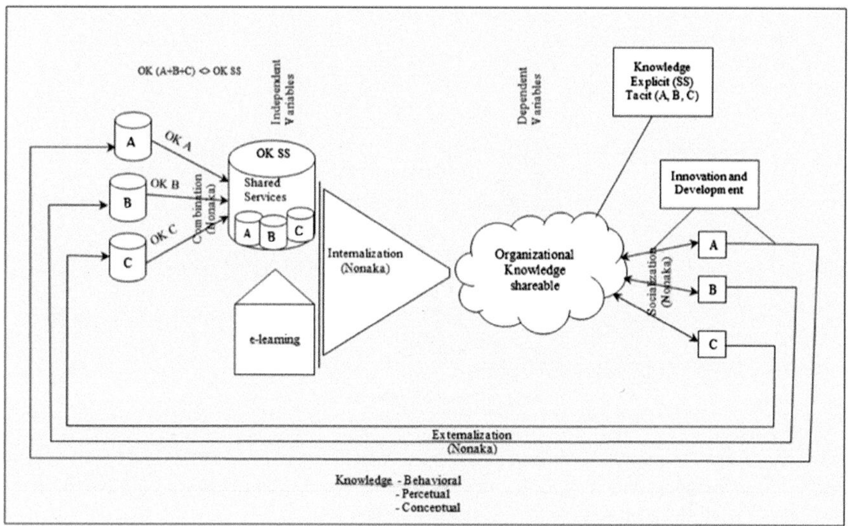

Fig. 2 Conceptual modeling for knowledge management using SSC and E-learning

becomes a dependent variable from the independent variables of SSC and E-Learning. The organizational learning is then dependant on these variables and results for its combinatory development and communication of learning.

The shared organizational knowledge results from the composition of two types of knowledge. First, from individual organizational knowledge, which is available at the SSC and which can be used and integrated by other organizations. Second, from the set of organizational knowledge that emerges from the combination of individual organizational knowledge that is only made possible by its co-existence in the SSC.

The conceptual model reflects also the concepts [3] that organizational knowledge can be arranged into three large classes:

- Behavioural knowledge—characterized as know how;
- Perceptual knowledge—characterized by the knowing; and
- Conceptual knowledge—characterized by its use.

Moreover, the Delta Model assumes characteristics of crowdsourcing because it is able to make a contact between different agents who seek and consume the best solutions, also contributing to the decision taken by the organization. In this case, the organization is the Shared Services Centre and the public the various collaborators, framed in their organizations and spread over large geographical areas.

Crowdsourcing is a model that uses the collective intelligence, collaborative culture and the formation of communities to solve problems, create content or to support innovation. Their existence depends directly on the involvement and participation of people.

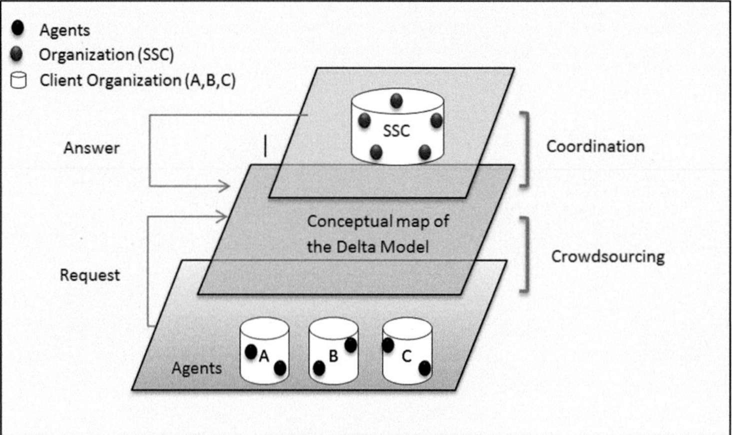

Fig. 3 The delta model as a crowdsourcing solution. Based on interagency map of harnessing the crowdsourcing power of social media for disaster relief

Crowdsourcing comes this new context, producing collaborative actions which are based on collective intelligence and training communities to create content and problem solving, so coveted by the Shared Services, Fig. 3.

The Shared Services Centers can then take actions, share and receive information and coordinate with each other using the information on the map those results from applying the Delta Model.

5 Results

It is fundamental, due to the component of joining SSC and E-Learning that the developed conceptual model intends to add value, through demonstrating that there exists a possible Δ (Delta) factor. This Δ (Delta) factor result from the difference between every organizational memory that the SCC has and the resulting combined knowledge. The Δ (Delta) factor is described according to Eq. 1. In Eq. 1, Kssc represents the organizational shareable knowledge resulting from the sum of all the different organizational memories.

$$\Delta = Kssc - \sum_{i-1}^{n} K_i \quad (1)$$

Equation 2 describes the composition of the *Kssc*. In Eq. 2, *K* represents the amount of sharable knowledge; *P* represent the amount of perceptual knowledge; *C* represents the amount of conceptual knowledge; *Ce* represents the amount of explicit behavioural knowledge and *K′* represents the combination of knowledge.

This combination results from the existing in the individual organizational memories of the SSC according to Eq. 3.

$$Kssc = \sum_{i=1}^{n} K_i(P + C + Ce) + K' \tag{2}$$

$$K' = C(K, p) \tag{3}$$

Equation 3 characterises K'. K' represents the patterns of knowledge that result from the combination, C, of the different knowledge, K, of different organizations, p. Replacing in Eq. 1 the $Kssc$ and Kp by the equivalents in Eqs. 2 and 3, respectively, results in Eq. 4.

$$\Delta = K' \tag{4}$$

This measure, the Δ (Delta) factor, allows the SSC manager to develop strategies for justifying and improving SSC. The Δ (Delta) factor determines that this is possible through the adding of the amount of diversity that exist at the SSC and the efficiency of E-Learning, as a communication for the learning tool.

It is assumed and demonstrated that the from the delta factor emerging from the application of the delta model is, thus, demonstrated the possibility of creating knowledge using the application of this complex adaptive system. Alongside comes a theoretical reference model, which is represented by a set of classes for improvement, which should be applied in organizations.

From an analysis of the data, there emerges a set of classes for improvement, which serve as a reference for organizations to improve knowledge management and thereby promote their efficiency.

In addition, was possible to accept a set of recommendations that were aggregated into categories or classes. We present a set of recommendations designed Improvement Classes, which aim to guide a successful implementation of the Delta Model.

- Exercising and exploiting intellectual capital.
- Specializing the level of service and this increases the effectiveness of the organization, through the enhancement of the tacit knowledge of incorporation.
- Promoting Innovation as something more than an intention but, in this context, an organizational process.
- Increasing the control and accuracy of the governance of organizations
- The combination or recombination of processes or products, from different organizations, known via Shared Services, provide, or may give the appearance of new processes, services or products.
- There is some additional motivational effect in that if you share what is best made in each organization.
- The need to share implies systematization and coding, which itself externalized implicit organizational knowledge.

- In a shared services environment, the standardization, homogenization, and even professionalism, learning times tend to be lower, ensuring e-learning one customization of this learning.
- The Shared Services and E-learning are combined drive or artifacts that can propel intentionally achieve the goals.
- Only by knowing the intention for which it creates, we can give real value to create information and knowledge.
- Shared Services in the use of E-learning fosters, by their nature and availability, autonomous learning.

6 Conclusion

This conceptual model should be seen and evaluated, as a model focused on the management of organizational knowledge. When we talk about shared services, we focus on economy of scale, standardisation, re-engineering of processes and control. When we think about E-Learning after overcoming the convenient temptation to associate it with traditional education, we think in technological solutions for continuous learning.

With the present model and its adoption in a context of integrating SSC and E-Learning, we are able to integrate the process of organizational knowledge management. At the same time this incorporates the process of knowledge discovery that happens in first place, through the sharing of diversity and in second place, as a result of the explicit activities of standardisation and re-engineering.

It can also support the discovery of new knowledge by developing the potential of infinite growth, through the use of analytical tools that are used to discover patterns. This discovery often happens and is sustained in the model by the elicitation of knowledge, in particular behavioural knowledge, which by other ways would be no more than tacit knowledge.

Thereby the amount of knowledge that potentially exists in the knowledge bases and organizational memory of the SSC and which evolves continuously for the stated reason becomes sharable knowledge. This knowledge can then be used and re-used by all those intervening, thus sustaining innovation and promoting the development of the organization.

References

1. Bradley, S.P., Hausman, J.A., Nolan R.A.: Globalization, technology, and competition: the fusion of computers and telecommunications in the 1990s, p. 392 (1993)
2. Brownlee, J.: Complex Adaptive Systems. Complex Intelligent Systems Laboratory (2007)

3. Carvalho, J.A., Morais, M.P.: Sistemas Informáticos e Conhecimento Organizacional: Uma Reinterpretação dos Papeis Desempenhados pelos Sistemas Informáticos nas Organizações, pp. 1–16 (2004)
4. Chunhua, Z.: E-learning: the new approach for knowledge management (KM). In: International Conference on, Presented at the Computer Science and Software Engineering, vol. 5, pp. 291–294 (2008)
5. Davenport, T.H.: Conhecimento empresarial (1998)
6. Davenport T.H., Prusak, L.: Working knowledge: how organizations manage what they know (2000)
7. Liu, Y., Bai, Y., Chen, Z.: Research on e-learning supporting system: a review from the perspective of knowledge management, presented at the e-learning, e-business, enterprise information systems, and e-government, pp. 329–332 (2009)
8. Maier, R.: Knowledge management systems: information and communication technologies for knowledge management (2004)
9. Mazzocchi, F.: Exceeding the limits of reductionism and determinism using complexity theory (2008)
10. Nonaka, I., Toyama, R.: The knowledge-creating theory revisited: knowledge creation as a synthesizing process. Knowl. Manage. Res. # 38 Pract. **1**(1), 2–10 (2003)
11. Nonaka, I.: Perspective-tacit knowledge and knowledge conversion: controversy and advancement in organizational knowledge creation theory. Organ. Sci. (2009)
12. Popper, K.R.: Conhecimento objetivo: uma abordagem evolucionária (1975)
13. Popper, K.R.: Truth, rationality and the growth of scientific knowledge (1979)
14. Santos M.Y., Ramos, I.: Business Intelligence: Tecnologias da informação na gestão de conhecimento (2006)
15. Schneider, M., Somers, M.: Organizations as complex adaptive systems: implications of complexity theory for leadership research. Leadersh. Quart. 351–365 (2006)
16. Strikwerda, H., Zee, J.: Fostering execution-how to leverage leadership in a confused business environment (2003)
17. Su, N., Akkiraju, R., Nayak, N., Goodwin, R.: Shared services transformation: conceptualization and valuation from the perspective of real options. Decis. Sci. **40**(3), 381–402 (2009)
18. Sutherland, J., van den Heuvel, W.J.: Enterprise application integration and complex adaptive systems. portal.acm.org (2002)

Online Communities of Practice: Social or Cognitive Arenas?

Azi Lev-On and Nili Steinfeld

Abstract This study examines whether interactions between members of communities of practice typically have cognitive or social character. Content analysis of more than 7000 posts, automatic words frequencies analyses as well as interviews with community members demonstrate that the interactions between members of the Israeli Ministry of Social Affairs' communities of practice, the subject of the present study, emphasize the cognitive rather than social aspects. This emphasis is reflected in the content of posts, the avoidance from discussing personal cases or offering emotional support and more. The findings are particularly interesting given the nature of these communities as a space for social workers whose work requires and is characterized by a high degree of social and emotional interactions.

Keywords Online communities of practice · Content analysis · Digital government · Organizational systems · Social services · Cognitive exchange

1 Introduction

Online communities of practice are groups of professionals who share a common interest in the examination of professional issues related to their work [1, 2]. In some communities, the social aspect of communication between members is particularly important, and fulfills needs that are impossible to similarly satisfy in offline contexts. For example, people with unique needs and problems, or those who have social difficulties, may find it difficult to create peer groups in their offline social environment. People who belong to marginalized groups, as well, typically seek connections to people similar to them, who may not always be found in their

A. Lev-On · N. Steinfeld (✉)
Ariel University, Ariel, Israel
e-mail: nilisteinfeld@gmail.com

A. Lev-On
e-mail: azilevon@gmail.com

S. Kunifuji et al. (eds.), *Knowledge, Information and Creativity Support Systems*, Advances in Intelligent Systems and Computing 416,
DOI 10.1007/978-3-319-27478-2_10

immediate environment. The Internet creates novel opportunities for social contacts in such cases [3].

Due to intensive and stressful work, both physically and emotionally, social workers often require support and sympathy [4]. A closed online group where colleagues and partners converge can function as place for ventilation, expression of support, solidarity, and creation of social ties among participants. Thus, online communities of practice of social workers can be expected to carry a more social character than comparable communities of professionals.

However, the literature about user behavior in online communities focuses less on the social aspects and more on the cognitive aspects of the use of the communities, such as the amount and quality of the information shared, information overload, uses of and gratifications from the information in the communities [5, 6]. The focus on the cognitive aspects is not surprising since these are online communities of practice, i.e. groups of professionals who are interested in a joint examination of professional issues related to their work. For this they share professional knowledge, tools, resources and experiences. The amount of shared knowledge in communities, its flow and its uses are therefore important aspects of such environments.

The research on the cognitive aspects of online communities of practice focuses on the various motivations to contribute information [7–10]. Yet, although motivations to contribute information have been investigated in depth, there are no prospective studies that analyzed the content of communities of practice to learn whether cognitive uses are indeed central, and if the content uploaded to the communities tends to be social or cognitive. This is the main contribution of the current study.

2 Research Environment and Methodology

This study analyzes the case of Israeli social workers' online communities of practice established by the Ministry of Social Affairs.

The research on communities of practice presents a unique case in Israel, where a government office established online discussion groups to allow interaction between office staff and the wider community of practitioners and professionals in related fields. Such communities may very well have advantages in terms of exposure of tacit and local knowledge, improving the flow of knowledge between professionals and even improving professional acquaintanceship and solidarity among workers [11].

The communities of practice project of the Ministry of Social Affairs was established in 2006, aiming to promote the development of methodologies of organizational learning among groups of social workers in the Israeli welfare system [2, 12]. Since the establishment of the project, more than 7,700 members have joined one or more of 31 communities. Registration to communities requires the approval of the communities' managers. All communication within the communities is identified using the real names of the members. The list of members is available to all members of the community, and they know who might read the content they upload and comment on. Community management is voluntary, and the managers receive a small reward in the form of vouchers [2]. While the communities are hosted on governmental platforms, less than 30 % of their members are government employees, and the rest are employed in NGOs, municipalities and elsewhere [12].

This study used three research methods. First, content analysis of 11 communities of practice that were selected for the study was performed. The communities were chosen to constitute a representative sample of the various communities in the project in terms of the date of establishment (older communities vs. newer ones), the scope of activity within the community (measured by the percentage of active members out of all members of the community), the size of the community (measured by the number of members in the community) and the areas of practice of the community (therapeutic communities compared to communities engaged in formal issues and procedures). We categorized communities as "therapeutic" if the topic of the community relates directly to work with clients or specific populations that receive welfare services, and "non-therapeutic" if the topic of the community is more related to organizational needs, programs creation and general work related issues. Therefore, the communities included in the study were composed of two groups.

- The first group included communities that were categorized as therapeutic: Intellectual Disability, Children at Risk, Immigrants and Inter-Cultural Issues, Blind and the Visually Impaired, Domestic Violence, Foster care and Juvenile Delinquency.
- The second group included non-therapeutic communities, which are more organizational in nature (less concerned with clients and more concerned in estimations and programs creation): Community Work, Policy and Performance, Welfare Management at Municipalities and Organizational Learning.

In each community of practice all posts available at the time of data collection (beginning of 2012) were analyzed. A total of 7248 posts were coded using a coding sheet developed for the study.

The study involves two units of analysis: Single posts and threaded discussions (a first post and at least one additional comment related to it). Thus, some of the categories in the coding sheet relate to posts and others to discussions.

Altogether, the coding sheet comprised 24 quantitative categories, which included the identity of the author (manager or community member), time of publication, the name of the community where the post was published, the post's order (is it a first post, opening a new discussion or a response to a previous post), and a number of categories relevant to the content of the post, as will be described later on. The posts were coded by 13 coders after an intensive training and a reliability test, which repeated until an agreement rate of 90 % between the coder and the leading researcher was obtained in each of the categories of the coding sheet. In addition, the first set of posts each coder has coded (usually 50–120 posts) were examined by the lead researcher, who went over the coding to make sure that reliability is maintained.

The rationale behind looking at first posts distinctively is the assumption that when community members initiate a new discussion, they dictate the topic of discussion. Reply to messages being raised in previous posts will mainly deal with the issues brought up in the first post. Therefore examining the first posts only allows us to study the issues that community members decide to initiate discussions around.

In particular, the following categories were used to code the content of messages:

- Practical advice, which is directly related to daily work with clients, for example: what is the impact of certain kinds of interventions?
- Organizational advice, related to employees' daily work unrelated to working with clients, for example concerning forms, procedures, programs and courses.
- Statements about the community's theme, which are statements that relate not to employees' daily work, but to more general issues related to the community's main theme, for example: How to improve service for patients? How to improve the status of blind people in the Israeli society?
- Emotional support—addressing community members' manifestations of charged emotions (anger, frustration, fear, sadness, etc.) that are related to their work.
- Additional categories were: academic advice (references to academic literature, relevant research etc.); informing on an event or conference; greetings and gratitude; publication of a project or organization, submitting contact details, and finally—other topics.

Another section in the coding sheet examined whether the message included sharing of personal experiences. It was also examined whether posts included provision and/or supply of assistance, and at the level of the entire thread, it was examined whether questions raised in the discussion were answered.

Secondly, word frequency analysis (conducted with Corsis, formerly Tenka Text, an open-source corpus analysis class library [13],[1] and visualized using Wordle—an online open-source tag cloud generator [14][2]) was used for linguistic analysis of texts in the discussions of the various communities, to examine what are the main issues discussed within the communities and whether they stay "on-topic", dealing with the main topic of the community.

Finally, interviews were conducted with 71 community members. The interviewees selected were sampled based on the level of involvement measured by the number of logins to the community and the number of times they contributed content within the community, so that the study involved members with low, medium and high involvement in communities. The interviews were conducted by

[1]We thank the developer, Cetin Sert, for the use of the tool.

[2]We thank the developer, Jonathan Feinberg, for the use of the tool.

five interviewers across the country, and the average length of an interview was about 45 min.

3 Hypotheses

The communities of practice of the Ministry of Social Affairs is a professional project, established and managed by the Ministry, hosted on the Ministry's website and its stated purpose is to promote organizational learning processes among employees. Consequently, we expect that the perception of the project as will be reflected in the approaches of members of the communities and the content of their posts, would be that it is a place of cognitive interactions between colleagues, rather than a place of social, emotional or personal engagements, in spite of the social and even emotional character of the profession.

Accordingly, the following hypotheses are:

3.1 Subjects of Posts

H1. Posts will include mostly professional and organizational advice and involve less emotional support.
H2. Posts that open new discussions will focus mostly on organizational and professional advice and less on emotional support.
H3. In therapeutic communities, due to the nature of work in the community, and the mental challenges members of these communities are faced with, a higher percentage of posts will contain emotional support. Still, they will be less prominent than professional and organizational advice.

3.2 Subjects of Discussions

H4. The discourse in discussions of the various communities tends to remain within the practice field of the community (on-topic) and doesn't drift to other areas, personal or social. The common words in each community will be words relevant to the community's field of activity.

3.3 Sharing Personal Experiences

H5. Community members will rarely share personal experiences from their work. The subjects of posts will be more general and members will rarely discuss particular cases.

H6. This tendency will also characterize the first posts in discussions. Community members will rarely initiate discussions dealing with personal work cases.
H7. In therapeutic communities a higher percentage of messages containing sharing of personal cases will be found. But such posts will still be rare.

4 Findings

4.1 General Findings

The dataset includes 7,248 posts. 22.1 % of the posts in the dataset contained messages requesting assistance from community members and 42.1 % of the posts contained provision of assistance. This indicates that for any question, a few answers were typically provided. In more than half of the discussions (55.1 %) a relevant response to a question was received during a discussion. When looking only at discussions in which questions were raised, in 76.7 % a relevant response was given. This portrays the communities as having lively dynamics of knowledge circulation, where questions are raised and answers provided.

4.2 Subjects of Posts

What are the issues raised in discussions in the various communities of practice? Table 1 presents the percentage of posts containing reference to the ten topics included in the coding book. Note that many posts include a number of topics, therefore the percentages total to more than 100 %.

The findings support hypotheses H1–H3.

Table 1 Distribution of the posts' and first posts' subjects in the sample

Topic of post	% of all posts	% of first posts
Practical advice	39.3	35.1
Organizational advice	25.7	26
Academic advice	6.4	10.4
Emotional support	3.4	3.4
Informing on an event or conference	8.1	12.4
Greetings and gratitude	10.9	3.7
Publication of a project or organization	8.5	16.8
Submitting contact details	7.6	9.3
Statements about the community's theme	23	12.5
Other topics	8.6	7

Of all the posts in the sample, the main topic (found in 39.3 % of the posts) is professional advice. The second most common topic (25.7 %) is organizational advice. Third prominent (23 %) was statements about the community's theme.

With the prominence of these issues in the posts, it is noticeable that emotional support is uncommon among community members' posts. Only 3.4 % of the posts in the sample included emotional support.

Further analysis referred only to first posts that were published in the community. There are 1509 first posts in the sample. The findings indicate similarities between the first posts and all posts in the sample, where the two prominent topics in the first posts are professional advice (35.1 %) and organizational advice (26 %). The third most prominent topic among 16.8 % of the first posts is a publication of a project or organization.

The percentage of the first posts that included emotional support is identical to the percentage of all posts in the sample that included this topic—3.4 %. This is the least common topic among all posts as well as among first posts.

Due to the therapeutic nature of some of the communities of practice, it can be expected that a higher percentage of messages containing emotional support will be found in communities dealing with more therapeutic areas in nature, in comparison to communities who's main concern is more organizational. In communities engaged in therapeutic areas, where workers are faced with complex personal cases, we would expect a higher demand from the community members for ventilation of feelings, sharing or leaning on other members' shoulders for support.

A chi-square test was performed to examine the relation between the type of community (therapeutic or non-therapeutic) and degree of emotional support in members' posts. The relation between these variables was significant ($\chi^2 = 22.35$, $p < 0.01$), supporting hypothesis H3. The effect size was calculated using Cramer's v and was found to be weak ($r = 0.06$). In non-therapeutic communities, only 1.6 % of the posts included emotional support, while in therapeutic communities, 4 % of the posts included such aspect. Still, the percentage of only 4 % within therapeutic communities strengthens the conclusion that emotional support is rarely present in posts published in the communities.

Interviews with community members complement the findings suggesting that the main use of the community is cognitive, and not social or emotional. Interviewees emphasized the exchange of knowledge within the framework of the discussions in the communities as the most fundamental aspect to them, an exchange of knowledge which allows them to learn from the experience of others:

> [The advantage of] transferring information at the most basic level, that at the push of a button [information] gets to so many professionals.
> The possibilities are endless in terms of flow of information.

An additional cognitive use that was brought up is stimulation of creativity and original thought, with group discussions helping to come up with creative ways to solve professional problems. Interviewees mentioned how the use of the communities encouraged them to think "outside of the box", or outside of the narrow framework they are used to:

> "The discussion is a good stimulation, which takes us out of the Sisyphean every-day work. I'm all for renewal, for new materials" says one respondent.
> Another respondent describes how learning about other professionals' decisions can be beneficial: "People have really creative solutions. The outcome is often similar, but the way to get there—people have wonderful ideas, it's very nice to read… to see how they, what path they took".

Interviewees also see differences between communities in terms of emotional support, but insist that emotional support is not a main function of most communities. Interviews revealed differences between respondents about the role of the communities as a place of support among members. One respondent expressed a firm stand regarding emotional support:

> Not in any form! It should be something unmediated-by personal conversations and not by interactive means. It gets lost through the computer […] something emotional has to be personal and not interactive.

Another member sees emotional support in contrast to the professional goals of the communities:

> Oh boy… if this becomes a main source of support… Communities need to be professional, and if it becomes the source of support-that's a sign that the situation is difficult….

On the other hand, other members see the importance of support within the community:

> It's ventilation of emotions… Someone brings up something that happened to him […] and then it triggers reactions and sharing around emotions. Or someone who was attacked […] she was reinforced by all of us and… A lot of support.
> Once there are many others that experience with me the same frustration it puts things in perspective.

However, the opinion expressed in the last quote does not represent the majority of interviewees, and indeed we see that in some cases, and especially in therapeutic communities, there was a small degree of emotional support in the posts analyzed, but these were rare in comparison to posts discussing professional or organizational matters.

4.3 Subjects of Discussions

Word frequency analysis was used to determine the most common words in each of the communities.[3] The word clouds below—show examples of the most frequent words that appeared in all posts published in the communities analyzed, with stop-words in Hebrew (such as "I", "it", "not", "of" etc.) excluded as customary (Fig. 1). The size of the word indicates the extent of its incidence in the texts, so that

[3]The texts are originally in Hebrew and the frequent words were translated to English by the researchers.

Foster Care

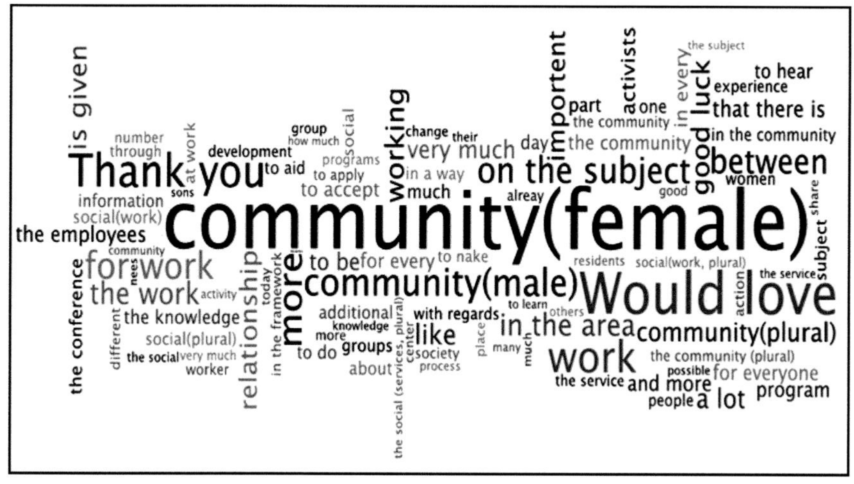

Community Work

Fig. 1 Word clouds of the 100 most common words in the communities *Foster Care* and *Community Work* (*Source* Wordle)

the most common words are also the largest. Data from the analysis software, Corsis, includes information on the frequency and percentage of each word in the text. The analysis clearly shows that the texts that comprise the posts tend to deal with issues relevant to the community, supporting hypothesis H4:

1. In the Foster Care community the words “the child” (appears 123 times in the texts, 0.66 % of the text) or “children” (51 times, 0.27 % of the text), “the family” (65 times, 0.35 % of the text, and “family”—51 times, 0.27 %), “the

foster", "foster", "in foster" and "foster care taker", "adoption", "the guardian" and "the parents" were the most frequent words in the text.

2. In the Domestic Violence community the most common words were "violence" (appears 910 times in the texts, 0.62 % of the text), "the violence", "in violence", "women" (732 times, 0.50 %), "for therapy" or "therapy" and "in the family".
3. In the Children at Risk community common words were variations of "children" (such as "children"—108 times, 0.43 %, "the child"—60 times, 0.24 %, "the children", "for children", "child" etc.) and "parents" (such as "parents", "the parents", "the family").
4. The community Policy and Performance is characterized by a high frequency of the words "the welfare" (27 times, 0.28 %) and "organization" in various forms (such as "organizational", "organizations", "the office", "organization").
5. In the Blind and the Visually Impaired community especially prevalent are variations of the word "blind" ("for the blind"—34 times, 0.25 %, "blind" in plural form, "the blind"), variations of "community" ("the community", "in the community") and "the service".
6. In the community Organizational Learning most common were variations of the word "knowledge" ("the knowledge"—117 times, 0.52 %, "knowledge"—109 times, 0.48 %,), "learning" and "community" ("the community", "communities", "in the community", "for the community", "community" etc.).
7. The community Welfare Management at Municipalities was characterized by the prevalence of the word "welfare" (51 times, 0.51 %), variations of the word "manager" ("manager"—22 times, 0.22 %, "the managers", "managerial"), and the words "social", "services" and "departments".
8. In the Community Work community most common were variations of the words "work" ("work"—229 times, 0.26 %, "for work"—196 times, 0.22 %, "the work", "working", "the workers"), "community" ("community", "communities", "the community" etc.), "activists", and it is interesting to note the frequency of the words "thank you" in this community.
9. The community Juvenile Delinquency is characterized by the prevalence of variations of the word "youth" ("The youngster"—236 times, 0.24 %, "youth", "the youth", "teens" etc.), and "probation" ("the probation" 398 times, 0.40 %, "probation"—218 times, 0.22 %), probably as part of the pair of words "probation officer", while the words "officer" and "officers" ware also frequent.
10. In the Immigrants and Inter-Cultural Issues community, the most common words were "immigrants" (11 times, 0.39 %), "Ethiopia" (10 times, 0.35 %), "in Israel", variations of the word "social", and it is interesting to note the frequency of the words "good luck" in this community.
11. The community Intellectual Disability was characterized by the prevalence of the words "retardation" (545 times, 0.41 %), "mental" (394 times, 0.30 %), "in the community", "the tenants", "at the framework" and "at the center".

4.4 *Sharing Personal Experiences*

The findings support hypotheses H5–H7:

The communities of practice at the Ministry of Social Affairs were created to serve as a safe space to raise professional dilemmas, and as a platform for discussions on methods of conduct that should be performed with different clients. However, it appears that the members of the communities rarely share information related to personal cases they encounter in their work. Among all posts in the sample, only 6.6 % share a personal case related to work. A slightly higher percentage can be found in first posts—9.7 % include sharing of a personal case, but this is still a fairly low percentage considering that these are communities of practice of social workers.

In this case as well, it is natural to assume that therapeutic communities will include a higher percentage of messages containing sharing of personal cases than communities revolving around issues that are more organizational or principle in nature. Therefore, a chi-square test was performed to examine the relation between the type of community (therapeutic or non-therapeutic) and degree of sharing personal cases in posts published in the community. The relation between these variables was significant ($\chi^2 = 7.39$, $p < 0.01$). The effect size was calculated using Cramer's v and was found to be weak ($r = 0.03$). In non-therapeutic communities, 5.1 % of the posts included sharing of a personal case, while in therapeutic communities, 7 % of the posts included this aspect. Still, 7 % is a pretty low degree of sharing when dealing with communities of practice of care-givers in the social services.

5 Discussion

The findings clearly indicate that the communities of practice are first and foremost a place of professional advice, exchange of knowledge and a place of general discussion of organizational, practical and general community-related issues. The study also demonstrates that the discourse within communities of practice tends to remain within the occupational field of the community-as expressed in the analysis of the most common words in each community.

Despite the importance of social interactions for the success of online communities, as indicated by the literature, analysis of the content of the posts in the communities suggests that community members deal primarily with practical and organizational issues, both when they are initiating new discussions and when they are participating in existing discussions, and they rarely deal with more personal or emotional issues. Sharing personal experiences and cases is also quite rare.

The interviews with community members as well as content analysis of the posts show that communities are perceived and used primarily as a space of cognitive and professional interactions, and as an organizational learning platform. Although these are communities of practice that serve the employees of the Ministry of Social Affairs, which by virtue of their role deal with emotional challenges and complex social situations which obviously can lead sometimes to distress, it is clear that they do not see these spaces as an appropriate place for emotional support or social interactions. Even within communities dealing with therapeutic issues they do not often turn to community members for assistance, sharing or reinforcement on a personal level, and from the interviews it is also apparent that the idea of these communities being a platform for sharing and support was seen by many of the members as misplaced. When members need a comforting shoulder, they will choose other platforms for this purpose and maintain the communities of practice as areas of professional discussions among colleagues.

What may be the implications of introducing an emotional or social discourse into the communities' discussions? One possible reason for the avoidance of community members from engaging with each-other on an emotional level may be that the communities are perceived as an extension of the workplace, being hosted on the ministry's servers, managed and to some degree supervised by the ministry. The employees and professionals who make use of the platform "know their place" and keep clear boundaries in their interactions within the communities.

Future work can analyze other online spaces used by social workers (such as Facebook groups) to learn whether the cognitive emphasis of the discussions in the online communities of practice occurs there as well, or whether the interactions in these other spaces have a more social or emotional character, due to the social and informal nature of the platform, and the lack of organizational involvement or supervision. Studies can also analyze other communities of practice that are hosted on a government platform, to see whether government platforms lead to a more cognitive character of conversations in the social media platforms they host. If this is indeed the case, interviews with policy experts can suggest whether this is beneficial to the community and if not, which interventions may be necessary to bring the conversation to a more social direction.

References

1. Lave, J., Wenger, E.: Situated Learning: Legitimate Peripheral Participation. Cambridge University Press, Cambridge (1991)
2. Sabah, Y.: Online learning groups and communities of practice in the social services. Society and Welfare (in Hebrew). **30**, 111–130 (2010)
3. McKenna, K.Y., Green, A.S.: Virtual group dynamics. Group. Dyn.-Theory Res. Pract. **18**, 116–127 (2002)
4. Meier, A.: An online stress management support group for social workers. J. Tech. Hum. Serv. **20**, 107–132 (2002)
5. Johnson, C.M.: A survey of current research on online communities of practice. Internet. High. educ. **4**, 45–60 (2001)
6. Kraut, R., Wang, X., Butler, B., Joyce, E., Burke, M.: Beyond information: developing the relationship between the individual and the group in online communities. Unpublished

7. Kankanhalli, A., Tan, B.C., Wei, K.K.: Contributing knowledge to electronic knowledge repositories: an empirical investigation. MIS. Quart. **29**, 113–143 (2005)
8. Lev-on, A., Hardin, R.: Internet-based collaborations and their political significance. J. Inform. Tech. Polit. **4**, 5–27 (2007)
9. Ma, M., Agarwal, R.: Through a glass darkly: Information technology design, identity verification, and knowledge contribution in online communities. Inform. Syst. Res. **18**, 42–67 (2007)
10. Wasko, M.M., Faraj, S.: Why should I share? Examining knowledge contribution in electronic networks of practice. MIS. Quart. **29**, 1–23 (2005)
11. Cook-Craig, P.G., Sabah, Y.: The role of virtual communities of practice in supporting collaborative learning among social workers. Brit. J. Soc. Work. **39**, 725–739 (2009)
12. Fein, T.: Online communities of practice in the social services (in Hebrew). MA Thesis. The Hebrew University. Israel (2011)
13. Corsis, http://sourceforge.net/projects/corsis/
14. Wordle, http://www.wordle.net/

Knowledge Extraction and Annotation Tools to Support Creativity at the Initial Stage of Product Design: Requirements and Assessment

Julia Kantorovitch, Ilkka Niskanen, Anastasios Zafeiropoulos, Aggelos Liapis, Jose Miguel Garrido Gonzalez, Alexandros Didaskalou and Enrico Motta

Abstract The initial stage of the conceptual product design requires creativity and is characterized by an intensive knowledge exploration process. To this purpose annotated resources have to be available, a need which introduces requirements on knowledge extraction tools. In this paper, we assess the current state of this technology vis-à-vis the demands identified to support the work of designers.

Keywords Conceptual product design · Creativity · Knowledge extraction and annotation · Semantic technologies

---

J. Kantorovitch (✉) · I. Niskanen
VTT—Technical Center of Finland, Vuorimiehentie 3, Espoo, Finland
e-mail: julia.kantorovitch@vtt.fi

I. Niskanen
e-mail: ilkka.niskanen@vtt.fi

A. Zafeiropoulos · A. Liapis
Intrasoft International, Markopoulou-Peania Avenue, Athens, Greece
e-mail: anastasios.zafeiropoulos@intrasoft-intl.com

A. Liapis
e-mail: aggelos.liapis@intrasoft-intl.com

J.M.G. Gonzalez
Atos, Albarracín, Madrid, Spain
e-mail: jose.garridog@atos.net

A. Didaskalou
DesignLab, Garyttou St.150, Athens, Greece
e-mail: alexandros.didaskalou@designlab.gr

E. Motta
The Open University, Walton Hall, Milton Keynes, UK
e-mail: enrico.motta@open.ac.uk

S. Kunifuji et al. (eds.), *Knowledge, Information and Creativity Support Systems*,
Advances in Intelligent Systems and Computing 416,
DOI 10.1007/978-3-319-27478-2_11

1 Introduction

Semantic technologies and tools has undergone significant development in recent years. The value of these technologies has been demonstrated in several application fields, including healthcare, logistics, retail, transport and finance [2]. Another field which may potentially benefit from the modelling, reasoning and decision support ability of semantic technologies is the knowledge-intensive process of product design—in particular during the initial design stage. Here, to help in the generation of new ideas, designers rely both on existing practices and resources available from company databases, such as documents and sketches produced in the course of previous designs, as well as external information sources. These may include electronic books, images, music, online design journals, as well as sources of domain specific design knowledge. During the initial stage of the design process designers are typically required to solve ill-defined problems. This means that the requirements to be addressed are not completely known and that there is no objective true-or-false evaluation of a solution [6]. However solutions can be assessed as good or bad, appropriate, innovative, creative, and so on, during the collaborative design process. The designers therefore need to go through the process of data gathering and exploration to clarify the vision they have in mind to deconstruct and specify the problem and solution for the task in hand. Accordingly the initial stage of the conceptual design is the most creative and knowledge-exploration intensive process. Extending French's design model [10] the role of knowledge exploration in the overall process of conceptual design is framed in Fig. 1. As the designer is proceeding from the stage of problem analysis to a working conceptual design, the knowledge search space is narrowing and becoming more structured and domain specific. To facilitate the process of knowledge exploration by designers, the availability of large quantities of annotated resources is essential.

Methods for knowledge exploration based on semantic annotation using ontologies are recognised as a powerful approach, which can make the processing of information resources more "intelligent"—i.e. machine interoperable, effective and meaningful [4, 14, 17, 18]. Ontologies can provide elegant mechanisms to organise multimedia content in logically contained groups while linking them with other related concepts [16]. Annotations with well-defined semantics are a requirement to ensure the interoperability of available information supporting knowledge sharing and collaboration across design teams. Semantic annotation can enhance the ability to perform effective customised searches, exploring the knowledge contained within annotations in order to access heterogeneous content. The recently introduced Linked Data technology [1] has the potential to facilitate the interlinking of unconnected documents and other data sources (e.g., internal enterprise data assets) to generate large interlinked data ecosystems.

While knowledge extraction (KE) and semantic annotation research has been undertaken for several decades and an extensive number of tools and frameworks have been proposed [11], our analysis shows that there is still some effort required before semantic tools will be able to address the knowledge exploration needs of

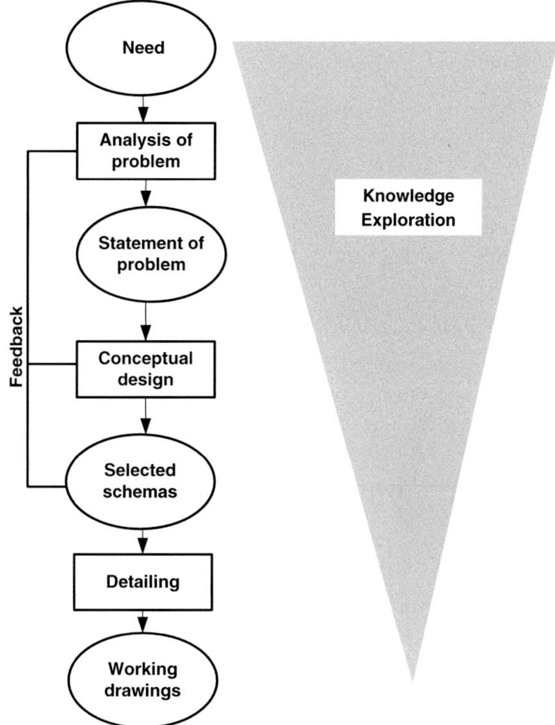

Fig. 1 French's model of design process extended with knowledge exploration process

designers in the context of a collaborative creative design process. While semantic approaches are potentially promising, we argue that further developments are needed to address particular use cases, application tasks and user groups.

In the following section, we discuss the identified requirements for knowledge extraction and annotation, which are essential from the perspective of applications in order to support design teams. The identified requirements are utilised as a basis for the assessment of semantic tools which is presented in Sect. 3. Conclusions are presented in Sect. 4.

2 Requirements for Knowledge Extraction and Annotation Tools

The requirements have been acquired through a multi-faceted approach, which included an analysis of the dimensions of creativity, a literature review which focused on what is needed to support collaboration and creativity across a design team, as well as extensive discussions with product designers from the COnCEPT

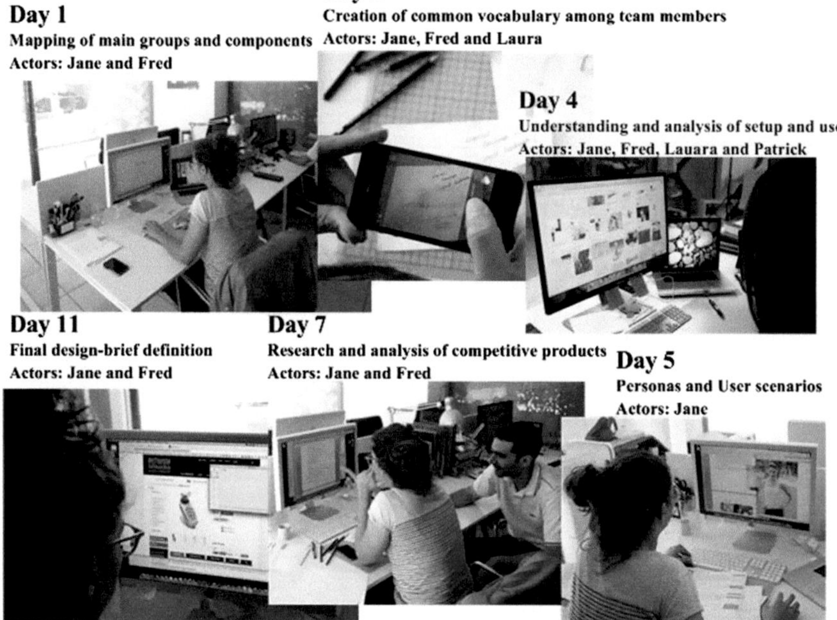

Fig. 2 Working activities of design team

consortium.[1] A crucial element of this process was the analysis of design teams in action and in what follows we will discuss how the design team of a small design company approached a sample project, for a product called "smart breathing trainer".

Generally, the initiation of new product design can start in several different ways. The design studios may get design briefs from their clients, but it is also often the case that they generate design briefs of their own. In the latter case, for example, they may come up with a new ground-breaking idea and want to make a real product out of it, or they may participate in a design competition. In the case of the initial design brief being defined by a client, it may contain a description of the product to be designed, a user profile, context of use, available technologies, competitive products, market positioning for the new product, target cost, functional and aesthetic features, design requirements, etc.

The design process partly visualised in Fig. 2 consists of two phases. During Discovery the designers are searching for information about the latest technologies on measuring breathing, medical information about training the respiratory system and about relevant, competitive products. The Discovery phase concludes with a

[1] http://www.concept-fp7.eu/.

detailed definition (or redefinition) of the design brief. During the Vision phase the team then focuses on generating new ideas and concepts.

The design group may include several designers and design engineers. In our example we assume four designers, Jane, Fred, Laura, and Patrick. They may have different design experience and areas of expertise. The process of conceptual design moves back and forth from Discovery to Vision and might take several days during which:

- Depending on the initial setup, if for example the requirements for a new breathing trainer are not specified extensively, Jane and Fred would start researching by using the Internet to search on Google or other online sources for specific information and inspiration. Alternatively, in the ideal case of a detailed design brief and after a useful discussion with the client, the received information is carefully analyzed and enriched by performing additional search on the Internet, on existing design company databases and other possible sources of information. Relevant data including images, drawings, notes, specifications, reviews, internet links, etc. are collected and saved to the design project space.
- Jane also creates sketches using pencil and paper, which are photographed and uploaded to the project space. Fred creates a *mood board*, pinning images, videos and other material and adds annotations in the form of notes and keywords, before storing it to the project space. The mood board contributes to the creation of a shared vocabulary for the design project. Furthermore, Laura looks for the appropriate tools to store interesting bookmarks she has found, adding comments and keywords where required.
- Jane may create a set of personas for the product using a word processor. For this task she complies a short description of the end-user's profile and searches on Google for images to illustrate these personas.
- Fred and Jane look for similar products online. Fred creates a document where he groups the features of similar products and shares it with Jane. The competitive products are also categorized and rated according to their functionality and usability.
- At certain points during the design process, Jane, Patrick and Fred have meetings to discuss their findings and development of the design concepts. All the stored material can be easily accessed by the designers, and it can also be easily reorganized, filtered and interconnected while interesting data can be flagged in a direct, visual manner.
- Looking for additional ideas while finalising the product concept, the designers continue to search on the Internet and locally stored databases for inspiring stuff from various possible sources—e.g., natural shapes, universal design principles (such as, symmetries, geometry, proportions etc.), existing objects with interesting features that could be useful when composing new forms, materials, processes, behaviours, etc. In addition, they engage in brainstorming, sketch solutions, compare ideas, evaluate proposals, produce detailed designs in 2D and 3D, make mockups, model prototypes, etc.

The collection of requirements has been an iterative process. Attention has been paid to quality of extraction and annotation means; the type of resources to be annotated—these include both external and company resources; and the types of annotation that might be required to support the designer, including visual features (colour, shape, texture), content level (tags, location, description/meaning, etc.), and functional (who, where, access rights); as well as the format of annotations and requirements for the interoperability of the system. The requirements have also been assessed analysing the role of the designer in the knowledge extraction process and the possible support that might be needed. The compatibility with existing resources on the internet (multimedia databases, virtual museums, etc.) as well as support needed for the porting of annotation tools to the existing company databases, has been considered. Accordingly several main requirements were identified. These are discussed in the following section.

2.1 Versatility and Conceptual Expressivity of Semantic Annotations

When considering products and the role of the designer in its creation, it is useful to refer to Offenbach's Theory of Product Language [22]. Offenbach makes the distinction between a product's function (functions and services) and the product's semantics (product language), which refers to the way in which the product is understood by users (see Fig. 3).

Looking further on the product language, the formal aesthetic functions of an object are referred to as the "grammar" of the design concept, which includes shapes, colour, texture, material, etc. The indicating functions specify the way in which the product could be used. Symbolic functions refer to impressions and

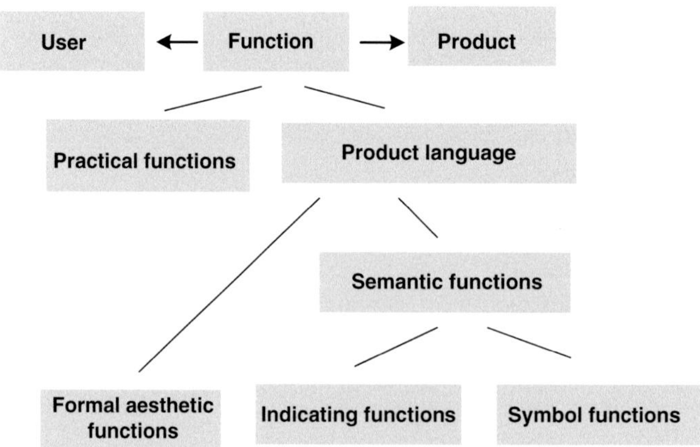

Fig. 3 Conceptual model of Offenbach theory of product language [22]

associations within the mind of the observer (designer and end user) using the product—for example, historical, economic aspects as well as associations such as cold/warm, strong/weak. From Offenbach's point of view, the product language and the communication of the product's message and formal meaning is considered to be within the domain of the designer, while engineers, ergonomists, etc. are responsible for the product's practical aspects (in other words, technical and technological functions). New technology can provide a source of innovation for new products. An innovation in product language means the creation of semantically coherent product concepts where product technical functions are framed by communicative functions of the product. The design of a semantically coherent product concept is the great source of innovation where the creativity is most needed.

In order to support innovation, to help designers to access useful resources for their creative tasks, the need for resources has to be identified. This may for example include the identification of properties of the product under design. Then these properties can be potentially mapped to some visual resources on the web or in company databases to bring more useful information about available styles, colour schemas and shapes or information from distant but still relevant domains based on the product language discussed above. The mapping and 'context'-aware knowledge delivery indicates the level of semantic annotation required and consequently the knowledge extraction technology used by the system. Levels of expression are achieved by introducing a versatility of concepts and richness of relationships between them. Moreover problem formalization requires changing the context and viewpoints surrounding the problem in hand, where the associated knowledge plays a critical role. Knowledge organization and annotation patterns should be able to make the information search and delivery in a context-aware and effective way, supporting the designers whilst considering the task in hand from alternative perspectives.

2.2 *Support for Various Types of Knowledge Extraction*

The wide variety of information available on the web and locally in company's repositories can spark creativity and support collaboration. Required information can be available as part of text embedded in documents, or delivered by image or sketch or music lyric or composed of multimedia content of various types. Knowledge extraction tools may provide the required information using subject-based metadata of documents as well as the metadata for the multimedia content, such as images (keywords, tags and also low level visual properties).

Looking at a typical model for the creative design, this process consists of several iterative steps such as incubation, creative insight and evaluation [7]. During the incubation phase, association, analogy, metaphors, etc., are techniques which are often used to facilitate the process of creative thinking and inspiration, which shape new product design. Metaphor is an association between two concepts or artefacts when their properties are common. Analogous thinking is often triggered by visual stimulation using images to stimulate creativity. Many studies have demonstrated that designers make intensive use of analogies, adapting design features from other fields to their own design problem [20]. Moreover it was shown that often the most creative analogies are those that are made between the most distant domains [3, 21]. Because multimedia content based knowledge extraction plays an important role in creativity support, this sets the requirement to multimedia knowledge extraction and annotation tools to bring together low-level visual features and high level meaning of multimedia objects.

The system should also consider the location of required content which may reside in enterprise databases, personal collections and catalogues as well as existing on the Web. Ideally the knowledge extraction and annotation tools should cope with both situations.

2.3 Open Standards

Any IT system and in particular a knowledge management system can only benefit by making use of open standards as far as possible. Open standards facilitate the porting of various applications to knowledge extraction platforms where guidance, rules and interfaces are known. Moreover the integration of various components developed by different parties (e.g., information extraction, ontology and annotation management) is easier, resulting in smooth interoperation with one another. For annotation systems in particular, open standards may efficiently bridge heterogeneous resources to be accessed by collaborating users and organizations to share annotations [23]. It's important for semantic knowledge extraction tools to be compatible with the most used open Semantic Web standards, such as RDF and OWL metadata models and, if available, Linked Data concepts as "control vocabularies" (i.e., collections of classes and properties used to describe resources in a particular domain) to tag resources unambiguously.

Knowledge extraction components and extensions should also be compliant with widely used middleware and communication standards such as Web Services, REST-like interfaces and data formats.

2.4 Support for the Extension of Semantic Concepts

While the product design from a process point of view can be considered as domain independent, from a knowledge management perspective, there are still aspects that need to be considered such as: various creativity techniques can be applied which may require different knowledge organization patterns; various aspects in the product design domain may require different depths of knowledge (marketing, technology, application domain) and accordingly make use of different knowledge bases; the product being designed may belong to different application domains (e.g., sport, health, work, leisure), which would require domain specific 'control vocabularies' to ensure sharing the understanding of domain concepts.

It is obvious that knowledge extraction tools can't address all the needs of product designers for various application domains, thus a knowledge extraction and annotation frameworks should support a clear separation between domain/task independent and domain/task specific knowledge and allow easy-to-use creation of knowledge bases and patterns for task specific needs (e.g., to facilitate different existing creativity-supporting techniques from a knowledge management point of view).

2.5 Support for the Automation of Knowledge Extraction and Annotation

In general, automatic approaches to knowledge extraction and annotation are desirable compared to manual, as they promise scalability to annotate the considerable amount of Web resources and organisational databases. Once they are available, they are extremely useful to the designer. Resources such as documents and various types of multimedia content can be used by the system. Thus we anticipate that the applications to support the work of a designer will need various types of annotation (technical, content- and visual-related). While automatic document annotation systems are more mature in their state of development, the image annotation methods remain a challenge for many real world applications [5, 8, 9].

In addition, we can't expect designers to expend effort on the annotation of material they use during the design process, unless of course a natural and intuitive interface is offered. Therefore to become useful, the annotation tools should provide as much as possible a degree of automation for knowledge extraction in order to reduce the burden of annotation.

2.6 Usability and Collaboration Support

Taking into account that end users are designers who may not necessarily have a 'technical' background or in-depth experience of knowledge management systems, the usability aspect which is related to human involvement in the generation of semantic metadata, is an important consideration in the design of the user interface.

Ideally the complexity of creating semantic data should be hidden behind an intuitive user interface and the vocabularies used for annotation as much as possible should be naturally aligned with the world of the designers. In addition the user interface technology should be designed to be portable and interoperable with existing web based interface standard (such as HTML5, JavaScript, etc.) and web applications. As regards to offering annotation concepts, some learning features might be useful to facilitate the suggestion of earlier used concepts by the designer, and aligned with widely agreed categories such as, for example, those defined by the open Linked Data community.

The design and deployment of advanced collaborative functions is considered crucial to support creativity. Knowledge sharing and re-use of existing content among the product designers has to be based on interactions between individuals that work on a specific project. Interaction between designers making use of shared design vocabularies supported by Knowledge Engineering technology should be carefully considered. In addition interconnection with available resources or relevant ideas online as well as options for tools that may be used to ensure seamless activities flow (e.g., collaborative editing and content annotation, content synchronization etc.) should be provided through the application of a semantic backbone.

3 Discussion

Semantic knowledge extraction and annotation tools have been searched through the literature in several domains including knowledge management, information retrieval, and semantic resource annotations using the digital libraries of ACM, IEEE and Springer. Analysing designers' needs, several KE systems have been found to be the most relevant and potentially promising candidates for further assessment. A selection was made based on the initial test experience and reported functionalities to support knowledge management tasks. Apart from functions provided by tools, the ease of installation, and the availability of suitable APIs in addition to web applications, as well as suitable licencing options and active development community around tools have been considered.

Various design briefs, HTML pages and documents are used to assess the tools. The results of the assessment in relation to the requirements discussed early are summarised in the Table 1. The Quality Function Deployment (QFD) concept [12] is used to express the overall impression about how the requirements are supported in each particular case. According to the QFD model, in the process of defining the product concept the design requirements (WHAT's) served as input to establish components characteristics (HOW's) of product design. Then HOW MUCH represents the amount that can be achieved. Adopting the QFD approach, in our case, WHAT's refers to the requirements specified in Sect. 2 and the HOW's are the knowledge extraction and annotation tools that have been assessed. HOW MUCH is rated as follows: **S**—requirement is Strongly supported; **M**—requirement is Moderately supported; **W**—requirement is Weakly supported; **N—**requirement is NOT supported.

It can be seen from the table that most of tools are compliant with open and widely used standards (2.3). Technically speaking most of the tools provide support for Web Service and REST-like interfaces as well as RDF/OWL and linked data to

Table 1 The assessment of tools against requirements

KE tools	2.1	2.2	2.3	2.4	2.5	2.6
LMF[3]	M	M	S	M	M	M
Stanbol[4]	M	M	S	M	M	W
KIM[5]	S	M	S	M	S	W
OpenCyc[6]	M	W	M	W	M	N
OpenCalais[7]	S	W	M	N	M	M
Pundit[8]	S	M	S	M	M	M
GoNTogle[9]	M	W	S	M	S	W
Annomation[10]	M	W	M	M	M	W
Annotator[11]	W	M	M	M	W	W
DBPediaSpotlight[12]	S	W	S	M	S	W
TextRazor[13]	S	W	S	M	S	W
FRED[14]	M	W	S	M	S	N
Zemanta[15]	M	W	S	M	S	N

2.1-conceptual expressivity; 2.2-heterogeneous content; 2.3-open standards; 2.4-extendability of semantic concepts; 2.5-automatic knowledge extraction; 2.6-usability and collaboration

[3]http://code.google.com/p/lmf/
[4]http://dev.iks-project.eu:8081/
[5]http://www.ontotext.com/kim
[6]http://www.cyc.com/platform/opencyc
[7]http://www.opencalais.com
[8]https://thepund.it/
[9]http://web.imis.athena-innovation.gr/projects/gontogle/
[10]http://annomation.open.ac.uk/annomation
[11]http://annotatorjs.org
[12]https://github.com/dbpedia-spotlight/dbpedia-spotlight/wiki
[13]https://www.textrazor.com/
[14]http://wit.istc.cnr.it/stlab-tools/fred/
[15]http://www.zemanta.com/

frame KE in an unambiguous manner. The exceptions are KIM and OpenCyc. In addition, in spite of the conceptual richness and expressiveness of annotations, the encyclopaedic ambitions of the OpenCyc knowledge base and the fact that it can only be extended by hand, increases the complexity of the system (2.4; 2.6).

It was found that many tools also fail to support collaborative activity in the process of knowledge annotation and that the usability of knowledge extraction and annotation tools is an area which is still in its infancy (2.6). The available technology is mostly developed keeping in mind the needs of an application developer or, in the case of semi-automatic annotation, providing support for individuals performing web authoring tasks or simple tag based functions in a non-professional context. Considering that product designers are neither data exploitation professionals nor application developers, existing user interfaces need considerable effort to hide the complexity of existing knowledge extraction technology. Moreover many tools are stand-alone. Their integration with popular content management platforms, such as Drupal or WordPress, would facilitate the collaboration and sharing content among designers (2.6).

As evidenced by the assessment undertaken for this report, the support for heterogeneous resource formats is not as developed as one may anticipate (2.2). A majority of tools focuses on knowledge extraction from text. This area of technology is more advanced compared with multimedia content processing. This is in line with other analyses made earlier [9]. Whilst prototype deployments as well as research initiatives are available and a plethora of image and video retrieval tools have been proposed, the process of information extraction and annotation is still much reliant on human involvement, and is semi-automatic in nature (2.5). On the contrary the support for annotation interoperability utilising ontologies and open linked data is widely provided [13, 15]. On the other hand, it should be acknowledged that these tools have facilitated the providers and community of web based online multimedia collections with the ability to semantically annotate their content (e.g., Europeana[2]), hereby enhancing knowledge extraction outcome.

Overall, keeping in mind that knowledge extraction from text is a much more mature technology, it's reasonable to take this aspect into consideration in developing applications to support collaboration and creativity in particular. It's not a surprise that during the past years a number of Semantic Web start-ups such as Zemanta, OpenCalais and others have emerged to take advantage of knowledge extraction technology from text, and provide APIs to enable bloggers and businesses to link published text with other content.

From the tools considered, only the LMF framework addresses the needs for multimedia based knowledge extraction and annotation extending the functionality developed in Stanbol. It was found that the LMF platform development was an attractive technology because of its interoperability and content interlinking features and ease of integration of knowledge management functions with enterprise databases. An additional advantage of Stanbol is that its components are already integrated with CMS systems such as Drupal which is a popular content management platform. However, compared to LMF, Stanbol does not support annotation of multimedia content. Perhaps the biggest deficiencies of both tools are related to their complexity and their ability to analyze text fragments. In order to efficiently exploit the functionalities of the systems, developers have to simultaneously understand many technologies including OSGI. Additionally, the ability to configure the frameworks to better serve domain specific use cases (2.4) requires versatile skills from developers. Finally, semantic content processing in LMF and Stanbol is still behind some of the more established text analysis tools [11]. Whilst LMF is more advanced with respect to knowledge extraction and annotation from multimedia, still images and videos can be automatically annotated with only technical information (date, owner, etc.) (2.5).

TextRazor and DBpedia Spotlight provide lightweight efficient solutions for Entity Recognition from text. While they are not open source, Zemanta and TagIt

[2]http://www.europeana.eu/.

are well developed services. Providing portable and interoperable APIs, these tools can potentially facilitate the content enrichment functions. It should be mentioned that FRED is the only tool able to automatically produce RDF/OWL ontologies and link data from natural language sentences. It can be potentially used to facilitate the analysis of design briefs or to support the creation of control vocabularies as well as to enrich syntactic keyword based searches that may be used by designers. These would however require an additional effort to extend the existing knowledge base with domain specific knowledge as well as with reasoning functions depending on the objectives of the use cases targeted by the application–in particular, the user interface which will be required in order to address the needs of designers.

The literature review [21] and the brainstorming sessions and interviews with both designers from the COnCEPT consortium and external experts have revealed that designers are the active users of standard web tools such as Google, Getty, and Flickr, specific design portal such as Fotolia and Corbis, mood boards (e.g., Pinterest) and other online resources (e.g., co.design, yatzer, designboom, designobserver, etc.). These are used by designers on a daily basis to search for new product concepts and inspiration. In addition, to support the work of application developers, various APIs are provided to utilize the content of the web. For example, with recent extensions, Google APIs allow for the extraction of images based on various parameters which relate to image features such as colour or face recognition, as well as images retrieved from particular web sites. The Flickr API can be used through REST or Web Service interfaces and public photos can be extracted using various parameters, such as a combination of tags, text as well as technical metadata such as location, date, owner, etc. In addition, there are many open datasets and online collections along with APIs available for developers, which are provided by museums and galleries nowadays. There are various configuration options offered for developers to perform context-aware content retrieval. An example is the Victoria and Albert museum's API, which allows users to filter retrieved content based on its type, materials, techniques and categories related to colour and the semantic content of the image and similarity to particular concepts/artefacts.

The advances of Internet tools discussed above, combined with recent developments in information extraction from text, used for example to analyse text-based design briefs or the result of brainstorming sessions, can bring real benefit to product designers in background information search, inspiration and the overall process of assessing product requirements. Accordingly, a better alignment of semantic knowledge extraction and annotation tools with the advances of classic web and application specific solutions is desirable. The Zemanta service discussed earlier, which was developed to support bloggers, is an example of such application-specific enhancement. A similar support for the initial stage of product design would be of great help, by providing the means to bring associations to stimulate creativity or tools to facilitate the development of various mood boards as well as the technology to ease the access, filtering, selection, comparison, interconnection, grouping and presentation of project related information.

4 Conclusion

Semantic knowledge extraction and annotation technologies have the potential to support designers in knowledge exploration tasks during the initial stage of conceptual product design. While an extensive number of KE tools have been proposed by the research community, our analysis shows that there is still effort required before the tools will be able to address the needs of designers with respect to collaboration and creativity support. The areas of essential improvements include the usability of knowledge extraction technology. Also taking into account that multimedia is one of the main sources of inspiration for designers, further work on knowledge extraction and annotation of images is required. Moreover, we argue that future developments ought to address particular use cases/application tasks (e.g. definition of requirements, brainstorming, creative thinking, creation of design-views/visualizations) and user groups (designers, developers). In addition the alignment of knowledge extraction technology with classical web support, from the point of view of both end users and application developers is very desirable.

In this paper we have revised and extended the research results presented earlier in [19]. In particular further attention has been paid to the context of application domain, work practices and activities of design teams.

Acknowledgments This research is funded by the European Commission 7th Framework ICT Research Programme. Further details can be found accessed at: http://www.concept-fp7.eu

References

1. Bizer, C., Heath, T., Berners-Lee, T.: Linked data—the story so far. Int. J. Semant. Web Inf. Syst. **5**(3), 1–22 (2009)
2. Blomqvist, E.: The Use of Semantic Web Technologies for Decision Support—a Survey. Semantic Web Journal, pp. 177–201, IOS Press (2012)
3. Bonnardel, N., Marmèche, E.: Towards supporting evocation processes in creative design: a cognitive approach. Int. J. Hum. Comput. Stud. **63**, 422–435 (2005)
4. Carbone, F., et al.: Enterprise 2.0 and semantic technologies for open innovation support. In: Trends in Applied Intelligent Systems Lecture Notes in Computer Science, vol. 6097, pp. 18–27 (2010)
5. Caputa, B., et al.: ImageCLEF 2013: the vision, the data and the open challenges. In: Information Access Evaluation. Multilinguality, Multimodality, and Visualization Lecture Notes in Computer Science, vol. 8138, pp. 250–268 (2013)
6. Cross, N.: Engineering Design Methods Strategies for Product Design. Willey, New York (2000)
7. Csikszentmihalyi, M.: Creativity: Flow and the Psychology of Discovery and Invention. HarperCollins, New York (1996)
8. Datta, R., et al.: Image retrieval: ideas, influences and trends of new age. ACM Comput. Surv. **40**(2), Article 5, pp. 5:1–60 (2008)
9. Dasiopoulou, S., Giannakidou, E., Litos, G., Malasioti, P., Kompatsiaris, Y.: A survey of semantic image and video annotation tools. In: Paliouras, G., Spyropoulos, C.D., Tsatsaronis,

G. (ed.) Knowledge-Driven Multimedia Information Extraction and Ontology Evolution, Springer, pp. 196–239 (2011)
10. French, M.J.: Conceptual Design for Engineers. Springer, New York (1999)
11. Gangemi, A.: A comparison of knowledge extraction tools for the semantic web. ESWC 351–366 (2013)
12. Govers, C.P.: What and how about quality function deployment (QFD). Int. J. Prod. Econ. **46**, 575–585 (1996)
13. Grassi, M., Morbidoni, C., Nucci, M., Fonda, S., Di Donato, F.: Pundit: creating, exploring and consuming semantic annotations. In: Proceedings of the 3rd International Workshop on Semantic Digital Archives, pp. 65–72 (2013)
14. Hare, J.S., Lewis, P.H., Enser, P.G.B., Sandom, C.J.: Mind the gap:another look at the problem of the semantic gap in image retrieval. In: Multimedia Content Analysis, Management and Retrieval, SPIE, vol. 6073, pp. 1–12 (2006)
15. Haslhofer, B., Momeni, E., Gay, M., Simon, R.: Augmenting Europeana content with linked data resources. In: Proceedings of 6th International Conference on Semantic Systems (I-Semantics) (2010)
16. Halaschek-Wiener, C., Schain, A., Grove, M., Parsia, B., Hendler, J.: Management of digital images on the semantic web. In: Proceedings of the International Semantic Web Conference (2005)
17. Hollink, L., Worring, M.: Building a visual ontology for video retrieval. In: Proceedings of the 13th International ACM Conference on Multimedia (MM), ACM Press, New York, pp. 479–482 (2005)
18. Kobilarov, G., et al.: Media meets semantic web—how the bbc uses dbpedia and linked data to make connections. In: The Semantic Web: Research and Applications. Lecture Notes in Computer Science, vol. 5554, pp. 723–737 (2009)
19. Kantorovitch, J. et al.: Knowledge extraction and annotation tools to support creativity in the initial stage of product design: requirements and assessment. In: Proceedings of the KICSS Conference, pp. 129–140 (2014)
20. Leclercq P., Heylighen A.: 5.8 analogies per hour—a designer's view on analogical reasoning. In: AID'02 Artificial Intelligence in Design, Cambridge, pp. 1–16 (2002)
21. Mougenot, C., Bouchard, C., Aoussat, A.: Creativity in design—how designers gather information in the "Preparation" phase. In: ASDR, pp. 1–16 (2007)
22. Steffen, D.: Design semantics of innovation, product language as a reflection on technical innovation and socio-cultural change. In: Proceedings of Design Semiotics in Use Workshop, Held as a Part of World Congress in Semiotics Communication: Understanding/Misunderstanding (2007)
23. Uren, V., et al.: Semantic annotation of knowledge management: requirements and a survey of the state of the art. Wb Semant. Sci. Serv. Agents World Wide Web **4**(1), 14–28 (2006)

Web Board Question Answering System on Problem-Solving Through Problem Clusters

Chaveevan Pechsiri, Onuma Moolwat and Rapepun Piriyakul

Abstract This paper aims to work on the Question Answering (QA) system within online web boards, especially the Why-question, How-question, and Request-Diagnosis-question types approach for solving problems. The research QA system benefits for the online communities in solving their problems, especially on health-care problems of symptoms. Both question and answer expressions are based on multiple EDUs (Elementary Discourse Units) where each EDU is equivalent to a simple sentence or a clause. The research involves two main problems: how to identify the question types of Why, How, and Request-Diagnosis and how to determine the corresponding answer from the knowledge source after solving the question focuses. Thus, the research applies different machine learning techniques, Naïve Bayes and Support Vector Machine, to solve the reasoning question type identification. The knowledge source contains several symptom-treatment vector pairs and several cause-effect vector pairs. Therefore, we propose clustering symptoms/problems of the knowledge source before determining an answer based on top-down levels of determining similarity scores between a web board question and the knowledge source. The research achieves 83 % correctness of the answer determination with potentially saving amounts of search time.

Keywords Why question · How question · Similarity score · Problem solving

C. Pechsiri (✉) · O. Moolwat
Department of Information Technology, Dhurakij Pundit University, Bangkok, Thailand
e-mail: itdpu@hotmail.com

O. Moolwat
e-mail: moolwat@hotmail.com

R. Piriyakul
Department of Computer Science, Ramkhamhaeng University, Bangkok, Thailand
e-mail: rapepunnight@yahoo.com

S. Kunifuji et al. (eds.), *Knowledge, Information and Creativity Support Systems*,
Advances in Intelligent Systems and Computing 416,
DOI 10.1007/978-3-319-27478-2_12

1 Introduction

The objective of this research is to develop a Question Answering (QA) system based on explanation knowledge or reasoning knowledge for primary problem solving, especially on the health care problems, through the online community web boards (e.g. http://haamor.com/webboard/). When people have problems of their disease symptoms, some of them direct to a certain clinic or hospital whilst some others with their own reasons (e.g. a time conflict, a tight budget, a location problem, and etc.) prefer to post their problems of disease symptoms including queries or questions on the health-care-community web boards. Then, they wait for several minutes to a couple days depending on their topics of queries to receive the recommended answers of solving their problems on the web boards posted by the experts. However, it is time consuming for them to know how to solve their problems. Moreover, some common accidents can be occurred on trip, in family, or at work as burns, bleeding wounds, dislocations, or sprains. If someone knows how first aid from the health-care-community web boards through the QA system, it can limit the damage, maybe even save the lives of victims. Thus, it is necessary to develop an automatic web board QA system, based on Why and How questions, to solve their problems. The corresponding answers of the web board QA system can be determined from the knowledge source containing question-answer pairs of the previous extracted Symptom-Treatment Relation from the medical-care-consulting documents [10] and also the extracted causality from the health-care documents [11]. Thus, our research of the web board QA system can provide the explanation knowledge or the reasoning knowledge for people to make their own decisions. In addition, the posted health care problems of disease symptoms on the web boards are expressed in the form of several Elementary Discourse Units or several EDUs (where each EDU is defined as a simple sentence or a clause, [4]) And, the posted problems always consist of multiple EDUs of the symptom expression followed by 1–3 questions as shown in the following question-pattern.

$$EDU_{ct-1}EDU_{ct-2}\ldots EDU_{ct-n}EDU_{q-1}\ldots EDU_{q-end} \quad \text{where } 0 < \text{end} < 4$$

EDU_{q-i} is the question EDU containing a question word (qw_{q-i}) and $qw_{q-i} \in$ QW where QW is a question-word set.

QW = {‘ทำไม/*Why*’ ‘อย่างไร/*How*’ ‘อะไร/*What*’ ‘บอกคำวินิจฉัย/*RequestDiagnosis*’ ‘ใช่หรือไม่/*yes-no*’…}

EDU_{ct-a} is a content EDU expressing a problem content of EDU_{q-i} based on ‘*Why-Question*’ (Why-Q), ‘*How-Question*’ (How-Q), and ‘บอกคำวินิจฉัย/*RequestDiagnosis*’ where a = 1, 2, … n and i = 1, 2, 3, … n is an integer number and is greater than 0.

EDU_{ct-a} and EDU_{q-i} have the following Thai linguistic patterns where the problem content of EDU_{q-i} for this research is based on a symptom event expressed by a verb phrase.

$$\mathrm{EDU}_{\mathrm{ct-a}} \rightarrow \mathrm{NP1VP} \mid \mathrm{conj\ NP1VP}$$
$$\mathrm{VP} \rightarrow \mathrm{V}_{\mathrm{sym}}\,\mathrm{NP2}$$
$$\mathrm{EDU}_{\mathrm{q-i}} \rightarrow qw_{\mathrm{q-i}}\mathrm{NP1\ V_q\ NP2} \mid \mathrm{NP1\ V_q\ NP2}\, qw_{\mathrm{q-i}}$$

(where $\mathrm{V}_{\mathrm{sym}}$ is a symptom verb concept set having $v_{\mathrm{ct-a}} \in \mathrm{V}_{\mathrm{sym}}$; $\mathrm{V_q}$ is a verb set expressed on $\mathrm{EDU}_{\mathrm{q-i}}$ having $v_{\mathrm{q-i}} \in \mathrm{V_q}$; conj is a conjuction word; NP1 and NP2 are noun phrases.)

For example:

Example 1

$\mathrm{EDU}_{\mathrm{ct-1}}$ "เมื่อผมไอออกมา/***When I cough***,"
(เมื่อ/When ผม/I ไอออกมา/cough)

$\mathrm{EDU}_{\mathrm{ct-2}}$ "เสมหะเป็นเลือด/***phlegm contains blood***."
(เสมหะ/phlegm (sputum) เป็นเลือด/contain blood)

$\mathrm{EDU}_{\mathrm{ct-3}}$ "แต่ว่า[ผม]ไม่เป็นไข้/***But*** [***I***] ***have no fever***."
(แต่ว่า/But [ผม/I] ไม่เป็นไข้/have no fever)

$\mathrm{EDU}_{\mathrm{ct-4}}$ "[ผม]เป็นมาประมาณ *2* วันครับ/[***I***] ***have the symptoms about 2 days***."
([ผม/I] เป็นมา/have the symptoms ประมาณ 2 วันครับ/about 2 days)

$\mathrm{EDU}_{\mathrm{q-1}}$ "ทำไม[ผม]ไอออกมาเป็นเลือดครับ/***Why do*** [***I***] ***cough up blood?***"
(ทำไม/Why [ผม/I] ไอออกมาเป็นเลือดครับ/cough up blood)

$\mathrm{EDU}_{\mathrm{q-2}}$ "[ผม/*I*] ควรจะทำอย่างไรครับ/***How should*** [***I***] ***remedy?***"
([ผม/I] ควรจะ/should ทำ/remedy อย่างไร/How)

(where [...] means ellipsis.)

This research emphasizes on the problem-solving questions especially the disease-symptom questions on the health-care domain. According to the knowledge source, the Symptom-Treatment relation is represented by a symptom-treatment vector pair (which consists of a symptom-EDU vector, $\langle \mathrm{EDU}_{\mathrm{sym-1}}\mathrm{EDU}_{\mathrm{sym-2}}\cdots \mathrm{EDU}_{\mathrm{sym-m}}\rangle$, as a question, and a treatment-EDU vector, $\langle \mathrm{EDU}_{\mathrm{treat-1}}\mathrm{EDU}_{\mathrm{treat-2}}\cdots \mathrm{EDU}_{\mathrm{treat-p}}\rangle$, as an answer for How-Q). And, the causality relation is also represented by a cause-effect vector pair (which consists of a cause-EDU vector, $\langle \mathrm{EDU}_{\mathrm{cause-1}}\mathrm{EDU}_{\mathrm{cause-2}}\ldots\mathrm{EDU}_{\mathrm{cause-s}}\rangle$ as an answer for Why-Q of solving causes, and an effect-EDU vector or the symptom-EDU vector, $\langle \mathrm{EDU}_{\mathrm{sym-1}}\mathrm{EDU}_{\mathrm{sym-2}}\cdots \mathrm{EDU}_{\mathrm{sym-m}}\rangle$, as a question).

In addition to the question-pattern, $\mathrm{EDU}_{\mathrm{ct-a}}$ expresses the problem content such as a symptom event based on a verb phrase. Thus, $\mathrm{EDU}_{\mathrm{ct-a}}$ can be represented by a word co-occurrence of two adjacent words (called "contentWord-CO", $v_{\mathrm{ct-a}}\ w_{\mathrm{ct-a}}$) after stemming words and skipping the stop words in between two adjacent words. The first word of contentWord-CO is a verb expression, $v_{\mathrm{ct-a}}$, where $v_{\mathrm{ct-a}} \in \mathrm{V}_{\mathrm{sym}}$ and $\mathrm{V}_{\mathrm{sym}}$ approaches to the symptom concept. The second word of contentWord-CO is a co-occurred word, $w_{\mathrm{ct-a}}$ ($w_{\mathrm{ct-a}} \in \mathrm{W}_{\mathrm{sym}}$ where $\mathrm{W}_{\mathrm{sym}}$ is a co-occurred word set inducing the $v_{\text{ct}-\text{a}}$ $w_{\text{ct}-\text{a}}$ co-occurrence to have the symptom concept [5], based on Wordnet and MeSH after using Thai-To-English dictionay).

V_{sym} = {'ถ่าย/*defecate*','เบ่ง/*push*', 'ปวดท้อง/*have an abdomen pain*','ปวด/*pain*', 'อึดอัด/*be uncomfortable*','รู้สึกไม่สบาย/*be uncomfortable*','มี[อาการ]/*have* [*symptom*]',}
W_{sym} = {'', 'ยาก/*difficultly*', 'ถ่าย/*stools*', 'เชื้อ/*germ*', 'เหลว/*liquid*', 'ประจำเดือน/*period*', 'แน่นท้อง/*fullness*', 'ท้องเฟ้อ/*flatulence*', 'ไข้/*fever*',...}

$\text{EDU}_{\text{q}-\text{i}}$ can also be represented by a word co-occurrence of two adjacent words where the first word is $v_{\text{q}-\text{i}}$ and the second word is the co-occurred word, $w_{\text{q}-\text{i}}$ after stemming words and the stop word removal.

There are several techniques [2, 7, 9, 12, 15] having been used for reasoning QA system especially on a Why and How QA system (see Sect. 2). However, the Thai documents have several specific characteristics, such as zero anaphora or the implicit noun phrase, without word and sentence delimiters, and etc. All of these characteristics are involved in two main problems. The first problem is how to classify 'Why' 'How' and 'Request-Diagnosis', from $\text{EDU}_{\text{q}-\text{i}}$ having the $qw_{\text{q}-\text{i}}$ ambiguity where Request-Diagnosis requires the answer from both Why-Q and How-Q. The second problem is how to determine a corresponding answer from the knowledge source for each question, $\text{EDU}_{\text{q}-\text{i}}$, after the question focuses of $\text{EDU}_{\text{q-i}}$ have been solved, where the knowledge source contains several question-answer pairs of the symptom-treatment vector pairs and the cause-symptom vector pairs (see Sect. 3.2). According to these problems, we need to develop a framework which combines the machine learning technique and the linguistic phenomena to learn the several EDU expressions of the question-pattern on the health-care-community web boards. Therefore, we apply Naïve Bayes (NB) and Support Vector Machine (SVM) to learn the posted-problem question types especially the Why, How, Request-Diagnosis, and Other questions from two EDUs, $\text{EDU}_{\text{ct}-\text{n}}$ and $\text{EDU}_{\text{q}-\text{i}}$ where i = 1, 2, 3. Since each symptom-treatment vector pair and each cause-symptom vector pair mostly contain several symptoms/problems, we propose clustering the symptoms/problems from the knowledge source before applying an IR (Information Retrieval) technique of determining the corresponding answer from the knowledge source. The applied IR technique of this research is based on top-down levels of determining similarity scores between the content-EDU vector ($\langle \text{EDU}_{\text{ct}-1}\text{EDU}_{\text{ct}-2}\ldots\text{EDU}_{\text{ct}-\text{n}} \rangle$ of the posted-problem question on the web board) and each symptom-EDU vector, $\langle \text{EDU}_{\text{sym-1}}\text{EDU}_{\text{sym-2}}\cdots \text{EDU}_{\text{sym-m}} \rangle$, of the knowledge source through the certain symptom clusters.

Our research is organized into 5 sections. In Sect. 2, related work is summarized. Problems in web board QA system are described in Sect. 3 and Sect. 4 shows our framework for web board QA system. In Sect. 5, we evaluate and conclude our proposed model.

2 Related Work

Several strategies [2, 7, 9, 12, 15] have been proposed to solve their QA systems including Why questions or How questions.

Girju [7] worked on the Why question with the answer based on the lexico-syntactic pattern as 'NP1 Verb NP2' (where NP1 and NP2 are the noun-phrase expressions of a causative event and an effect event, respectively), i.e. "What causes Tsunami? → Earthquake causes Tsunami". However, it is not suitable for our research mostly based several effect-event explanations which express by verbs/verb phrases.

Schwitter et al. [12] worked on the procedural questions/How questions with their answers being extracted from technical documents by the ExtrAns system. Their procedural answer is often expressed in a procedural writing style with guidelines. The high performance in their QA system is best achieved through logic-based and pattern-matching techniques.

Verberne et al. [15] proposed using RST (Rhetorical Structure Theory) structures to approach Why questions by matching the question topic with a nucleus in the RST tree while yielding the answer from the satellite. The RST approach to the Why-QA system achieved the answer correctness of 91.8 % and a recall of 53.3 %.

Baral et al. [2] developed a formal theory of answers to why and how questions by developing the biological-graph model having event nodes and compositional edges as the knowledge-base with corresponding to why and how questions on the biology domain. Their questions are based on the forms: "How are X and Y related in the process Z?" and "Why is X important to Y.

Oh et al. [9] used intra- and inter-sentential causal relations between terms or clauses as evidence for answering Why-questions. They ranked their candidate answers (from documents retrieval Japanese web texts) with the ranking function including re-ranking the answer candidates done by a supervised classifier (SVM). Their why-QA system achieves 83.2 % precision.

However, most of previous researches on a Why-QA system/a How-QA system [2, 7, 12] are based on a single sentence/one EDU of a Why question/a How question, except [15] and [9] based on two EDUs of Why question, whereas our Why-Q and/or How-Q is based on several EDUs (see Sect. 1).

3 Problems of Web Board QA System

To develop the web board QA system, there are two main problems that must be solved: how to identify the reasoning questions (especially a Why question and a How question) from the posted problems with multi-questions having the $qw_{\mathrm{q-i}}$ ambiguity and how to determine the corresponding answer of each question from the knowledge source.

3.1 How to Identify Why Question and How Question with qw_{q-i} Ambiguity

The problem of identifying the question expression without having the question mark symbol ('?') is solved by using a question-word set, QW, {'ทำไม/Why', 'อย่างไร/How' 'อะไร/What', …} having $qw_{q-i} \in$ QW. Where a 'ทำไม/Why' function is a reasoning question, a 'อะไร/What' function is asking for information about something (http://www.englishclub.com/vocabulary/wh-question-words.htm). However, there is a question word's function ambiguity, e.g. 'อะไร/What' as in reasoning (as shown in the following example on EDU_{q-1}).

Example 2

EDU_{ct-1} "[ผม] ไม่มีน้ำมูก/[*I*] *do not have mucus*."
EDU_{ct-2} "[ผม] ไม่ไอ/[*I*] *do not cough*."
EDU_{ct-3} "แต่[ผม] คัดจมูกบ้าง/*But* [*I*] *have a congested nose*."
EDU_{ct-4} "[ผม] เจ็บคอ/[*I*] *have sore throat*."
EDU_{q-1} "เป็นเพราะอะไร/***What*** *are the reasons?*"
EDU_{q-2} "มีวิธีรักษาอาการอย่างไรคะ/*Show how to treat symptoms*."

Therefore, we solve the qw_{q-i} ambiguity by apply NB and SVM to learn the posted-problem question type from EDU_{q-i} and EDU_{ct-n} where EDU_{ct-n} is the last content EDU and i = 1, 2, 3. There are three features used for NB learning and SVM learning, $v_{ct-n}\ w_{ct-n}$ co-occurrence of EDU_{ct-n}, qw_{q-i} from EDU_{q-i}, and another verb-noun co-occurrence, $v_{q-i}\ w_{q-i}$, of EDU_{q-i} after stemming words and eliminating the stop words.

3.2 How to Determine Corresponding Answer from Knowledge Source

The knowledge source contains the several question-answer pairs of several symptom-treatment vector pairs and several cause-symptom vector pairs. Each symptom-EDU of a symptom-EDU vector and each treatment-EDU of a treatment-EDU vector are represented by the $v_{sym-\beta}\ w_{sym-\beta}$ co-occurrence (with the symptom concept, called 'symptomWord-CO', where $v_{sym\text{-}\beta} \in V_{sym}$ and $w_{sym\text{-}\beta} \in W_{sym}$) and the $v_{treat-\gamma}\ w_{treat-\gamma}$ co-occurrence (with the treatment concept, called 'treatmentWord-CO') respectively [10]. Where $\beta = 1, 2, \ldots, m$; $\gamma = 1, 2, \ldots, p$; m is the number of symptom-EDUs on the symptom-EDU vector; p is the number of treatment-EDUs on the treatment-EDU vector (see Sect. 1); $v_{treat-\gamma} \in V_{treat}$; $w_{treat-\gamma} \in W_{treat}$; V_{treat} is the treatment-verb concept set; W_{treat} is co-occurred word sets inducing the $v_{treat-\gamma}\ w_{treat-\gamma}$ co-occurrence to have treatment concept. Then, $v_{sym-\beta}\ w_{sym-\beta} \in VW_{sym}$ (where VW_{sym} is a symptomWord-CO set) and $v_{treat-\gamma}\ w_{treat-\gamma} \in$

VW_{treat} (where VW_{treat} is a treatmentWord-CO set). A contentWord-CO of EDU_{ct-a} is also an element of VW_{sym}(v_{ct-a} $w_{ct-a} \in VW_{sym}$). Thus, the content-EDU vector, $\langle EDU_{ct-1}EDU_{ct-2}...EDU_{ct-n}\rangle$, of the question-pattern can be represented by a contentWord-CO vector, $\langle v_{ct-1}w_{ct-1}\ v_{ct-2}\ w_{ct-2}\ \ldots\ v_{ct-n}\ w_{ct-n}\rangle$ where SubSym, $\{v_{ct-1}w_{ct-1}\ v_{ct-2}\ w_{ct-2}\ \ldots\ v_{ct-n}\ w_{ct-n}\}$, is a set containing all elements of the contentWord-CO vector. SubSym can occur on several symptom-treatment vector pairs/several cause-symptom vector pairs of the knowledge source (where SubSym $\subseteq VW_{sym}$) as shown in the following example.

Example 3

EDU_{ct-1} "Yesterday, I felt dizzy."
EDU_{ct-2} "I vomited frequently."
EDU_{ct-3} "Now I am very nauseated."
EDU_{q-1} "What can I do?"
EDU_{q-2} "What could have caused this?"

From Example 3, SubSym, {'feel_dizzy' 'vomit_ frequently' 'be_ nauseated'}, occurs on the symptom-treatment vector pairs of the following diseases, Brain Tumor, Food Poisoning, and Vestibular Imbalance. It is challenge to determine the answer of this question of the posted problem on the web board. Moreover, each symptom-treatment vector pair and also each cause-symptom vector pair mostly contain several symptom EDUs. Therefore, we propose clustering the symptom feature of the knowledge source before using top-down levels of determining similarity scores among the content-EDU vector (the posted problem) and the symptom-EDU vectors from the knowledge source clusters which involve with stemming words and skipping the stop words. The first/top level of determining similarity scores is at the cluster level which is the conceptual symptom level. This first level potentially causes of saving amounts of all levels' search time to match the symptom-EDU of the knowledge source. The second level of determining similarity scores is at the EDU level of each EDU pair (EDU_{ct-a} $EDU_{sym-\beta}$; a = 1, 2, …, n; $\beta = 1, 2, \ldots, m$). The EDU pair is marked 'match' if its similarity score is ≥ 0.5 otherwise 'un-match'. The third level of similarity scores is determined from the second level as the vector level based on the 'match' marks of each content-symptom vector pair,$\langle EDU_{ct-1}EDU_{ct-2}...EDU_{ct-n}\rangle$ $\langle EDU_{sym-1}EDU_{sym-2}...EDU_{sym-m}\rangle$. The answer of How-Q/Why-Q is selected from the symptom-treatment vector pairs or the cause-symptom vector pairs having the top 3 ranked final similarity scores.

4 A Framework for Web Board Question Answering System

There are four steps in the web board QA system, Question-Corpus Preparation, Question-Type Learning, Question-Type Identification, Knowledge Source Preparation and Answer Determination as shown in Fig. 1.

4.1 Question-Corpus Preparation

This step is the preparation of the question corpus in the form of EDU from the medical-care-consulting documents on the hospital's web-boards of the non-government organization (NGO) website. The step involves with using Thai word segmentation tools [13], including Name entity [5]. After the word segmentation is achieved, EDU segmentation is then to be dealt with [6]. These annotated EDUs will be kept as an EDU corpus. This corpus contains 6000 EDUs of 580 questions from four disease categories, gastrointestinal tract diseases, brain-never diseases, ear-nose-throat diseases, and bone diseases including arthritis. The corpus is separated into 2 parts; one part is 400 questions for learning the question type concepts based on ten folds cross validation. And, the other part of 180 questions is for testing the question type identification. In addition to this step of corpus preparation, we semi-automatically annotate the question-word concepts, the question-focus group, and the contentWord-CO concepts of symptoms, as shown in Fig. 2. All concepts of annotation are referred to Wordnet (http://wordnet.princeton.edu/obtain) and MeSH after translating from Thai to English, by using Lexitron (the Thai-English dictionary) (http://lexitron.nectec.or.th/).

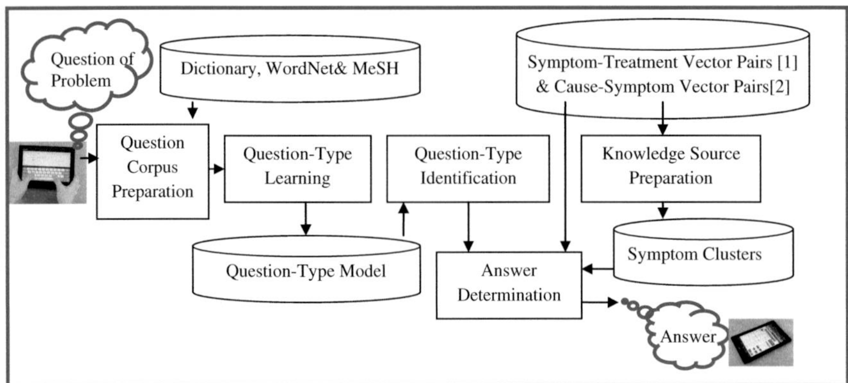

Fig. 1 System overview

<EDUct1> "[ผม] รู้สึก คลื่นไส้" ([I] *feel nauseated.*)
[ϕ = ผม/I]/NP <Qfocus> [<WordCo concept= be nauseated><vct1 =feel> รู้สึก/vi</vct1><wct1= nauseated> คลื่นไส้/vi </wct1></WordCo>]/VP</Qfocus></EDUct1>
<EDUct2> "[ผม] อาเจียน" ([I] *vomit.*)
[ϕ = ผม/I]/NP <Qfocus>[<WordCo concept= vomit><vct1 =vomit> อาเจียน/vi</vct1><wct1=null> </wct1> </WordCo>]/VP</Qfocus></EDUct2>
<EDUct3> "[ผม]เวียนศีรษะ" ([I] *am dizzy.*)
[ϕ = ผม/I]/NP <Qfocus> [<WordCo concept= be dizzy><vct1 =be dizzy> เวียนศีรษะ/vi</vct1><wct1= null></wct1> </WordCo>]/VP</Qfocus></EDUct3>
<EDUct4>"[ผม]นอนไม่หลับ" ([I] *can't sleep.*)
[ϕ = ผม/I]/NP <Qfocus>[<WordCo concept= cannot sleep><vct1 =sleep>นอน/vi </vct1> <wct1=cannot>ไม่หลับ/adv </wct1> </WordCo>]/VP</Qfocus></EDUct4>
<EDUq1> "[ผม]ควรจะทำอย่างไร" (*How should* [*I*] *do*?)
[ϕ=ผม/I]/NP [*should*-ควรจะ/prev <vq1 : concept=do>ทำ/vt </vq1><wq1=null> </wq1> <Qword concept=how> *how*-อย่างไร/pint </Qword>]/VP </EDUq1>

Where: a 'Qfocus' tag is a question focus tag. A WordCo tag is a word co-occurrence tag. A 'vct#' tag is a verb tag of a content EDU and the other verb concept set. A wct# tag is a co-occurred word tag. A 'vq#' tag is a verb tag of an EDU containing the question word. A 'Qword' tag is a question word tag. An EDUct tag is an EDU content tag. An EDUq tag is a tag of an EDU having the question word. And, ϕ stands for a zero anaphora or ellipsis.

Fig. 2 Examples of question annotation

4.2 Question-Type Learning

This step is using Weka (http://www.cs.waikato.ac.nz/ml/weka/) to learn Why-Q, How-Q, Request-Diagnosis, and Other-Q by NB, and SVM from EDU_{q-i} and EDU_{ct-n} (where i = 1, 2, 3) of the question-pattern from the annotated-question corpus. The features used for these learning consist of three feature sets; VW_{sym}, QW, and $VW_{question}$. $VW_{question}$ is a questionWord-CO set where each element of the set is formed by a verb and a co-occurred word from a question EDU, EDU_{q-i}, after stemming words and eleminating stop words. Where EDU_{ct-n} contains $v_{ct-n}w_{ct-n} \in VW_{sym}$, and EDU_{q-i} contains $qw_{q-i} \in QW$ and $v_{q-i}w_{q-i}, \in VW_{question}$ after stemming words and skipping the stop words.

Naïve Bayes (NB) According to [8], the NB learning is a generic classification to determine the feature probabilities of four classes (class1, class2, class3, and class4) of the question types; Why-Q, How-Q, Request-Diagnosis, and Other-Q respectively. The features of NB classifiers consist of three features, $v_{ct-n}w_{ct-n}$, qw_{q-i}, and $v_{q-i}w_{q-i}$ of VW_{sym}, QW, and $VW_{question}$ respectively, from the annotated corpus of two EDUs (EDU_{q-i} and EDU_{ct-n} where i = 1, 2, 3).

Support Vector Machine The linear binary classifier, SVM, applies in this research to classify the four question types, Why-Q, How-Q, Request-Diagnosis, and Other-Q based on pairwise classification. According to [14] this linear function, f(x), of the input $x = (x_1x_2 \ldots x_n)$ assigned to the positive class if $f(x) \geq 0$, and otherwise to the negative class if $f(x) < 0$, can be written as:

$$
\begin{aligned}
f(x) &= \langle \mathrm{w} \cdot \mathrm{x} \rangle + b \\
&= \sum_{j=1}^{n} w_j x_j + b
\end{aligned}
\tag{1}
$$

where x is a dichotomous vector number, w is weight vector, b is bias, and (w,b) $\in R^n \times R$ are the parameters that control the function. The SVM learning results are w_j and b for x_j which consists of three feature sets, VW_{sym}, QW, and $VW_{question}$, for the corresponding elements, $v_{ct-n}w_{ct-n}$, qw_{q-i}, and $v_{q-i}w_{q-i}$ where i = 1, 2, 3.

4.3 Question-Type Identification

All probabilities and weights from the previous learning step by NB and SVM respectively are used to identify the question types.

Naïve Bayes According to [8], Eq. 2 and the feature-probabilities determined by the previous step of NB are used to identify the question type class of the question-pattern.

$$
\begin{aligned}
Question\ Class &= \underset{class \in class}{\arg\max}\ P\left(class \middle| v_{ct-n}w_{ct-n}, v_{q-i}w_{q-i}, qw_{q-i}\right) \\
&= \underset{class \in class}{\arg\max}\ P(v_{ct-n}w_{ct-n}|class)P\left(v_{q-i}w_{q-i}|class\right)P\left(qw_{q-i}|class\right)P(calss)
\end{aligned}
\tag{2}
$$

$$
\begin{aligned}
&where\ v_{ct-n}w_{ct-n} \in VW_{sym};\ VW_{sym}\ is\ a\ symptom\ word - CO\ set \\
&\quad v_{q-i}w_{q-i} \in VW_{question};\ VW_{question}\ is\ a\ word - CO\ set\ on\ EDU_{q-i}\ and\ i = 1, 2, 3 \\
&\quad qw_{q-i} \in QW;\ QW\ is\ a\ question\ word\ set
\end{aligned}
$$

Support Vector Machine The results from SVM learning are weight, w_j, and bias, b, of each feature (x_i). According to Eq. (2), the input vector (x) consisting of contentWord-CO features, $v_{ct-n}w_{ct-n}$, questionWord-CO features, $v_{q-i}w_{q-i}$, and question-word features, qw_{q-i}, including their weights, w_j, and bias, b, are used to determine the question types based on pairwise classification.

4.4 Knowledge Source Preparation

This step is to organize the knowledge source, especially the symptom/problem feature, by clustering the symptoms/problems for enhancing the symptom/problem search which most users have questions about this feature. We cluster ρ symptom-vector samples from the knowledge source by using k-mean as shown in Eq. (3) [1]

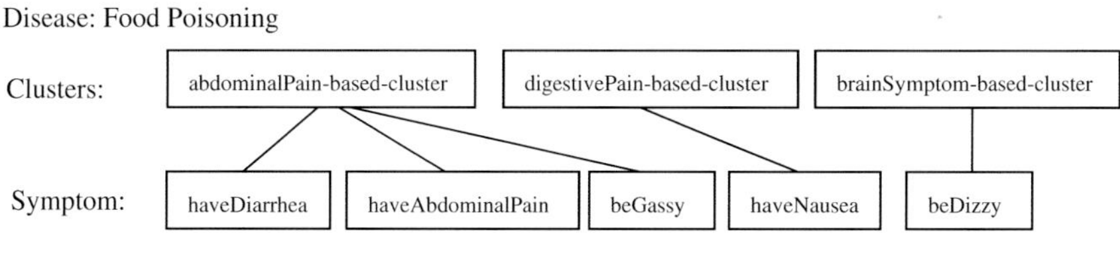

Fig. 3 An example of symptom clusters from knowledge source [10, 11]

$$cluster(\mathrm{x}_j) = \underset{1 \le k \le K}{\arg\min} \left\| \mathrm{x}_j - \mu_k \right\|^2 \tag{3}$$

where xj is a symptom vector of an object $\langle v_{\text{sym}-1}\, w_{\text{sym}-1}, v_{\text{sym}-2}\, w_{\text{sym}-2}, \ldots, v_{\text{sym}-m}\, w_{\text{sym}-m} \rangle$ (which represents a symptom-EDU vector, $\langle \text{EDU}_{\text{sym-1}}\text{EDU}_{\text{sym-2}}..\text{EDU}_{\text{sym-m}} \rangle$) and j = 1, 2, ..., ρ. μ_k is the mean vector of the kth cluster. The highest number of $v_{\text{sym}-\beta}\, w_{\text{sym}-\beta}$ occurrences in each cluster is selected to its cluster representative by. Thus, we have a symptom cluster set (Y) {rhinorrhoea-based-cluster, abdominalPain-based-cluster, brainSymptom-based-cluster, ….., k-Symptom-basedcluster} where $\text{Element}_r \in \text{Y}$ and r is an integer number; for example: Element_1 = rhinorrhoea-based-cluster, Element_2 = abdominalPain-based-cluster, Element_3 = brainSymptom-based-cluster,…, Element_k = k-Symptom-based-cluster, as shown in Fig. 3.

$$\begin{aligned}
\text{Element}_1 &= \left\{ v_{\text{sym}-11} w_{\text{sym}-11}, v_{\text{sym}-12} w_{\text{sym}-12}, \ldots, v_{\text{sym}-1\alpha} w_{\text{sym}-1\alpha} \right\} \\
\text{Element}_2 &= \left\{ v_{\text{sym}-21} w_{\text{sym}-21}, v_{\text{sym}-22} w_{\text{sym}-22}, \ldots, v_{\text{sym}-2\delta} w_{\text{sym}-2\delta} \right\} \\
\text{Element}_3 &= \left\{ v_{\text{sym}-31} w_{\text{sym}-31}, v_{\text{sym}-32} w_{\text{sym}-32}, \ldots, v_{\text{sym}-3\eta} w_{\text{sym}-3\eta} \right\} \\
\text{Element}_k &= \left\{ v_{\text{sym}-k1} w_{\text{sym}-k1}, v_{\text{sym}-k2} w_{\text{sym}-k2}, \ldots, v_{\text{sym}-k\varphi} w_{\text{sym}-k\varphi} \right\}
\end{aligned}$$

where α, δ, η, and φ are integer numbers.

4.5 Answer Determination

According to the correct question-type identification from the previous step, we randomly select 120 correct questions which consist of 30 correct questions of each disease category, gastrointestinal-tract diseases, brain-never diseases, ear-nose-throat-diseases, and bone-arthritis-diseases. The selected questions consist of 50 questions of Why-Q, 50 questions of How-Q, and 20 questions of Request-Diagnosis. Since the focuses of Why-Q, How-Q, and Request-Diagnosis from the content-EDU vector of the question-pattern are based on the symptom events expressed by verb phrases, the question focuses of this research can be solved by the contentWord-CO vector with $v_{ct-a}\ w_{ct-a} \in VW_{sym}$. The answers can be solved by using top-down levels of determining similarity scores between the content-EDU vector (which contains the content Word-CO vector as focuses) and the symptom-EDU vector of the following vector pairs of the knowledge source.

- a symptom-treatment vector pair for How-Q where the symptom-EDU vector is equivalent to the question and the treatment-EDU vector is the answer.
- a cause-symptom vector pair for a Why-Q where the symptom-EDU vector is equivalent to the question and the cause-EDU vector is the answer.
- a cause-symptom vector pair and a symptom-treatment vector pair for a question of Request-Diagnosis where the symptom-EDU vector is equivalent to the question and the cause-EDU vector and the treatment-EDU vector are the answers.

The top/first level of determining similarity scores starts by determining Element_r of the conceptual cluster from $v_{ct-a}w_{ct-a}$ as $v_{ct-a}w_{ct-a} \in \text{Element}_r$.

The second level of determining similarity scores from Element_r as shown in Eq. 4 [3] is at the EDU level between EDU_{ct-a} and $EDU_{sym-\beta}$

$$Similarity\_Score = \frac{|S1 \cap S2|}{\sqrt{|S1| \times |S2|}} \tag{4}$$

where

S1 is an EDU_{ct-a} of the content-EDU vector (having a = 1, 2, …, n) after stemming words and skipping stop words

S2 is an $EDU_{sym-\beta}$ of the symptom-EDU vector (having β = 1, 2, …, m) after stemming words and skipping stop words

All word concepts of a S1 EDU and a S2 EDU are based on WordNet and MeSH after using the Thai-to-English dictionary. The number of words in the S1 EDU and the number of words in the S2 EDU are not significantly different. According to the similarity score determination at EDU level, both EDU_{ct-a} and $EDU_{sym-\beta}$ are marked 'match' if its similarity score is ≥ 0.5 otherwise 'un-match'. The final similarity scores based on the vector level is calculated by Eq. 5 from the content-symptom vector pair, $\langle EDU_{ct-1}\ EDU_{ct-2}\ \ldots EDU_{ct-n}\rangle\ \langle EDU_{sym-1} EDU_{sym-2} \ldots EDU_{sym-m}\rangle$ based on the 'match' marks. After ranking the final similarity scores, the answer of How-Q/Why-Q is selected from the symptom-treatment vector pairs or the cause-symptom vector pairs from the knowledge source having the top 3 ranked similarity scores at the vector level as shown in Fig. 4.

$$Similarity\_Score = \frac{|Vec1 \cap Vec2|}{\sqrt{|Vec1| \times |Vec2|}} \tag{5}$$

Assume that each EDU is represented by (NP V P). L is a list of EDUs.
S1 is an $EDU_{ct\text{-}a}$ of the content-EDU vector. S2 is an $EDU_{sym\text{-}\beta}$ of the symptom-EDU vector. Vec1 is a content-EDU vector of the question-pattern on a certain web-board. Vec2 is a symptom-EDU vector of a symptom-treatment vector pair or a cause-symptom vector pair. D is a matrix of Vec2. $v_{ct\text{-}a}\ w_{ct\text{-}a}$ is a Word-CO of $EDU_{ct\text{-}a}$.Y is a symptom cluster set of the Knowledge Source.

```
ANSWER_DETERMINATION(D_vec2,L_s1,L_s2)
x=1;
while x<= row_of_matrix(D) do
{ j=1, count_intersect_Vector=0;
  while j<= length[L_s1] do
    { found = true;
      Find Element_r from v_ct-a w_ct-a
      if found
      {   while each_member_of_Element_r
          {   i=1;flag=true;
              where i<= length[L_s2] and flag do
              {  sim_score1 = similarity_score(S_1-i,S_2-i) (eq.4)
                 if sim_score1>0.5 then
                 {   count_intersect_Vector++;
                     intersect_Vector = intersect_Vector+S_1-i; //match
                     i++; flag = true; }
                 else flag = false; } } }
      else found = false; j++; }
  sim_score2=abs(count_intersect_vector)/sqrt(summation(abs(length(Vec1))+
  abs(length(Vec2))))
  Vec2_x = sim_score2;
  x++;
  Sorting Vec2 as descending;
  Select top three Vec2;
```

Fig. 4 Answer determination algorithm

where

Vec1 is a content-EDU vector of the question-pattern on a certain web-board
Vec2 is a symptom-EDU vector of a symptom-treatment vector pair or a cause-symptom vector pair

5 Evaluation and Conclusion

There are two evaluations of the proposed research, the question type identification and the answer determination. The evaluation of the question type identification performance is based on the precision and the recall with the testing question corpus of 180 questions. The results of precision and recall are evaluated by three expert judgments with max win voting. The average precision and the average recall of the question type identification by the proposed machine learning is 0.918 and 0.863

Table 1 Show precision and recall of question type identification by NB and SVM

180 Web board questions	NB		SVM	
	Precision	Recall	Precision	Recall
Why-Q	0.864	**0.877**	0.923	0.907
How-Q	0.891	0.865	0.932	0.922
Request-Diagnosis	**0.952**	0.819	**0.960**	**0.941**
Other	0.885	0.755	0.936	0.818

respectively (see Table 1). SVM yields the better precision and recall results than NB whereas some features are dependency.

And, the evaluation of the answer determination performance of the proposed method is based on the percent correctness of the answer based on the knowledge source as the answer source. The results of answer correctness are evaluated by three expert judgments with max win voting. We drew a random sample of 120 correct questions from the question type identification for the answer evaluation. The answer correctness for Why-Q, How-Q, and Request-Diagnosis questions at the top 3 ranked similarity scores is 83 %. However, our proposed methodology including the clustering technique has a potential of taking less search time to find an answer than our previous methodology [1] without applying the clustering one because our current algorithm is based on top-down levels (which is similar to the tree structure) of determining similarity. The correctness of answer determination results should be increased if the noun ellipsis or the zero anaphora has been solved and also the major symptoms of each disease have been solved from its symptom-EDU vectors before clustering. Therefore, the web board QA system of this research provides the explanation knowledge or reasoning knowledge for solving problems and also for the feasibility of self-study.

Acknowledgments This is supported by Thai Research Fund (MRG5580030). The medical-care knowledge and the pharmacology knowledge applied in this research are provided by Prof. Puangthong Kraipiboon, a clinician of Division of Medical Oncology, Department of Medicine, Ramathibodi Hospital, and Uraiwan Janviriyasopak, a pharmacist of RexPharmcy, respectively.

References

1. Aloise, D., Deshpande, A., Hansen, P., Popat, P.: NP-hardness of Euclidean sum of squares clustering. Mach. Learn. **75**, 245–249 (2009)
2. Baral, C., Vo, N.H., Liang, S.: Answering why and how questions with respect to a frame-based knowledge base: a preliminary report. In: Proceedings of ICLP 2012, Hungary (2012)
3. Biggins, S., Mohammed, S., Oakley, S.: University of Sheffield: two approaches to semantic text similarity. In: Proceedings of First Joint Conference on Lexical and Computational Semantics, Montréal, Canada (2012)

4. Carlson, L., Marcu, D., Okurowski, M.E.: Building a discourse-tagged corpus in the framework of rhetorical structure theory. Curr. New Dir. Discourse Dialogue **22**, 85–112 (2003)
5. Chanlekha, H., Kawtrakul, A.: Thai named entity extraction by incorporating maximum entropy model with simple heuristic information. In: IJCNLP' 2004 Proceedings (2004)
6. Chareonsuk, J., Sukvakree, T., Kawtrakul, A.: Elementary discourse unit segmentation for Thai using discourse cue and syntactic information. In: NCSEC 2005 Proceedings (2005)
7. Girju, R.: Automatic detection of causal relations for question answering. In: Proceedings of 41st Annual Meeting of the Association for Computational Linguistics, Workshop on Multilingual Summarization and Question Answering-Machine Learning and Beyond, Japan (2003)
8. Mitchell, T.M.: Machine Learning. The McGraw-Hill Companies Inc. and MIT Press, Singapore (1997)
9. Oh, J-H., Torisawa, K., Hashimoto, C., Sano, M., Saeger, S.D., Ohtake, K.: Why-Question answering using intra- and inter-sentential causal relations. In: Proceedings of the 51st Annual Meeting of the Association for Computational Linguistics, Bulgaria (2013)
10. Pechsiri, C., Moolwat, O., Piriyakul, R.: Symptom-treatment relation extraction from web-documents for construct know-how map. In: KICSS' 2013 Proceedings (2013)
11. Pechsiri, C., Piriyakul, R.: Explanation knowledge graph construction through causality extraction from texts. J. Comput. Sci. Technol. **25**(5), 1055–1070 (2010)
12. Schwitter, R., Rinaldi, F., Clematide, S.: The importance of how-questions in technical domains. In: Proceedings of TALN-04, Workshop Question—Réponse, Fez, Morocco (2004)
13. Sudprasert, S., Kawtrakul, A.: Thai word segmentation based on global and local unsupervised learning. In: NCSEC'2003 Proceedings (2003)
14. Vapnik, V.N.: The Nature of Statistical Learning Theory. Springer, USA (1995)
15. Verberne, S., Boves, L., Coppen, P.-A., Oostdijk, N.: Discourse-based answering of Why-Questions. Traitement Automatique des Langues **47**, 2 (2007)

A PHR Front-End System with the Facility of Data Migration from Printed Forms

Atsuo Yoshitaka, Shinobu Chujyou and Hiroshi Kato

Abstract Systems for PHR (Personal Health Record) data management are expected to be supplemental infrastructure for improving health care environment in addition to the improvement of medical care environment. One of the major issues in current PHR environment is that only a very limited data is distributed from hospital or medical institution to citizens as electric data; most of the data is still distributed as printed document. In this paper, we propose a PHR system which enables to migrate from PHR data of printed document to the database by photography. Since it is estimated to take years to offer full electric data distribution online, we think this is one of the practical solutions for the issue. In addition to this feature, we designed the user interface suitable for touch panel display for more intuitive data manipulation. The result of preliminary user test showed the proposed system is more preferable for users.

Keywords PHR (Personal Health Record) · Data migration by photography · OCR · Adaptive dictionary configuration

1 Introduction

Japan is known as a country where the average life expectancy of citizen is long, and the importance of health care is recognized in accordance with aging society. Because of the growing ratio of aged person, Japanese government recognizes that

A. Yoshitaka (✉)
School of Information Science, Japan Advanced Institute of Science and Technology, 1-1 Asahidai, Nomi, Ishikawa 923-1292, Japan
e-mail: ayoshi@jaist.ac.jp

S. Chujyou
goowa inc, 3-4 Unetanaka, Kanazawa, Ishikawa 920-0343, Japan

H. Kato
Life Care on Demand, 3-13-5 Asahimachi, Kanazawa, Ishikawa 920-0941, Japan

S. Kunifuji et al. (eds.), *Knowledge, Information and Creativity Support Systems*, Advances in Intelligent Systems and Computing 416,
DOI 10.1007/978-3-319-27478-2_13

policies for improving healthy life expectancy are becoming more important. These include not only the development of medical care environment but also that of the environment which enables effective use of medical data for both of medical personnel and citizen. The development of health-care environment toward both of the above-mentioned two directions is expected to improve the quality of life especially for aged persons.

For the next step for matured aged society, Japanese government has put more focus on developing infrastructure for the citizens to maintain health by their effort. In accordance with this context, various aspects of EHR (Electric Health Record) [16] or PHR (Personal Health Record) systems have been studied. Topics related to PHR systems include standardization, security [2, 6, 9], access control [10, 15], user interface [4], hardware implementation, CCD (Continuity of Care Document) [1] or CCR (Continuity of Care Record) [13], and user study for PHR can be found in [17].

Currently, sharing of electric medical data among hospitals or clinics is not established yet. One of the reasons is that the widely accepted standard of medical data format is still in progress. HL7 CDA [5] is one of the examples of such standard which is similar to a study managing PHR based on XML [18]. Data sharing/exchanging frameworks based on HL7 are widely studied as seen in [3, 20]. Current status of medical data sharing is so-called "inter-sharing", and "intra-sharing" is expected as the next step to improve medical service. Therefore, EHR or PHR system is considered to be a breakthrough for the above-mentioned issue. The EHR infrastructure includes the idea of data sharing among hospitals, clinics, medical inspection institutions, local governments, and citizen. Practical experiments of PHR system are carried out, and some of them are under the cooperation with local government. When we mention PHR system, we put more emphasis is on how the citizen takes care of his/her own health record by his/her own effort.

In this paper, we focus on PHR systems, since our motivation is based on how we can offer more effective platform for PHR data management. PHR systems [7, 11, 12] consolidates health information such as height, weight, BMI (Body Mass Index), blood type, blood pressure, hepatic function indices, uric protein, medication, history of disease, and so on. Currently, such health information is managed independently in each hospital/clinic or institution for medical examination, and therefore such data is distributed as many places as a person had been seen. The purpose of the development of a PHR system is to offer an environment for citizens to consolidate distributed information, and contribute to watch his/her health condition by his/her own effort. A PHR system makes it possible to refer to all his/her personal health information anytime and anywhere it needed: it is expected to contribute to diminish unintentional duplication of basic medical check.

An example of the report of annual health checkup is shown in Fig. 1. Note that the form shown in Fig. 1 is currently used for distributing the report to each person who had a medical checkup, but the data printed on the form is fictional for demonstration. The report includes the result of the measurement of height, weight, BMI (body mass index), abdominal circumference, glucose, protein, and occult

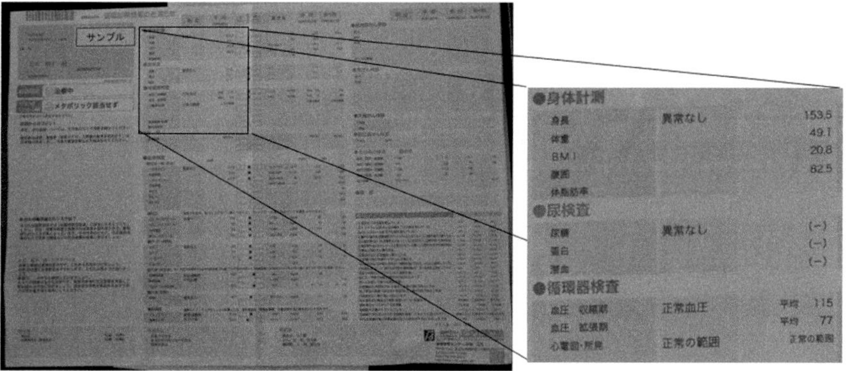

Fig. 1 An example of the report of annual health checkup

blood in urine, and so on. Though the number of items to be measured depends on the age, gender, and/or category of labor in case of an employee in a company, there are approximately 40 items.

There are studies and demonstration experiments on PHR system in Japan. One of the examples of PHR system is called Lico which is implemented by "Life care on demand" project. The Lico is a web-based PHR service and is aimed at maintaining his/her PHR by his/her own effort. One of the functions of Lico is to maintain the information of annual health checkup. There are approximately 40 items to enter manually with keyboard. As a result of operating the "Life care on demand" project, it is revealed that better interface in migrating printed data is awaited.

One of the possible solutions for this issue is to exploit an image scanner to capture a printed data of health check-up, and to apply OCR (Optical Character Recognition) processing for data migration. A general OCR engine reads out characters in each row, without recognizing logical structure of the table. Therefore, simple installation of a general image scanner with OCR engine will not be a promising solution. Alternative solution for this issue is to utilize a camera device to capture a form. For minimizing additional cost for equipping devices which may be devoted only for PHR data migration and for employing intuitive user interface, we think using a tablet PC is one of the promising solutions. Our scenario is as follows. A user takes photos of a printed report of health checkup with the built-in camera of a tablet PC in which a PHR front-end system is installed. Then the captured images are sent to PHR front-end system for OCR processing to migrate printed PHR data.

One of the major issues in data migration from printed form with OCR is how we assure acceptable accuracy in OCR process. The accuracy of character recognition is not always satisfactory considering the burden for checking and correcting errors. This issue is more serious in the scenario of migrating printed data by photography, because the images shot with a hand-held tablet may be blurred, which results to deteriorate the accuracy of character recognition. In addition to this, we need to consider how we cope with geometric normality issue because of the flexibility of condition in photography.

In this paper, we describe a front-end system which enables a user to migrate PHR data by taking photography of printed document. We discuss the issue of improving the accuracy of character recognition by OCR engine for PHR data migration. Though this is considered to be a transient technique, however, we believe this is significant in order to offer better PHR platform at earlier convenience. In addition to the user evaluation we reported in [19], further analysis is described later in this paper.

2 A PHR System Lico

2.1 System and Interface Design in Lico

As introduced in Sect. 1, there are several demonstrational experiments for PHR systems. An example of them is led by a local government of Uchinada town in Ishikawa prefecture, Japan, which is called "Lico". The Lico system is in operation which is implemented by the project called "Life care on demand" [8]. Lico offers a web based PHR service, which enables a user to enter PHR data by filling values into a form. Lico enables a registered user to store PHR data such as a report of annual health checkup, and to view stored PHR data.

Figure 2 shows an example of screen snapshot when a user enters PHR data using Lico system. As mentioned, current implementation of Lico is a web-based system with simple data migration method. Lico data migration interface is designed using text boxes, and a user need to enter the values on the printed report of annual checkup one by one with keyboard and mouse. In this example, a user is

Fig. 2 User interface for entering PHR data in current Lico system

prompted to enter the date of health checkup, height, weight, abdominal circumstance, and body fat percentage.

Though the numbers of items to be measured during an annual health checkup depends on age, gender, and company/institution (i.e., employer or health insurance system) which offers an annual medical check, there are approximately 40 items to enter. It is not difficult to imagine that entering all of such data into the form as shown in Fig. 2 is time-consuming, stressful, and error-prone for the user. Note that all the reports of annual health checkup are distributed as a printed form, but no electric data is distributed.

2.2 Data Migration Issues in Lico PHR System

Up to now, there are over 600 registered users of Lico system. As introduced in the previous section, current implementation of PHR data migration interface of Lico system is simple, text-box based user interface. A user needs to enter the values of measured items one by one into the fields manually. According to the experience of operating Lico system and feedback from the Lico user, insufficient operability of data migration is revealed in the current system. That is, the task of migrating printed data by filling the values into tens of fields in a Web page one by one discourages the user to keep using the system. One reason is that the task of typing all the data manually is time-consuming, and confirming if there is a miss-typing is stressful. Another reason relates to the unfamiliarity of specific terms such as MCV, MCH, GOT, or GPT, all of which are the indices on blood quality or soundness of liver. Since it is difficult for most of the users to understand the meaning of such values, it is stressful for them to enter such values without errors because it is not obvious whether a certain value is erroneous or not.

Since electric data distribution from the hospitals/clinics or medical institutions to end users is not established, one of the most practical solutions is to migrate PHR data printed on a paper into electric data by OCR. Two kinds of devices are considered to capture printed data; one is an image scanner and another is a digital camera. We think the image scanner is rather special and less versatile device, and the device such as a camera module in tablet PC or smart phone is more general. Moreover, general users are more familiar with using a camera than an image scanner. Since most tablet PCs and smart phones equip camera modules for taking photos, we chose a tablet PC as a device for capturing forms of annual health checkup. Another advantage of assuming a tablet PC as the device for PHR front-end, we can provide a user with intuitive user interface with direct manipulation on LCD display with touch panel.

As far as we experimented, the accuracy of character recognition by OCR processing is not always perfect, in spite of the maturity of character recognition technique. Therefore, when we design a PHR data migration facility by OCR, what we need to take into consideration is whether the accuracy of OCR is acceptable or not, in the sense that the amount of errors that need to be corrected manually by the user. One reason of imperfect accuracy in OCR is in the insufficient resolution of source image or bleeding and/or blur of character printed on a paper. Another is the case where an image is captured with blur, especially in a case where an image is captured by holding the camera device by hands. In addition to the above mentioned issues, there is also an issue where a character is misclassified into the different character whose visual appearance is quite similar to the character to be recognized. One example is the case where a single "l" (the lower case of the alphabet "L") is misrecognized as a single "1" (the letter which corresponds to the number "one"). This problem is inevitable if the OCR engine does not hold a dictionary of the letters specific to the printed fonts.

Moreover, the fundamental problem in applying general purpose OCR engine is difficulty in proper recognition of structured data such as a table. If a table which includes items with null values is processed, output of recognition will be a sequence of characters in horizontal order, which may loose the correspondence between item name and value and/or the logical structure of the table.

3 PHR Data Management Front-End

3.1 System Organization

Figure 3 illustrates the system organization of the PHR system which we propose in this paper. The PHR system consists of a PHR front-end and PHR manager. The PHR front-end is implemented as application software on a tablet PC. Form capturing module assists a user to capture printed form by showing guide frames. Captured image is then transferred to OCR processor so as to carry out OCR processing. Each item name is associated with the value which is indicated in the same row, and 'null' case is also detected when no character is found. At the final

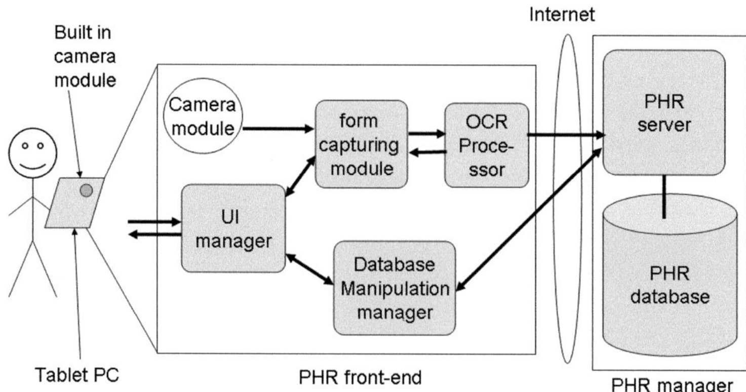

Fig. 3 The organization of the PHR front-end

process of data migration, PHR data received from the PHR front-end is stored into PHR database by PHR server. When a user accesses his/her own PHR data, it is transferred to the PHR front-end via PHR server. It is then shown on the display of PHR front-end. In the succeeding sections, we will describe the detail of form capturing assistance and the method for improving OCR accuracy which we call adaptive dictionary configuration.

3.2 Form Capture Assistance in Data Migration by Photography

Data migration by photography is the idea of migrating printed information into a system as electric data by OCR process, where the input image is obtained by photography. If we allow a user much freedom in shooting a form, many kinds of pre-processing such as geometric normalization are mandatory. In addition to this, there appear cases where image captured by photography is not usable because of low resolution or blur that is not negligible for realistic OCR processing. Therefore, proper shooting assistance is mandatory which guides a user to capture images properly. The detail of form capture module and OCR processor is shown in Fig. 4. Prior to the data migration, the user is prompted to choose the institution which conducted his/her annual checkup, since the form layout is not standardized throughout institutes. (Note that the items to check are defined by government.) After that, guide frames are superimposed on the image captured by the build-in camera module, and is shown on the display. The image is updated frame by frame, and all the user has to do is simply to adjust the position of the tablet, that he/she holds in hands, so that the borders of each table component overlap with guide

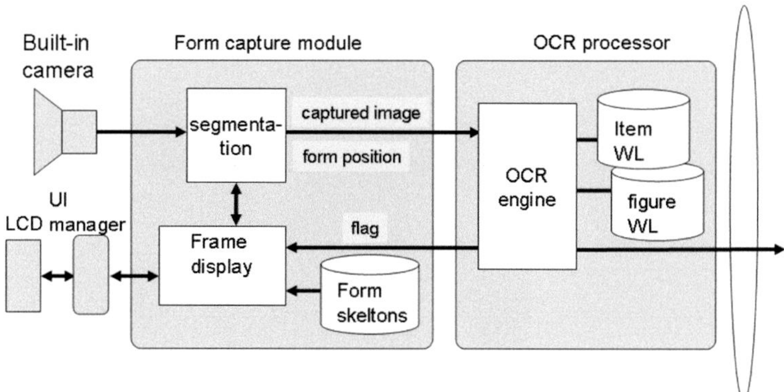

Fig. 4 The organization of capture module and OCR processor

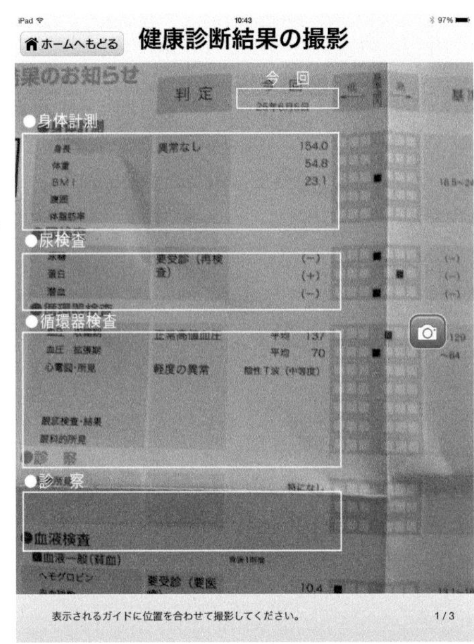

Fig. 5 Displaying guide frames for capturing printed report

frames. Finally, he/she touches a shutter icon to capture the tables. An example of displaying guide frames is shown in Fig. 5. There are 5 guide frames displayed as white rectangles, which correspond to the rectangular areas of "checkup date", "anthropometric", "urine test", "circulatory organ test", and "inquiry", respectively. A snapshot in the process of data migration by photography is shown in Fig. 6. Figure 6 is an example when a user adjusts the position of his tablet so as to migrate printed information of"anthropometric", "urine test", and "circulatory organ test".

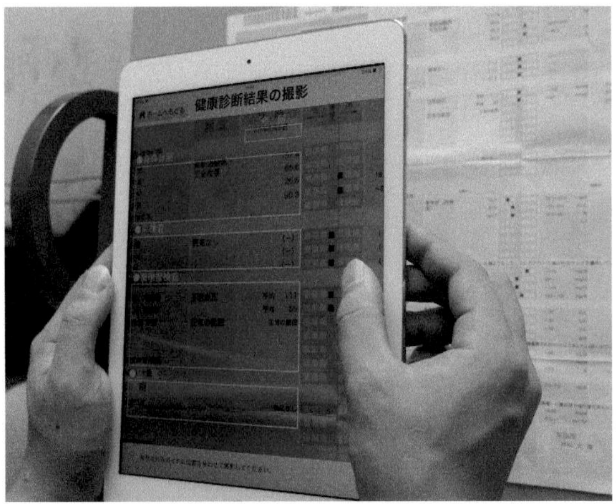

Fig. 6 Data migration by photography

3.3 OCR Processing with Adaptive Dictionary Configuration

Our preliminary experiment showed that OCR processing for images captured by handheld tablet with improper image results more recognition errors with standard dictionary, compared with OCR process with image scanner. Figure 7 shows an example of OCR processing where an image is captured by a handheld tablet. The entire image is 1223 × 1631 pixels, and we clipped two columns as shown in red and green rectangle. The result of OCR process with standard Japanese dictionary is shown in the rectangles on the right. Red characters in the figure correspond to recognition error. Capturing condition in the sense of image quality such as resolution or blur largely affects the accuracy of character recognition.

In order to improve the accuracy of OCR processing, adaptive dictionary configuration is carried out in OCR processor as shown in Fig. 4. In case of recognizing characters in an annual health checkup report, we can expect a word or figure within a limited number of instances. For example, the possible words appeared in a report of annual health check-up only include item names such as ‘height’, ‘weight’, or ‘BMI’, but no other words. The number of possible word is quite limited compared with the case of general documents.

Adaptive dictionary configuration is based on this observation. Since the number of kanji characters (i.e., Chinese letters used in Japan) designated for everyday use is more than 2000, excluding Chinese letters that are not used in annual health check-up report from OCR dictionary contributes to improve the accuracy. Figure 8 illustrates how adaptive dictionary configuration is performed in OCR process. Candidate set of character, i.e., Chinese letter, Japanese character, and/or figures is associated with each sub-region in the guide frames as shown in Fig. 8. In this Figure, the left side part for the area of anthropometric is associated with OCR

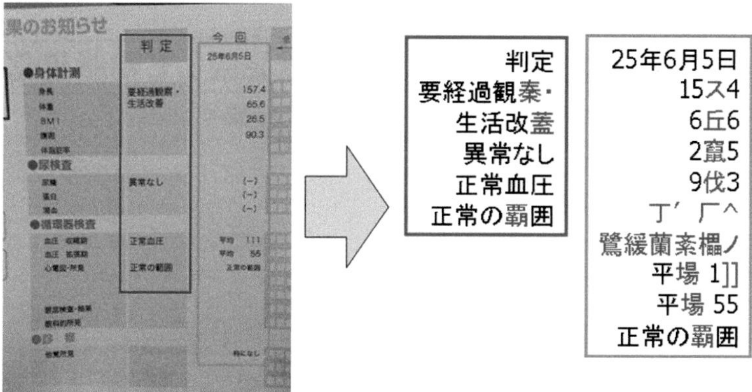

Fig. 7 An example of OCR processing

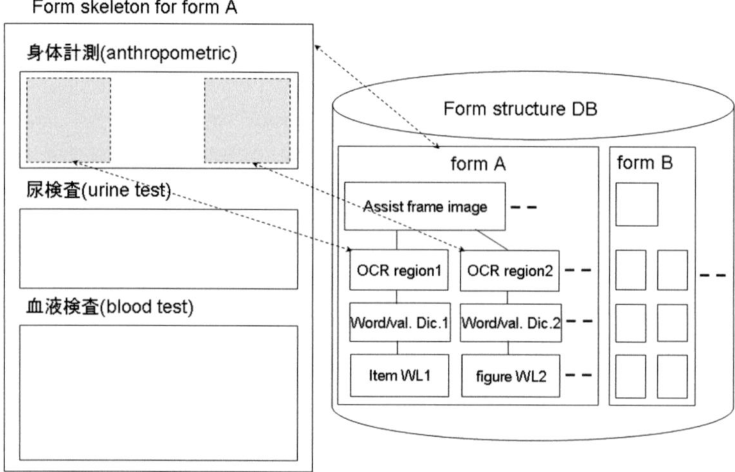

Fig. 8 Adaptive dictionary configuration for OCR processing

region 1, word/value dictionary1, and item WL1 (white list 1) which defines possible words/range of characters and values, respectively.

After an image is captured, it is segmented into sub-regions by following the guide frames, i.e., the spatial structure of the table. Then, each sub-image is sent to OCR processor with the meta-data as explained above. After that, the OCR engine switches the dictionary in accordance with the sub-region so as to converge to one of the characters that is supposed to appear in the region.

3.4 Data Confirmation and Correction by Display Synchronization

The adaptive dictionary configuration described in the previous section contributes to improve the OCR accuracy, however, it does not always make a user free from confirmation and correction of error. In other words, confirmation and correction process is mandatory step before registering the PHR data into the PHR database.

Since the current implementation of Lico offers a simple, Web-based migration interface, the confirmation process impose a user a large amount of eye movement between a field displayed on the computer screen and that corresponds to it on the printed form. Keeping correspondence between them is a stressful and error-prone task because of the long distance of eye movement.

Our PHR front-end takes this issue into consideration, and designed the user interface as shown in Fig. 9. There are two examples of aligning the captured image and the result of OCR processing; one is aligning them horizontally, and another is vertically. When a user drags either the captured image or the table that shows the

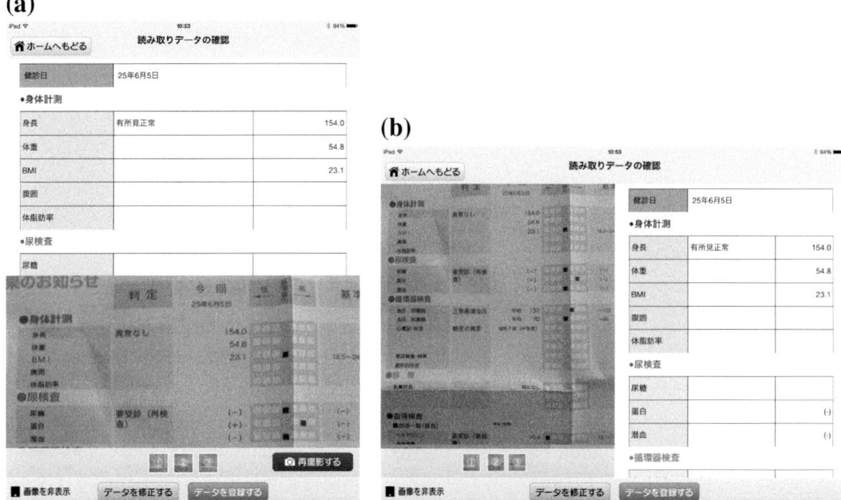

Fig. 9 Confirming migrated PHR data with display synchronization. **a** Vertical alignment. **b** Horizontal alignment

Table 1 Effect of adaptive dictionary configuration

	Standard config.	Adaptive config.
Item name	52 % (456/876)	79 % (692/876)
Figures	82 % (570/695)	91 % (630/695)

result of recognition, both of them is scrolled synchronously on the display so as to minimize burden on keeping recognizing the correspondence between an item in the image and that in the resultant table.

4 Experimental Evaluation

4.1 Evaluation OCR Accuracy Improvement

The result of an experiment to evaluate the effectiveness of adaptive dictionary configuration is shown in Table 1. We used Tesseract-OCR [14] as OCR engine, since it performs the most precisely among OCR engines available. The standard dictionary configuration for recognizing Japanese (including Hiragana, Katakana, Kanji) and figure is carried out by specifying an option as '-l jpn'. Another standard option of '-l eng' is preferable in case for recognizing figures, simply because there are less candidates of characters to be recognized. We regard the results obtained by these configurations as baseline. In adaptive dictionary configuration, we configured dictionary so that the OCR engine refers to only the characters that are expected to be appeared in item name in case of item name recognition, and configured it so that it refers to only the figure (i.e., number from 0 to 9) and decimal in case of recognizing the region of numeric values. The result of experiment is shown in Table 1. Fractions beside the percentage denote the number of correctly recognized characters out of characters being tested.

4.2 User Evaluation

We conducted usability test for the visitors of the technology exposition, which is called "e-messe 2014" held from May 15 to 17 at Kanazawa city in Ishikawa prefecture. We asked visitors who came to our booth to use our prototype system as well as current Lico system. We first explained each visitor purpose, issues and our solution (i.e., offering easier interface for data migration) for PHR system. After that, we explained how to manipulate our system as well as current (i.e., form based) Lico system. The total number of subjects is 68, which consists of 51 male subjects and 14 female subjects. To avoid order effect, we let 33 subjects to use proposed system first, and let 31 subjects to use current Lico system first. We collected questionnaires from all these visitors, where the profile of the subjects with regard to age and proficiency or familiarity of IT devices is summarized in Tables 2 and 3, respectively.

First inquiry of comparison was "Which platform was easier to use?" As the result of calculating summary math, about 78% of subjects evaluated that the proposed system is more useful, and 4.7 % of subjects evaluated the current implementation of Lico system is easier to use as shown in Table 4. Note that indication of "current Lico" and "proposed" were erratum in [19] and Table 4 in this paper shows the correct result of the user feedback.

Table 2 Age distribution of subjects

~10	20–29	30–39	40–49	50–59	60–69	70~	Total
4	19	14	14	4	12	1	68
5.9 %	27.9 %	20.6 %	20.6 %	5.9 %	17.6 %	1.5 %	

Table 3 Proficiency/familiarity to computers and communication device

Device	N/A	Up to once/week	2–3 times/week	Everyday	Total
Desktop PC	17	6	3	40	66
Note/laptop PC	5	6	13	44	68
Tablet PC	43	5	4	15	67
Mobile phone	30	0	4	34	68
Smart phone	20	1	1	46	68

Table 4 Comparison of usability between two platforms

Current Lico	Proposed (DMP)	No difference	Undecided	Total
3	50	3	8	64
4.7 %	78.1 %	4.7 %	12.5 %	100 %

Table 5 Evaluation on acceptability for daily use

Current Lico	Proposed (DMP)	None	Undecided	Total
3	53	6	5	67
4.5 %	79 %	9.0 %	7.5 %	100 %

Table 6 Acceptance ratio of subject with regard to OCR accuracy

Acceptable	Unacceptable	Undecided	Total
44	6	16	66
66.7 %	9.1 %	24.2 %	100 %

Second inquiry was "Which platform do you like to use for your dairy health check?" Feedback from the subjects is shown in Table 5. Comparing Table 4 with Table 5, we can see that the usability of a platform relates to the preference for daily use of a PHR system. However, 63 out of 68 (93 %) subjects rated that data migration by photography is more convenient than manual input into fields as the result of additional inquiry. (Note that 2 (3 %) of the subjects rated the current, form based data migration is easier to use.) This result shows that the usability of the platform with data migration by photography was rated less than the usability of the method of data migration by photography.

To explore the reason of the above-mentioned difference, we evaluated to see whether OCR accuracy is acceptable for the subjects. Concerning the evaluation by acceptance of the accuracy of character recognition which is presented in Sect. 4.1, about 67 % subjects evaluated the OCR accuracy is acceptable, and only 9 % of subject evaluated error correction is not negligible, as shown in Table 6. (Note that two subjects did not answer this inquiry.) However, none of the 6 subjects in Table 5 rated that the OCR accuracy is not acceptable, and 4 of 6 subjects in Table 6 rated that the proposed platform is preferable to use in spite of unacceptable OCR accuracy.

5 Conclusion and Future Work

In this paper, we described a PHR system for end users to maintain their PHR data by their own effort. Proposed front-end system assumes to use tablet PC with camera module. Data migration by photography, which enables a user to migrate printed data by taking photos, is employed. It enables a user to migrate printed data more easily, since guide frames are displayed so that the user can understand which part of the form to take intuitively. Adaptive dictionary configuration improves OCR accuracy even for images captured by handheld camera device. Data confirmation and correction by synchronized display offers less stressful confirmation task before registering the PHR data into the database. According to the result of evaluation of proposed system, though it is still a prototype system, about 78 % of subject rated that the proposed system is easier to use, and 93 % of subjects evaluated that the data migration by photography is more useful than form based manual data migration.

Acknowledgments The authors would like to express appreciation to Ms. Y. Nakada at the Industrial Collaboration Promotion Center, JAIST, for their management of this project. We would also like to acknowledge Mr. Y. Nishikawa, Mr. S. Urabe, and Mr. H. Yamashina for their effort of developing the system. This work is supported by the Strategic Information and Communications R&D Promotion Programme (SCOPE) supervised by the Ministry of Internal Affairs and Communications, Japan.

References

1. Andy, Y.-Y.L., et al.: Continuous, personalized healthcare integrated platform. In: Proceedings of IEEE Region 10 Conference on TENCON, pp. 1–6 (2012)
2. Bang, D.W., Jeong, J.S., Lee, J.H.: An implementation of privacy security for PHR framework supporting u-healthcare service. In: Proceedings of 6th International Conference on Networked Computing, pp. 1–4, (2010)
3. Bender, D., Sartipi, K.: HL7 FHIR: An Agile and RESTful approach to healthcare information exchange. In: Proceedings of IEEE 26th International Symposium on Computer-Based Medical Systems, pp. 326–331 (2013)
4. de Ridder, M., et al.: A web-based medical multimedia visualisation interface for personal health records. In: Proceedings of IEEE 26th International Symposium on Computer-Based Medical Systems, pp. 191–196 (2013)
5. Dolin, R.H., et al.: The HL7 clinical document architecture. J. Am. Med. Inform. Assoc. **8**, 552–569 (2001)
6. Gorp, P.V., Comuzzi, M., Fialho, A., Kaymak, U.: Addressing health information privacy with a novel cloud-based PHR system architecture. In: Proceedings of IEEE International Conference on Systems, Man, and Cybernetics, pp. 1841–1846 (2012)
7. Kahn, J.S., Aulakh, V., Bosworth, A.: What it takes: characteristics of the ideal personal health record. Health Aff. **28**(2), 369–376 (2009)
8. Lico PHR Service.: https://www.l-cod.com/
9. Nzanywayingoma, F., Huang, Q.: securable personal health records using ciphertext policy attribute based encryption. In: Proceedings of IEEE 14th International Conference on e-Health Networking, Applications and Services, pp. 502–505 (2012)
10. Perumal, B., Rajasekaran, M.P., Duraiyarasan, S.: An efficient hierarchical attribute set based encryption scheme with revocation for outsourcing personal health records in cloud computing. In: Proceedings of International Conference on Advanced Computing and Communication Systems, pp. 1–5 (2013)
11. Reti, S.R., et al.: Improving personal health records for patient-centered care. J. Am. Med. Inform. Assoc. **17**, 192–195 (2010)
12. Sittig, D.F.: Personal health records on the internet: a snapshot of the pioneers at the end of the 20th century. Int. J. Med. Inf. **65**(1), 1–6 (2002)

13. Sutanto, J.H., Seldon, H.L.: Translation between HL7 v2.5 and CCR message formats (For communication between hospital and personal health record systems). In: Proceedings of IEEE Conference on Open Systems, pp. 406–410 (2011)
14. Tesseract-ocr.: https://code.google.com/p/tesseract-ocr/
15. Thummavet, P., Vasupongayya, S.: A novel personal health record system for handling emergency situations. In: Proceedings of International Computer Science and Engineering Conference (ICSEC), pp. 266–271 (2013)
16. Tsiknakis, M., Katehakisa, D., Orphanoudakis, S.C.: A health information infrastructure enabling secure access to the life-long multimedia electronic health record. In: Proceedings of 18th International Congress and Exhibition on Computer Assisted Radiology and Surgery, vol. 1268, pp. 289–294 (2004)
17. Tulu, B., et al.: Personal health records: identifying utilization patterns from system use logs and patient interview. In: Proceedings of 45th Hawaii International Conference on System Science, pp. 2716–2725 (2012)
18. Xie, L., Yu, C., Liu L., Yao, Z.: XML-based personal health record system. In: Proceedings of 3rd International Conference on Biomedical Engineering and Informatics, pp. 2536–2540 (2010)
19. Yoshitaka, A. Chujyou, S., Kato, H.: Front end system for personal health record with data migration facility from printed information. In: Proceedings of 9th international conference on knowledge, information and creativity support systems, pp. 1–12 (2014)
20. Yuksel, M., Dogac, A.: Interoperability of medical device information and the clinical applications: an HL7 RMIM based on the ISO/IEEE 11073 DIM. IEEE Trans. Inf Technol. Biomed. **15**(4), 557–566 (2011)

Investigating Unit Weighting and Unit Selection Factors in Thai Multi-document Summarization

Nongnuch Ketui and Thanaruk Theeramunkong

Abstract Breaking down documents into small units, unit weighting and unit selection are two important factors in summarization of multiple related documents. This paper presents an investigation on performance of several variants of unit weighting and selection schemes on Thai multi-document summarization. Fifty sets of Thai news articles with their reference summaries are used to evaluate the performance of various weighting and selection methods. Compared to PageRank and Maximal Marginal Relevance (MMR) with four ROUGE measures, the results show that iterative weighting gets higher performance of traditional TF-IDF, the iterative node weighting, query relevance, centroid-based selection, and unit redundancy consideration can help improving summary quality.

Keywords Unit weighting · Unit selection · Summarization · Iterative node weighting · Query relevant · Centroid-based selection · Redundant removal

1 Introduction

Multi-document summarization (MDS) is one of the most challenging tasks in natural language processing community. Unlike single-document summarization, redundancy removal and reorganization are needed for summarizing multiple documents. In the past, some works defined an MDS process in terms of clustering problem. In such approach, the centroid-based techniques that generate a composite sentence from each cluster were proposed by [1, 10, 13]. They used features to

N. Ketui (✉)
Faculty of Science and Agricultural Technology,
Rajamangala University of Technology Lanna, Nan, Thailand
e-mail: nongnuchketui@rmutl.ac.th; nongnuchketui@gmail.com

T. Theeramunkong
School of Information, Computer and Communication Technology,
Sirindhorn International Institute of Technology, Thammasat University,
Bangkok, Thailand
e-mail: thanaruk@siit.tu.ac.th

S. Kunifuji et al. (eds.), *Knowledge, Information and Creativity Support Systems*,
Advances in Intelligent Systems and Computing 416,
DOI 10.1007/978-3-319-27478-2_14

modify sentence in each cluster while feature vectors were represented by a set of single words weighted by a kind of weighting systems such as TF-IDF. Carbonell and Goldstein [2] combined query relevance with information novelty as the topic and made a major contribution as topic-driven summarization, called Maximal Marginal Relevance (MMR) measure. Mani [9] presented an information extraction framework for summarization as well as a graph-based method to find similarities and dissimilarities in pairs of documents. Moreover, Mihalcea [11] introduced a new formula for a traditional graph-based ranking, called TexRank, that takes into account edge weights when computing the score associated with a vertex in the graph. The sentence scoring function was known as the PageRank algorithm [12].

In this work, we introduce three phrases of multi-document summarization. A Thai running text is segmented into Thai elementary discourse units (TEDUs) [5] as units in the unit segmentation phase. In the unit graph formulation phase, segmented units are assembled to form a graph where each node is assigned a weight, typically in the form of TF-IDF and the links among them (edges) are also associated with a weight, calculated by a form of similarity or associative measurement. In the unit selection phase, a so-call inclusion-based method is used to include the most important unit one by one with the consideration of PageRank factors, node weight, query relevance, centroid-based selection and redundancy removal. After selecting important non-redundant units, a summary is generated with factors of temporal factors and ordinal relations in original documents. To evaluate our method, a number of experiments are conducted using fifty sets of Thai news articles, the model summaries of which are given. Four measures of ROUGE-1, ROUGE-2, ROUGE-S, and ROUGE-SU4 are used as performance metrics.

In the rest, Sect. 2 describes the process of multi-document summarization. Unit weighting and unit selection factors are defined in Sect. 3. Experimental setup is shown in Sect. 4. Experimental results are discussed in Sect. 5. Error analysis is illustrated in Sect. 6. Finally, conclusion and future work are given in Sect. 7.

2 Multi-document Summarization

This section presents our multi-document summarization approach which is composed of three processes: (1) unit segmentation; (2) unit graph formulation; and (3) unit selection as shown in Fig. 1. In the unit segmentation, a Thai running text is segmented into a sequence of tractable units and tagged with part-of-speech (POS) and named entities (NEs). TEDUs are proposed for segmenting a Thai running text into units. In the unit graph formulation, a graph of units is constructed by conceptualizing a unit as a weighted node in a graph and a relationship of two nodes is formulated as a weighted link between the nodes. The weight of a node or a link is determined by considering its importance, that is, its contribution in the graph. In the unit selection, a number of important nodes and links are selected by considering importance level of nodes or links, together with redundancy among units (nodes), and focusing on the node weight recalculation.

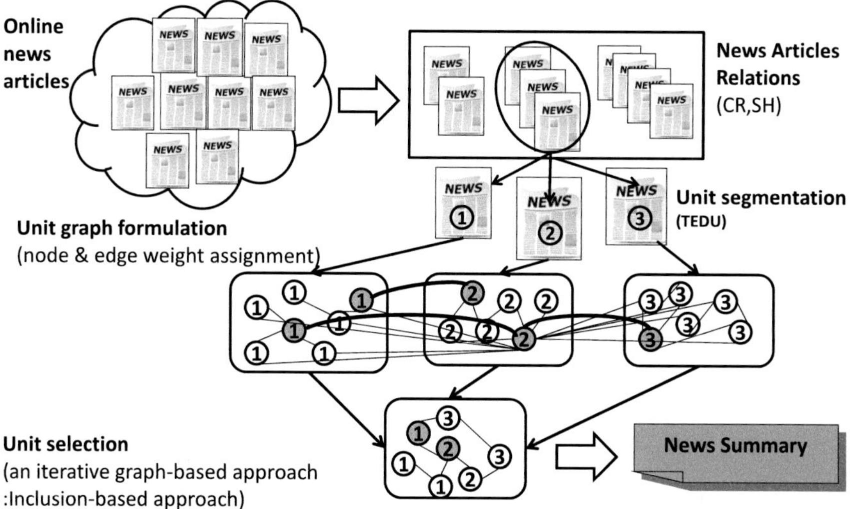

Fig. 1 The multi-document summarization framework

2.1 Unit Segmentation

Since Thai language has no sentence boundary (or even word boundary), it is difficult to define a unit for summarization. The most straightforward approach for segmenting blocks of Thai documents are paragraphs, as done [3]. However, using paragraphs as units for Thai document summarization is inflexible since a paragraph may contain heterogeneous contents and such paragraph-based approach does not allow us to exclude some unimportant parts in a paragraph. In this work, one alternative for unit segmentation in Thai language is to apply Thai Elementary Discourse Units (TEDUs) and COMmon Phrases (COMPs) [5]. A Thai running text is segmented into TEDUs+COMPs by using context free grammar rules with a bottom-up chart parser and the longest matching technique [8]. TEDUs in a text are detected by together with their structures.

2.2 Unit Graph Formulation

After a running text is split into tractable unit as TEDUs+COMPs, we need a model to determine importance of units and select the most suitable ones for a summary. Towards this, a graph model is proposed to express units and their relations extracted from multiple documents targeted for summarization. In our approach, a node in a graph corresponds to a unit while a link in a graph expresses connections between two units. Two subprocesses are implemented: (1) Node Weight Calculation and (2)

Link Weight Calculation. For the node weight calculation process, we assign the node weight either TF-IDF (Term Frequency times Inverse Document Frequency) or the iterative weighting [4]. For the link weight calculation, it is worth investigating relations between two units (nodes). A link weight (relation strength) between two nodes (units) describes how much two units are identical or related. Although there have been several possibilities on definition of relations between units, this work simply uses a common method, called cosine similarity. The cosine similarity ranges from 0 to 1 and the highest value indicates high similarity. It implies that two units are duplicates or highly related.

2.3 Unit Selection

Given the unit graph derived from the process described in the previous subsection, unit selection is a task to select a set of suitable units for constructing a summary. In the past, several works [2] applied a straightforward method to select units based on their weights. In this work, we utilize a variant of the inclusion-based summarization method proposed in [4]. In the algorithm, the number of units to be selected is set. The most important nodes are repeatedly added one-by-one into the summary band and deleted from a set of unselected units until the number of selected units reaches the predefined compression rate. After the node addition, the weight of each unselected unit is recalculated. When the number of selected units satisfies the predefined compression rate, the algorithm returns the graph of summary. In the detail, the variant of the inclusion-based summarization factors are discussed in next section.

3 Unit Weighting and Unit Selection Factors

In this section, we classify the variant factors into two groups: (1) unit weighting factors; and (2) unit selection factors as shown in the next section.

3.1 Unit Weighting Factors

After segmenting a document into units to form a set of nodes in a graph, each node is assigned a weight for expressing its importance. In our work, we investigate two weighting alternatives as follows. However, the links between nodes are given a weight by similarity. In this work, it is cosine similarity.

TF-IDF: As the most naive method, it is possible to apply TF-IDF (Term Frequency times Inverse Document Frequency) to weight words in a document and weight a unit by calculating the summation of weights of all words in that unit.

Iterative Weighting: As a more sophisticated weighting method, we have proposed a so-called iterative weighting [4] to obtain more accurate weights of units by considering importance of documents, units, and words and reflect them when we weigh units.

3.2 Unit Selection Factors

After constructing a graph of units obtained from documents to be summarized, units are selected one by one to form a summary graph. In the work, the selection criterias are five types as follows.

PageRank consideration: A traditional graph-based ranking model applies the sentence scoring function for finding the relevance information [12] as shown in Eq. 1. Let $PR(s_i)$ is the ranking score of unit. s_j represents the unit which relates to the considered unit s_i while s_k is the neighbour of s_j. Here, the constant d equals 0.001.

$$PR(s_i) = (1 - d) + \left(d \times \sum_{s_j \in Neighbour(s_i)} \frac{sim(v_{s_j}, v_{s_i})}{\sum_{s_k \in Neighbour(s_j)} sim(v_{s_j}, v_{s_k})} \right) \tag{1}$$

Node weighting: A node weight is assigned by either TF-IDF or the iterative weighting. It is possible to the high weight has a important information.

Query relevance: A unit combines query relevance with information as the title of a set of related news articles. The overlapped content between the unit and query may improve the text summarization since the query guides to find the main content.

Centroid-based selection: A preferable unselected unit is a node with high similarity to all other unselected units. As well as, the unit close to the centroid of the unselected units should be included in the summary. In contrast of the centroid-based selection, the unit far from the centroid of the unselected units should be considered to include in the summary.

Redundancy removal: Two units with an identical content or a highly similar content should not be selected simultaneously. In order words, it is reasonable to eliminate content redundancy in order to have a good short summary.

Reflecting the first concepts on weighting units, the unit selection can be formulated as follows. Here, the original weighting of a unit is modified to reflect the linkage page factor, query relevance factor, centroid-related factor, and redundancy-related factor as shown in Eq. (2). The best unit ($\hat{u}$) can be selected by maximizing the value in the equation, where the five terms indicate the PageRank consideration, its node weight, the query relevance, the centroid-based selection and the redundancy removal factors.

$$\hat{u} = \left(\begin{array}{l} \lambda \times \left(PR(s_i)\right)^{\beta} \times \left(W(s_i)\right)^{\phi} \times \left((\varepsilon + sim(v_q, v_{s_i}))\right)^{\varphi} \\ \times \left[\left(1 - \frac{\sum_{k \neq 1, v \neq v_{s_i}}^{|V|} sim(v, v_{s_i})}{|V|-1} \right)^{\delta} \times \left(\frac{\sum_{k \neq 1, v \neq v_{s_i}}^{|V|} sim(v, v_{s_i})}{|V|-1} \right)^{1-\delta} \right]^{\alpha} \end{array} \right) - \left((1-\lambda) \max_{s_j \in R} (sim(v_{s_j}, v_{s_i})) \right)^{\omega} \quad (2)$$

In the past, TF-IDF was usually used to express such importance levels of words by using term (word) frequency and inverse document frequency. As an alternative mentioned in [4], iterative weighting may be used. Such weight helps selecting suitable units for summarization. In Eq. 2, the constants (λ and ε) are 0.003 since the current unit weight may equal to 1. For multi-criteria terms, the first term expresses the original weight $W(s_i)$ which is assigned by either TF-IDF or iterative weighting ($\beta = 1$). The second term displays the PageRank consideration $PR(s_i)$ ($\phi = 1$) while the query relevance indicates in the third term ($\varphi = 1$). Let v_q is the word vector of query and v_{s_i} is the vector of considered unit. The fourth term explores the centroid/discentroid/non-consideration of the unselected units. If δ equals 0 that we focus on the centroid-based selection and do not consider the discentroid-based selection. Let $|V|$ is the total number of unselected units. While α is 0, both the centroid and discentroid-based selection will not work. The fifth term represents the level of content redundancy ($\omega = 1$). In the extreme case, if the unit has similar content with any in the set of selected units, the maximum is one and the term will become 0.

4 Experimental Setup

4.1 Datasets

This work utilizes the THAI-NEST corpus developed in [14] which comprises 10,000 news articles in seven categories: crimes (CR); sports (SP); foreign affairs (FO); politics (PO); entertainment (EN); economics (EC); and education (ED), gathered from seventeen on-line news sources Later, a method for discovering document relations in [6] is applied to find relation between news documents and to group highly related news documents into a data set for summarization. While most previous works focused on finding document relations judged to be either relevant or non-relevant, this work classified documents into three main types of relations; (1) 'completely related' (CR), and (2) 'somehow related' (SH), and (3) 'unrelated' (UR). In this work, we randomly selected 50 sets of related documents with CR and SH relations for testing our proposed graph-based summarization approach. Given each set of related documents, the documents were tagged with POSs/NEs by Thai E-Class [15].

Later a Thai running text with POS and NEs tagging is segmented into TEDUs, COMPs, and TEDU-LPs by using 446 Context-Free Grammar rules (CFG rules) with a bottom-up chart parser [5] with the longest matching technique [8] to detect TEDUs in a text, together with their structures. Here, the CFG rules are built based on the syntactic categories defined by a Thai E-Class [15]. We applied three groups of CFG rules: 342 rules for TEDUs; 95 rules for COMPs; and 9 rules for TEDU-LPs. To form a reference summary (i.e., model summary) for evaluation, we have asked a number of Thai language experts in the Faculty of Liberal Art, Thammasat University to manually summarize the prepared set of news articles as gold standard for evaluating system results. The summarizers were instructed to construct an abstractive-based summary with the size of 20–100 words for each of the fifty datasets. The summaries contain main contents in the original documents. Some discourse markers are added to connect clauses. These reference summaries are used for evaluating a summary obtained from the system.

4.2 Experimental Settings

To examine effect of weighing factor and unit selection factors on multi-document summarization, we have conducted two experiments using 50 sets of related news documents, containing 23,781 words. The first experiment targets to compare performance of top 20 ranked combination methods of the unit weighting factors and the five unit selection factors. Moreover, the performances of twenty combinations are summarized and ranked to clarify which factor combination is optimal. The type of unit is based on TEDUs+COMPs. The two unit weighting factors we consider are (1) TF-IDF ('TF'), or (2) the iterative weighting ('IT') for assigning to the node weights. The five unit selection factors we explore are (1) the PageRank consideration ('PR'), (2) the node weighting ('NW'), (3) the query relevance ('QR'), (3) the centroid-based selection ('CS'), and (4) the redundancy removal ('RR'). For the compression rate, we examine ten rates of 0.1 to 1.0. Here, the compression rate is defined as the ratio of the number of units in a summary to the number of units in the original documents. Moreover, we compare our summarization methods with a basic extractive summary, called the first n character which the summary is composed of the ranked units by position in the original document, a traditional graph-based ranking model, called the PageRank algorithm [12] and a query-based model, called the Maximal Marginal Relevance (MMR) measure [2]. The second experiment aims to investigate summarization performance according to TEDUs+COMPs unit type, weighing factor, and five unit selection factors, and ten compression rates.

4.3 Evaluation Method

To evaluate a summary output from a system, we use the reference summaries as described in Sect. 4.1. A reference summary is an abstractive-based summary (with consideration of semantics; i.e., what, where, who, whom, why, and how) constructed by requesting Thai language experts to manually summarize a set of related news articles into 20–100 words. In this work, we utilize a standard metric, called ROUGE [7], to evaluate a system's summarization result by comparing it with its reference summary. Among various types of ROUGE, this work uses ROUGE-1 (unigram-based co-occurrence statistics), ROUGE-2 (bigram-based co-occurrence statistics), ROUGE-S (skip-bigram-based co-occurrence statistics), and ROUGE-SU4 (skip-bigram plus unigram-based co-occurrence statistics). Originally developed by NIST, the ROUGE we used is a new variant of ROUGEs that considers precision (R-1_P, R-2_P, R-S_P, R-$SU4_P$), recall (R-1_R, R-2_R, R-S_R, R-$SU4_R$), and F-score (R-1_F, R-2_F, R-S_F, R-$SU4_F$).

5 Experimental Results

5.1 *Effect of Unit Weighting and Unit Selection Factors*

Table 1 shows the effect of unit weighting factors and unit selection factors to the summarization methods. R-1_F, R-2_F, R-S_F, R-$SU4_F$, and the average of four $ROUGE_F$ values of top-20 summarization methods are illustrated. Our methods are compared with the first n character, PageRank algorithm, and MMR method. For the unit weighting factors, the performance of iterative weighting ('IT') (the top-9 summarization methods) is higher than the standard TF-IDF weighting ('IF'). Since 'IT' weighting considers the important documents, units, and words in a set of related document, TF-IDF only focuses on the word occurrence in documents. For the unit selection factors, the summarization does not need the PageRank factor ('PR') since the extractive summarization considers the importance information (not focus on the linkage unit). While the top-10 summarization methods include the node weight factor ('NW'), they achieve the high R-1_F, R-2_F, R-S_F, and R-$SU4_F$ values. For the query relevance factor ('QR'), it can be improve the performance of summarization methods (the MDS1-MDS7 except MDS2) because the query as title of a set of news article represents the main content. To consider the centroid-based selection factor ('CS'), even though the MDS1 method having this factor achieves the highest average of four ROUGEs performance, the discentroid selection and non-consideration have occurred in the top-10 summarization methods. In the contrast of the previous factors, the redundancy removal ('RR') absolutely improves the performance of summarization methods since the elimination of duplicated information should be considered in multi-document summarization.

To conclude the effect of unit selection factors, four factors (the node weight, query relevance, centroid-based selection, and redundancy removal) can affect to get the performance of summarization method highly. Comparing the performance of summarization methods and the baselines, the average of four ROUGEs values of the MDS1-MDS14 methods are higher than the first n character ('FNC') while PageRank ('PR') and MMR get the lowest performance.

Table 1 Classification accuracy of the 20 best unit weighting and unit selection factors

Methods	Unit weighting factors	Unit selection factors					R-1$_F$	R-2$_F$	R-S$_F$	R-SU4$_F$	AVG
		PR	NW	QR	CS	RR					
MDS1	IT	N	Y	Y	C	Y	0.2946	0.1473	0.1159	0.2137	0.1929
MDS2	IT	N	Y	N	C	Y	0.2924	0.1477	0.1155	0.2126	0.1921
MDS3	IT	N	Y	Y	N	Y	0.2910	0.1457	0.1146	0.2111	0.1906
MDS4	IT	Y	Y	Y	N	Y	0.2913	0.1449	0.1136	0.2105	0.1901
MDS5	IT	Y	Y	Y	D	Y	0.2912	0.1448	0.1135	0.2105	0.1900
MDS6	IT	N	Y	Y	D	Y	0.2903	0.1452	0.1137	0.2104	0.1899
MDS7	IT	Y	Y	Y	C	Y	0.2909	0.1449	0.1131	0.2105	0.1899
MDS8	IT	N	Y	N	D	Y	0.2891	0.1459	0.1129	0.2092	0.1893
MDS9	IT	N	Y	N	N	Y	0.2889	0.1456	0.1127	0.2088	0.1890
MDS10	TF	N	N	N	C	Y	0.2879	0.1433	0.1122	0.2103	0.1884
MDS11	IT	N	N	Y	C	Y	0.2882	0.1417	0.1124	0.2107	0.1882
MDS12	IT	Y	N	Y	N	Y	0.2874	0.1421	0.1128	0.2105	0.1882
MDS13	IT	Y	Y	N	C	Y	0.2880	0.1438	0.1120	0.2086	0.1881
MDS14	IT	Y	N	Y	D	Y	0.2863	0.1422	0.1123	0.2099	0.1877
MDS15	IT	Y	N	Y	C	Y	0.2868	0.1404	0.1119	0.2108	0.1875
MDS16	TF	N	N	Y	C	Y	0.2880	0.1410	0.1086	0.2081	0.1864
MDS17	TF	N	Y	N	D	Y	0.2800	0.1451	0.1133	0.2051	0.1859
MDS18	TF	N	Y	N	N	Y	0.2800	0.1451	0.1133	0.2050	0.1858
MDS19	IT	Y	Y	N	D	Y	0.2866	0.1409	0.1095	0.2057	0.1857
MDS20	IT	Y	Y	N	N	Y	0.2865	0.1408	0.1095	0.2056	0.1856
FNC	–	–	–	–	–	–	0.2802	0.1438	0.1150	0.2098	0.1872
PR	TF	Y	N	N	N	N	0.2450	0.1211	0.0968	0.1931	0.1640
MMR	TF	N	N	Y	N	Y	0.2767	0.1382	0.1098	0.2021	0.1817

5.2 *Performance of Multi-document Summarization Methods*

In this experiment, we investigate the performance of summarization methods, and ten compression rates (0.1-1.0). R-1_F, R-2_F, R-S_F, and R-SU4_F values of the best summarization method ('MDS1') and three baselines, that is, the first n character ('FNC'), the PageRank algorithm ('PR'), and the query-based approach ('MMR') are shown in Table 2. Here, the bold font shows the highest performance of ROUGE value. For R-1_F value, TEDUs+COMPs with MDS1 gets the highest R-1_F performance at the compression rate of 0.2 (0.3368). Both FNC and MMR are the best performance at the compression rate of 0.4 (0.3263 and 0.3261, respectively). PR is the lowest performance at the compression rate of 0.3 (0.2641). For R-2_F value, even though R-2_F value of MDS1 (0.1596) is lower than FNC and MMR (0.1632 and 0.1637, respectively), our method gains the high R-2_F performance at the lower compression rate (at the compression rate 0.4). Normally, the good summary is less than 50 % of original text. As well as R-2_F value, TEDUs+COMPs with FNC achieves the highest performance of R-S_F at the compression rate 0.5 (0.1307) while R-S_F value of MDS1 is higher than the baseline (PR) at the compression rate 0.4. For R-SU4_F value, TEDUs+COMPs with MDS1 is superior to other methods with R-SU4_F performance of 0.2314 (at the compression rate 0.4). Both MMR and FNC performance are similar (0.2295 and 0.2294, respectively) at the compression rate 0.4-0.5 while PR achieves the lowest performance of R-SU4_F at the compression rate 0.6 (0.2093).

We conclude that MDS1 obtains the best performance of R-1_F and R-SU4_F at the compression rate 0.2 and 0.4, respectively while R-2_F and R-S_F values of MDS1 are lower than MMR and FNC as shown in Fig. 2. Another baseline ('PR') gets the lowest performance of all ROUGEs values.

6 Error Analysis

In order to analyze errors in the proposed methods, we rank 50 datasets based on their performances (average F-score over all ROUGEs and compression rates). The worst-15 datasets are sorted and shown in Table 3. Here, the column 'No.' displays the index of each dataset, 'S' represents the size of the reference summary (Ref.) and original documents (Org.) in a kilobyte unit (KB), '#W' is the number of words in the dataset, '#D' is the number of documents in the dataset, '#P' is the number of paragraphs in the dataset, and 'RE' shows the type of dataset, where 'CR' means completely related and 'SH' means somehow related.

From the table, some observations can be drawn as follows. Firstly, datasets with a very low ratio of document size between the reference summary and the original documents (i.e., 37, 43, 48 and 50) seems to obtain low performance in summarization. This implies the difficulty in selecting suitable phrases/words for a summary due to a large number of candidates. Secondly, datasets with heterogeneous contents

Table 2 ROUGE performance of MDS methods and compression rates

ROUGE	Method	Compression rate									
		0.1	0.2	0.3	0.4	0.5	0.6	0.7	0.8	0.9	1
R-1	MDS1	0.3075	**0.3368**	0.3361	0.3351	0.3216	0.3042	0.2813	0.2608	0.2406	0.2218
	FNC	0.2532	0.3105	0.3206	0.3263	0.3189	0.2941	0.2711	0.2504	0.2351	0.2218
	PR	0.2366	0.2594	0.2641	0.2595	0.2553	0.2516	0.2439	0.2316	0.2259	0.2218
	MMR	0.2186	0.2820	0.2993	0.3261	0.3216	0.3094	0.2839	0.2615	0.2428	0.2218
R-2	MDS1	0.1316	0.1499	0.1550	0.1596	0.1584	0.1567	0.1514	0.1452	0.1371	0.1283
	FNC	0.1165	0.1500	0.1529	0.1623	0.1632	0.1528	0.1455	0.1360	0.1306	0.1283
	PR	0.0846	0.1019	0.1188	0.1234	0.1299	0.1343	0.1332	0.1279	0.1288	0.1283
	MMR	0.0987	0.1152	0.1227	0.1523	0.1603	**0.1637**	0.1551	0.1464	0.1388	0.1283
R-S	MDS1	0.1047	0.1164	0.1210	0.1259	0.1237	0.1228	0.1189	0.1138	0.1090	0.1030
	FNC	0.0929	0.1195	0.1250	0.1297	**0.1307**	0.1201	0.1153	0.1088	0.1045	0.1030
	PR	0.0684	0.0794	0.0961	0.0995	0.1026	0.1077	0.1069	0.1014	0.1025	0.1030
	MMR	0.0778	0.0908	0.0966	0.1222	0.1276	0.1292	0.1227	0.1167	0.1119	0.1030
R-SU4	MDS1	0.1929	0.2180	0.2237	**0.2314**	0.2265	0.2231	0.2150	0.2077	0.2009	0.1977
	FNC	0.1652	0.2120	0.2221	0.2294	0.2278	0.2211	0.2152	0.2069	0.2006	0.1977
	PR	0.1452	0.1753	0.1910	0.2024	0.2052	0.2093	0.2069	0.1989	0.1986	0.1977
	MMR	0.1348	0.1792	0.1945	0.2272	0.2295	0.2282	0.2181	0.2093	0.2027	0.1977

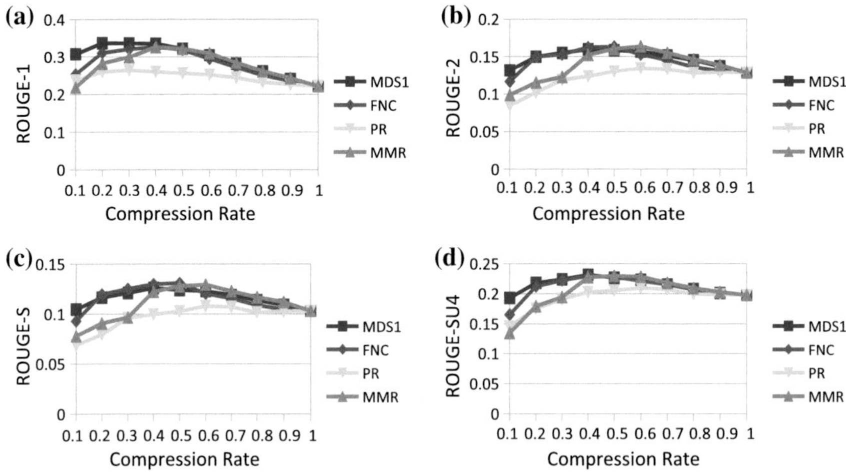

Fig. 2 ROUGE performance comparison between the MDS1 with baselines

(somehow related = 'SH' in Table 3) (e.g., 7, 9, 34, 37, 43, and 48) tends to get low performance. For instance, while the dataset number 43 consists of fifteen documents, only five documents are strongly related but the others are trivially related. As another evidence, the dataset number 7 and 34 each contains three documents but only two documents are strongly related and one document is weakly related. The task of summarizing heterogeneous-contented documents are difficult.

Thirdly, datasets that occupy documents with a small number of paragraphs (i.e., 6, 15, 18, 23, 28, 33, and 35), tend to get low performance for paragraph-based units (PARA). When unit selection is done based on paragraph, it is high possibility to include words/phrases that are not suitable to be in the summary since the paragraph-based method will select the whole paragraph as one unit. Fourthly, datasets with improper TEDU+COMP or CTEDU segmentation (i.e., 4, 9, 15, 33, and 35), may trigger lower performance in ROUGE-2 and ROUGE-SU4. Selection on mis-segmented unit is not efficient, resulting in low performance.

7 Conclusion and Future Work

This paper presented a multi-document summarization approach which is composed of three phases: unit segmentation; unit graph formulation; and unit selection. As unit definition, we utilized a method in [5] to segment a documents into a number of Thai Elementary Discourse Units and COMmon Phrase units (TEDUs+COMPs). With such units, a graph was formed with each node representing a unit and each edge expressing similarity among two units (nodes). This work investigated several unit weighting and unit selection factors in Thai Multi-document Summarization. Using

Table 3 Average ROUGE-based (R-1, R-2, R-S, and R-SU4) F-scores (F) (calculated from performances of all methods and compression rates)

No.	Ref.		Org.				% of	$R\text{-}1_F$	$R\text{-}2_F$	$R\text{-}S_F$	$R\text{-}SU4_F$	AVG
	S	#W	S	#W	#D	RE	Ref/Org					
43	0.92	43	31.0	1,628	15	SH	2.64	0.0445	0.0076	0.0044	0.0217	0.0195
34	0.84	41	6.0	299	3	SH	13.71	0.1404	0.0391	0.0187	0.0740	0.0681
50	1.00	54	25.5	1,385	10	CR	3.90	0.1112	0.0552	0.0308	0.0780	0.0688
37	1.28	77	21.3	1,133	7	SH	6.80	0.1631	0.0684	0.0370	0.1019	0.0926
7	1.06	44	9.3	424	3	SH	10.38	0.1925	0.0572	0.0361	0.1158	0.1004
33	0.97	45	5.2	226	2	CR	19.91	0.2062	0.0544	0.0336	0.1185	0.1032
18	1.10	50	5.3	271	2	CR	18.45	0.1986	0.0607	0.0374	0.1220	0.1047
4	1.04	46	5.6	288	2	SH	15.97	0.2142	0.0640	0.0316	0.1227	0.1081
28	0.68	24	5.0	223	2	SH	10.76	0.2074	0.0553	0.0471	0.1338	0.1109
48	1.81	80	21.1	1,046	5	SH	7.65	0.1689	0.0914	0.0688	0.1182	0.1118
9	1.07	52	8.6	412	3	SH	12.62	0.1996	0.0814	0.0529	0.1318	0.1164
15	1.30	65	6.4	276	2	SH	23.55	0.2373	0.0677	0.0357	0.1345	0.1188
29	0.86	35	5.7	236	2	CR	14.83	0.2164	0.0766	0.0504	0.1318	0.1188
23	0.66	35	4.0	248	2	CR	14.11	0.1579	0.0724	0.1241	0.1272	0.1204
35	1.04	43	8.1	352	3	CR	12.22	0.2580	0.0946	0.0452	0.1533	0.1378
5	1.08	57	6.7	345	2	CR	16.52	0.2448	0.1042	0.0562	0.1480	0.1383
8	1.92	86	10.7	491	2	CR	17.52	0.2755	0.0873	0.0618	0.1673	0.1480
14	1.68	88	11.3	543	2	CR	16.21	0.2555	0.1294	0.0805	0.1802	0.1614
11	0.85	54	4.7	275	2	CR	19.64	0.3086	0.0915	0.0477	0.1997	0.1619
13	2.67	134	17.1	880	2	SH	15.23	0.2670	0.1103	0.1113	0.2096	0.1746
6	0.86	46	4.2	190	2	CR	24.21	0.2704	0.1475	0.0891	0.1917	0.1747
39	1.09	55	6.2	339	2	CR	16.22	0.2791	0.1222	0.1138	0.1862	0.1753

(continued)

Table 3 (continued)

No.	Ref.		Org.				% of	R-1$_F$	R-2$_F$	R-S$_F$	R-SU4$_F$	AVG
	S	#W	S	#W	#D	RE	Ref/Org					
47	2.79	149	25.7	1,373	4	CR	10.85	0.2517	0.1545	0.1250	0.2084	0.1849
22	1.27	69	6.2	332	2	SH	20.78	0.3314	0.1067	0.0796	0.2244	0.1855
27	0.85	39	5.9	208	2	CR	18.75	0.2779	0.1419	0.1321	0.1959	0.1869
16	1.38	62	7.1	331	2	SH	18.73	0.2930	0.1644	0.0981	0.2126	0.1920
40	2.63	137	15.2	532	2	CR	25.75	0.3342	0.1180	0.0908	0.2253	0.1921
12	1.04	53	5.2	257	2	CR	20.62	0.2720	0.1627	0.1228	0.2136	0.1928
32	1.69	76	7.5	354	2	SH	21.47	0.3030	0.1322	0.1190	0.2237	0.1945
41	1.81	89	10.2	492	2	CR	18.09	0.2983	0.1571	0.0988	0.2303	0.1961
20	1.73	86	15.7	724	4	SH	11.88	0.2833	0.1578	0.1223	0.2212	0.1962
1	1.44	81	6.0	325	2	CR	24.92	0.3261	0.1547	0.1030	0.2094	0.1983
10	1.62	81	7.4	369	2	CR	21.95	0.3130	0.1545	0.1030	0.2241	0.1987
3	2.12	115	10.6	563	2	CR	20.43	0.3072	0.1344	0.1207	0.2423	0.2012
49	1.18	68	12.0	621	4	CR	10.95	0.2591	0.1922	0.1547	0.2282	0.2085
25	1.44	72	9.1	459	3	CR	15.69	0.3176	0.1533	0.1271	0.2464	0.2111
19	1.38	83	7.5	419	2	SH	19.81	0.3191	0.1502	0.1287	0.2615	0.2148
38	0.86	43	4.9	250	2	SH	17.20	0.3314	0.1693	0.1158	0.2429	0.2149
31	0.88	41	6.0	221	2	SH	18.55	0.2950	0.1789	0.1480	0.2418	0.2159
26	1.43	67	8.6	338	2	SH	19.82	0.3465	0.2006	0.1397	0.2631	0.2375
36	1.08	52	5.9	279	2	CR	18.64	0.3171	0.2250	0.2153	0.2955	0.2632
17	1.67	84	6.8	347	2	SH	24.21	0.3753	0.2327	0.1810	0.3216	0.2777
2	1.37	77	6.0	338	2	CR	22.78	0.3187	0.2538	0.2430	0.3183	0.2835
21	1.90	97	9.6	492	2	SH	19.72	0.3741	0.2522	0.1820	0.3427	0.2877

(continued)

Table 3 (continued)

No.	Ref.		Org.				% of	$R\text{-}1_F$	$R\text{-}2_F$	$R\text{-}S_F$	$R\text{-}SU4_F$	AVG
	S	#W	S	#W	#D	RE	Ref/Org					
45	1.74	86	16.9	758	4	CR	11.35	0.3467	0.2626	0.2118	0.3573	0.2946
46	1.51	81	14.0	704	4	CR	11.51	0.3299	0.2769	0.2573	0.3150	0.2948
30	1.57	84	6.7	332	2	CR	25.30	0.4015	0.2347	0.2177	0.3470	0.3002
24	1.13	61	6.5	319	2	CR	19.12	0.3446	0.2643	0.2588	0.3637	0.3079
42	1.17	73	5.7	302	2	CR	24.17	0.4624	0.3066	0.2444	0.3641	0.3444
44	1.01	47	5.2	232	2	CR	20.26	0.4232	0.3274	0.2949	0.3676	0.3533

fifty sets of Thai news articles, the results showed that TEDUs+COMPs with the MDS1 method (composed of iterative weighting, the PageRank non-consideration, the node weight inclusion, the query relevance, the highest-weight priority with centroid preference, and the redundancy removal) yielded the best performance in R-1_F, R-S_F, and R-$SU4_F$ evaluations while the MDS2 method (having the same factors as MDS1 except the query relevance) achieved the highest R-2_F value. Finally, an error analysis was provided to find a way to improve the method in the future. Some main sources of errors are document heterogeneity, unit size effect and incorrect segmentation. We will analyze the relation between TEDUs in order to form a more suitable set of combined TEDU with consideration of semantic.

Acknowledgments This work was supported by the National Research University Project of Thailand Office of Higher Education Commission, Thammasat Center of Excellence in Intelligent Informatics, Speech and Language Technology and Service Innovation, and Rajamangala University of Technology Lanna Nan. We would like to thank to all members at KINDML laboratory at Sirindhorn International Institute of Technology for fruitful discussion.

References

1. Barzilay, R., McKeown, K.R., Elhadad, M.: Information fusion in the context of multi-document summarization. In: Proceedings of the 37th Annual Meeting of the ACL, pp. 550–557 (1999)
2. Carbonell, J., Goldstein, J.: The use of MMR, diversity-based reranking for reordering documents and producing summaries. Res. Dev. Inf. Retr. 335–336 (1998)
3. Jaruskulchai, C., Kruengkrai, C.: A practical text summarizer by paragraph extraction for Thai. In: Proceedings of the sixth international workshop on Information retrieval with Asian languages (AsianIR '03), vol. 11, pp. 9–16 (2003)
4. Ketui, N., Theeramunkong, T.: Inclusion-based and exclusion-based approaches in graph-based multiple news summarization. In: Knowledge, Information and Cre-ativity Support Systems, LNCS 6746, pp. 91–102 (2010)
5. Ketui, N., Theeramunkong, T., Onsuwan, C.: Thai elementary discourse unit analysis and syntactic-based segmentation. Inf. Int. Interdisc. J. (INFORMATION-TOKYO) **16**(10B), 7423–7436 (2013)
6. Kittiphattanabawon, N., Theeramunkong, T., Nantajeewarawat, E.: News relation discovery based on association rule mining with combining factors. IEICE Trans. **94**(D(3)), 404–415 (2011)
7. Lin, C.-Y.: Rouge: a package for automatic evaluation of summaries. In: Proceedings of ACL Workshop on Text Summarization, pp. 74–81 (2004)
8. Maier, D.: The complexity of some problems on subsequences and supersequences. J. Assoc. Comput. Machin. **25**(2), 322–336 (1978)
9. Mani, I.: Multi-document summarization by graph search and matching. In: Proceedings of the Fifteenth National Conference on Artificial Intelligence (AAAI-97), pp. 622–628 (1997)
10. McKeown, K., Klavans, J., Hatzivassiloglou, V., Barzilay, R., Eskin, E.: Towards multidocument summarization by reformulation: progress and prospects. AAAI/IAAI 453–460 (1999)
11. Mihalcea, R.: Graph-based ranking algorithms for sentence extraction, applied to text summarization. In: Proceedings of the ACL: on Interactive poster and demonstration sessions, ACLDEMO '04. Association for Computational Linguistics, Stroudsburg, PA, USA (2004)
12. Page, L., Brin, S., Motwani, R., Winograd, T.: The PageRank Citation Ranking: Bringing Order to the Web. Technical report, Stanford University (1998)

13. Radev, D.R., Jing, H., Budzikowska, M.: Centroid-based summarization of multiple documents: sentence extraction, utility-based evaluation, and user studies. In: Proceedings of NAACL-ANLP 2000 Workshop on Automatic Summarization, pp. 21–30 (2000)
14. Theeramunkong, T., Boriboon, M., Haruechaiyasak, C., Kittiphattanabawon, N., Kosawat, K., Onsuwan, C., Siriwat, I., Suwanapong, T., Tongtep, N.: Thai-nest: a framework for Thai named entity tagging specification and tools. In: Proceedings of the 2nd International Conference on Corpus Linguistics (CILC '10), pp. 895–908. University of A Coruna, Spain (2010)
15. Tongtep, N., Theeramunkong, T.: Multi-stage automatic NE and POS annotation using pattern-based and statistical-based techniques for Thai corpus construction. IEICE Trans. Inf. Syst. **E96–D**(10), 2245–2256 (2013)

On the Use of Character Affinities for Story Plot Generation

Gonzalo Méndez, Pablo Gervás and Carlos León

Abstract One of the aspects that is used to keep the reader's interest in a story is the network of relationships among the characters that take part in that story. We can model the relationship between two characters using their mutual affinities, which allow us to define which interactions are possible between two characters. In this paper we present a model to represent characters' affinities and we describe how we have implemented this model using a multi-agent system that is used to generate stories. We also present the result of one experiment to measure the evolution of the affinities between two characters throughout a story.

Keywords Computational creativity · Narrative · Story generation · Multi-agent simulation · Character affinities

1 Introduction

In the book *The Thirty-Six Dramatic Situations* [17] Polti explores the assertion made by Gozzi (author of *Turandot*) saying that there can only be thirty-six tragic situations. Polti analyses what these thirty-six situations are, their variations, and what characters are involved. At the end of the book, he begins his conclusions by saying that, to obtain the nuances of the situations, the first thing he did was to "*enumerate the ties of friendship or kinship between the characters*". A century before that, Goethe had already proposed his theory (maybe just metaphorical) of *elective affini-*

G. Méndez (✉) · P. Gervás · C. León
Facultad de Informática, Universidad Complutense de Madrid, Madrid, Spain
e-mail: gmendez@ucm.es
URL: http://nil.fdi.ucm.es

P. Gervás
e-mail: pgervas@ucm.es

C. León
e-mail: cleon@ucm.es

S. Kunifuji et al. (eds.), *Knowledge, Information and Creativity Support Systems*, Advances in Intelligent Systems and Computing 416,
DOI 10.1007/978-3-319-27478-2_15

ties [20] to depict human relations, specially marriages, and he showed how affinities between characters can be represented by a topological chart.

Even in modern TV shows which expand for several seasons, one of the aspects that create more engagement with spectators are the relationships that exist between characters and the way in which they evolve from the beginning of the first season to end of the last one.

Through all of theses examples, we can see that the affinity between characters is an important factor to take into account when generating stories, and one that can help us to maintain the necessary narrative tension to keep the reader interested in the story.

In the following sections, we present a model of character affinities and the way in which we have implemented it using a multi-agent system that is used to generate stories based on the relationships between characters.

2 Related Work

The first story telling system for which there is a record is the Novel Writer system developed by Sheldon Klein [7]. Novel Writer created murder stories within the context of a weekend party. It relied on a micro-simulation model where the behaviour of individual characters and events were governed by probabilistic rules that progressively changed the state of the simulated world (represented as a semantic network). The flow of the narrative arises from reports on the changing state of the world model. A description of the world in which the story was to take place was provided as input. The particular murderer and victim depended on the character traits specified as input (with an additional random ingredient). The motives arise as a function of the events during the course of the story. The set of rules is highly constraining, and allows for the construction of only one very specific type of story.

Overall, Novel Writer operated on a very restricted setting (murder mystery at weekend party, established in the initial specification of the initial state of the network), with no automated character creation (character traits were specified as input). The world representation allows for reasonably wide modeling of relations between characters. Causality is used by the system to drive the creation of the story (motives arise from events and lead to a murder, for instance) but not represented explicitly (it is only implicit in the rules of the system). Personality characteristics are explicitly represented but marked as "not to be described in output". This suggests that there is a process of selection of what to mention and what to omit, but the model of how to do this is hard-wired in the code.

TALESPIN [13], a system which told stories about the lives of simple woodland creatures, was based on planning: to create a story, a character is given a goal, and then the plan is developed to solve the goal. TALESPIN introduces character goals as triggers for action. Actions are no longer set off directly by satisfaction of their conditions, an initial goal is set, which is decomposed into subgoals and events. The systems allows the possibility of having more than one problem-solving character in the story (and it introduced separate goal lists for each of them). The validity of a story is established in terms of: existence of a problem, degree of difficulty in solving the problem, and nature or level of problem solved.

Lebowitz's UNIVERSE [9] modelled the generation of scripts for a succession of TV soap opera episodes (a large cast of characters play out multiple, simultaneous, overlapping stories that never end). UNIVERSE is the first storytelling system to devote special attention to the creation of characters. Complex data structures are presented to represent characters, and a simple algorithm is proposed to fill these in partly in an automatic way. But the bulk of characterization is left for the user to do by hand.

UNIVERSE is aimed at exploring extended story generation, a continuing serial rather than a story with a beginning and an end. It is in a first instance intended as a writer's aid, with additional hopes to later develop it into an autonomous storyteller. UNIVERSE first addresses a question of procedure in making up a story over a fictional world: whether the world should be built first and then a plot to take place in it, or whether the plot should drive the construction of the world, with characters, locations and objects being created as needed. Lebowitz declares himself in favour of the first option, which is why UNIVERSE includes facilities for creating characters independently of plot, in contrast to Dehn [3] who favoured the second in her AUTHOR program (which was intended to simulate the author's mind as she makes up a story).

The actual story generation process of UNIVERSE [9] uses plan-like units (plot fragments) to generate plot outlines. Treatment of dialogue and low-level text generation are explicitly postponed to some later stage. Plot fragments provide narrative methods that achieve goals, but the goals considered here are not character goals, but author goals. This is intended to allow the system to lead characters into undertaking actions that they would not have chosen to do as independent agents (to make the story interesting, usually by giving rise to melodramatic conflicts). The system keeps a precedence graph that records how the various pending author goals and plot fragments relate to each other and to events that have been told already. To plan the next stage of the plot, a goal with no missing preconditions is selected and expanded. Search is not depth first, so that the system may switch from expanding goals related with one branch of the story to expanding goals for a totally different one. When selecting plot fragments or characters to use in expansion, priority is given to those that achieve extra goals from among those pending.

The line of work initiated by TALESPIN, based on modelling the behaviour of characters, has led to a specific branch of storytellers. Characters are implemented as autonomous intelligent agents that can choose their own actions informed by their internal states (including goals and emotions) and their perception of the environment. Narrative is understood to emerge from the interaction of these characters with one another. While this guarantees coherent plots, Dehn pointed out that lack of author goals does not necessarily produce very interesting stories. However, it has been found very useful in the context of virtual environments, where the introduction of such agents injects a measure of narrative to an interactive setting.

MEXICA [16] is a computer model designed to study the creative process in writing in terms of the cycle of engagement and reflection [18]. It was designed to generate short stories about the MEXICAS (also wrongly known as Aztecs). MEXICA was a pioneer in that it takes into account emotional links and tensions between the characters as means for driving and evaluating ongoing stories. It relies on certain structures to represent its knowledge: a set of *story actions* (defined in terms of preconditions and post-conditions) and a set of *previous stories* (stated in terms of story actions).

The reflection phase revises the plot so far, mainly checking it for coherence, novelty and interest. The checks for novelty and interest involve comparing the plot so far with that of previous stories. If the story is too similar to some previous one, or if its measure of interest compares badly to previous stories, the system takes action by setting a guideline to be obeyed during engagement. These guidelines are a low level equivalent of author goals, driving which types of action can be chosen from the set of possible candidates. The check for coherence is only carried out over the final version of the story, and it involves inserting into the text actions that convey explicitly either character goals or tensions between the characters that are necessary to understand the story. Unless they are explicitly added during this check, goals and tensions are not included in the discourse.

The Virtual Storyteller [19] introduces a multi-agent approach to story creation where a specific director agent is introduced to look after plot. Each agent has its own knowledge base (representing what it knows about the world) and rules to govern its behaviour. In particular, the director agent has basic knowledge about plot structure (that it must have a beginning, a middle, and a happy end) and exercises control over agent's actions in one of three ways: environmental (introduce new characters and object), motivational (giving characters specific goals), and proscriptive (disallowing a character's intended action). The director has no prescriptive control (it cannot force characters to perform specific actions). Theune et al. report the use of rules to measure issues such as surprise and "impressiveness".

Comme il Faut (CiF) [12] is a knowledge-based system that models the interplay between social norm, social interactions, character desires, and cultural background. The underlying model of social interaction covers a range of aspects, from cultural static knowledge relevant to social interaction to fleeting desires of characters, with models for intervening factors like social exchanges, microtheories for significant concepts (such as friendship), and set of complex rules capturing likely behaviours of characters when faced with particular social circumstances.

Stella [10] performs story generation by traversing a conceptual space of partial world states based on narrative aspects. World states are generated as the result of non-deterministic interaction between characters and their environment. This generation is narrative agnostic, and an additional level built on top of the world evolution chooses the most promising ones in terms of their narrative features. Stella makes use of objective curves representing these features and selects world states whose characteristics match the ones represented by these curves. Stella is aligned to the current approach in the sense that simulation is also the base for generation. Stella, however, does not address characters' interactions as a key feature in the creative process.

In general, approaches to Interactive Storytelling have some degree of simulation as conceived in this work [1, 2, 11]. While every approach models the problem of story generation in a specific way, there exist some degree of similarity in the way they perform, namely by chaining sequential states that are driven or selected by an implicit or explicit model of plot quality.

3 A Model of Character Affinity

When running intelligent agents in simulations, and specially when they are in the form of intelligent virtual agents within virtual environments, some authors report the impossibility to run but a few of them at the same time [5, 6], since all the artificial intelligence involved in making them intelligent implies a high computational cost. One of our concerns when designing the current model has been for it to be as light-weight and cost-effective as possible, so its combination with other artefacts to create intelligent characters with personality traits, emotions and complex decision making maintains a low computational cost.

One of the most relevant research works on the subject is *Thespian* [14], the social behaviour framework used in [6]. In this work, the authors describe the use of an affinity factor to model social interaction which affects how characters can behave with each other. In this case, affinity is affected by other factors, such as social obligations and characters goals.

The first approach we used was to model affinity as a set of symbolic values that would be subsequently used to reason about the character's actions. The advantage of this approach is that it is easier to understand and reason about what is happening in the simulation. However, it is more difficult to operate with these values, a certain semantic has to be added to the code to understand how these values change and, on the long run, symbolic reasoning tends to be slow when combined with other processes.

Therefore, we have opted for a numeric representation that allows us to use common arithmetic operators to modify the degree of affinity between characters. The main drawback of this approach is that it is more difficult to calibrate the model and interpret what is happening in the simulation. To reduce this drawback, we have opted for a representation similar to the fuzzy concepts proposed in [22], as shown in Fig. 1, an approach that has already been used by other authors to model cognitive architectures [4, 5].

We have modelled four levels of affinity according to four different kinds of affinity: foe (no affinity), indifferent (slight affinity), friend (medium affinity) and mate (high affinity). These four levels of affinity overlap on their limits, which allows for relationships not to change constantly when moving around the limits of two different levels. Therefore, the change from *indifferent* to *friend*, takes place when the affinity value is 70, and changing from *friend* to *indifferent* is done with an affinity value of 50.

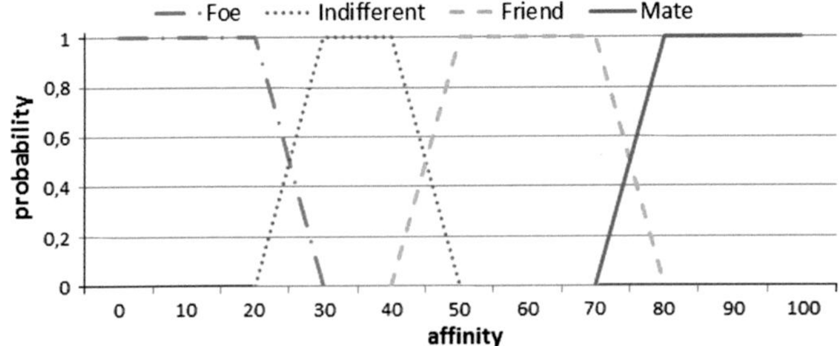

Fig. 1 Model of affinity

An additional aspect of affinity is that it is not symmetrical. Given two characters, their mutual affinity is likely to have different values and it may even be situated in different levels, with the exception of mates: *character A* considers *character B* as its mate only if *character B* considers *character A* as its mate, too. However, if they are not mates, *character A* may think *character B* is a friend, while *character B* may think *character A* is a foe.

There are two ways in which the affinity value can change. The first one is by lack of interaction, in which case the affinity value moves towards the indifferent level. The second one is through interactions among characters, that obey a few simple rules. There is a set of interactions that is appropriate for each affinity level, so when dealing with a friend a character may only propose to carry out friend actions, but not mate actions. In addition, characters ignore proposals that do not correspond to their perceived affinity level, and receiving such proposals may penalise the affinity with the character proposing them. The exception to this rule are foes, who carry out what they intend to do irrespective of what the other character may want. When receiving a proposal, a character may decide to either accept or reject it. If the proposal is accepted, both characters increase their mutual affinity. If it is rejected, the proposer will penalise its affinity with the receiver. Actions for the same level of affinity have different impact on it. For example, a romantic dinner has a higher effect on affinity than watching tv together. Similarly, the negative effect of rejecting an invitation is opposite to the positive effect of accepting it.

4 Implementation of the Model

The described model has been implemented by means of a multi-agent system developed using JADE.[1]

The main objectives of the implementation were: to test the model apart from other factors such as the environment in which the story takes place or the personality traits and emotional state of the characters, which cause them to make different

[1] http://jade.tilab.com

decisions in the same situation; and to implement the model as independent as possible from the domain of the story, so it can be easily used to generate different kinds of stories.

The system consists of two types of agents: a Director Agent, which is in charge of setting up the execution environment and creating the characters; and Character Agents, one for each character of the story, which are the ones that interact to generate the story. In the current case, the story consists of a set of interactions that make the affinity between characters change accordingly.

The information that the Director Agent needs to set up the execution environment is written in a text file that contains the number and names of the characters that have to be created. Subsequently, the information needed to configure each character is also included in a text file that currently contains the name and gender of the character, the name and affinity with its mate, a list of friend names and affinities and a list of foe names and affinities.

Each Character Agent is endowed with three different behaviours: one to interact with its mate, another one to interact with its friends and the last one to interact with its foes. Each behaviour is independent from the others, and they can all be added, blocked and removed dynamically to keep the system as lightweight as possible. These behaviours run the interaction protocols needed to implement and test the affinity model. The information needed by these behaviours, mainly the actions that characters can perform when executing them, along with the degree in which these actions affect the affinity between characters, is also stored in text files, so it is easy to add and remove actions and modify their influence on the affinity without having to change the code and recompile the system. This also means that, as far as the affinity model is concerned, actions have no semantic apart from their influence on the affinity value.

In Fig. 2 we can see how the MateBehaviour works. When a character receives a message from its mate, it checks whether it is a proposal to do something together or not. If it is, it may accept it, in which case it increases its affinity with its mate, or decline. In both cases, the decision is notified to its mate. When a character receives an acceptance, it increases its affinity with its mate, whereas if is a rejection, it checks its affinity with its mate and, if it is already below a given threshold, it will decide to break up with its mate. When an agent receives a break-up notification, it decreases

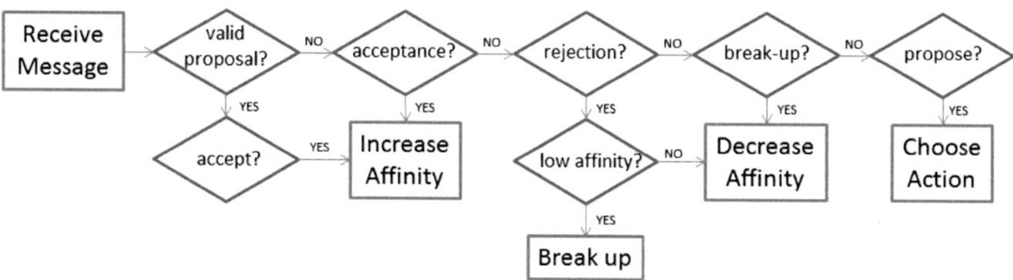

Fig. 2 Interaction protocol for mate characters

its affinity with its mate and decides whether to make it its friend or its foe. If none of this has happened, the character may then decide to propose its mate to do some activity together.

In the points where the characters should make a decision, such as whether to accept a proposal to do something or not, a random probabilistic decision has been made in order to be able to test the implementation of the affinity model by itself, without the interference of other processes. Thus, for example, the probability of accepting a proposal of the character's mate has been empirically established in 0.6. The reason for this value in our experiments is that it is high enough for couples to remain fairly stable, but it is low enough to keep things happening, so that stories don't turn boring.

Running the implemented system with 15 characters (8 females and 7 males forming 7 couples), we have chosen the couple formed by two of them, Betty and Clark, to show the evolution of their affinity over time, as shown in Fig. 3. The image shows how their affinity varies between 80 and 100 over the execution, but both affinities evolve separately (although they are not completely independent). In general, affinity increases at a lower speed than it decreases. This is due to two causes: the heavier impact that rejections have on the affinity than acceptances; and the fact that, if no other action is taken, affinity slowly fades over time, which affects the overall decreasing speed. The most remarkable fact that can be appreciated in the image is how, at the end of the execution, the affinity between both characters falls dramatically due to the final break-up of the couple. This break-up is caused by Clark's rejections of several of Betty's proposals, which in turn causes Betty's decision to put an end to the relationship, once the affinity level has gone below the threshold of mate affinity.

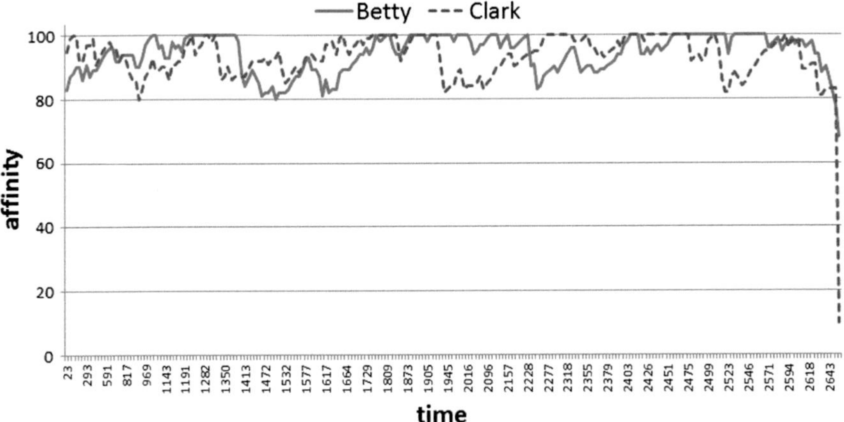

Fig. 3 Evolution of the affinity between characters Betty and Clark

5 System Test

Although a more exhaustive testing is still needed, we have run some tests with different configurations in order to check empirically that the model is consistent and that it allows us to generate different stories by just modifying the thresholds that are used for the characters to make decisions.

In Figs. 4 and 5 we can observe the results of two of these tests, where we show three of the most representative affinities in each of the generated stories. In both tests, the initial situation was the same: 15 characters, 14 mate relationships (i.e. 7 couples), 28 friend relationships (with all characters having at least one friend) and 4 foe relationships (only two characters had foes).

The graphics displayed in Figs. 4 and 5 show how the affinities between two given characters change over time. They have been selected because they represent relationships that have persisted over time and have a high degree of variability, so they are likely to give rise to interesting stories.

One of the aspects that has drawn our attention is that the affinities between two characters tend to evolve in parallel, and although at some points they may diverge, they keep on moving in the same direction and they eventually converge, showing different values but a similar tendency.

In Fig. 4 we can see the result of running the system with high thresholds for foe interaction (0.75 over 1) which means there is only a probability of 0.25 that a character will decide to act against a foe or to respond to an act from a foe. Higher values for these thresholds (i.e. less probability for action) often produce stories with little or no interaction among foes. In general, the stories we have obtained produce some stable relationships, a few relationships that change from friend to foe and back again, and a fair amount of relationships that fade into indifference over time.

The first two graphics displayed in Fig. 4 show a similar evolution of the affinity value. In both cases, one of the character's affinity value decreases enough for the character to decide to break up the relationship and, as a response, their mate decides to consider them as foes. At this point, the values of the affinities start diverging, and one of the affinities moves to the lower range of foe affinity, while the other moves to the upper values. In both cases, the affinities evolve separately, showing the same tendency but with separated values. As it can be seen, the affinities eventually converge and start showing the same evolution.

The third graphic shows a different starting situation, since Meredith considers Violet as an enemy, while Violet is initially indifferent about Meredith. When Meredith acts against Violet, she immediately responds in the same way and considers Meredith as a foe, but using a higher affinity value as a starting point. As in the other cases, the affinity values eventually converge and start evolving in parallel.

In all three cases, we can see how the affinity values tend to separate from the lowest limit, due to the fact that, since the threshold for action is set to a high value, the characters don't act too much against their foes, so there is a chance for the affinity values to increase and for the relationships to evolve in different ways.

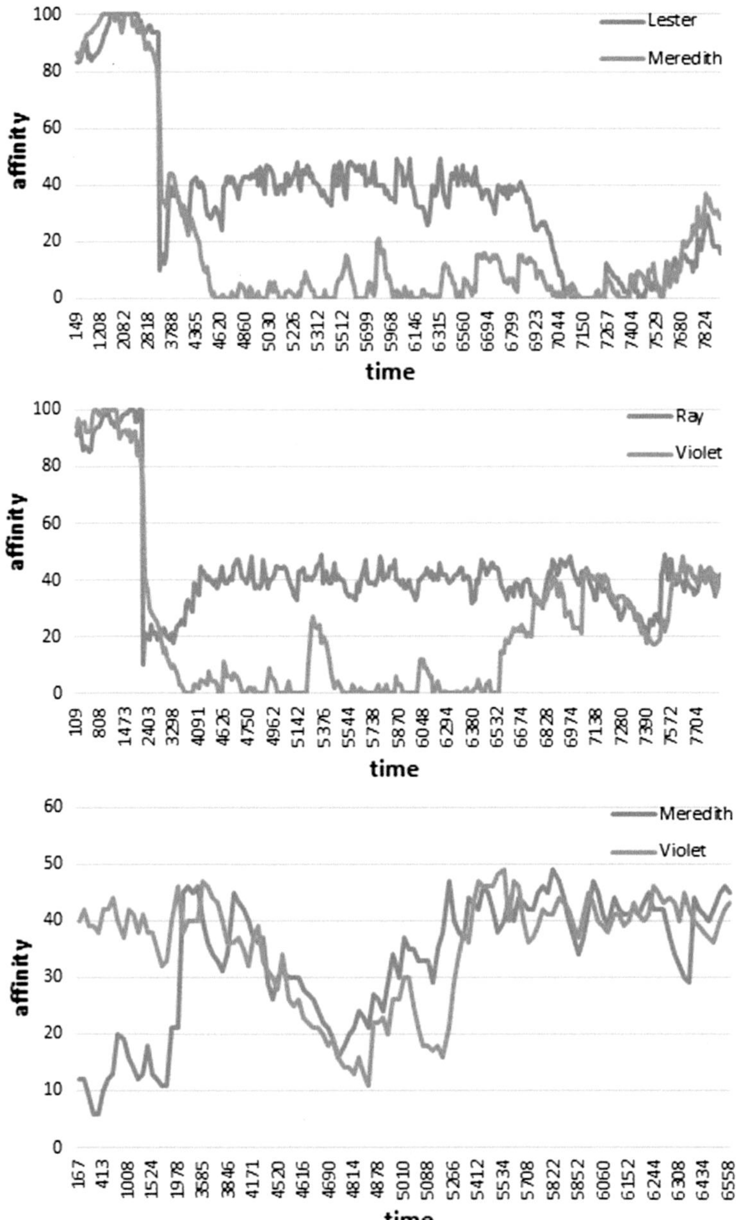

Fig. 4 Evolution of the affinity with a high threshold for foe interaction

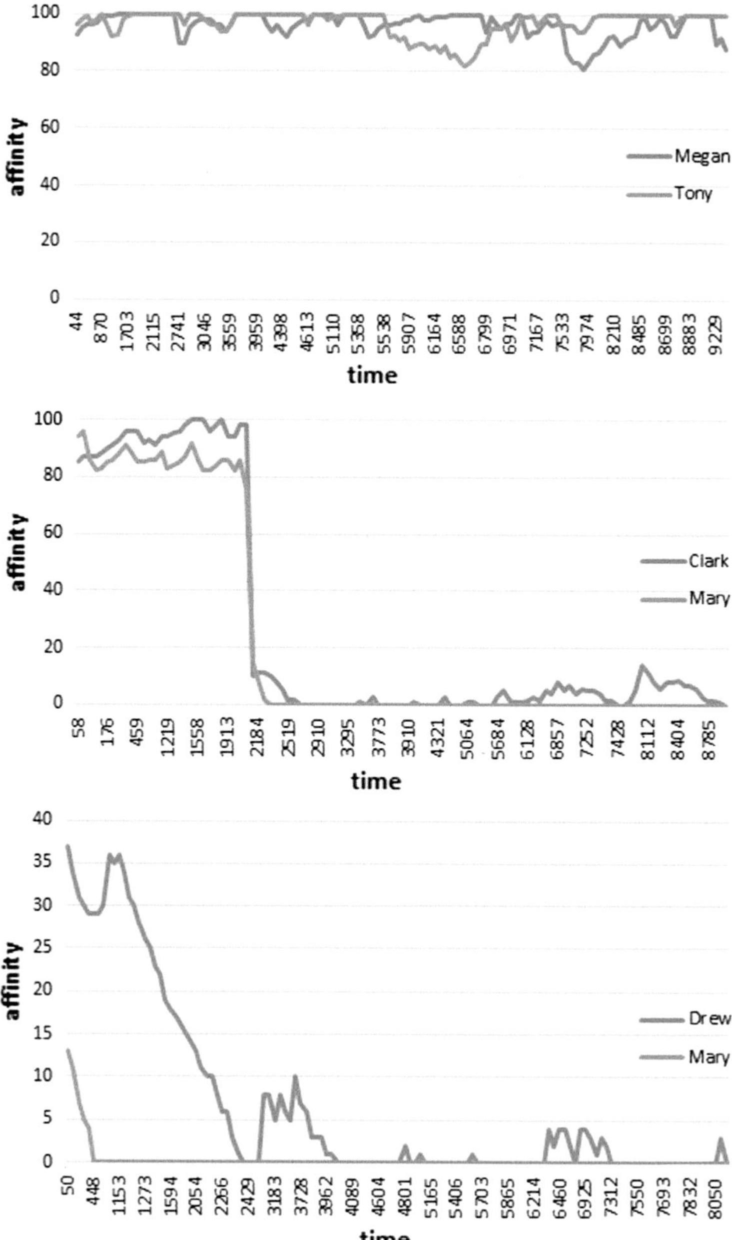

Fig. 5 Evolution of the affinity with a low threshold for foe interaction

In contrast to the described graphics, the ones displayed in Fig. 5 show what happens when the threshold for action and reaction to foe events are set to lower values (in this case, 0.4 to initiate an action and 0.5 to react to another character's action). The probability of acting against a foe is considerably higher than in the former case, and the graphics show how, in similar initial situations, the affinity values present much less variation and tend to be much closer to the lower limit of the foe affinity values.

However, as it can be seen in the first of the three graphics, it is still possible to obtain stable relationships in different ranges of the affinity spectrum, as long as they do not fall into the foe category, from which it is difficult to get out due to the high activity imposed by the thresholds. This behaviour is useful when our aim is to generate stories that do not have a happy ending.

6 Discussion

There is room for improvement in the current model. A number of design decisions have been taken in order to prove the plausibility of the hypotheses, but these can be further developed.

One of the most relevant aspects of the current system is the description of the domain, namely the number of actions that can take place among the characters, the values for the parameters, and so on. The current prototype proposes a straightforward implementation meant to serve as placeholder for a richer definition of the domain. More specifically, the interaction protocols used to drive the calculation of the affinities has been defined by hand, being driven only by common sense intuition. While this well serves the purpose of deploying a working prototype, following a model of affinities grounded on a psychological model is necessary for the conclusions to have a general validity.

From the point of view of creativity, two important aspects are worth the discussion: the evaluation process and the exploration and comparison of different evolution sequences for affinity and their relation with the quality of the generated narratives. Wiggins proposes a model in which every creative artifact—story in this case—is a member of a more general set of artifact, and the overall objective of creativity is to find the best ones [21]. From this perspective, it makes sense to compare the output of different executions in terms of the perceived quality of the generated narratives. While the design of a full evaluation procedure is still planned as part of the future work, the current work makes it possible to highlight some particular points to be addressed in this evaluation.

First, the evaluation must focus on the emotional aspects of the stories and not on a general value of a narrative. It would not make sense to assess other aspects as literary quality, for instance. While this might sound trivial, it is crucial to acknowledge that isolating particular aspects when evaluating a narrative is not simple. As humans, we tend to provide overall ratings that are not properly bound to any concrete parameter of stories.

Second, if we assume the previous point solved, we would have to go the other way around and discover the influence of the evolution of the affinity on the overall perception of the story, which is exactly the general objective for this kind of creative systems.

7 Conclusions

In the previous sections we have described a model to express characters relations based on their affinity, and we have shown how this model has been implemented using a multi-agent system that generates character-based plots.

We have run the implemented system with up to 15 characters and the results show that the possible interactions are rich enough to generate a high variety of stories, and the variability between executions is such that, for the same initial situation, the output story is always different. For this reason, it is necessary to develop evaluation mechanisms that allow us to decide when a story is good or when we need to keep on trying to generate a story that meets certain quality standards.

In addition, it is possible to change most of the information needed to generate the stories through configuration files, which makes it easy to produce new stories with different situations almost effortlessly. However, these situations are still generated by hand, so for non-trivial set-ups, an automated way to check and correct the consistency of the initial situations would be useful.

In particular, we have seen that the model can be configured in such a way that it keeps relationships stable, but it allows enough flexibility for unexpected events to happen and make the plot more interesting, and it can also be tuned so that we can easily obtain a story with the desired type of ending (either happy or not).

8 Future Work

The model can be further improved. More relations can be included in the system and a more refined selection of them can be tried and evaluated. The results on how the selection of features affects the complexity and the number of generated stories can shed light on what the set of relevant aspects of affinity are.

There is still much work to be done in order to generate stories that are not only based on character relationships. We will start by integrating the work described in [15] (which this paper extends) with the generator presented in [8], which will allow us to situate characters within a map and a context, giving us the chance to generate interactions only when proximity makes them possible.

We intend to endow characters with personality traits and emotions, in order to complement the affinity model and give characters the possibility to make decisions in a more cognitive way. We plan to use an approach similar to the one described in [5] to model emotions, so that it can be easily integrated with the present model and the implementation can maintain a low computational cost. This work line is specially relevant, since it will allow characters to make consistent decisions according to the personality and state of mind, instead of random ones, and it will also allow them to behave in a different way from each other.

Finally, since we are capable of generating a high variety of different stories, we need to develop a mechanism to evaluate these stories in order to discard those that lack interest and to refine the generation mechanism so that less non-interesting stories are generated. The first hypotheses have been pinpointed (as show in Sect. 6), but a concrete computational model is still to be created.

Acknowledgments This paper has been partially supported by the project WHIM 611560 funded by the European Commission, Framework Program 7, the ICT theme, and the Future Emerging Technologies FET program.

References

1. Aylett, R.S., Louchart, S., Dias, J., Paiva, A., Vala, M.: Fearnot!: an experiment in emergent narrative. In: 5th International Working Conference on IntelligentVirtual Agents, pp. 305–316. LNCS, Springer, Sep 2005
2. Cavazza, M., Charles, F., Mead, S.: Planning characters behaviour in interactive storytelling. J. Vis. Compu. Anim. **13**, 121–131 (2002)
3. Dehn, N.: Story generation after tale-spin. In: Proceedings of the International Joint Conference on Artificial Intelligence, pages 16–18 (1981)
4. El-Nasr, M.S., Yen, J., Ioerger, T.R.: Flame—fuzzy logic adaptive model of emotions. Auton. Agents Multi-agent Syst. **3**(3), 219–257 (2000)
5. Imbert, R., de Antonio, A.: An emotional architecture for virtual characters. In: International Conference on Virtual Storytelling. pp. 63–72 (2005)
6. Johnson, W.L., Valente, A.: Tactical language and culture training systems: using artificial intelligence to teach foreign languages and cultures. In: AAAI, pp. 1632–1639 (2008)
7. Klein, S., Aeschliman, J.F., Balsiger, D., Converse, S.L., Court, C., Foster, M., Lao, R., Oakley, J.D., Smith, J.: Automatic novel writing: a status report. Technical Report 186, Computer Science Department, The University of Wisconsin, Dec 1973
8. Laclaustra, I.M., Ledesma, J.L., Méndez, G., Gervás, P.: Kill the dragon and rescue the princess: Designing a plan-based multi-agent story generator. In: 5th International Conference on Computational Creativity, Ljubljana, Slovenia (2014)
9. Lebowitz, M.: Storytelling as planning and learning. Poetics **14**, 483–502 (1985)
10. León, C., Gervás, P.: Creativity in story generation from the ground up: non-deterministic simulation driven by narrative. In: 5th International Conference on Computational Creativity, ICCC 2014, Ljubljana, Slovenia, 06/2014 (2014)
11. Mateas, M., Stern, A.: Structuring content in the faade interactive drama architecture. In: Proceedings of AIIDE, pp. 93–98 (2005)
12. McCoy, J., Treanor, M., Samuel, B., Reed, A.A., Mateas, M., Wardrip-Fruin, N.: Social story worlds with comme il faut. IEEE Trans. Comput. Intell. AI in Games, **6**(2), 97–112 (2014)
13. Meehan, J.R.: Tale-spin, an interactive program that writes stories. In: Proceedings of the Fifth International Joint Conference on Artificial Intelligence, pp. 91–98 (1977)
14. Mei, S., Stacy, M., Pynadath, C., David, V.: Thespian: Modeling Socially Normative Behavior in a Decision-Theoretic Framework. In: Intelligent Virtual Agents, LectureNotes in Computer Science, vol. 4133, pp. 369–382. Springer (2006)

15. Méndez, G., Gervás, P., León, C.: A model of character affinity for agent-based story generation. In: 9th International Conference on Knowledge, Information and Creativity Support Systems, Limassol, Cyprus, 11/2014 (2014)
16. Pérez, R.P.: MEXICA: A Computer Model of Creativity in Writing. Ph.D. thesis, The University of Sussex (1999)
17. Polti, G.: The Thirty-Six Dramatic Situations (1917)
18. Sharples, M.: How We Write. Routledge (1999)
19. Theune, M., Faas, E., Nijholt, A., Heylen, D.: The virtual storyteller: story creation by intelligent agents. In: Proceedings of the Technologies for Interactive Digital Storytelling and Entertainment (TIDSE) Conference, pp. 204–215 (2003)
20. von Goethe, J.W.: Elective Affinities/Kindred by Choice (1809)
21. Wiggins, G.:. A preliminary framework for description, analysis and comparison of creative systems. Knowl.-Based Syst. **19**(7) (2006)
22. Zadeh, L.A.: A computational approach to fuzzy quantifiers in natural languages. Comput. Math. Appl. **9**(1), 149–184 (1983)

Vicarious: A Flexible Framework for the Creative Use of Sensed Biodata

Paul Tennent, Joe Marshall, Brendan Walker, Paul Harter, Steve Benford and Tony Glover

Abstract In this paper we discuss *vicarious*, a flexible, extensible, distributed framework for capturing, processing, visualising, recording and generally *handling* sensed biodata. We outline six specific creative needs: heterogeneity of sensor inputs, liveness, high quality video, synchronisation, scalability, usability. Next, by outlining the architecture and features of vicarious, we show how the system meets each of those expectations. We then provide four examples of creative experiences developed using vicarious. The paper thus contributes the tool vicarious as support for creative experience design, as well as concrete examples of creative, artistic or performative applications of the system.

1 Introduction

In recent years there has been a proliferation of computing systems that make use of biosensing and biodata. Traditionally these data have primarily been used for study as part of the process of capturing and analysing a user's experience. More recently, we have seen the emergence of so called 'wellbeing' systems to the commercial

---

P. Tennent (✉) · J. Marshall · B. Walker · P. Harter · S. Benford · T. Glover
Mixed Reality Laboratory, Department of Computer Science, University of Nottingham, Nottingham NG8 1BB, UK
e-mail: paul.tennent@nottingham.ac.uk

J. Marshall
e-mail: joe.marshall@nottingham.ac.uk

B. Walker
e-mail: brendan.walker@nottingham.ac.uk

P. Harter
e-mail: paul@cleverplugs.com

S. Benford
e-mail: steve.benford@nottingham.ac.uk

T. Glover
e-mail: tony.glover@nottingham.ac.uk

S. Kunifuji et al. (eds.), *Knowledge, Information and Creativity Support Systems*, Advances in Intelligent Systems and Computing 416,
DOI 10.1007/978-3-319-27478-2_16

market with peripherals such as fitbit[1] and Nike+[2] becoming a common sight in fitness circles. In academia we have seen a move towards systems which make active and creative use of so called biodata (or physiological data) either for control or to in some way adapt an experience. These take many forms: for example there are several examples of biodata-driven games: Brainball [5], Perping [13], and others e.g. [8, 10] to cover but a few. Other, less game-oriented but similarly playful systems such as themepark-style rides like the broncomatic [7] and breathless [1], or more artistically oriented experiences such as 'ere be dragons [3] or Donnarumma's muscle-movement based music generation [4] are amongst a host of biodata-themed experiences punctuating the HCI literature.

One common feature of these systems is that to construct them requires the use of one or more biofeedback sensor, and the application or development of software to make use of the data from that sensor. Choosing appropriate sensors when designing an experience turns out to be something of a challenge. The range of available kit is extensive and diverse; in most cases one is confronted with a staggering array of choices. Should my sensors be able to transmit live or simply record the data to local memory? What format might that data be in: EDF(+), CSV, XML or some proprietary standard? What about the quality of the equipment and data? Do I need consumer or medical grade monitoring? The next challenge relates to how that equipment may be used in coordination. What if we want one sensor from one company and another from somewhere else? Later in the paper we will discuss this more, but this should serve to demonstrate the myriad possibilities for which we have to account when choosing equipment.

After several years of designing experiences which require a new system built essentially from the ground up each time, or the integration of several different systems, it was concluded it was necessary to construct a framework through which heterogeneous sensors might be integrated. This framework developed into the eponymous *vicarious*, and in the rest of the paper we will explore the system itself, then present a series of examples where we were able to successfully apply the system in creative or performative experiences.

2 Motivation

HCI and computer science in general has a history of creating general purpose data visualisation tools. Many of these focus on post-experimental data analysis e.g. [2, 6, 9] Virtually none of them however support any kind of live data display. Of course there are commercial systems which do exactly that; *Biotrace*[3] is one such example, however these tools tend to be developed to work with a particular system: Mind Media's *nexus* family in the case of Biotrace. Similarly Affectiva's *Q Live* and Cam-

[1] http://www.fitbit.com/.

[2] https://secure-nikeplus.nike.com/plus/.

[3] http://www.humankarigar.com/biotrace.htm.

NTech's *Actiwave* Suite[4] provide comprehensive analysis tools as long as one has recorded the data using their *Q* and *actiwave* products. There is nothing intrinsically wrong with this, except that when creating experiences that use a fusion of different sensors a designer looses the ability to perform data integration. A combination of proprietary data formats and closed-source software make expansion challenging at best and impossible at worst.

There are then three specific challenges relating to the sensor data: liveness, biodata-specifc tools and heterogeneity of input sensors—and no currently available tools meet all three of these challenges.

Changing focus slightly, let us consider video. Synchronisation of video and data is challenging. Indeed synchronisation of video and video is challenging, and this is reflected in the relative paucity of support for it in many of these tools, despite the incredibly rich context it can offer to the data. Even when video is included in such systems it is rarely of a quality suitable for broadcast. And yet as a creative medium, video is incredibly familiar and often serves a key role in the delivery of experiences involving biodata—as we will see in several of the examples later in this paper.

Next we have questions of scale. The vast majority of these tools are designed to handle a single, or at most a handful of participants for a short duration experiment. What do we do if we want to scale a system to handle biodata from a larger group or over a larger period of time? What if we wish to broadcast our experience to an audience of millions. In an era of multi-screen, cross media design, the local area network may represent a creative prison.

One difficulty creative designers often face when building experiences that make use of biodata is a simple lack of knowledge about what to do with the data and how to get it into a quantifiable state. Signal processing is something of an arcane art to those without relevant training—and as creative experience developers, that state is reasonable to assume for at least a significant portion of potential users. Most data wrangling systems provide a plethora of choice in how one might wish to process the data, but making those choices meaningful is often challenging.

All this has led us to develop a set of key challenges for building a biodata handling system that is flexible, useful and usable. These are as follows:

- *Heterogeneity of sensors and data sources*: Can a system be designed to handle inputs from a whole range of possible sensors, and indeed a wider corpus of data sources?
- *Liveness and pre-recorded data*: Can we design a system able to usefully display, or make use of in some other way, live data, as well as allow for the playback and review of data recorded either within our system or from some other source or collection of sources, such as non-connected sensors?
- *Handling of broadcast quality video*: Can a system be designed in such a way as to support very high quality video suitable for use in broadcast media and can the video and biodata be usefully and attractively combined?

[4]http://www.camntech.com/.

- *Effective data synchronisation*: If we have several videos and several feeds of data can these be effectively synchronised to allow for accurate analysis, and account for different data rates, framerates and dropouts?
- *Scalability*: Can we design a system that is able to scale up to handle lots of data sources over extended periods of time?
- *Usability*: Is it possible to create a system that can be essentially picked up and used by creative designers who may know little about the complexities of biodata signals, and can those signals be visualised in such a way as to be understandable to lay persons when broadcast?

In the remainder of this paper we will describe the vicarious system with reference to these questions, before giving some examples of experiences or performances delivered using vicarious as a backbone.

3 Architecture

Vicarious has been developed in such a way as to allow it to be both robust and scalable. It is built on the premise that each part of the system should be an autonomous, atomic unit and that those units should communicate in a system of publish and subscribe via network protocols. This supports robustness by allowing any single component to fail and be restarted, or be added, removed, modified or moved without interfering with the rest of the system and allows complex work to be distributed across several devices if necessary. From a users' point of view this is virtually transparent, but from a design point of view, as we will see, it allows for a very flexible system. Each program, henceforth referred to as a component is typically one of six types: *producers*, *translators*, *processors*, emphconsumers, *gateways* and *datastores*.

These components are used to construct chains whereby data ultimately gets from producer to consumer. Once data has passed into the system all the data is transformed to a consistent format and thus any component can (generally) subscribe to and consume data from any other. This allows for very large ranges of configurations from relatively small numbers of components.

Another significant benefit of this approach is platform and language independence. The vast majority of vicarious is written in python and thus roughly platform agnostic. However, Because each component is atomic and thus not dependant on any other, it can potentially be written in any language that supports network sockets. In practice this has been implemented on multiple occasions to communicate with e.g. MaxMSP in *The Experiment Live* (see Sect. 4.1). This inbuilt architectural flexibility in particular allows for creatives to construct visualisations etc. using whichever platforms and tools they are comfortable with and still have access to vicarious' data capturing and processing resources.

3.1 Communications

At its heart, Vicarious is a publish and subscribe system where each component publishes zero or more streams of data and any other component can subscribe to and consume zero or more of those streams. Streams can be consumed by as many subscribers as desired, allowing the construction of some fairly complex chains. In most cases a component publishes only one stream, however if multiple streams are required a single port listens for numerical queries which define which stream is being requested, then set up and service that logical connection as required. Streaming is handled over TCP/IP and consists of a series of newline-separated strings. In practice, the vast majority of what is sent through the system is numerical data, but there are some cases where one might wish to send more complex data. By using strings as our base medium we create implicit support for json encoding—however, the onus is then on the consumer to expect the data in that format.

Treating all our data as network ready streams supports both robustness and scalability. Individual components can be run on any system on the same subnet, allowing process heavy components to have dedicated CPU time; the whole system can be very effectively multithreaded on one machine letting the CPU sort out priority in most cases; and it supports the graceful stopping and starting of data at any point in the system. If we wish to leave the local subnet, one of the *gateways* discussed in Sect. 3.7 can be applied.

3.2 Producers

We use the term producer to refer to any data source. In most cases this will be a single sensor such as a heart rate monitor, or sensor hardware platform such as the Nexus. It might however be something which programmatically generates data such as our person simulator (below). Much of the time producers are third party hardware/software and as such the produced data are likely to be in any one of a myriad of often proprietary formats. This leads us the need for *translators*. Some producers are also instances of *gateways* see Sect. 3.7.

3.2.1 The Person Simulator

Is a test harness, used to generate a facsimile of one person's physiological data, as if it were being sensed. It includes electrodermal activity, an electrocardiogram and a respiration trace. It can generate data well beyond normal human ranges, a feature which has been creatively co-opted as we shall see in the examples section. While EDA, ECG and Respiration are by no means a complete set of signals, they have proved a reasonable analogue for the types of sensors typically applied in such systems, and we have used equivalent third party software to do a similar job for electroencephalography.

3.3 Translators

A *translator* serves as a bridge between the data format and communication medium used by any one of a vast array of possible third party producers and that used internally by vicarious. It is the job of a translator to serve streams of data in a standardized, simple form, (usually floats, but occasionally json objects) to any other component which may the consume that stream or pass it further down the chain. For vicarious to succeed as intended, there must be an expandable pool of translators, and they must be easily creatable. Any time one wishes to make use of a sensor or system that lacks a translator one has to be able to write a new one, and so a template is provided and the complexity required is dictated by the data format and connection method of the producer. Translators have been created for several popular sensor systems including *nexus*, *Q* and *emotiv* to name but a few. The open source framework encourages the sharing of additional translators as they are created.

3.4 Processors

By *processor* we mean any component that subscribes to one or more streams and publishes one or more streams, usually performing some kind of signal processing on the incoming data. Processors will be of one of four types:

- Linear: one stream in, one stream out.
- Combinatorial: Multiple streams in, one stream out.
- Multiplexing: One stream in, multiple streams out.
- Multilinear: Multiple Streams in, multiple streams out.

The vast majority of familiar signal processors are linear: bandpass filters, smoothing, rate of change etc. Similarly more specific data oriented ones such as ECG to heart rate or RSP to breathing rate are also typically linear. The processor group is also the place for threshold counters, concordancers, and a myriad of other possible processing tasks.

As with the translators, there is an ever increasing set of processor modules and a very simple template for creating new ones. This means that with only a little programming skill it is straightforward to create new processors to make project-sepcific calculations, combine incoming streams, or any one of countless other operations without the need for learning new and arcane scripting languages. We hope we will be able to encourage the community to upload any created processors for the benefit of others, but we will moderate these for one key practical reason: usability. One of our key motivating factors in this development was that everything should be at least usable without field specific knowledge (medical or signal processing). Thus every processor in vicarious includes a clear description of what inputs it requires, what outputs it produces, what it does, and what it is intended for. A user browsing for appropriate processors or wishing to build a system to use a particular type of data may the search the database of processors to find exactly what is required and understand exactly what it does rather than having to use trial and error to get the necessary results. Of course this is not an issue for users who are experts in signal processing, but the target user group for vicarious is technical creative designers, so that knowledge cannot be assumed.

3.5 Consumers

Consumers, in this context refer to any component which subscribes to a stream and does not publish. Most consumers in our example systems are ultimately contained within the *visualisation suite*, but it is possible to create standalone programs that consume data from vicarious. In many cases these will be *gateways* to some other system.

3.6 Datastores

Datastores are special cases of consumers which write incoming data streams into databases. They can then be used to replay those streams as live—taking the role of producers. Incoming data is timestamped to allow for playback synchronisation and post-hoc analysis.

3.7 Gateways

A *gateway* is a special case of either a *translator* or a *consumer*. They allow vicarious to pass data into other systems and formats. For example, there exists a gateway to the massive-scale messaging system pubnub[5] which buffers up some amount of some collection of streams, encodes them as json and delivers them to a pubnub channel. This can then be either handled by some external resource, or an incoming pubnub-vicarious gateway can be used to consume the stream from another internet connected computer and feed it into another vicarious setup. This allows us to sense at a distance, and is particularly desirable for massively mobile systems. Similarly, there is a gateway that serves as a websocket-javascript API for doing web based visualisation of streams of sensed data, and another that delivers accessible streams into the Arduino sketch environment for more physical computing-based experiences.

---

[5]http://www.pubnub.com.

3.8 Component Management

The distributed nature of the vicarious framework makes for an uncomfortable number of separate processes. Starting each of these manually in a console window or similar, is hardly ideal, so vicarious includes a component manager, that reads an XML specification for a particular system and starts and manages all the necessary processes from within one local application (per device). A drag and drop visual-programming approach to building these XML files is currently under development to further simplify this process.

3.9 Visualisation Suite

Thus far we have examined the *backbone* of the vicarious framework—looking at tools for collecting, processing and storing data, but it is also necessary to provide tools for interacting with that data. While the range of creative uses to which sensed biodata might be put are far without our scope as technical designers to produce, we have created a suite through which incoming data can be visualised and combined with video, and which is in itself widely flexible and extensible. The system provides line charts, bar charts (in various flavours), heatmaps, alpha-shapes, videos and a plethora of other basic visualisation tools, which can be combined to create more complex visualisations.

The nature of the system is such that creatives can of course make use of the data outside vicarious using either *consumers* in their own system or vicarious' *gateways*, but the *visualisation suite*, which is written in OpenGL and has been designed to allow new visualisation components to be quickly added. The system is based on the idea of *active textures*, each of which constitutes one or more consumers (either of video or of data), and which can be laid out with suitable alpha-blending on the screen to create quite complex visualisation systems. Each of the example systems shown below (all of which were creatively directed by artists or performers) make use of the visualisation suite as the principle delivery method for data.

4 Examples of Successful Applications

In this section we give four examples of experiences which have made use of vicarious to deliver creative and interesting uses of biodata. The first is a live performance, the second is a marketing campaign, the third is a live game at a music festival run by an environmental NGO, and the last is a dynamic sculpture. The diversity of these experience scratches the surface of what vicarious is capable of supporting. In the discussion section we will outline some future directions.

4.1 The Experiment Live

The experiment live was a live event run on Halloween and delivered on the screen of a local arts cinema. It centred around group of four ‘ghost hunters’ who, with their biodata being displayed to all, were exploring the ‘haunted’ basement of a local tea-room, to a soundtrack of music generated from their biosignals. A story was told in which the data-recorders became inactive and were then ‘possessed’ (unbeknownst to the ghost hunters themselves) after being removed. The experience is more fully documented in [12]. In system terms, each ‘ghost hunter’ was equipped with a nexus measuring ECG, EDA and RSP, each of which were passed though a processor to generate heart rate, excitement, and breathing rate respectively, and combined in another processor to create a ‘fear factor’. The streams were consumed by the visualisation suite along with video data from a series of static and roaming cameras. Similarly, the music—generated in MaxMSP consumed the streams straight out of vicarious. All this was delivered live to the cinema, then when it was time for the ‘possession’ the nexus data streams were swapped for the *person simulator* (Fig. 1).

4.2 Juke: Built to Thrill

Unlike the experiment live, *Juke: Built to Thrill* was not exclusively a live performance. Instead it was part of an advertising campaign for the Nissan Juke in coordination with the advertising company TBWA, with a series of events and short films

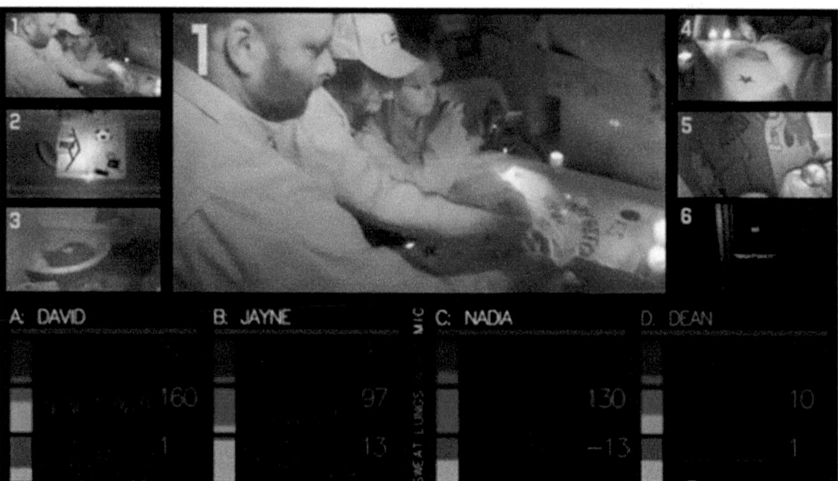

Fig. 1 The experiment live as presented on a cinema screen. Six cameras with one selected for main display at any given time, four ‘ghost hunters’ performing a seance with biodata displayed for ECG, EDA and RSP

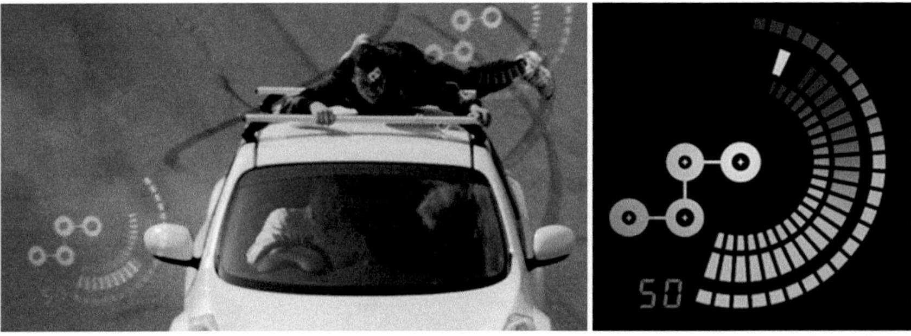

Fig. 2 The visualisation of biodata in the Juke: Built to thrill campaign. *Right* as rendered in vicarious, *Left* as applied to the film in production

made for web consumption. There were four events: a pre-recorded stunt film to introduce the campaign, that required the data to be 'staged' post-hoc then overlaid. A skydiving simulator running live at the Goodwood festival of speed which showed live 'thrill data' processed from heart rate and EDA, and an adventure trip in which four participants had their 'thrill' data recorded to create a montage of their experience including an on screen comparison of their thrill data.

All three cases used the same visualisation (see Fig. 2) which was developed as a new active texture for the visualisation suite. In the case of the stunt film, this was driven with 'acted' biodata—a series of clips were made attempting to match a scripted 'bio-story' of the experience. These were then added into the final film by an external production team, using Adobe After Effects. At Goodwood, the data was captured live using a vilistus blood volume pulse sensor and an EDA sensor. This was processed and displayed live at the event using the visualisation suite, then these visualisations were sent to the same video production team and were again added to the final film in post. The adventure trip participants had their ECG and EDA data recorded using actiwave cardios and Q sensors. The data was not monitored live, but downloaded each evening and processed, inspected and visualised for inclusion in the final video.[6]

4.3 Man Versus Turtle

Man versus turtle was an interactive experience designed to help an environmental NGO called Medasset raise its public profile and create awareness regarding its aims. It consisted of an interactive biodata driven game which then led to a reflective experience in the form of a short, but carefully staged interview and was developed to create conversation time between the players (general public) and the NGO representatives. Practically this was delivered live as a public two-player game in a marquee

[6]The videos can be viewed at https://www.youtube.com/playlist?list=PL2C59049F28D8B580.

Fig. 3 Players of the Man versus Turtle game at the Plissken music festival in Athens

at the Plissken music festival in Athens. Each player of the game had their EDA captured using a nexus, this was then processed to rate of change and the differing values were used in the visualisation suite to drive an alpha-blended 'tug-o-war' between two videos: one of serene turtles, and another of a beach music festival (Fig. 3).

The game itself was a spin on the classic relax to win mechanic established by brainball [5], wherein one player, representing the turtles needed to relax, and the other player, representing the humans, needed to get excited. This was supposed to be representative of the differing goals of turtles and humans and the relative success of each player served to drive the discussion after the game.

4.4 Duality

Duality: one body, two brains was an artefact-based performance around halloween at a local arts cinema. It centred on the classical cinema trope of reanimation, and took the form of a sculpture of a brain (actually made of jelly) in a jar with a number of lights throughout which activated based on the real neural activity of somebody watching a horror film in the auditorium—Fig. 4. A 3D rendering of the brain was simultaneously shown online, again with live data causing different areas of it to"light up".

An Emotiv Epoc EEG headset was used in the auditorium to capture 14 channels of raw EEG data from a participant, and these data were streamed live, using vicarious, to both the brain sculpture and the web representation. Before each film was shown, a theatrical trailer for "the experiment" was shown in both the auditorium (where the data was being captured) and the bar (where it was being presented through the sculpture)—creating a story about capturing participants' personalities and "downloading" them to another brain. This event was in part active performance—that is, the data were being captured live and very publicly by team members in the auditorium, and simultaneously passive—that is, the data as presented to the audience are displayed through the sculpture which "sits on the bar blinking" and was designed to become a curio.

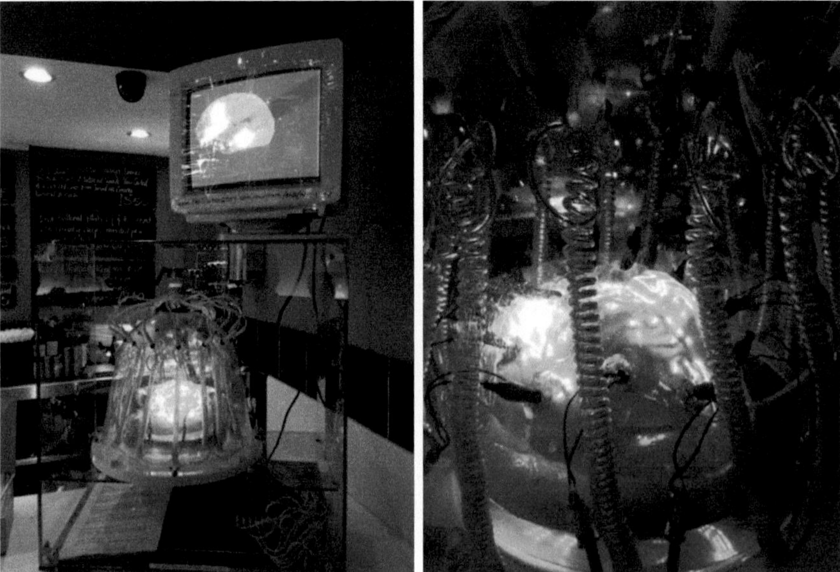

Fig. 4 The brain sculpture from duality: one body two brains in context (*left*) and close up (*right*)

While the audience experience of this event is quite subtle compared to the other case studies here, the system used to deliver it is relatively complex: Data are gathered from the emotiv using a translator, then processed to be manageable for controlling lights, the data are then buffered up into a large JSON object representing one second of data, along with timestamps to ensure good synchronisation between the brain and the website. Those data are then published over the web, where in one instance they are consumed by a Vicarious-WebGL bridge and used to drive the light objects in the rendered brain, and in the other instance are consumed by an arduino output node, which is then wired to the lights in the physical brain. Recorded data were also played back through the system between performances and an extra light on the sculpture indicated whether the data was live or not.

Duality makes an important step in the data performance, beyond the other experiences described so far by adding the physical element of the brain, rather than depending on-screen visualisations, It also did not provide any interpretation of the meaningfulness of the data, rather serving as a seemingly scientistic talking piece.

5 Discussion

Here we revisit the design challenges outlined in the motivation section with reference to the design of the system as discussed in the previous sections. In each case we outline how vicarious attempts to meet the challenge.

Heterogeneity of sensors and data sources: Vicarious uses an extensible system of translators and gateways to be able to talk to potentially anything that exposes a stream of data.

Liveness and pre-recorded data: The system allows for the monitoring, recording and playback of data and videos, and includes import and export for a variety of standard data formats such as CSV and EDF+.

Handling of broadcast quality video: As long as the hardware it is running on has the necessary support, e.g. Blackmagic *Decklinks* or similar, vicarious' visualisation suite can support multiple full HD video feeds and blend these with complex visualisations. As we saw with Juke: Built to thrill—these visualisations can also be exported to be combined with video in post production using more familiar tools and workflows.

Effective data synchronisation: Vicarious' playback system uses absolute time stamps for every data point and video frame to ensure that everything remains perfectly in sync all the time when playing. This leads to generally excellent synchronisation, but is limited by the latency of the producer's communication medium as the timestamps are applied once the data enters the vicarious framework.

Scalability: The distributed nature of vicarious makes it scalable up to a point. Gateways to systems such as *Pubnub* dramatically increase the scalability. Generally, vicarious has been designed with a broadcast view of the world, that is, data are generated by a small number of people and consumed by a larger audience. It remains to be seen whether the reverse would be feasible within this framework—and this stands as a remaining research question.

Usability: Vicarious as it stands is usable up to a point. Several areas of its interface—including a visual system generator for the backbone are currently under development. However, it still requires a significant degree of technical skill to operate and extend. The gateways mitigate this to some extent—delivering the data into familiar formats for visualisation, and the visualisation suite itself may be suitable for many applications, as demonstrated in the examples section, but for now, Vicarious remains something of a technical undertaking. Our local observations suggest that these techno-creative experiences are often composed by teams of people—with a designer/performer/artist at the helm and a supporting technical advisor or developer assisting them. For a team such as this then, vicarious is indeed a usable system, but perhaps not universally so. We hope to further develop the interface to the system to make it more accessible in the future, but the complexity of these systems will likely always necessitate tech-savvy support in their development.

5.1 Interpreting Data

In our first three case studies, the data are interpreted for the audience. In some cases this is a trivial interpretation as part of a video screening, as with *The Experiment Live* and *Juke: Built to Thrill*, or an expert there at the time, as with *Man versus Turtle*. However in the case of *Duality* the audience is left to draw its own conclusions about the meaningfulness of the data. This is an important distinction. It remains an open question whether it is necessary when presenting data that are not easily lay-interpretable, whether more explanation is necessary. In the case of Duality, we wanted to give the sense of brain activity in the sculpture, and the sense that "science was being done", without going into much detail about what the data meant. This was in part to do with the "story" of the performance, and in part to see what an audience would do with the data. We know that there are biodata that audiences can relate to, such as heart beats, at least in part because they have become narrative tropes. The PQRST wave complex is familiar to most people by shape if not by name. This may however not be the case with less relatable data such as EDA or EEG. This likely down to familiarity—we can literally feel our heart beating, but have no real sense of our EDA or EEG activity. We hope to further explore these questions in future work.

6 Conclusions and Future Work

Generally vicarious as a framework has achieved much of what we set out to do. We have created a system for ourselves and our artistic collaborators that allows us to rapidly prototype, develop and deploy interesting and creative experiences based on biodata. In this paper we have outlined the vicarious system structure and provided case studies of four example projects that make use of it. We have discussed the requirements for systems to support data performance, and how vicarious might meet them. We have also examined the question of interpretation: Should these data be interpreted for the audience? We conclude for now that it depends on the intended purpose of the data within the performance, but we aim to further explore this question in future work. We are now also looking beyond biodata as a source, to other data types such as environmental sensors, and looking at other forms of delivery such as web based data explorers or second screens to live broadcasts. Much remains to be done, but vicarious may just prove to be an effective backbone with which to do much.

Acknowledgments This paper was published in an earlier form as [11]. This work is supported by Horizon Digital Economy Research, RCUK grant EP/G065802/1.

References

1. Benford, S., Greenhalgh, C., Giannachi, G., Walker, B., Marshall, J., Rodden, T.: Uncomfortable interactions. In: Proceedings of the 2012 ACM Annual Conference on Human Factors in Computing Systems—CHI '12, p. 2005. ACM Press, New York, New York, USA, May 2012
2. Crabtree, A., Benford, S., Greenhalgh, C., Tennent, P., Chalmers, M., Brown, B.: Supporting ethnographic studies of ubiquitous computing in the wild. In: Proceedings of the 6th ACM conference on Designing Interactive systems—DIS '06, p. 60, ACM Press, New York, New York, USA, June 2006

3. Davis, S.B., Moar, M., Jacobs, R., Watkins, M., Riddoch, C., Cooke, K.: Ere be dragons: heartfelt gaming. Digit. Creativity **17**(3), 157–162 (2006)
4. Donnarumma, M.: Music for Flesh II: informing interactive music performance with the viscerality of the body system. In: New Interfaces for Musical Expression Conference (NIME), Ann Arbor, Michigan, USA (2012)
5. Hjelm, S.I., Browall, C.: Brainball—using brain activity for cool competition. In: Proceedings of the NordiCHI, Stockholm, Sweden (2000)
6. Kipp, M.: Anvil—A generic annotation tool for multimodal dialogue. In: Proceedings of the 7th European Conference on Speech Communication and Technology (Eurospeech), pp. 1367–1370, Aalborg, Sep 2001
7. Marshall, J., Harter, P., Longhurst, J., Walker, B., Benford, S., Tomlinson, G., Rennick Egglestone, S., Reeves, S., Brundell, P., Tennent, P., Cranwell, J.: The gas mask. In: Proceedings of the 2011 annual conference extended abstracts on Human factors in computing systems—CHI EA '11, p. 127. ACM Press, New York, USA, May 2011
8. Masuko, S., Hoshino, J.: A fitness game reflecting heart rate. In: Proceedings of the 2006 ACM SIGCHI International Conference on Advances in Computer Entertainment Technology, ACE '06, ACM, New York, NY, USA (2006)
9. Morrison, A., Tennent, P., Chalmers, M.: Coordinated visualisation of video and system log data. In: Fourth International Conference on Coordinated & Multiple Views in Exploratory Visualization (CMV'06), pp. 91–102. IEEE, July 2006
10. Tadeusz Stach, T.C., Graham, N., Yim, J., Rhodes, R.E.: Heart rate control of exercise video games. pp. 125–132, May 2009
11. Tennent, P., Marshall, J., Walker, B., Harter, P., Benford, S.: Vicarious: a flexible framework for the creative use of sensed biodata. In: Proceedings of the 9th International Conference on Knowledge, Information and Creativity Support Systems, KICSS2014, Limassol, Cyprus. University of Cyprus (2014)
12. Tennent, P., Reeves, S., Benford, S., Walker, B., Marshall, J., Brundell, P., Meese, R., Harter, P.: The machine in the ghost. In: Proceedings of the 2012 ACM annual conference extended abstracts on Human Factors in Computing Systems Extended Abstracts—CHI EA '12, p. 91. ACM Press, New York, USA, May 2012
13. Tennent, P., Rowland, D., Marshall, J., Rennick-Egglestone, S., Harrison, A., Jaime, Z., Walker, B., Benford, S.: Breathalising games: understanding the potential of breath control in game interfaces. In: Proceedings of the 8th International Conference on Advances in Computer Entertainment Technology, p. 58. ACM (2011)

Application of Rough Sets in *k* Nearest Neighbours Algorithm for Classification of Incomplete Samples

Robert K. Nowicki, Bartosz A. Nowak and Marcin Woźniak

Abstract Algorithm *k*-nn is often used for classification, but distance measures used in this algorithm are usually designed to work with real and known data. In real application the input values are imperfect—imprecise, uncertain and even missing. In the most applications, the last issue is solved using marginalization or imputation. These methods unfortunately have many drawbacks. Choice of specific imputation has big impact on classifier answer. On the other hand, marginalization can cause that even a large part of possessed data may be ignored. Therefore, in the paper a new algorithm is proposed. It is designed for work with interval type of input data and in case of lacks in the sample analyses whole domain of possible values for corresponding attributes. Proposed system generalize *k*-nn algorithm and gives rough-specific answer, which states if the test sample may or must belong to the certain set of classes. The important feature of the proposed system is, that it reduces the set of the possible classes and specifies the set of certain classes in the way of filling the missing values by set of possible values.

Keywords Rough sets · k-nn · Missing values

R.K. Nowicki (✉) · B.A. Nowak
Institute of Computational Intelligence, Czestochowa University of Technology, Al. Armii Krajowej 36, 42-200 Czestochowa, Poland
e-mail: robert.nowicki@iisi.pcz.pl
URL: http://www.iisi.pcz.pl

B.A. Nowak
Department of Mathematical Methods in Computer Science, University of Warmia and Mazury, Ul. Słoneczna 54, 10-710 Olsztyn, Poland
e-mail: bartosz.nowak@iisi.pcz.pl

M. Woźniak
Institute of Mathematics, Silesian University of Technology, Kaszubska 23, 44-101 Gliwice, Poland
e-mail: Marcin.Wozniak@polsl.pl

S. Kunifuji et al. (eds.), *Knowledge, Information and Creativity Support Systems*, Advances in Intelligent Systems and Computing 416,
DOI 10.1007/978-3-319-27478-2_17

1 Introduction

The problem of imperfect, especially incomplete, input data is inherent in many real applications of decision support systems. In the industry, some information could be unavailable due to e.g. measuring instrument failure or temporary exceeding the measure range in some part of monitoring process. In medical diagnosis some tests procedures are omitted because of the patient state, unacceptable cost, lack of reagents or they are rejected by the community because of beliefs. Moreover, it could be deemed unnecessary by a doctor. The decision support system cannot remain idle in such cases. Generally, regardless of applied methodology (neural networks, fuzzy systems, k-nn classifiers, svm systems etc.) there are two general methods to process data with missing values—marginalization and imputation—as well as its hybrids and modifications.

Methods that belong to the first group boil down to temporary reduce the dimensionality of considered space to the features of known values. Thus, some elements of the system are just turned off. Moreover, sometimes the elimination of whole incomplete samples is also treated as the marginalization. However, it is eventually accepted only in developing time.

When we would like to use imputation, the unknown values are replaced by estimated ones. The palette of available methods is generally unlimited. The most primitive ones are confined to insertion of random, average or most common values. More sophisticated ones apply EM (Expectation Maximization) or k nearest neighbour algorithms, neural networks, and various fuzzy systems [5, 8, 24, 26]. Promising results are obtained by multiple imputation and interval imputation. If we know the probability density distribution, we can use the Bayesian solution [6, 9, 22, 23]. Then, if we know the possibility distribution, we can use fuzzy imputation.

A specific approach to the problem comes from the rough set theory [19]. An object can be classified to a positive region of the class (i.e. the object certainly belongs to the class), to a negative region of the class (i.e. the object certainly does not belong to the class) or to a boundary region of the class (i.e. it is not possible to determine if the object belongs to the class or not). Membership to these regions depends on the quality of object description. If this description is good enough, the object belongs either to the positive or negative regions. If the description is too weak, then the object belongs to the boundary region. In the rough set theory [19] as well as in the theory of evidence [27], we do not use the individual elements of the consideration but some granules [21]. The granules contain elements which are indistinguishable basing on knowledge that we dispose. Thus, the size and the shape of granules depend on the used (known) knowledge about the elements. Hence, the many hybrid approaches apply rough sets together with other methods mentioned above.

In this paper, we focus on the hybrid system merging the rough set theory with one of the, probably, most popular classification method, i.e. k nearest neighbours [4]. The solutions based on this both ideas are already present in the literature. Let us give some examples. In [2] in order to improve the performance of k-nn classifier, the specific (tolerant) rough relation has been applied to build the similarity function. The quite popular idea is to apply rough set theory to reduction of dimensionality in nearest neighbour classifier. We can find various realizations of it in [10–12]. The rough sets and nearest neighbour algorithm are combined also with fuzzy sets (logic) and genetic (evolutionary) algorithms, see [13, 25, 29, 30] or [32].

The aim of solution presented in this paper is different from cited above. We propose to rebuild the k-nn classifier to add support for incomplete input data. However, we do not use either the imputation or marginalization. In general, the idea could be described as follow. The new solution is built according to rough set theory [19] and is extension of idea discussed in [17]. When all input features have known and available values then the classifier works as basic k-nn system. In the case, when some value is missing the all possible output of classifier are estimated. When only one decision is possible, the feature with missing value could be considered negligible and the classified sample is assigned to lower approximation (positive region) of classes indicated by the classifier. When more than one decision is possible, the classified sample is assigned to upper approximation of each class indicated by the classifier. Obviously, it is not possible to test all features with missing values. In presented solution we use the intervals and specific procedure shown below.

2 Proposed Algorithm

The proposed algorithm can operate with interval-type value of attributes and missing values. In our approach lacks are replaced by interval $[v_i^{min}, v_i^{max}]$, where v_i^{min}, v_i^{max} are minimal and maximal values of the attribute v_i.

The first part of k-nn algorithm requires to compute distance from current sample ($x^{(t)}$) to each sample in the reference set. In our approach distance similar to Manhattan metric is used, i.e.

$$\overline{d}_s = \overline{d}(x^{(t)}, x_s) = \sum_{i=1\ldots n} \begin{cases} |\bar{v}_i^{(t)} - \bar{v}_{s,i}| & \text{if } v_{s,i} \in P \wedge v_{t,i} \in P \\ \max\begin{pmatrix} |\bar{v}_i^{(t)} - v_i^{max}|, \\ |\bar{v}_{t,i} - v_i^{min}| \end{pmatrix} & \text{if } v_{s,i} \in G \wedge v_{t,i} \in P \\ \max\begin{pmatrix} |v_i^{max} - \bar{v}_{s,i}|, \\ |v_i^{min} - \bar{v}_{s,i}| \end{pmatrix} & \text{if } v_{s,i} \in P \wedge v_{t,i} \in G \\ v_i^{max} - v_i^{min} & \text{if } v_{s,i} \in G \wedge v_{t,i} \in G, \end{cases} \tag{1}$$

$$\underline{d}_s = \underline{d}(x^{(t)}, x_s) = \sum_{\substack{i=1\ldots n \\ v_{s,i} \in P \wedge v_{t,i} \in P}} |\bar{v}_i^{(t)} - \bar{v}_{s,i}|, \tag{2}$$

where $\bar{v}_i^{(t)}$ is a value of ith attribute in the test sample, $\bar{v}_{s,i}$ is a value of ith attribute of sth reference sample, P is set of attributes with available values, G is set of attributes with unavailable values, $[\underline{d}_s, \overline{d}_s]$ is interval-type distance of test sample $x^{(t)}$ to reference sample x_s.

Let the vector $\mathbf{d}^{unique}$ contains all sorted and unique values of $\overline{d}_s$ and $\underline{d}_s$, where $s = 1 \ldots M$ and M is an amount of reference samples. Algorithm uses matrices $\overline{\Psi}$ and $\underline{\Psi}$. Both have $m \times \overline{\overline{d^{unique}}}$ elements, where $\overline{\overline{d^{unique}}}$ is dimension of space of $\mathbf{d}^{unique}$, m is number of classes. Values $\overline{\Psi}_{j,c}$, $\underline{\Psi}_{j,c}$ define how many reference samples of the class ω_j are in the distance not greater than d_c^{unique} using $\underline{\mathbf{d}}$ for $\overline{\Psi}_{j,c}$ or $\overline{\mathbf{d}}$ for $\underline{\Psi}_{j,c}$, where

$$\overline{\Psi}_{j,c} = \sum_{\substack{s=1\ldots M: \\ \underline{d}_s \le d_c^{unique} \\ \wedge x_s \in \omega_j}} 1 , \tag{3}$$

$$\underline{\Psi}_{j,c} = \sum_{\substack{s=1\ldots M: \\ \overline{d}_s \le d_c^{unique} \\ \wedge x_s \in \omega_j}} 1 . \tag{4}$$

Let vectors $\overline{\psi}$, $\underline{\psi}$ contain decimal values. Values $\overline{\psi}$, $\underline{\psi}$ are equal to number of samples, which are in the distance not greater than d_c^{unique} using $\underline{\mathbf{d}}$ for $\overline{\psi}_{j,c}$ or $\overline{\mathbf{d}}$ for $\underline{\psi}_{j,c}$. So, they are defined as follows

$$\overline{\psi}_c = \sum_{\substack{s=1\ldots M: \\ \underline{d}_s \le d_c^{unique}}} 1 , \tag{5}$$

$$\underline{\psi}_c = \sum_{\substack{s=1\ldots M: \\ \overline{d}_s \le d_c^{unique}}} 1 . \tag{6}$$

It is always true, that $\overline{\psi}_c \ge \underline{\psi}_c$. When all distances are singletons then $\overline{\psi}_c = \underline{\psi}_c$ and the system operates the same as k-nn.

Example values of $\overline{\Psi}_{c,j}$, $\underline{\Psi}_{c,j}$, $\overline{\psi}_c$, $\underline{\psi}_c$ are presented in Figs. 1 and 2. They were calculated for Wine Recognition Database from UCI (more detail in Sect. 3). The last sample was used as test one and the rest of samples as reference set. The test sample had also artificially removed value of the second attribute.

Additionally, for greater clarity, the symbols $\overline{\Psi}_{\sim j,c}$ and $\underline{\Psi}_{\sim j,c}$ have been introduced and defined as follows

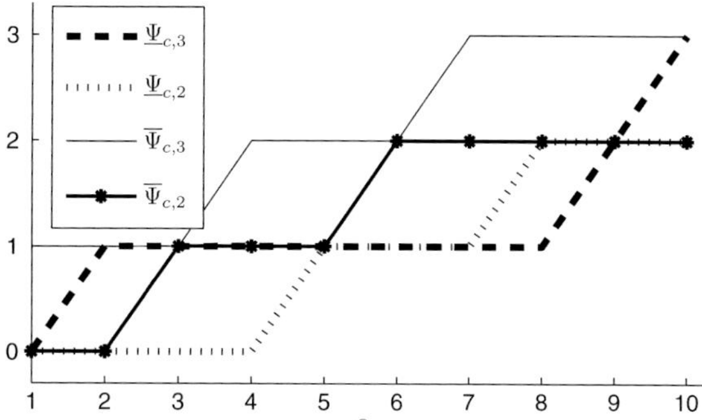

Fig. 1 Values of $\overline{\Psi}_{c,j}$ and $\underline{\Psi}_{c,j}$ for the first sample in WR, with induced lack in one attribute value, rest of samples from the database were chosen as reference, $k = 5$, $c_{min} = 7$, $c_{max} = 10$

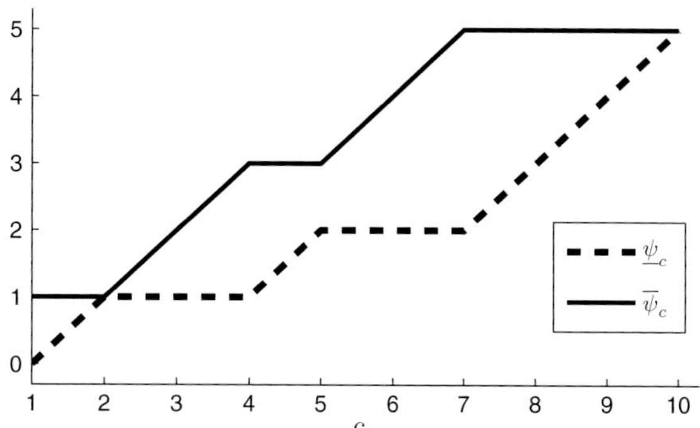

Fig. 2 Values of $\overline{\Psi}_c$ and $\underline{\Psi}_c$ for the same conditions as in the previous figure

$$\overline{\Psi}_{\sim j,c} = \sum_{\substack{s=1\ldots M: \\ \underline{d}_s \le d_c^{unique} \\ \wedge x_s \notin \omega_j}} 1 = \overline{\psi}_c - \overline{\Psi}_{j,c} \,, \tag{7}$$

$$\underline{\Psi}_{\sim j,c} = \sum_{\substack{s=1\ldots M: \\ \overline{d}_s \le d_c^{unique}, \\ \wedge x_s \notin \omega_j}} 1 = \underline{\psi}_c - \underline{\Psi}_{j,c} \,. \tag{8}$$

Let c_{min} equals index of the first element in $\overline{\psi}_c$, which value is not lower than parameter k, and c_{max} equals index of the first element in $\underline{\psi}_c$ not lower than k. Because $c_{min} \le c_{max}$ always occurs, we can write down as follows

$$c_{min} = \arg\min_c \left(\overline{\psi}_c \ge k\right), \tag{9}$$

$$c_{max} = \arg\min_c \left(\underline{\psi}_c \ge k\right). \tag{10}$$

Proposed algorithm examines only distances $\{d_c^{unique} : c_{min} \le c \le c_{max}\}$. Because only for these distances the main idea of k-nn algorithm is fulfilled: exactly k-nn nearest reference samples (or more in case of a tie) are considered (in our case for classification).

Let $\Psi^{pot}_{1,c}, \Psi^{pot}_{2,c} \ldots \Psi^{pot}_{m,c}$ equal some potential numbers of samples that belong to an appropriate class and are in the distance not greater than d^{unique}_c. Authors assumed that for each distance $\{d^{unique}_c : c_{min} \leq c \leq c_{max}\}$ all possible combinations of the values $\{\Psi^{pot}_{j,c} : 1 \leq j \leq m\}$ meet k-nn conditions if $\forall j = 1 \ldots m, \underline{\Psi}_{j,c} \leq \Psi^{pot}_{j,c} \leq \overline{\Psi}_{j,c}$ and $\sum_{j=1\ldots m} \Psi^{pot}_{j,c} \geq k$.

The test sample ($x^{(t)}$) may belong to an upper approximation of some class ($x^{(t)} \in P^*\omega_j$) only if for any possible combination of $\{\Psi^{pot}_{j,c} : 1 \leq j \leq m\}$ the most of sample elements belong to that class (ω_j), i.e.

$$x^{(t)} \in P^*\omega_j \Leftrightarrow \left(\exists c : c_{min} \leq c \leq c_{max}\right) \begin{pmatrix} \exists \Psi^{pot}_{1,c}, \Psi^{pot}_{2,c} \ldots \Psi^{pot}_{m,c} : \\ (\underline{\Psi}_{1,c} \leq \Psi^{pot}_{1,c} \leq \overline{\Psi}_{1,c}) \wedge \\ (\underline{\Psi}_{2,c} \leq \Psi^{pot}_{2,c} \leq \overline{\Psi}_{2,c}) \wedge \\ \ldots \\ (\underline{\Psi}_{m,c} \leq \Psi^{pot}_{m,c} \leq \overline{\Psi}_{m,c}) \wedge \\ (\sum_{j'=1\ldots m} \Psi^{pot}_{j',c} \geq k) \end{pmatrix} \begin{pmatrix} \forall j' : 1 \leq j' \leq m \\ \wedge j' \neq j \end{pmatrix} \left(\Psi^{pot}_{j,c} > \Psi^{pot}_{j',c}\right). \tag{11}$$

Similarly, test sample ($x^{(t)}$) must belong to a lower approximation of some class ($x^{(t)} \in P_*\omega_j$) only if for all possible combination of $\{\Psi^{pot}_{j,c} : 1 \leq j \leq m\}$ the most of sample elements belong to that class (ω_j), what we can write as follows

$$x^{(t)} \in P_*\omega_j \Leftrightarrow \left(\forall c : c_{min} \leq c \leq c_{max}\right) \begin{pmatrix} \forall \Psi^{pot}_{1,c}, \Psi^{pot}_{2,c} \ldots \Psi^{pot}_{m,c} : \\ (\underline{\Psi}_{1,c} \leq \Psi^{pot}_{1,c} \leq \overline{\Psi}_{1,c}) \wedge \\ (\underline{\Psi}_{2,c} \leq \Psi^{pot}_{2,c} \leq \overline{\Psi}_{2,c}) \wedge \\ \ldots \\ (\underline{\Psi}_{m,c} \leq \Psi^{pot}_{m,c} \leq \overline{\Psi}_{m,c}) \wedge \\ (\sum_{j'=1\ldots m} \Psi^{pot}_{j',c} \geq k) \end{pmatrix} \begin{pmatrix} \forall j' : 1 \leq j' \leq m \\ \wedge j' \neq j \end{pmatrix} \left(\Psi^{pot}_{j,c} > \Psi^{pot}_{j',c}\right). \tag{12}$$

In practice usage these two formulas—(11), (12)—may be difficult and time consuming. Therefore in the paper, the following classification rules are proposed.

To state that **the sample belongs to upper approximation of class** ω_j at least for one $c : c_{min} \leq c \leq c_{max}$ the following two conditions must be fulfilled:

- If $\overline{\Psi}_{j,c} + \underline{\Psi}_{\sim j,c} \geq k$, then this sample belongs to upper approximation of ω_j, if for all other classes $\omega_{j'}$, $\underline{\Psi}_{j',c} < \overline{\Psi}_{j,c}$, i.e.

$$\begin{gathered}\left(\exists c : c_{min} \leq c \leq c_{max}\right) \\ \left(\left(\overline{\Psi}_{j,c} + \underline{\Psi}_{\sim j,c} \geq k\right) \wedge \left(\forall_{\substack{j':1\leq j\leq m \\ \wedge j'\neq j}}(\overline{\Psi}_{j,c} > \underline{\Psi}_{j',c})\right)\right) \Rightarrow \\ \Rightarrow x^{(t)} \in P^{*}\omega_j \; .\end{gathered} \tag{13}$$

- If $\overline{\Psi}_{j,c} + \underline{\Psi}_{\sim j,c} < k$, then this sample belongs to upper approximation of ω_j, if for all other classes $\omega_{j'}$, $\underline{\Psi}_{j',c} < \overline{\Psi}_{j,c}$. However, more samples (than $\underline{\Psi}_{j',c}$) must be chosen from other classes until k samples will be selected. Of course number of chosen samples from other class cannot exceed its current maximum ($\overline{\Psi}_{j',c}$) and should be lower than $\overline{\Psi}_{j,c}$, i.e.

$$\begin{gathered}\left(\exists c : c_{min} \leq c \leq c_{max}\right) \\ \left(\begin{array}{c}\left(\overline{\Psi}_{j,c} + \underline{\Psi}_{\sim j,c} < k\right) \wedge \left(\forall_{\substack{j':1\leq j\leq m \\ \wedge j'\neq j}}(\overline{\Psi}_{j,c} > \underline{\Psi}_{j',c})\right) \\ \wedge \left(\overline{\Psi}_{j,c} + \sum_{\substack{j':1\leq j\leq m \\ \wedge j'\neq j}} \left(\min\{\overline{\Psi}_{j,c} - 1, \overline{\Psi}_{j',c}\}\right) \geq k\right)\end{array}\right) \Rightarrow \\ \Rightarrow x^{(t)} \in P^{*}\omega_j \; .\end{gathered} \tag{14}$$

These equations may be presented in following, easy to implementation formula

$$\begin{gathered}x^{(t)} \in P^{*}\omega_j \Leftrightarrow \\ \left(\exists c : c_{min} \leq c \leq c_{max}\right) \\ \left(\forall_{\substack{j':1\leq j\leq m \\ \wedge j'\neq j}}(\overline{\Psi}_{j,c} > \underline{\Psi}_{j',c})\right) \wedge \\ \left((\overline{\Psi}_{j,c} + \underline{\Psi}_{\sim j,c} < k) \Rightarrow \sum_{\substack{j':1\leq j\leq m \\ \wedge j'\neq j}} \left(\min\{\overline{\Psi}_{j,c} - 1, \overline{\Psi}_{j',c}\}\right) \geq k - \overline{\Psi}_{j,c}\right) \; .\end{gathered} \tag{15}$$

To state that **the sample belongs to lower approximation of class** ω_j, for all $c : c_{min} \leq c \leq c_{max}$ the following conditions must be fulfilled:

- When $\underline{\Psi}_{j,c} + \overline{\Psi}_{\sim j,c} \geq k$ and sample belongs to the lower approximation of ω_j, then $\underline{\Psi}_{j,c}$ must be greater than $\overline{\Psi}_{j',c}$ for all other classes, i.e.

$$\begin{gathered}\left(\forall c : c_{min} \leq c \leq c_{max}\right) \\ \left(\underline{\Psi}_{j,c} + \overline{\Psi}_{\sim j,c} \geq k \wedge x^{(t)} \in P_{*}\omega_j\right) \Rightarrow \\ \Rightarrow \left(\forall_{\substack{j':1\leq j\leq m \\ \wedge j'\neq j}}(\underline{\Psi}_{j,c} > \overline{\Psi}_{j',c})\right) .\end{gathered} \tag{16}$$

- When $\underline{\Psi}_{j,c} + \overline{\Psi}_{\sim j,c} < k$ and sample belongs to the lower approximation of ω_j, more sophisticated condition must be fulfilled than before. In this case more (than $\underline{\Psi}_{j,c}$) samples from current class must be chosen until k samples will be selected. Nevertheless, number of chosen samples from current samples must be greater than $\overline{\Psi}_{j',c}$ for all other classes, i.e.

$$\begin{gathered}\left(\forall c : c_{min} \leq c \leq c_{max}\right)\\ \left(\underline{\Psi}_{j,c} + \overline{\Psi}_{\sim j,c} < k \wedge x^{(t)} \in P_*\omega_j\right) \Rightarrow\\ \Rightarrow \left(\forall_{\substack{j' : 1 \leq j \leq m\\ \wedge j' \neq j}} (k - \overline{\Psi}_{\sim j,c} > \overline{\Psi}_{j',c})\right).\end{gathered} \tag{17}$$

Previous equations may be combined into one, easy to implementation formula

$$\begin{gathered}x^{(t)} \in P_*\omega_j \Leftrightarrow\\ \left(\forall c : c_{min} \leq c \leq c_{max}\right)\\ \left((\underline{\Psi}_{j,c} + \overline{\Psi}_{\sim j,c} \geq k) \Rightarrow \left(\forall_{\substack{j' : 1 \leq j \leq m\\ \wedge j' \neq j}} (\underline{\Psi}_{j,c} > \overline{\Psi}_{j',c})\right)\right) \wedge\\ \left((\underline{\Psi}_{j,c} + \overline{\Psi}_{\sim j,c} < k) \Rightarrow \left(\forall_{\substack{j' : 1 \leq j \leq m\\ \wedge j' \neq j}} (k - \overline{\Psi}_{\sim j,c} > \overline{\Psi}_{j',c})\right)\right),\end{gathered} \tag{18}$$

what can be simplify to the following form

$$\begin{gathered}x^{(t)} \in P_*\omega_j \Leftrightarrow \left(\forall c : c_{min} \leq c \leq c_{max}\right)\\ \left(\forall_{\substack{j' : 1 \leq j \leq m\\ \wedge j' \neq j}} (\max\{\underline{\Psi}_{j,c}, k - \overline{\Psi}_{\sim j,c}\} > \overline{\Psi}_{j',c})\right).\end{gathered} \tag{19}$$

Using these equations algorithm states that for example presented in Figs. 1 and 2 the test sample belongs only to upper and lower approximation of ω_3.

3 Performed Simulations

Proposed algorithm was tested using databases (Table 1) from well-known UCI Repository [3]. All the research results were concluded according to 10-fold cross validation [1]. The main parameter of k-nn algorithm was set empirically. Authors used Manhattan distance instead of Euclidean, because it gave better results. In the course of experimental research following datasets have been applied:

Glass Identification (GI)—samples were prepared by B. German from Home Office Forensic Science Service in Aldermaston. This database was composed for construction of the system to classify the type of glass for forensic purpose. The database contains 214 samples described by 9 attributes, which are non-negative real numbers.

Table 1 Properties of used databases from UCI repository

Database name	#samples (M)	#attributes (n)	#classes (m)
Glass identification (GI)	214	9	2
Ionosphere	351	34	2
Iris	150	4	3
Wine recognition (WR)	178	13	3
Breast cancer wisconsin (BCW)	699	9	2

The number of classes was changed from original 7 to 2. The first one contains 163 samples and concerns window-type of glass, while the second one describes other type of glass.

Ionosphere—samples were gathered by Space Physics Group from Johns Hopkins University. This database may be used for classification type in ionosphere analysing radar [28]. It contains 351 samples, each described by 34 real numbers. The sample belongs to one of two classes of 225 or 126 items.

Iris—data was prepared by Fisher [7]. This database is probably the most popular for benchmarking of classification algorithms. It contains 150 samples, each one is described by 4 attributes of real numbers. They represent characteristics of one of three types (classes) of iris flowers, containing 50 items each.

Wine Recognition (WR)—database contains results of chemical analysis of 178 samples of wine, which can be divided into three classes. Every sample is described by 13 attributes.

Breast Cancer Wisconsin (BCW) Dataset—samples were prepared by William H. Wolberg from University of Wisconsin Hospitals. The samples may be used to build classifier which recognizes type of breast cancer (malignant or not) [31]. The database contains 699 samples described by 9 attributes, which are positive decimals. Each sample belongs to one of two classes, containing 458 and 241 samples. Value of one of the attributes is missing for 16 samples.

Proposed algorithm indicates membership of sample to upper and lower approximations, what requires more complex evaluation model that used in case typical classification. The system was tested using 4 different measures:

1. Lower approximation of correct classification rate—the first criterion counts how often test sample belongs to the lower approximation of an appropriate class, i.e.

$$CCR_* = \frac{1}{M^{(t)}} \sum_{s=1}^{M^{(t)}} \sum_{j=1}^{m} \begin{cases} 1 & \text{if } x_s^{(t)} \in P_*\omega_j \wedge \omega\left(x_s^{(t)}\right) = \omega_j \\ 0 & \text{otherwise} \end{cases}, \tag{20}$$

where $M^{(t)}$ is a number of test samples, $x_s^{(t)}$ is sth test sample, $x_s^{(t)} \in P_*\omega_j$ denotes that sample $x_s^{(t)}$ has been classified to the lower approximation of class ω_j and $\omega\left(x_s^{(t)}\right) = \omega_j$—that the proper class of sample $x_s^{(t)}$ is class ω_j.

2. Lower approximation of incorrect classification rate—this criterion computes how often test sample belongs to the lower approximation, but improper class, i.e.

$$ICR_* = \frac{1}{M^{(t)}} \sum_{s=1}^{M^{(t)}} \sum_{j=1}^{m} \begin{cases} 1 & \text{if } x_s^{(t)} \in P_*\omega_j \wedge \omega\left(x_s^{(t)}\right) \neq \omega_j \\ 0 & \text{otherwise} \end{cases}, \tag{21}$$

where $\omega\left(x_s^{(t)}\right) \neq \omega_j$ denotes that the proper class of sample $x_s^{(t)}$ is other than class ω_j.

3. Upper approximation of correct classification rate—the third measure defines how often test sample belongs to upper approximation of appropriate class, i.e.

$$CCR^* = \frac{1}{M^{(t)}} \sum_{s=1}^{M^{(t)}} \sum_{j=1}^{m} \begin{cases} 1 & \text{if } x_s^{(t)} \in P^*\omega_j \wedge \omega\left(x_s^{(t)}\right) = \omega_j \\ 0 & \text{otherwise} \end{cases}, \tag{22}$$

where $x_s^{(t)} \in P^*\omega_j$ denotes that sample $x_s^{(t)}$ has been classified to the upper approximation of class ω_j.

4. Upper approximation of incorrect classification rate—the last criterion measures how often test sample belongs to the upper approximation of improper class, i.e.

$$ICR^* = \frac{1}{(m-1) \cdot M^{(t)}} \sum_{s=1}^{M^{(t)}} \sum_{j=1}^{m} \begin{cases} 1 & \text{if } x_s^{(t)} \in P^*\omega_j \wedge \omega\left(x_s^{(t)}\right) \neq \omega_j \\ 0 & \text{otherwise} \end{cases}. \tag{23}$$

It is worth to note that in the denominator of criterion (23) occurs part $(m-1)$. This is due to the sample may belongs to upper approximations of more than one class. This measure can be also called "wastefulness" of the algorithm.

It is obvious, that

$$CCR_* + ICR^* = CCR^* + ICR_* = 100\,\%. \tag{24}$$

Proposed algorithm was tested using mentioned databases and measures. Parameter k was different for each dataset and was empirically defined: 11 for GI, 9 for Ionosphere, 5 for Iris, 11 for WR and 13 for BCW. For evaluation purposes, values of specified attributes in test samples were removed (considered as missing). The algorithm was tested in two different scenarios: in the first one (Tables 2, 5, 7 and 9) each attribute was erased separately, and in the second scenario (Tables 3, 4, 6, 8 and 10) set of "removed" attributes was growing until the algorithm gave completely useless answer ($CCR_* = 0, ICR^* = 100\,\%$). In each step the test important attribute was removed.

Table 2 Results of proposed algorithm for GI database and 8 available attributes

List of available attributes	Results (%)			
	CCR_*	ICR_*	CCR^*	ICR^*
$v_1, v_2, v_3, v_4, v_5, v_6, v_7, v_8$	65.8	0.5	99.5	32.7
$v_1, v_2, v_3, v_4, v_5, v_6, v_7, v_9$	20.6	0.0	100.0	78.1
$v_1, v_2, v_3, v_4, v_5, v_6, v_8, v_9$	58.8	0.0	100.0	40.8
$v_1, v_2, v_3, v_4, v_5, v_7, v_8, v_9$	30.8	0.0	100.0	68.2
$v_1, v_2, v_3, v_4, v_6, v_7, v_8, v_9$	76.2	0.5	99.5	23.4
$v_1, v_2, v_3, v_5, v_6, v_7, v_8, v_9$	62.7	0.0	100.0	35.5
$v_1, v_2, v_4, v_5, v_6, v_7, v_8, v_9$	4.7	0.0	100.0	94.4
$v_1, v_3, v_4, v_5, v_6, v_7, v_8, v_9$	71.4	0.0	100.0	27.1
$v_2, v_3, v_4, v_5, v_6, v_7, v_8, v_9$	64.9	0.5	99.5	34.7

Table 3 Results of proposed algorithm for GI database and chosen set of available attributes

List of available attributes	Results (%)			
	CCR_*	ICR_*	CCR^*	ICR^*
$v_1, v_2, v_3, v_4, v_5, v_6, v_7, v_8, v_9$	92.0	8.0	92.0	8.0
$v_1, v_2, v_3, v_4, v_5, v_6, v_7, v_9$	20.6	0.0	100.0	78.1
$v_1, v_3, v_4, v_5, v_6, v_7, v_9$	0.0	0.0	100.0	100.0

Table 4 Results of proposed algorithm for Ionosphere database and chosen set of available attributes

List of removed attributes	Results (%)			
	CCR_*	ICR_*	CCR^*	ICR^*
–	86.9	13.1	86.9	13.1
v_{29}	78.7	7.7	92.0	21.3
v_{18}, v_{29}	70.4	4.5	95.5	28.5
v_{10}, v_{18}, v_{29}	47.1	0.8	99.2	49.2
$v_{10}, v_{18}, v_{29}, v_{33}$	22.8	0.0	100.0	76.6
$v_8, v_{10}, v_{18}, v_{29}, v_{33}$	13.7	0.0	100.0	86.3
$v_8, v_{10}, v_{15}, v_{18}, v_{29}, v_{33}$	8.8	0.0	100.0	90.3
$v_7, v_8, v_{10}, v_{15}, v_{18}, v_{29}, v_{33}$	2.0	0.0	100.0	98.0
$v_7, v_8, v_{10}, v_{15}, v_{18}, v_{27}, v_{29}, v_{33}$	0.0	0.0	100.0	100.0

In case of no missing values, proposed algorithm returns the same answer as typical k-nn method, because $c_{min} = c_{max}$ and identical samples used for voting as in standard k-nn algorithm. Usually, the level of ICR_* was very low in cost of low CCR_*, conversely CCR^* was very high in cost of high ICR^*. In the first scenario, when values of single attribute were removed, there were significant differences (Tables 2, 7, 9 and particularly 5) in efficiency between various cases.

Table 5 Results of proposed algorithm for Iris database and 3 available attributes

List of available attributes	Results (%)			
	CCR_*	ICR_*	CCR^*	ICR^*
v_1, v_2, v_3	0.0	0.0	100.0	61.0
v_1, v_2, v_4	0.0	0.0	100.0	56.3
v_1, v_3, v_4	26.7	0.0	100.0	40.0
v_2, v_3, v_4	26.7	0.0	100.0	36.3

Table 6 Results of proposed algorithm for Iris database and chosen set of available attributes

List of available attributes	Results (%)			
	CCR_*	ICR_*	CCR^*	ICR^*
v_1, v_2, v_3, v_4	94.7	5.3	94.7	2.7
v_2, v_3, v_4	26.7	0.0	100.0	36.3
v_3, v_4	0.0	0.0	100.0	90.0
v_3	0.0	0.0	100.0	100.0

Table 7 Results of proposed algorithm for WR database and 12 available attributes

List of available attributes	Results (%)			
	CCR_*	ICR_*	CCR^*	ICR^*
$v_1, v_2, v_3, v_4, v_5, v_6, v_7, v_8, v_9, v_{10}, v_{11}, v_{12}$	42.7	0.0	100.0	30.3
$v_1, v_2, v_3, v_4, v_5, v_6, v_7, v_8, v_9, v_{10}, v_{11}, v_{13}$	54.4	0.0	100.0	24.8
$v_1, v_2, v_3, v_4, v_5, v_6, v_7, v_8, v_9, v_{10}, v_{12}, v_{13}$	58.3	0.0	99.4	19.4
$v_1, v_2, v_3, v_4, v_5, v_6, v_7, v_8, v_9, v_{11}, v_{12}, v_{13}$	50.0	0.0	100.0	25.8
$v_1, v_2, v_3, v_4, v_5, v_6, v_7, v_8, v_{10}, v_{11}, v_{12}, v_{13}$	60.6	0.0	100.0	18.8
$v_1, v_2, v_3, v_4, v_5, v_6, v_7, v_9, v_{10}, v_{11}, v_{12}, v_{13}$	65.1	0.0	100.0	16.9
$v_1, v_2, v_3, v_4, v_5, v_6, v_8, v_9, v_{10}, v_{11}, v_{12}, v_{13}$	56.6	0.0	100.0	22.2
$v_1, v_2, v_3, v_4, v_5, v_7, v_8, v_9, v_{10}, v_{11}, v_{12}, v_{13}$	64.1	0.0	100.0	17.7
$v_1, v_2, v_3, v_4, v_6, v_7, v_8, v_9, v_{10}, v_{11}, v_{12}, v_{13}$	55.1	0.0	100.0	21.4
$v_1, v_2, v_3, v_5, v_6, v_7, v_8, v_9, v_{10}, v_{11}, v_{12}, v_{13}$	66.3	0.0	100.0	15.2
$v_1, v_2, v_4, v_5, v_6, v_7, v_8, v_9, v_{10}, v_{11}, v_{12}, v_{13}$	68.0	0.0	100.0	15.7
$v_1, v_3, v_4, v_5, v_6, v_7, v_8, v_9, v_{10}, v_{11}, v_{12}, v_{13}$	52.8	0.0	100.0	23.3
$v_2, v_3, v_4, v_5, v_6, v_7, v_8, v_9, v_{10}, v_{11}, v_{12}, v_{13}$	49.5	0.0	100.0	24.4

Table 8 Results of proposed algorithm for WR database and chosen set of available attributes

List of available attributes	Results (%)			
	CCR_*	ICR_*	CCR^*	ICR^*
$v_1, v_2, v_3, v_4, v_5, v_6, v_7, v_8, v_9, v_{10}, v_{11}, v_{12}, v_{13}$	97.7	2.3	97.7	1.1
$v_1, v_2, v_3, v_5, v_6, v_7, v_8, v_9, v_{10}, v_{11}, v_{12}, v_{13}$	66.3	0.0	100.0	15.2
$v_1, v_3, v_5, v_6, v_7, v_8, v_9, v_{10}, v_{11}, v_{12}, v_{13}$	6.1	0.0	100.0	61.7
$v_1, v_3, v_5, v_6, v_7, v_9, v_{10}, v_{11}, v_{12}, v_{13}$	0.0	0.0	100.0	96.9
$v_1, v_3, v_6, v_7, v_9, v_{10}, v_{11}, v_{12}, v_{13}$	0.0	0.0	100.0	100.0

Table 9 Results of proposed algorithm for BCW Dataset and 8 available attributes

List of available attributes	Results (%)			
	CCR_*	ICR_*	CCR^*	ICR^*
$v_1, v_2, v_3, v_4, v_5, v_6, v_7, v_8$	75.5	0.1	99.9	23.3
$v_1, v_2, v_3, v_4, v_5, v_6, v_7, v_9$	72.4	0.0	99.9	25.3
$v_1, v_2, v_3, v_4, v_5, v_6, v_8, v_9$	80.8	0.4	99.4	18.5
$v_1, v_2, v_3, v_4, v_5, v_7, v_8, v_9$	68.8	0.3	99.6	30.5
$v_1, v_2, v_3, v_4, v_6, v_7, v_8, v_9$	80.5	0.1	99.9	18.8
$v_1, v_2, v_3, v_5, v_6, v_7, v_8, v_9$	75.7	0.3	99.7	23.8
$v_1, v_2, v_4, v_5, v_6, v_7, v_8, v_9$	78.5	0.1	99.7	20.2
$v_1, v_3, v_4, v_5, v_6, v_7, v_8, v_9$	76.4	0.1	99.9	22.0
$v_2, v_3, v_4, v_5, v_6, v_7, v_8, v_9$	78.5	0.6	99.4	20.3

Table 10 Results of proposed algorithm for BCW Dataset and chosen sets of available attributes

List of available attributes	Results (%)			
	CCR_*	ICR_*	CCR^*	ICR^*
$v_1, v_2, v_3, v_4, v_5, v_6, v_7, v_8, v_9$	95.6	2.9	97.0	4.0
$v_1, v_2, v_3, v_4, v_5, v_6, v_7, v_9$	72.4	0.0	99.9	25.3
$v_1, v_2, v_3, v_4, v_5, v_6, v_7$	5.3	0.0	100.0	94.4
$v_1, v_3, v_4, v_5, v_6, v_7$	0.0	0.0	100.0	100.0

4 Conclusions

Proposed algorithm may be considered as non-trivial generalization of k-nn. The proposed improvement over k-nn allows to operate with interval-type of attribute value and as consequent interval-type of the distance. This modification can be also applied in the case of missing values, because missing value can be replaced be the whole spectrum (in the most cases interval) of possible values. Then, the range of input features variation should be known. The algorithm processes the input interval and calculates the lower and upper distances between samples. Thus, it can take into account all possible but unknown input states. This approach is inspired directly by Pawlak rough sets theory [18–20].

The experimental studies confirm the analytical consideration. The proposed classifier, alike as the rough-neuro-fuzzy classifiers presented in [14–16] does not give incorrect answers as the result of missing values. If the classification is proper for complete input vector, it is also proper in case of incomplete data. If the available information is insufficient to make certain decision, the proposed classifier indicates the additional classes corresponding to possible values of feature with missing values. It expressed in this way the uncertainty in the decision.

Acknowledgments The project was funded by the National Science Centre under decision number DEC-2012/05/B/ST6/03620.

References

1. Arlot, S., Celisse, A.: A survey of cross-validation procedures for model selection. Stat. Surv. **4**, 40–79 (2010)
2. Bao, Y., Du, X., Ishii, N.: Improving performance of the k-nearest neighbor classifier by tolerant rough sets. In Proceedings of the Third International Symposium on Cooperative Database Systems for Advanced Applications, pp. 167–171 (2001)
3. Collective work. Uci machine learning repository. http://archive.ics.uci.edu/ml/datasets.html
4. Cover, T., Hart, P.: Nearest neighbor pattern classification. IEEE Trans. Inf. Theory **13**(1), 21–27 (1967)
5. Cpałka, K., Rutkowski, L.: Flexible takagi-sugeno fuzzy systems. In: Proceedings of IEEE International Joint Conference on Neural Networks (IJCNN), vol. 3, pp. 1764–1769 (2005)
6. Duda, R.O., Hart, P.E., Stork, D.G.: Pattern Classification. A Wiley-Interscience Publication, Wiley, New York (2001)
7. Fisher, R.: The use of multiple measurements in taxonomic problems. Ann. Eugen. **7**, 179–188 (1936)
8. Gabryel, M., Korytkowski, M., Scherer, R., Rutkowski, L.: Object detection by simple fuzzy classifiers generated by boosting. In: Rutkowski, L., Korytkowski, M., Scherer, R., Tadeusiewicz, R., Zadeh, L., Zurada, J., (eds.), LNCS, vol. 7894, pp. 540–547. Springer, Berlin (2013)
9. Greblicki, W., Rutkowski, L.: Density-free Bayes risk consistency of nonparametric pattern recognition procedures. Proc. IEEE **69**(4), 482–483 (1981)
10. He, M., Du, Y.-P.: Research on attribute reduction using rough neighborhood model. In: Proceedings of International Seminar on Business and Information Management (ISBIM), vol. 1, pp. 268–270 (2008)
11. Ishii, N., Torii, I., Bao, Y., Tanaka, H.: Modified reduct: nearest neighbor classification. In: Proceedings of IEEE/ACIS 11th International Conference on Computer and Information Science (ICIS), pp. 310–315 (2012)
12. Ishii, N., Torii, I., Bao, Y., Tanaka, H.: Mapping of nearest neighbor for classification. In: Proceedings of IEEE/ACIS 12th International Conference on Computer and Information Science (ICIS), pp. 121–126 (2013)
13. Keller, J., Gray, M., Givens, J.: A fuzzy k-nearest neighbor algorithm. IEEE Trans. Syst. Man Cybern. **15**(4), 580–585 (1985)
14. Nowicki, R.: On combining neuro-fuzzy architectures with the rough set theory to solve classification problems with incomplete data. IEEE Trans. Knowl. Data Eng. **20**(9), 1239–1253 (2008)

15. Nowicki, R.: Rough-neuro-fuzzy structures for classification with missing data. IEEE Trans. Syst. Man Cybern.-Part B: Cybern. **39**(6), 1334–1347 (2009)
16. Nowicki, R.: On classification with missing data using rough-neuro-fuzzy systems. Int. J. Appl. Math. Comput. Sci. **20**(1), 55–67 (2010)
17. Nowicki, R.K., Nowak, B.A., Woźniak, M.: Rough k nearest neighbours for classification in the case of missing input data. In: Proceedings of the 9th International Conferenceon Knowledge, Information and Creativity Support Systems, pp. 196–207 (2014)
18. Pawlak, M.: Kernel classification rules from missing data. IEEE Trans. Inf. Theory **39**, 979–988 (1993)
19. Pawlak, Z.: Rough Sets: Theoretical Aspects of Reasoning About Data. Kluwer, Dordrecht (1991)
20. Pawlak, Z.: Rough sets, decision algorithms and bayes theorem. Eur. J. Oper. Res. **136**, 181–189 (2002)
21. Pedrycz, W., Bargiela, A.: Granular clustering: a granular signature of data. IEEE Trans. Syst. Man Cybern.-Part B: Cybern. **32**(2), 212–224 (2002)
22. Rutkowski, L.: On Bayes risk consistent pattern recognition procedures in a quasi-stationary environment. IEEE Trans. Pattern Anal. Mach. Intell. PAMI **4**(1), 84–87 (1982)
23. Rutkowski, L.: Adaptive probabilistic neural networks for pattern classification in time-varying environment. IEEE Trans. Neural Netw. **15**(4), 811–827 (2004)
24. Rutkowski, L., Cpałka, K.: Compromise approach to neuro-fuzzy systems. In: Sincak, P., Vascak, J., Kvasnicka, V., Pospichal, J. (eds.), Intelligent Technologies—Theory and Applications, vol. 76, pp. 85–90. IOS Press (2002)
25. Sarkar, M.: Fuzzy-rough nearest neighbors algorithm. In: Proceedings of IEEE International Conference on Systems, Man, and Cybernetics, vol. 5, pp. 3556–3561 (2000)
26. Scherer, R.: Neuro-fuzzy systems with relation matrix. In: Rutkowski, L., Scherer, R., Tadeusiewicz, R., Zadeh, L.A., Zurada, J.M. (eds.) LNAI, vol. 6113, pp. 210–215. Springer, Berlin (2010)
27. Shafer, G.: A Mathematical Theory of Evidence. Princeton University Press, Princeton (1976)
28. Sigillito, V., Wing, S., Hutton, L., Baker, K.: Classification of radar returns from the ionosphere using neural networks. Johns Hopkins APL Tech. Dig. 262–266 (1989)
29. Verbiest, N., Cornelis, C., Jensen, R.: Fuzzy rough positive region based nearest neighbour classification. In: Proceedings of IEEE International Conference on Fuzzy Systems (FUZZ-IEEE), pp. 1–7 (2012)
30. Villmann, T., Schleif, F., Hammer, B.: Fuzzy labeled soft nearest neighbor classification with relevance learning. In: Proceedings of Fourth International Conference on Machine Learning and Applications, pp. 11–15 (2005)
31. Wolberg, W., Mangasarian, O.: Multisurface Method of Pattern Separation for Medical Diagnosis Applied to Breast Cytology. In: Proceedings of the National Academy of Sciences, vol. 87, pp. 9193–9196. U.S.A. (1990)
32. Yager, R.: Using fuzzy methods to model nearest neighbor rules. IEEE Trans. Syst. Man Cybern. Part B: Cybern. **32**(4), 512–525 (2002)

An Application of Firefly Algorithm to Position Traffic in NoSQL Database Systems

Marcin Woźniak, Marcin Gabryel, Robert K. Nowicki and Bartosz A. Nowak

Abstract In this paper, an application of Computational Intelligence methods in positioning and optimization of traffic in NoSQL database system modeled with exponentially distributed service and vacation is discussed. Positioning of the system modeled with independent 2-order hyper exponential input stream of packets and exponential service time distribution is solved using firefly algorithm. Different scenarios of examined system operation are presented and analyzed.

Keywords Firefly algorithm · Queueing system · Evolutionary computation

1 Introduction

In modern computer systems more and more data is collected. The servers operate on large data sets. Service must be efficient and fast enough to work with many clients at the same time. In NoSQL systems we use efficient and dedicated solutions to increase performance, see [1]. Dedicated sorting algorithms help to organize large data sets and response to clients as fast as possible, see examples in [2–4]. However it is not easy to position these systems for processing large amounts of data and response to many clients in the service. Extended research on this situations are presented

M. Woźniak (✉)
Institute of Mathematics, Silesian University of Technology,
Kaszubska 23, 44-101 Gliwice, Poland
e-mail: Marcin.Wozniak@polsl.pl

M. Gabryel · R.K. Nowicki · B.A. Nowak
Institute of Computational Intelligence, Czestochowa University of Technology,
Al. Armii Krajowej 36, 42-200 Czestochowa, Poland
e-mail: Marcin.Gabryel@iisi.pcz.pl

R.K. Nowicki
e-mail: Robert.Nowicki@iisi.pcz.pl

B.A. Nowak
Department of Mathematical Methods in Computer Science,
University of Warmia and Mazury, ul. Słoneczna 54, 10-710 Olsztyn, Poland
e-mail: Bartosz.Nowak@iisi.pcz.pl

S. Kunifuji et al. (eds.), *Knowledge, Information and Creativity Support Systems*,
Advances in Intelligent Systems and Computing 416,
DOI 10.1007/978-3-319-27478-2_18

in [5, 6], where efficient methods of indexing and sorting in large NoSQL systems are discussed. Here, another important aspect of optimal service in large NoSQL systems will be discussed.

Traffic in the network and therefore efficient organization of the service are other aspects that increase efficiency and Quality of Service (QoS). Let us now think of the network, where NoSQL database server is serving various clients. In common service many clients send requests at the same time. The server collects them and proceeds actions. After processing knowledge in database server responses to the requests, but this goes according to the income queue. Earlier requests shall be answered first and so on, according to First In–First Out (FIFO) method. There are significant difficulties to position this system for the most efficient operation. One shall define optimal service, vacation and income parameters. In this paper NoSQL database system is positioned for the most efficient service and the lowest possible cost of work by the use of Evolutionary Computation (EC) methods.

For this type of object, various methods of modeling and simulation can be applied. One of them is analysis of provided stochastic characteristics, which describe operation of devices such as: IP routers, audio or video cards with buffer of packets waiting for displaying and various types of database servers with queue of requests from remote clients. In each of these objects, operation model is defined for applied Queueing System (QS).

Definition 1 Queueing System modeled for computer network with NoSQL database is a process, where requests from remote clients arrive to be served. Each request leaves the system after being served. However after arrival, some of them may wait for the service. These, that wait, are buffered in a queue.

Let us take situation in network with NoSQL database server. Various clients may want to find information and send requests to the server. Server finds the information and then sends it to the clients according to arrivals. The server answers, however there is limited number of requests to be served at the same time. Therefore, it gathers all the requests to put them in a FIFO queue. All the requests are serviced according to the order in this queue. Let us say, that the queue has defined capacity. Schematic situation is pictured in Fig. 1. This is sample description of the service in computer network with database, where a dedicated QS can be applied to optimize operation cost and increase QoS. The situation can be modeled for $T_{service}$, T_{income} and $T_{vacation}$, which describe average time of service for each request, average time of incoming requests and average time in which the system takes break (for backup, conservation and etc.), respectively. All inter-arrival times, service times and server vacations are totally independent random variables, where the symbols used in QS description in time t are:

- τ_1—the first busy period of the system (starting at $t = 0$);
- δ_1—the first idle time of the system (consisting of the first vacation time v_1 and the first server standby time q_1);
- $h(\tau_1)$—the number of packets completely served during τ_1;
- $X(t)$—the number of packets present in the system at time t.

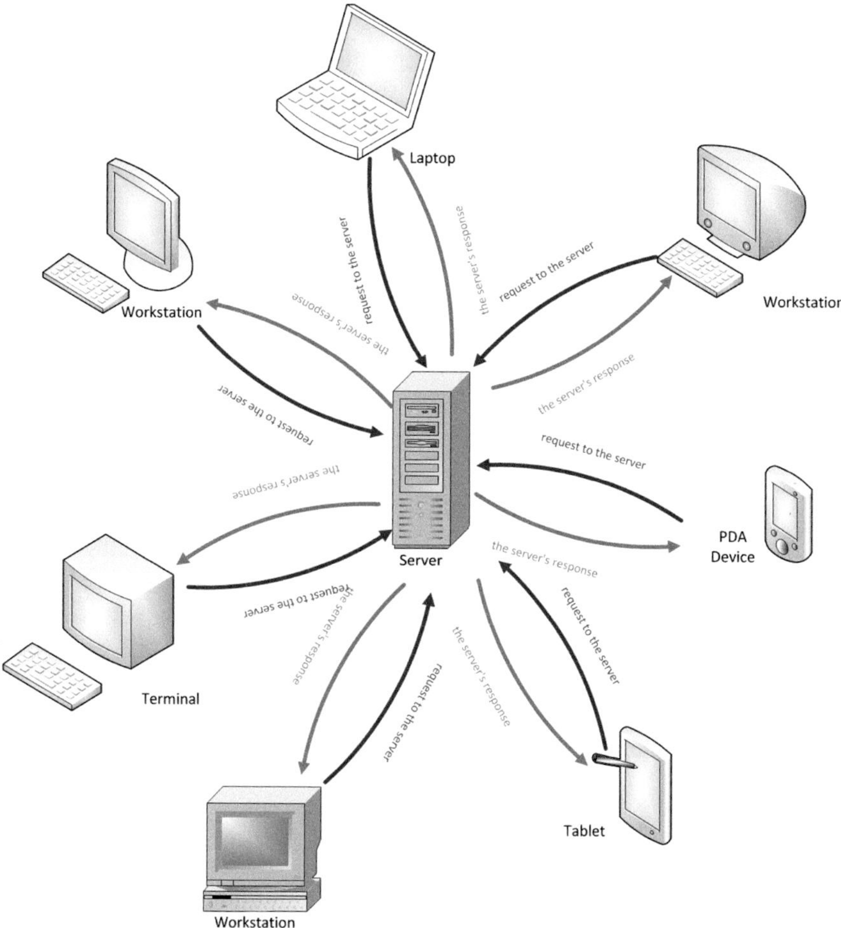

Fig. 1 Sample traffic in LAN with a single database server

In this paper we discuss a novel way to simulate and position NoSQL database traffic modeled with dedicated QS, where at the time there is only one request arrival or response departure. First attempt to this task was presented in [7]. Sample QS operation in this system is pictured in Fig. 2. This situation can be modeled with dedicated QS, which helps to position server operation, increases QoS and minimize costs of work. However this is non-trivial problem and many interesting papers are devoted to various positioned objects. Classical cost structure is considered in [8]. While in [9, 10] most important aspects of positioning and cost optimization using QS are presented. Various queueing models for applied server types are investigated in [11–14]. See also [15–17] for a review of important results on modeling and positioning.

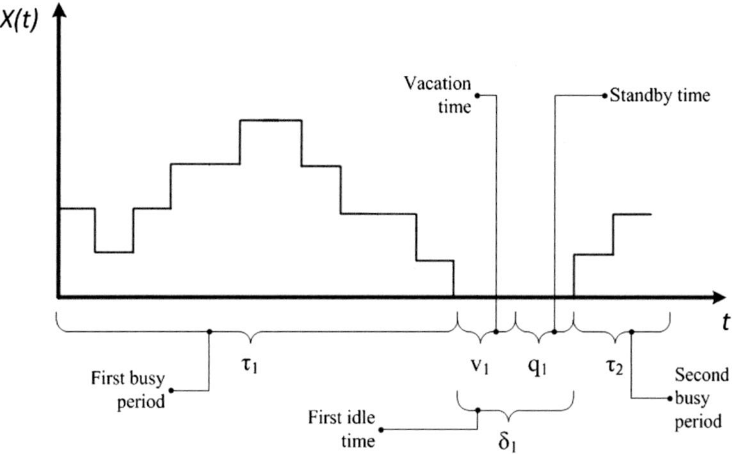

Fig. 2 Sample queue process in NoSQL database system over time t, when one request arrives or one after service departures

In this paper sample results of the research on similar QS are applied. In [18, 19] representation for joint transform of first busy period, first idle time and number of packets completely served during first busy period in $GI/G/1$-type system with batch arrivals and exponential single server vacation is discussed. Further characteristics of the system with single vacations, Poisson arrivals, generally distributed service times and infinite buffers can be found in [20, 21]. In [22] nonstationary behavior of waiting time distribution in a finite-buffer queue with single server vacations is investigated. All these research results can help to model and position traffic in NoSQL system types discussed in [1, 2, 23, 24]. Let us see the model of QS for NoSQL database traffic.

2 Queueing Model

Similar attempt to model and position QSs by dividing observation horizon on relatively short time intervals in which arrival process is recurrent with different distributions in different interval times is presented in [25–28]. To model NoSQL server operation a finite-buffer $H_2/M/1/N$-type QS was used, which the best simulate situation in this type of server traffic, for more details see [29, 30]. Here is presented only a brief description, just to help in understanding NoSQL positioning and simulation problem using Definition 1, a full analytical model is presented in details in [27].

In modeled service are used inter-arrival times of 2-order, hyper exponentially distributed random variables with distribution function

$$F(t) = p_1\left(1 - e^{-\lambda_1 t}\right) + p_2\left(1 - e^{-\lambda_2 t}\right), \quad t > 0, \tag{1}$$

where $\lambda_i > 0$ and $p_i \geq 0$ for $i = 1, 2$. Distribution of inter-arrival times is mixture of two exponential distributions with parameters λ_1 and λ_2, which are being "chosen" with probabilities p_1 and p_2 respectively. In the system, there are $(N - 1)$ places in a queue and one place for a packet in the service. System starts working at $t = 0$ with at least one packet present, see schematic situation in Fig. 2. After busy period the server begins compulsory single vacation in the service which is modeled with 2-order hyper exponentially distributed random variable with distribution function

$$V(t) = q_1\left(1 - e^{-\alpha_1 t}\right) + q_2\left(1 - e^{-\alpha_2 t}\right), \quad t > 0. \tag{2}$$

Interpretation of parameters α_i and q_i for $i = 1, 2$ is similar to that for λ_i and p_i. If at the end of vacation there is no packet present in the system, the server is on standby and waits for first arrival to start service. If there is at least one packet waiting for service in the buffer at the end of vacation, the service process starts immediately and new busy period begins.

2.1 Cost Optimization Problem

Functions $F(\cdot)$ and $V(\cdot)$ are implemented to investigate some various aspects of operation of the examined NoSQL system, where inter-arrival times and vacation period have distribution functions defined in (1) and (2), respectively. In the research optimal set of parameters λ_i, p_i, μ and α_i is found. The set defines possible service situations, for which optimal cost of work is found. Therefore $r_n(c_1)$ is defined to describe minimal amount of resources needed to perform all operations.

Definition 2 Optimal cost of work in modeled QS is a minimal amount of money, energy or any other resources that the system may need to perform all operations.

In scope of the Definition 2, optimal cost of work has a form of equation, used in optimization process

$$r_n(c_1) = \frac{Q_n(c_1)}{\mathbf{E}_n(c_1)} = \frac{r(\tau_1)\mathbf{E}_n\tau_1 + r(\delta_1)\mathbf{E}_n\delta_1}{\mathbf{E}_n\tau_1 + \mathbf{E}_n\delta_1}, \tag{3}$$

where the notations are: $r(\tau_1)$–fixed unit operation costs during busy period τ_1, $r(\delta_1)$–fixed unit operation costs during idle time δ_1, $\mathbf{E}_n\tau_1$–mean of busy period τ_1 and $\mathbf{E}_n\delta_1$–mean of idle time δ_1 on condition that the system starts with n packets present. The explicit formula with detailed information and description for conditional joint characteristic functions of τ_1, δ_1 and $h(\tau_1)$ is presented in [26–28]. Here let us briefly discuss modeling of applied QS. General equation to calculate this values is

$$B_n(s, \varrho, z) = \mathbf{E}\{e^{-s\tau_1 - \varrho\delta_1} z^{h(\tau_1)} \mid X(0) = n\} =$$

$$= \frac{D(s, \varrho, z) - G(s, \varrho, z)}{H(s, z)} R_{n-1}(s, z) + \sum_{k=2}^{n} R_{n-k}(s, z)\Psi_k(s, \varrho, z), \quad 2 \leq n \leq N. \tag{4}$$

where $s \geq 0, \varrho \geq 0$ and $|z| \leq 1, n \geq 1$. Moreover $a_n(s,z), \Psi_n(s,\varrho,z), D(s,\varrho,z), G(s,\varrho,z)$ and $H(s,z)$ are defined in (5), (6), (9), (10) and (11) respectively.

$$a_n(s,z) = \int_0^\infty \frac{(z\mu t)^n}{n!} e^{-(\mu+s)t} dF(t), \quad n \geq 0, \tag{5}$$

$$\Psi_n(s,\varrho,z) = -\frac{(z\mu)^n}{(n-1)!} \left[\int_0^\infty dF(t) \int_0^t x^{n-1} e^{-(\mu+s)x} \right.$$
$$\left. \times \left(e^{-\varrho(t-x)} V(t-x) + \int_{t-x}^\infty e^{-\varrho y} dV(y) \right) dx \right]. \tag{6}$$

Moreover, sequence $a_n(s,z)$ in (5) helps to recursively define

$$R_0(s,z) = 0, \quad R_1(s,z) = a_0^{-1}(s,z),$$
$$R_{n+1}(s,z) = R_1(s,z)(R_n(s,z) - \sum_{k=0}^{n} a_{k+1}(s,z) R_{n-k}(s,z)). \tag{7}$$

With introduced following function

$$f(s) = \int_0^\infty e^{-st} dF(t), \quad s > 0 \tag{8}$$

we finally have components of $r_n(c_1)$ operation costs defined in (3)

$$D(s,\varrho,z) = \sum_{k=1}^{N-1} a_k(s,z) \sum_{i=2}^{N-k+1} R_{N-k+1-i}(s,z) \Psi_i(s,\varrho,z), \tag{9}$$

$$G(s,\varrho,z) = \Psi_N(s,\varrho,z) + \big(1 - f(\mu+s)\big) \sum_{k=2}^{N} R_{N-k}(s,z) \Psi_k(s,\varrho,z) \tag{10}$$

and

$$H(s,z) = \big(1 - f(\mu+s)\big) R_{N-1}(s,z) - \sum_{k=1}^{N-1} a_k(s,z) R_{N-k}(s,z). \tag{11}$$

Indeed, since:

$$\mathbf{E}_n e^{-s\tau_1} = \mathbf{E}\{e^{-s\tau_1} \mid X(0) = n\} = B_n(s,0,1), \tag{12}$$

then

$$\mathbf{E}_n \tau_1 = -\frac{\partial}{\partial s} B_n(s,0,1)\Big|_{s=0} \tag{13}$$

and

$$\mathbf{E}_n\delta_1 = -\frac{\partial}{\partial\varrho}B_n(0, \varrho, 1)\Big|_{\varrho=0}. \tag{14}$$

The QS model presented above was solved using Wolfram Mathematica 9.0 in order to define analytical form of total cost Eq. (3). As You see QS has complicated mathematical model, which is not easy to position even using tailored analytical methods. Therefore an application of EC method to position presented NoSQL system for best QoS at lowest cost is presented.

3 Applied Firefly Algorithm

Firefly algorithm (FA) imitates behavior of flying insects that we all can see in the summer. FA method uses some characteristics of fireflies to search for optimum in examined object's solution space. As the research show, FA is efficient in various applications. In [31, 32] FA is applied to help in 2D images processing. In [33] FA is applied to minimum cross entropy threshold selection, while in [34] is described it's efficiency in image compression. This algorithm is also efficient in continuous optimization [35] or multi-modal optimization [36]. Let us now see dedicated mathematical model.

Firefly of given species can be described by several biological traits:

- Specific way of flashing,
- Specific way of moving,
- Specific perception of other individuals.

These features are modeled to map them in specific numerical values. In this way we translate natural characteristics in mathematical model used to develop EC algorithm. Thus, in implementation of firefly algorithm are:

γ–light absorption factor in given circumstances,
μ–factor for random motion of each individual,
β–factor for attractiveness of firefly species.

These features allow to implement behavior of different species or conditions of natural environment. Let us now see description of innate qualities:

- All fireflies are unisex, therefore one individual can be attracted to any other one regardless of gender.
- Attractiveness is proportional to brightness. Thus, for every two fireflies less clear flashing one will move toward brighter one.
- Attractiveness is proportional to brightness and decreases with increasing distance between individuals.
- If there is no clearer and more visible firefly within the range, then each one will move randomly.

Movement of individuals is conditioned by distance to other individuals surrounding it. Firefly will go to the most attractive one measuring intensity of flicker. Thus, using mathematical assumptions, individuals will be characterized by suitable metric. Therefore we can model natural identification of individuals and their attraction to each other depending on the distance separating them. In nature, fireflies that are closer, not only see better, but also seem to be more attractive to the others. Using these features calculations simulate natural behavior.

Distance between any two fireflies i and j situated at points $\mathbf{x_i}$ and $\mathbf{x_j}$ in object's solutions space can be defined as Cartesian metric

$$r_{ij}^t = \|\mathbf{x_i}^t - \mathbf{x_j}^t\| = \sqrt{\sum_{k=1}^{n}(x_{i,k}^t - x_{j,k}^t)^2}, \tag{15}$$

where the notations in t iteration are: $\mathbf{x}_i^t$; $\mathbf{x}_j^t$–points in R^n object's model space, $x_{i,k}^t$; $x_{k,j}^t$–kth components of the spatial coordinates $\mathbf{x_i}$ and $\mathbf{x_j}$ that describe each firefly position in the model space. In the research n is dimension of the object's fitness function in (3), what mainly is 9 as the operation cost depends on number of parameters calculated in Sect. 4 for different scenarios.

Attractiveness of firefly i to firefly j decreases with increasing distance. It is proportional to intensity of light seen by surrounding individuals defined as

$$\beta_{ij}(r_{ij}^t) = \beta \cdot e^{-\gamma \cdot (r_{ij}^t)^2}, \tag{16}$$

where the notations in t iteration are: $\beta_{ij}(r_{ij}^t)$–attractiveness of firefly i to firefly j, r_{ij}^t–distance between firefly i and firefly j defined in (15), γ–light absorption factor mapping natural conditions, β–attractiveness factor.

Finally, firefly i motions toward more attractive (clearer flashing) individual j using information about other individuals in the population denotes formula

$$\mathbf{x_i}^{t+1} = \mathbf{x_i}^t + (\mathbf{x_j}^t - \mathbf{x_i}^t) \cdot \beta_{ij}(r_{ij}^t) + \mu \cdot e_i, \tag{17}$$

where the notations in t iteration are: $\mathbf{x_i}^t$; $\mathbf{x_j}^t$–points in the object's model space, $\beta_{ij}(r_{ij}^t)$–attractiveness of firefly i to firefly j modeled in (16), r_{ij}^t–distance between fireflies i and j modeled in (15), γ–light absorption factor mapping natural conditions, μ–factor mapping natural random motion of individuals, e_i–randomized vector changing position of fireflies. Implementation of dedicated FA is presented in Algorithm 1.

Start,
Define all coefficients: γ–light absorption factor, β–attractiveness factor, μ–natural random motion, number of *fireflies* and *generation*–number of iterations in the algorithm,
Define fitness function for the algorithm: $r_n(c_1)$ in (3),
Create at random initial population P of *fireflies* in QS solution space,
$t = 0$,
while $t \leq$ *generation* **do**
  Calculate distance between individuals in population P using (15),
  Calculate attractiveness for individuals in population P using (16),
  Evaluate individuals in population P using (3),
  Create population O: move individuals towards closest and most attractive individual using (17),
  Evaluate individuals in population O using (3),
  Replace δ worst individuals from P with δ best individuals from O and the rest of individuals take at random,
  Next generation $t = t + 1$,
end
Values from population P with best fitness are solution,
Stop.

Algorithm 1: Dedicated FA to optimize defined QS with NoSQL database

4 Optimal NoSQL Database System Positioning

Research results help to position the system for lowest response time and optimize service cost considered in variants: under-load, critical load and overload. FA simulations were performed for $r(\tau_1) = 0.5$ and $r(\delta_1) = 0.5$. It means that according to Definition 2, modeled system service and vacation uses 0.5 energy unit each, which costs differ depending on the local energy billing rate. These values may be changed, what makes presented model flexible and easily applicable to any similar system. Presented results are averaged values of 100 samplings, 80 fireflies in 20 generations with $\beta = 0.3$, $\gamma = 0.3$, $\mu = 0.25$, $\delta = 25\,\%$, where:

- Average service time: $T_{service} = \frac{1}{\mu}$,
- Average time between packages income into the system: $T_{income} = \frac{p_1}{\lambda_1} + \frac{p_2}{\lambda_2}$,
- Average vacation time: $T_{vacation} = \frac{q_1}{\alpha_1} + \frac{q_2}{\alpha_2}$,
- Examined system size: $N =$ buffer size $+1$.

Scenario 1.
FA was performed to find set of parameters for lowest cost of work and the best QoS. In Table 1 are optimum values that affect NoSQL system operation.

FA was also arranged to position NoSQL in various scenarios. Hence system parameters must be set in peculiar way to optimize operation cost. In each scenario some possible situation of traffic is discussed.

Table 1 Parameters μ, λ_i, α_i, p_i, q_i for $i = 1, 2$ and lowest value of (3)

λ_1	λ_2	α_1	α_2	p_1	p_2	q_1	q_2
2.80	2.60	1.20	0.44	1.51	1.0	5.90	4.90
μ	0.51			$r_n(c_1)$	0.32		
	$T_{service}$		T_{income}		$T_{vacation}$		
$[sec]$	1.96		0.92		16.05		

Scenario 2.
NoSQL $T_{service} = 2[sec]$, what means that request service takes about $2[sec]$. This is similar to situation in small on-line shops or customer services. Research results are shown in Table 2.

Scenario 3.
NoSQL $T_{service} = 0.5[sec]$. This situation represents NoSQL business service with heavy traffic and efficient server machine. Research results with system positioning are shown in Table 3.

Using FA we can also position NoSQL system for given T_{income}. This will correspond to peculiar incoming situations.

Scenario 4.
NoSQL T_{income} was given as $2[sec]$, what means that requests are incoming to the server once in 2 seconds, similarly to scenario 2. Research results are shown in Table 4.

Scenario 5.
NoSQL T_{income} was given as $0.5[sec]$, what means that requests are incoming to the server twice in every second. This situation is describing an extensively used database, like these of business purpose. Research results are shown in Table 5.

Table 2 Parameters μ, λ_i, α_i, p_i, q_i for $i = 1, 2$ and lowest value of (3)

λ_1	λ_2	α_1	α_2	p_1	p_2	q_1	q_2
1.93	3.21	0.81	0.9	81	1	2.11	10.20
μ	0.5			$r_n(c_1)$	0.36		
	$T_{service}$		T_{income}		$T_{vacation}$		
$[sec]$	2.00		42.28		13.94		

Table 3 Parameters μ, λ_i, α_i, p_i, q_i for $i = 1, 2$ and lowest value of (3)

λ_1	λ_2	α_1	α_2	p_1	p_2	q_1	q_2
45.12	24.17	128.01	1.30	1.70	1.41	67.16	15.65
μ	2.00			$r_n(c_1)$	0.28		
	$T_{service}$		T_{income}		$T_{vacation}$		
$[sec]$	0.5		0.10		12.56		

Table 4 Parameters μ, λ_i, α_i, p_i, q_i for $i = 1, 2$ and lowest value of (3)

λ_1	λ_2	α_1	α_2	p_1	p_2	q_1	q_2
3.83	4.01	0.7	1.24	6.57	2.02	11.01	6.45
μ	0.3			$r_n(c_1)$	0.39		
	$T_{service}$		T_{income}		$T_{vacation}$		
$[sec]$	3.33		2.00		20.93		

Table 5 Parameters μ, λ_i, α_i, p_i, q_i for $i = 1, 2$ and lowest value of (3)

λ_1	λ_2	α_1	α_2	p_1	p_2	q_1	q_2
28.13	26.56	0.92	0.54	14.13	1.40	6.20	4.90
μ	0.92			$r_n(c_1)$	0.30		
	$T_{service}$		T_{income}		$T_{vacation}$		
$[sec]$	1.09		0.50		15.38		

Table 6 Parameters μ, λ_i, α_i, p_i, q_i for $i = 1, 2$ and lowest value of (3)

λ_1	λ_2	α_1	α_2	p_1	p_2	q_1	q_2
12.52	6.33	0.10	2.82	2.23	1.92	0.16	1.49
μ	0.87			$r_n(c_1)$	0.32		
	$T_{service}$		T_{income}		$T_{vacation}$		
$[sec]$	1.14		0.48		2.0		

In last two scenarios we simulate FA to position modeled system fo given $T_{vacation}$. These examples show simulation for predefined backup or conservation in the system.

Scenario 6.
This situation represents system operation, where the server has two seconds free of service after each busy period $T_{vacation} = 2[sec]$. Research results are shown in Table 6.

5 Conclusions

Evolutionary Computation methods are useful in simulation or positioning of various types of objects. They help to collect representative samples, which can be used by AI decision support systems. Computational Intelligence helps to simulate complicated objects and because of flexible design, calculations are possible even in discontinuous spaces [37].

Conducted experiments confirm efficiency in simulation of examined object in many possible scenarios representing common situations in reality. Positioned model was simulated in various situations with predefined time of service, income or vacation. These reflect situations when traffic is common or heavy and the system must serve many requests. Further work should be carried out to reduce time consuming operations, tentatively by using some knowledge prior to generate initial population. Moreover, modeled system could be non-stationary and parameters could change during work due to wear (expenditure) of elements (battery, voltage, etc.) or environment changes (temperature, air composition etc.). So, in future research it is important to take into account this aspects by i.e. fuzzyfication of system parameters.

Acknowledgments The project was funded by the National Science Centre under decision number DEC-2012/05/B/ST6/03620.

References

1. Marszałek, Z., Woźniak, M.: On possible organizing nosql database systems. Int. J. Inf. Sci. Intell. Syst. **2**(2), 51–59 (2013)
2. Woźniak, M., Marszałek, Z., Gabryel, M., Nowicki, R.K.: On quick sort algorithm performance for large data sets. In: Proceedings of the 8th International Conference on Knowledge, Information and Creativity Support Systems, Skulimowski, A.M.J. (ed.), pp. 647–656. Cracow, Poland, 7–9 Nov 2013
3. Woźniak, M., Marszałek, Z., Gabryel, M., Nowicki, R.K.: Triple heap sort algorithm for large data sets. In: Proceedings of the 8th International Conference on Knowledge, Information and Creativity Support Systems, Skulimowski, A.M.J. (ed.), pp. 657–665. Cracow, Poland, 7–9 Nov 2013
4. Woźniak, M., Marszałek, Z., Gabryel, M., Nowicki, R.K.: Preprocessing large data sets by the use of quick sort algorithm, pp. 36–72. Advances in Intelligent Systems and Computing, Springer International Publishing, Switzerland, (accepted-in press) (2014)
5. Woźniak, M., Marszałek, Z.: Extended Algorithms for Sorting Large Data Sets. Silesian University of Technology Press, Gliwice (2014)
6. Woźniak, M., Marszałek, Z.: Selected Algorithms for Sorting Large Data Sets. Silesian University of Technology Press, Gliwice (2013)
7. Woźniak, M., Gabryel, M., Nowicki, R.K., Nowak, B.A.: A novel approach to position traffic in nosql database systems by the use of firefly algorithm. In: Proceedings of the 9th International Conference on Knowledge, Information and Creativity Support Systems, Papadopoulos, G.A. (ed.), pp. 208–218. Limassol, Cyprus, 6–8 Nov 2014
8. Teghem, J.: Control of the service process in a queueing system. Eur. J. Oper. Res. **1**(23), 141–158 (1986)
9. Kella, O.: Optimal control of the vacation scheme in an $M/G/1$ queue. Oper. Res. J. **4**(38), 724–728 (1990)
10. Lillo, R.E.: Optimal operating policy for an $M/G/1$ exhaustive server-vacation model. Methodol. Comput. Appl. Probab. **2**(2), 153–167 (2000)
11. Gupta, U.C., Banik, A.D., Pathak, S.: Complete analysis of $MAP/G/1/N$ queue with single (multiple) vacation(s) under limited service discipline. J. Appl. Math. Stoch. Anal. **3**, 353–373 (2005)
12. Gupta, U.C., Sikdar, K.: Computing queue length distributions in $MAP/G/1/N$ queue under single and multiple vacation. Appl. Math. Comput. **2**(174), 1498–1525 (2006)
13. Niu, Z., Shu, T., Takahashi, Y.: A vacation queue with setup and close-down times and batch markovian arrival processes. Perform. Eval. J. **3**(54), 225–248 (2003)
14. Niu, Z., Takahashi, Y.: A finite-capacity queue with exhaustive vacation/close-down/setup times and markovian arrival processes. Queueing Syst. **1**(31), 1–23 (1999)

15. Takagi, H..: Queueing Analysis, vol. 1: Vacation and Priority Systems, vol. 2. Finite Systems. Amsterdam, North-Holland (1993)
16. Tian, N., Zhang, Z.G.: Vacation Queueing Models. Theory and Applications. Springer, Berlin (2006)
17. M. Woźniak.: On positioning traffic in nosql database systems by the use of particle swarm algorithm. In: Proceedings of XV Workshop DAGLI OGGETTI AGLI AGENTI—WOA'2014, Catania, Italy, 25–26 Sept, CEUR Workshop Proceedings (CEUR-WS.org), RWTH Aachen University Deutschland, vol. 1260 (2014)
18. Kempa, W.M.: $GI/G/1/\infty$ batch arrival queuing system with a single exponential vacation. Math. Methods Oper. Res. **1**(69), 81–97 (2009)
19. Kempa, W.M.: Characteristics of vacation cycle in the batch arrival queuing system with single vacations and exhaustive service. Int. J. Appl. Math. **4**(23), 747–758 (2010)
20. Kempa, W.M.: Some new results for departure process in the $M^X/G/1$ queuing system with a single vacation and exhaustive service. Stoch. Anal. Appl. **1**(28), 26–43 (200)
21. Kempa, W.M.: On departure process in the batch arrival queue with single vacation and setup time. Ann. UMCS Informatica **1**(10), 93–102 (2010)
22. Kempa, W.M.: The virtual waiting time in a finite-buffer queue witha single vacation policy. Lecture Notes in Computer Science, vol. 7314 pp. 47–60. Springer International Publishing, Switzerland (2012)
23. Ćwikła, G., Sękala, A., Woźniak, M.: The expert system supporting design of the manufacturing information acquisition system (MIAS) for production management. Adv. Mater. Res. **2014**, 852–857 (1036)
24. Marszałek, Z., Połap, D., Woźniak, M.: On preprocessing large data sets by the use of triple merge sort algorithm. In: Proceedings of International Conference on Advances in Information Processing and Communication Technologies, pp. 65–72. The IRED, Seek Digital Library, Rome, Italy, 7–8 June 2014
25. Gabryel, M., Nowicki, R.K., Woźniak, M., Kempa, W.M.: Geneticcost optimization of the $GI/M/1/N$ finite-buffer queue with a single vacation policy. Lecture Notes in ArtificialIntelligence, vol. 7895, pp. 12–23. Springer Publishing International, Switzerland (2013)
26. Woźniak, M.: On applying cuckoo search algorithm to positioning GI/M/1/N finite-buffer queue with a single vacation policy. In: Proceedings of 12th Mechican International Conference on Artificial Intelligence—MICAI'2013, pp. 59–64. IEEE, Mexico City, Mexico, 24–30 Nov 2013
27. Woźniak, M., Kempa, W.M., Gabryel, M., Nowicki, R.K.: A finite-buffer queue with single vacation policy—analytical study with evolutionary positioning. Int. J. Appl. Math. Comput. Sci. **24**(4), 887–900 (2014)
28. Woźniak, M., Kempa, W.M., Gabryel, M., Nowicki, R.K., Shao, Z.: On applying evolutionary computation methods to optimization of vacation cycle costs in finite-buffer queue. Lecture Notes in Artificial Intelligence, vol. 8467, pp. 480–491. Springer International Publishing, Switzerland (2014)
29. Hongwei, D., Dongfeng, Z., Yifan, Z.: Performance analysis of wireless sensor networks of serial transmission mode with vacation on fire prevention. In: Proceedings of ICCET'10, IEEE, pp. 153–155 (2010)
30. Mancuso, V., Alouf, S.: Analysis of power saving with continuous connectivity. Comput. Netw. **56**(10), 2481–2493 (2012)
31. Napoli, C., Pappalardo, G., Tramontana, E., Marszałek, Z., Połap, D., Woźniak, M.: Simplified firefly algorithm for 2d image key-points search. In: Proceedings of the 2014 IEEE Symposium on Computational Intelligence for Human-like Intelligence—CIHLI'2014, pp. 118–125. IEEE, Orlando, Florida, USA, 9–12 Dec 2014
32. Woźniak, M., Marszałek, Z.: An idea to apply fireflyalgorithm in 2D image key-points search. Communications in Computer and Information Science, vol. 465, pp. 312–323. Springer International Publishing, Switzerland (2014)
33. Horng, M.H.: Multilevel minimum cross entropy threshold selection based on the firefly algorithm. Expert Syst. Appl. **38**(12), 14805–14811 (2011)

34. Horng, M.H.: Vector quantization using the firefly algorithm for image compression. Expert Syst. Appl. **39**(1), 1078–1091 (2012)
35. Yang, X.S., Cui, Z.H., Xiao, R.B., Gandomi, A.H., Karamanoglu, M.: Swarm Intelligence and Bio-inspired Computation: Theory and Applications. Elsevier, Waltham, USA (2013)
36. Yang, X.S., Deb, S.: Cuckoo search via lévy flights. In: Proceedings of NaBIC'2009, pp. 210–214 (2009)
37. Woźniak, M.: Fitness function for evolutionary computation applied in dynamic object simulation and positioning. In: Proceedings of the 2014 IEEE Symposium on Computational Intelligence in Vehicle and Transportation Systems—CIVTS'2014, pp, 108–114. IEEE, Orlando, Florida, USA, 9–12 Dec 2014

Length of Hospital Stay and Quality of Care

José Neves, Vasco Abelha, Henrique Vicente, João Neves and José Machado

Abstract The relationship between Length Of hospital Stay (*LOS*) and *Quality-of-Care* (*QofC*) is demanding and difficult to assess. Indeed, a multifaceted intertwining network of countless services and *LOS* factors is available, which may range from organizational culture to hospital physicians availability, without discarding the possibility of lifting the foot on intermediate care services, to the customs and cultures of the people. On health policy terms, *LOS* remains a measurable index of efficiency, and most of the studies that have been undertaken show that QoC or health outcomes do not appear to be compromised by reductions in *LOS* times. Therefore, and in order to assess this statement, a *Logic Programming* based methodology to *Knowledge Representation* and *Reasoning*, allowing the modeling of the universe of discourse in terms of defective data, information and knowledge is used, being complemented with an *Artificial Neural Networks* based approach to computing, allowing one to predict for how long a patient should remain in a hospital or at home, during his/her illness experience.

Keywords Length of hospital stay · Logic programming · Knowledge representation and reasoning · Artificial neural networks

J. Neves (✉) · J. Machado
Centro Algoritmi/Departamento de Informática, Universidade do Minho, Braga, Portugal
e-mail: jneves@di.uminho.pt

J. Machado
e-mail: jmac@di.uminho.pt

V. Abelha
Departamento de Informática, Universidade do Minho, Braga, Portugal
e-mail: vascoabelha91@gmail.com

H. Vicente
Departamento de Química, Centro de Química de Évora, Escola de Ciências e Tecnologia, Universidade de Évora, Évora, Portugal
e-mail: hvicente@uevora.pt

J. Neves
DRS. NICOLAS & ASP, Dubai, United Arab Emirates
e-mail: joaocpneves@gmail.com

S. Kunifuji et al. (eds.), *Knowledge, Information and Creativity Support Systems*, Advances in Intelligent Systems and Computing 416,
DOI 10.1007/978-3-319-27478-2_19

1 Introduction

In light with recent events, our world is going through troubled social and economic times. In order to succeed and live under this scenery, one may cut on any kind of outgoing expenses, which means reducing costs, namely in education and healthcare. As a result, it was developed indicators that may assess the effectiveness of organizations and promote cost reduction. In the healthcare arena, the average *Length-Of-Stay* (*LOS*) in hospitals is often used as an indicator of efficiency. If all other parameters are equal, a shorter stay will reduce the costs per patient. However, it is important to note that shorter stays tend to be more service intensive and more costly on a day base. Furthermore, an extremely short length of stay may also lead to a greater readmission rate, and the effective costs per episode of illness may fall only slightly, or even rise. Some stated that reducing 1 day to the end of an inpatient stay has a minimal impact on the final costs [15]. According to these authors the bulk of health care expenses are incurred in the patients' early days of hospital stay.

In a recent study carried out in 12 (twelve) Dutch hospitals [2], it were identified a number of measures that reduce the *LOS* along the main phases of the care process, i.e., admission, stay and discharge. The authors identify, among other situations, the admission of patients on the day that the treatment actually starts and in the case of acute patients, to provide a plan of treatment immediately after admission. During the hospital stay the most common problems mentioned were the waiting time for diagnostic tests and interventions, as well as the improvement of the cooperation with paramedics to succeed in a expedite rehabilitation, in order to have a better and earlier clearance. Finally, at the clearance phase, an all-in-one transfer of patient care to a different health care facility is of utmost importance.

The average length of stay in hospitals has fallen over the past decade in nearly all *OECD* countries, from 8.2 days in 2000 to 7.2 days in 2009. Portugal has followed this pattern, and the average *LOS* of hospital patients has been decreasing from 7.3 to 5.9 days under this time period [13].

Indeed, a hospital should be regarded as a place where patients are treated to their illnesses and express and validate their health states. A patient *LOS* must be decided against some indicators, namely among those that are stated below:

- Age and Gender;
- Comorbidity;
- Context;
- Hospital Records;
- Treatments;
- Patient History; and
- Exams.

Therefore, and in order to fulfill this goal the variables of the universe of discourse are organized into two clusters, namely *Patient Source* and *Hospital Source*. *Patient Source* will include age, gender, history, context and comorbidity, i.e., the subjective information that is retrieved by a dialog between the individual and the physician, whose accuracy must be judged and evaluated. Undeniably, it can be compromised by factors such as memory, unawareness or fraud. On the other hand the patient may easily forget a given fact or situation, depending on his/her condition, details or even events that may be of great help to the physician. On top of that, they can also be lying and trying to deceive the clinic.

However, one has a reliable and objective source of information, which is provided by the healthcare unit. It includes vital signs and measurements (e.g. patient weight), findings from physical examination (e.g. results from laboratory and other diagnostics already completed), and healthcare diaries. The two groups of information, *Subjective—Patient*—and *Objective—Healthcare units*—were inspired by a method of documentation employed by the healthcare arena—*S.O.A. P. (Subjective, Objective, Analyses and Planning).* The aspects previously enumerated, are more than necessary, as presented in a wide-range of literature or among experts, to complete a risk assessment [4, 17], i.e., to determine, with a certain Degree of Confidence, if a patient should stay in the health care facility or not. The most prominent aspects can be linked to Patient History, Comorbidity and Context. To shed some light, the latter refers to the patient daily lives. If they have no conditions to be *set out in the open*—home hygienic conditions, no guidance—they ought to stay in the hospital in order not to jeopardize their health state. Indeed, *it is better to do it right the first time and not having the past come back to bite.* Nevertheless and in spite of these politics, this is exactly what happens most of the times. Patients tend to decrease their already short time span of admission in the hospital. With respect to the remaining aspects it is easy to infer their relevance and why they should to be taken into consideration. Such variables as *current treatments, exams and daily hospital diaries* have an association among them. Most of the time they share information or complement one another.

In order to explain the basics that are described below, namely the aspects needed to determine the length of stay in a hospital, the information available was clustered in 4 (four) groups, i.e., *Biological Factors*, *Patient History*, *Ambient Factors* and *Hospital Factors.* Patients' data such as age, sex, and comorbidity are linked to *Biological Factors*. *Patient History* refers to the detailed patient medical past. The information here comes from the patient good will, or any past records that the health system had recorded regarding his/her previous admissions. *Ambient Factors* are simply about the context embracing the patient. Anything that surrounds him/her is significant to determine the hospital *LOS*. *Hospital Factors* simply summarizes every exam, diagnosis he/she did at the healthcare facility. Indeed, after a briefly description of the relevance and meaning of the values and moments under inspection, their drawbacks lay on the incomplete and default information, noise and uncertainty to which they are attached. Our analysis aims to speculate and estimate the necessary length of stay in a hospital of a given patient, while reducing any mishap in their health. This is achieved by the use of a Logical Programming based approach to Knowledge Representation and Reasoning [11, 12], complemented with a computational framework based on Artificial Neural Networks [3], as it is shown below.

2 Knowledge Representation and Reasoning

Many approaches to knowledge representations and reasoning have been proposed using the Logic Programming (LP) paradigm, namely in the area of Model Theory [5, 7], and Proof Theory [11, 12]. In this work it is followed the proof theoretical approach in terms of an extension to the LP language to knowledge representations and reasoning. An Extended Logic Program is a finite set of clauses in the form:

$$p \leftarrow p_1, \ldots, p_n, \ not\ q_1, \ \ldots, not\ q_m \tag{1}$$

$$?\,(p_1, \ \ldots, p_n, \ not\ q_1, \ \ldots, \ not\ q_m)(n, m \geq 0) \tag{2}$$

where *?* is a domain atom denoting falsity, the p_i, q_j, and p are classical ground literals, i.e., either positive atoms or atoms preceded by the classical negation sign ¬ [12]. Under this emblematic formalism, every program is associated with a set of abducibles [7] given here in the form of exceptions to the extensions of the predicates that make the program. Once again, LP emerged as an attractive formalism for knowledge representations and reasoning tasks, introducing an efficient search mechanism for problem solving.

Due to the growing need to offer user support in decision making processes some studies have been presented [6, 8] related to the qualitative models and qualitative reasoning in Database Theory and in Artificial Intelligence research. With respect to the problem of knowledge representation and reasoning in Logic Programming (LP), a measure of the *Quality-of-Information* (*QoI*) of such programs has been object of some work with promising results [9, 10]. The *QoI* with respect to the extension of a predicate *i* will be given by a truth-value in the interval [0,1], i.e., if the information is *known* (*positive*) or *false* (*negative*) the *QoI* for the extension of $predicate_i$ is 1. For situations where the information is unknown, the *QoI* is given by:

$$QoI_i = \lim_{N \to \infty} \frac{1}{N} = 0 \ (N \gg 0) \tag{3}$$

where *N* denotes the cardinality of the set of terms or clauses of the extension of $predicate_i$ that stand for the incompleteness under consideration. For situations where the extension of $predicate_i$ is unknown but can be taken from a set of values, the *QoI* is given by:

$$QoI_i = 1/Card \tag{4}$$

where *Card* denotes the cardinality of the *abducibles* set for *i*, if the *abducibles* set is disjoint. If the *abducibles* set is not disjoint, the *QoI* is given by:

$$QoI_i = \frac{1}{C_1^{Card} + \cdots + C_{Card}^{Card}} \tag{5}$$

where C_{Card}^{Card} is a card-combination subset, with *Card* elements. The next element of the model to be considered is the relative importance that a predicate assigns to each of its attributes under observation, i.e., w_i^k, which stands for the relevance of attribute k in the extension of $predicate_i$. It is also assumed that the weights of all the attribute predicates are normalized, i.e.:

$$\sum_{1 \leq k \leq n} w_i^k = 1, \forall_i \tag{6}$$

where $\forall$ denotes the universal quantifier. It is now possible to define a predicate's scoring function $V_i(x)$ so that, for a value $x = (x_1, \ldots, x_n)$, defined in terms of the attributes of $predicate_i$, one may have:

$$V_i(x) = \sum_{1 \leq k \leq n} w_i^k * QoI_i(x)/n \tag{7}$$

allowing one to set:

$$predicate_i(x_1, \ldots, x_n) :: V_i(x) \tag{8}$$

that denotes the inclusive quality of $predicate_i$ with respect to all the predicates that make the program. It is now possible to set a logic program (here understood as the predicates' extensions that make the program) scoring function, in the form:

$$LP_{Scoring\ Function} = \sum_{i=1}^{n} V_i(x) * p_i \tag{9}$$

where p_i stands for the relevance of the $predicate_i$ in relation to the other predicates whose extensions denote the logic program. It is also assumed that the weights of all the predicates' extensions are normalized, i.e.:

$$\sum_{i=1}^{n} p_i = 1, \forall_i \tag{10}$$

where $\forall$ denotes the universal quantifier.

It is now possible to engender the universe of discourse, according to the information given in the logic programs that endorse the information about the problem under consideration, according to productions of the type:

$$predicate_i - \bigcup_{1 \leq j \leq m} clause_j(x_1, \ldots, x_n) :: QoI_i :: DoC_i \tag{11}$$

where $\bigcup$ and m stand, respectively, for *set union* and the *cardinality* of the extension of $predicate_i$. On the other hand, DoC_i denotes one's confidence on the attribute's values of a particular term of the extension of *predicate$_i$*, whose evaluation will be illustrated below. In order to advance with a broad-spectrum, let us suppose that the *Universe of Discourse* is described by the extension of the predicates:

$$f_1(\ldots), f_2(\ldots), \ldots, f_n(\ldots) \; where \; (n \geq 0) \tag{12}$$

Assuming that a clause denotes a happening, a clause has as argument all the attributes that make the event. The argument values may be of the type unknown or members of a set, or may be in the scope of a given interval, or may qualify a particular observation. Let us consider the following clause where the first argument value may fit into the interval [20,30] with a domain that ranges between 0 (zero) and 50 (fifty), where the second argument stands for itself, with a domain that ranges in the interval [0,10], and the value of the third argument being unknown, being represented by the symbol , with a domain that ranges in the interval [0,100]. Let us consider that the case data is given by the extension of predicate f_1, given in the form:

$$f_1 : x_1, x_2, x_3 \rightarrow \{0, 1\} \tag{13}$$

where "{" and "}" is one's notation for sets, "0" and "1" denote, respectively, the truth values *false* and *true*. Therefore, one may have:

$$
\begin{array}{l}
\{ \\
\quad \neg f_1(x_1, x_2, x_3) \leftarrow not \; f_1(x_1, x_2, x_3) \\
\quad f_1(\underbrace{[20, 30], \;\; 5, \;\;\; \perp}_{attribute's \; values \; for \; x_1, x_2, x_3}) :: 1 :: DoC \\
\quad \quad \underbrace{[0, 50][0, 10][0, 100]}_{attribute's \; domains \; for \; x_1, x_2, x_3} \\
\ldots \\
\}
\end{array}
$$

Once the clauses or terms of the extension of the predicate are established, the next step is to set all the arguments, of each clause, into continuous intervals. In this phase, it is essential to consider the domain of the arguments. As the third argument is unknown, its interval will cover all the possibilities of the domain. The second argument speaks for itself. Therefore, one may have:

$$
\begin{array}{l}
\{ \\
\quad \neg f_1(x_1, x_2, x_3) \leftarrow not\ f_1(x_1, x_2, x_3) \\
\quad f_1\Big(\underbrace{[20, 30], [5, 5], [0, 100]}_{attribute's\ values\ ranges\ for\ x_1, x_2, x_3}\Big) :: 1 :: DoC \\
\qquad \underbrace{[0, 50]\ \ [0, 10]\ [0, 100]}_{attribute's\ domains\ for\ x_1, x_2, x_3} \\
\quad \ldots \\
\}
\end{array}
$$

It is now achievable to calculate the *Degree of Confidence* for each attribute that make the term argument (e.g. with respect to the former attribute it denotes one's confidence that the attribute under consideration fits into the interval [20,30]). Next, we set the boundaries of the arguments intervals to be fitted in the interval [0,1] according to the normalization procedure given in the procedural form by $(Y - Y_{min})/(Y_{max} - Y_{min})$, where the Y_s stand for themselves. One may have:

$$
\begin{array}{l}
\{ \\
\quad \neg f_1(x_1, x_2, x_3) \leftarrow not\ f_1(x_1, x_2, x_3) \\
\quad x_1 = \left[\frac{20-0}{50-0}, \frac{30-0}{50-0}\right]\ x_2 = \left[\frac{5-0}{10-0}, \frac{5-0}{10-0}\right],\ x_3 = \left[\frac{0-0}{100-0}, \frac{100-0}{100-0}\right] \\
\quad f_1\ \ \underbrace{([0.4, 0.6], [0.5, 0.5], [0, 1])}_{\substack{attribute's\ values\ ranges\ for\ x_1, x_2, x_3 \\ once\ normalized}}\ \ :: 1 :: DoC \\
\qquad \underbrace{[0, 1]\quad\ [0, 1]\quad [0, 1]}_{\substack{attribute's\ domains\ for\ x_1, x_2, x_3 \\ once\ normalized}} \\
\quad \ldots \\
\}
\end{array}
$$

The *Degree of Confidence* (*DoC*) is evaluated using the equation $DoC = \sqrt{1 - \Delta l^2}$, as it is illustrated in Fig. 1. Here Δl stands for the length of the arguments intervals, once normalized. Therefore, one may have:

{

$$\neg f(x_1, x_2, x_3) \leftarrow not\ f_1(x_1, x_2, x_3)$$

$$f_1 \underbrace{(0.98, \quad 1, \quad 0)}_{\substack{attribute's\ confidence\\ values\ for\ x_1, x_2, x_3}} :: 1 :: 0.66$$

$$\underbrace{[0.4, 0.6][0.5, 0.5][0, 1]}_{\substack{attribute's\ values\ ranges\ for\ x_1, x_2, x_3\\ once\ normalized}}$$

$$\underbrace{[0, 1] \quad [0, 1] \quad [0, 1]}_{\substack{attribute's\ domains\ for\ x_1, x_2, x_3\\ once\ normalized}}$$

...

}

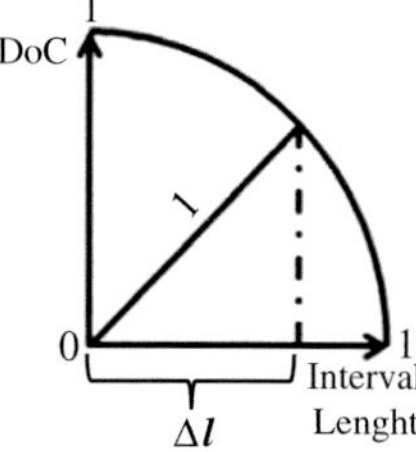

Fig. 1 Evaluation of the degree of confidence

where the *DoC's* for $f_1(0.98, 1, 0)$ is evaluated as (0.98 + 1 + 0)/3 = 0.66, assuming that all the argument's attributes have the same weight.

3 A Case Study

In order to exemplify the applicability of our model, we will look at the relational database model, since it provides a basic framework that fits into our expectations, and is understood as the genesis of the *Logic Programming* (*LP*) approach to *Knowledge Representation and Reasoning* [12].

As a case study, consider a scenario where a relational database is given in terms of the extensions of the relations depicted in Fig. 2, which stands for a situation where one has to interpret the available information about a patient state with a given disease, and set a timetable to his/her stay at the hospital or home. Under this scenario some incomplete and/or default data is also available. For instance, relatively to *Ambient Factors* the presence/absence of underlying problems for case 2 is unknown, while in case 1, it varies from 5 to 10. The *Length of Stay* relation (Fig. 2)

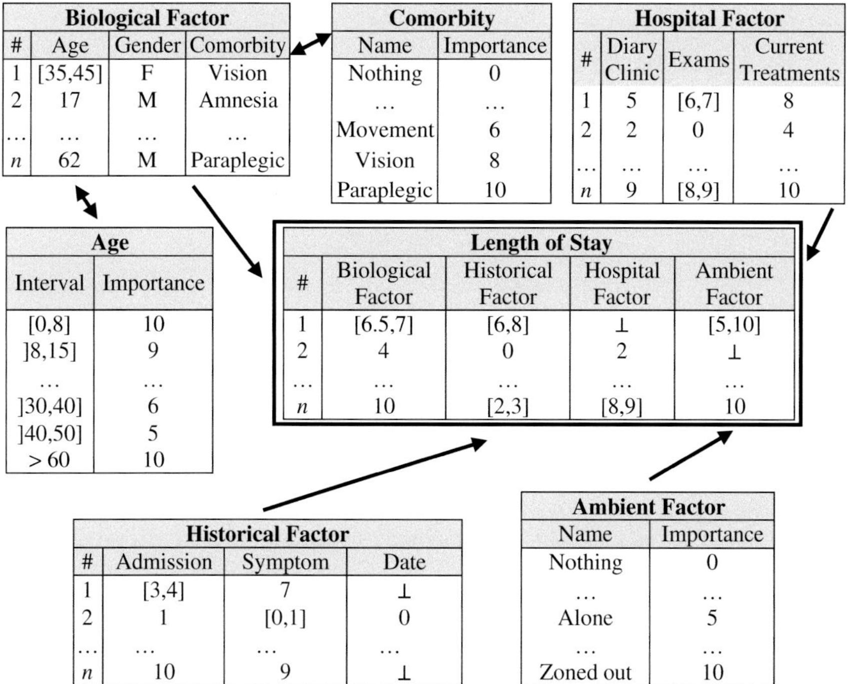

Biological Factor			
#	Age	Gender	Comorbity
1	[35,45]	F	Vision
2	17	M	Amnesia
…	…	…	…
n	62	M	Paraplegic

Comorbity	
Name	Importance
Nothing	0
…	…
Movement	6
Vision	8
Paraplegic	10

Hospital Factor			
#	Diary Clinic	Exams	Current Treatments
1	5	[6,7]	8
2	2	0	4
…	…	…	…
n	9	[8,9]	10

Age	
Interval	Importance
[0,8]	10
]8,15]	9
…	…
]30,40]	6
]40,50]	5
> 60	10

Length of Stay				
#	Biological Factor	Historical Factor	Hospital Factor	Ambient Factor
1	[6.5,7]	[6,8]	⊥	[5,10]
2	4	0	2	⊥
…	…	…	…	…
n	10	[2,3]	[8,9]	10

Historical Factor			
#	Admission	Symptom	Date
1	[3,4]	7	⊥
2	1	[0,1]	0
…	…	…	…
n	10	9	⊥

Ambient Factor	
Name	Importance
Nothing	0
…	…
Alone	5
…	…
Zoned out	10

Fig. 2 A summary of the relational database model

is populated according to various results of the different columns, namely *Biological Factors*, *Historical Factors*, *Hospital Factors* and *Ambient Factors*.

These are all the necessary variables to estimate the safety risk that a patient may incurs if he/she stay in the hospital/home more than the expected time, as well as their intervals' boundaries.

It should be reminded and noted that every single value is classified from level 0 to 10. The higher the level of stay energises the more important it is. As seen in the Fig. 2 every set has the equivalent result. These will and may be, later, calibrated by the medical staff according to their needs. Then, one may abstract on the subject and describe the extension of the *length* predicate in the form:

$$length\colon Bio_{logical\ Factor}, Hist_{orical\ Factor}, Hosp_{ital\ Factor}, Amb_{ient \cdot Factor} \rightarrow \{0, 1\}$$

where 0 (zero) and 1 (one) denote, respectively, the situation of *not need* (in this case the patient stays at home), and *need,* i.e., the patient stays at the health care facility. It is now possible to give the extension of the predicate *length*, in the form:

$$\{$$

$$\neg length(Bio, Hist, Hosp, Amb) \leftarrow not\ length(Bio, Hist, Hosp, Amb)$$

$$length\left(\underbrace{[6.5,7],\ [6,8],\ \perp,\ [5,10],}_{attribute's\ values}\right) :: 1 :: DoC$$

$$\underbrace{[0,10]\ [0,10][0,10][0,10]}_{attribute's\ domains}$$

$$\ldots$$

$$\}$$

In this program, the former clause denotes the closure of predicate *length*. The next clause corresponds to patient 1, taken from the extension of the *length* relation presented in Fig. 2. Patient 1 is a female, with an uncertain age, somehow between 35 and 45 years old, with vision problems, i.e., she is eyeless. Given that her age varies between 35 and 45, it may correspond to the *significance factor* 6 or *significance factor* 5, according to the *Age* Table. *Comorbidity* Table shows that vision corresponds to *significance* factor 8, i.e., the average of the values of her biological factors will return a variable whose values range in the interval [6.5,7]. We keep doing these kinds of operations for all the aspects that are not unknown. If they are unknown we just assume that the level ranges in the interval 1…10. To eliminate any doubt, case 1 refers to a patient that should no longer remain in the Health Care facility, while in case 2 one has the opposite, i.e., the patient should remain held for further examination and care. For further enlightenment, it is essential to explain that in the *Historical Factor* Table, *Admission, Symptom* and *Date* columns is relative to the severity of his/her past status. *Date* is the number of times he/she has been the hospital/home, given in an explicit form in terms of the time periods he/she stay at the healthcare facility/home. The latter, *Admission* and *Symptom* is referent to his/her health condition and how it evolved during the admission and subsequent period of time in the medical facility/home.

Moving on, the next step is to transform all the argument values into continuous intervals, and then move to normalize the predicate's arguments, as it was shown above. One may have:

$\{$

$$\neg length(Bio, Hist, Hosp, Amb) \leftarrow not\ length(Bio, Hist, Hosp, Amb)$$

$$length\left(\underbrace{0.99, \quad 0.98, \quad 0, \quad 0.87}_{attribute's\ confidence\ values}\right) :: 1 :: 0.71$$

$$\underbrace{[0.65, 0.7][0.7, 0.8][0, 1][0.5, 1]}_{attribute's\ values\ ranges\ once\ normalized}$$

$$\underbrace{[0, 1] \quad [0, 1] \quad [0, 1] \quad [0, 1]}_{attribute's\ domains\ once\ normalized}$$

$\cdots$

$\}$

where its terms make the training and test sets of the *Artificial Neural Network* given in Fig. 3.

4 Artificial Neural Networks

Several studies have demonstrated how *Artificial Neural Networks* (*ANNs*) could be successfully used to model data and capture complex relationships between inputs and outputs [14, 16]. *ANNs* simulate the structure of the human brain, being populated by multiple layers of neurons. As an example, let us consider the last case presented in Fig. 2, where one may have a situation in which a prolonged hospital stay is needed, which is given in the form:

$\{$

$$\neg length(Bio, Hist, Hosp, Amb) \leftarrow not\ length(Bio, Hist, Hosp, Amb)$$

$$length\left(\underbrace{10, \quad [2, 3], \quad [8, 9], \quad 10}_{attribute's\ values}\right) :: 1 :: DoC$$

$$\underbrace{[0, 10][0, 10][0, 10][0, 10]}_{attribute's\ domains}$$

$\cdots$

$\}$

According to the formalism presented above, once the transition to continuous intervals is completed, it is possible to have the arguments of the predicate extension normalized to the interval [0,1] in order to obtain the *Degree of Confidence* of the *length* predicate:

$$\{$$

$$length\left(\underbrace{1, \quad 0.99, \quad 0.99, \quad 1}_{attribute's\ confidence\ values}\right) :: 1 :: 0.995$$

$$\underbrace{[1,1][0.2,0.3][0.8,0.9][1,1]}_{attribute's\ values\ ranges}$$

$$\underbrace{[0,1]\ \ [0,1]\ \ \ [0,1]\ \ [0,1]}_{attribute's\ domains}$$

$$\ldots$$

$$\}$$

In Fig. 3 it is shown how the normalized values of the interval boundaries and their *DoC* and *QoI* values work as inputs to the *ANN*. The output translates the patient *Local-of-Stay* (0, if it is at home, 1 if it is at the healthcare facility), and *DoC* the confidence that one has on such a happening. In addition, it also contributes to

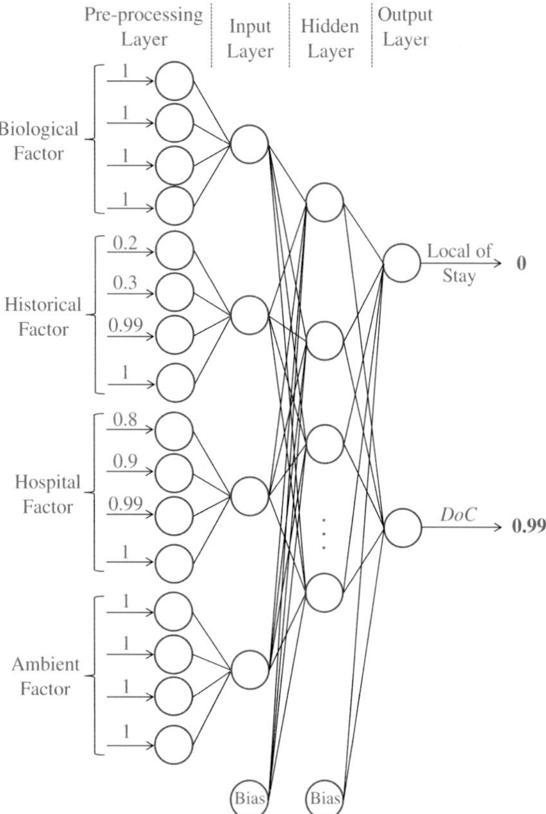

Fig. 3 The artificial neural network topology

build a more complete database of study cases that may be used later to retrain and test the *ANN*.

In this study were considered 538 patients with an age average of 57.6 years, ranging from 34 to 92 years old. The gender distribution was 47.8 and 52.2 % for female and male, respectively. The data came from a main health care center in the north of Portugal.

To ensure statistical significance of the attained results, the *ANN's* runs *stop-condition* was set to a value that must be less or equal to the modulus between the *DoC's* value evaluated taking into consideration the arguments' values of the former term in the extension of the predicate *length* given below (i.e., 0.995), and the one return by the *ANN* (i.e., 0.99). In each simulation, the available data were randomly divided into two mutually exclusive partitions, i.e., the training set with 67 % of the available data and, the test set with the remaining 33 % of the cases. The back propagation algorithm was used in the learning process of the *ANN*. As the output function in the pre-processing layer it was used the identity one. In the others layers it was used the sigmoid one.

The model accuracy was 95.3 % for the training set (344 correctly classified in 361) and 94.3 % for test set (167 correctly classified in 177).

5 Conclusions and Future Work

A true *LOS* performance measurement and improvement model would not only be an unending process or practice, but something that has to have their roots or become part of the in-structure of medicine. When the practice of medicine is intertwined with its simultaneous evaluation as to its impact on restoring health or improving functional status and *Quality-of-Life*, or even the costs of healthcare provision, then we may have a true discussion about quality, accounting and accountability. After all, the term *accountability* requires an inherent characteristic of measurement, in addition to the physician social and ethical responsibilities. It is at this point that one as to look to accounting. Indeed, the functioning of individuals and organizations should be continuously measured, not in a desire to censure or punish but to improve. As a matter of fact, value-added performance and improved management are seen as parallel tracks to the practice of medicine, where other incoming practices or studies may have much less ability to succeed, like alter culture, mind set, or perceived responsibilities. Indeed, the criterion of temporality is one of the most important if one is considering establishing a causal relationship between *LOS* and the *Quality-of-Care*, i.e., the sine qua non for causality is temporality. The cause must precede the effect. How can these criteria be used in assessing whether a relationship found between *Quality-of-Care* and *LOS* is causal? The strength of the relationship is not at all clear. Studies have been published which suggest an increase in *Quality-of-Care* with both a shorter and a longer *LOS*, and this finding does not easily comply with the consistency or biological gradient criteria. Plausible reasons for the relationship between *LOS* and *Quality-of-Care* can

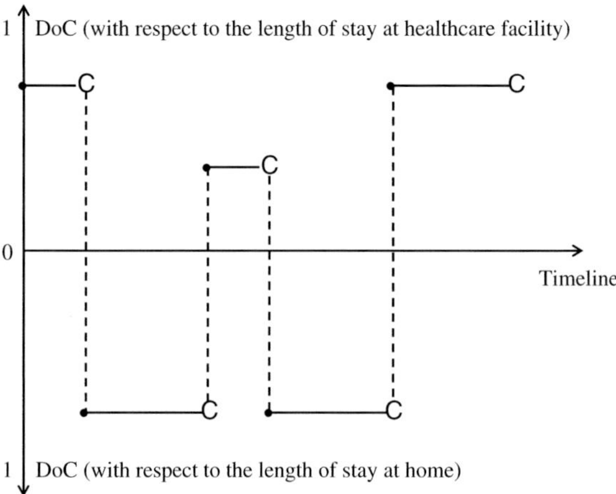

Fig. 4 Patient timeline for a given disease

be put forward to support either a longer or shorter *LOS* (e.g., a longer *LOS* might be thought of as allowing more time for appropriate investigation and treatment, while a shorter *LOS* may be consistent with a rapid, ordered and systematic care pathway). It is at this point that enters the work developed by Abelha et al. [1], and presented at the 9th International Conference on *Knowledge, Information and Creativity Support Systems* (*KICSS 2014*). Indeed, instead of speculating about the relationship between *LOS* and *Quality-of-Care* (i.e., who comes first?), the problem must be put in terms that must focus on the patient current condition, which may change with time. One's answer is given throughout Fig. 4, where a patient timeline (here understood as the disease signature) is provided. The corresponding logic program is given below:

$$\{$$

$$\neg length(Bio, Hist, Hosp, Amb) \leftarrow not\ length(Bio, Hist, Hosp, Amb).$$

$$length\left(\underbrace{1, \quad 0.99, \quad 0.99, \quad 1}_{attribute's\ confidence\ values}\right) :: 1 :: 0.995$$

$$\underbrace{[1,1][0.2,0.3][0.8,0.9][1,1]}_{attribute's\ values\ ranges}$$

$$\underbrace{[0,1] \quad [0,1] \quad [0,1] \quad [0,1]}_{attribute's\ domains}$$

$$\cdots$$

$$\}$$

which stands for the patient signature for a given disease.

Acknowledgments This work is funded by National Funds through the FCT—Fundação para a Ciência e a Tecnologia (Portuguese Foundation for Science and Technology) within project PEst-OE/EEI/UI0752/2014.

References

1. Abelha, V., Vicente, H., Machado, J., Neves, J.: An assessment on the length of hospital stay through artificial neural networks. In: Papadopoulos, G. (ed.) Proceedings of the 9th International Conference on Knowledge, Information and Creativity Support Systems (KICSS 2014), pp. 219–230. Cyprus Library, Cyprus (2014)
2. Borghansa, I., Koola, R.B., Lagoec, R.J., Westert, G.P.: Fifty ways to reduce length of stay: an inventory of how hospital staff would reduce the length of stay in their hospital. Health Policy **104**, 222–233 (2012)
3. Cortez, P., Rocha, M., Neves, J.: Evolving time series forecasting ARMA models. J. Heuristics **10**, 415–429 (2004)
4. Fry, J.: General practice: the facts. Radcliffe Medical Press, Oxford (1993)
5. Gelfond, M., Lifschitz, V.: The stable model semantics for logic programming. In: Kowalski, R., Bowen, K. (eds.) Logic Programming—Proceedings of the Fifth International Conference and Symposium, pp. 1070–1080 (1988)
6. Halpern, J.: Reasoning about uncertainty. MIT Press, Massachusetts (2005)
7. Kakas, A., Kowalski, R., Toni, F.: The role of abduction in logic programming. In: Gabbay, D., Hogger, C., Robinson, I. (eds.) Handbook of Logic in Artificial Intelligence and Logic Programming, vol. 5, pp. 235–324. Oxford University Press, Oxford (1998)
8. Kovalerchuck, B., Resconi, G.: Agent-based uncertainty logic network. In: Proceedings of the IEEE International Conference on Fuzzy Systems—FUZZ-IEEE, pp. 596–603 (2010)
9. Lucas, P.: Quality checking of medical guidelines through logical abduction. In: Coenen, F., Preece, A., Mackintosh A. (eds.) Proceedings of AI-2003 (Research and Developments in Intelligent Systems XX), pp. 309–321. Springer, London (2003)
10. Machado, J., Abelha, A., Novais, P., Neves, J., Neves, J.: Quality of Service in healthcare units. Int. J. Comput. Aided Eng. Technol. **2**, 436–449 (2010)
11. Neves, J., Machado, J., Analide, C., Abelha, A., Brito, L.: The halt condition in genetic programming. In: Neves, J., Santos, M.F., Machado, J. (eds.) Progress in Artificial Intelligence. LNAI, vol. 4874, 160–169. Springer, Berlin (2007)
12. Neves, J.: A logic interpreter to handle time and negation in logic databases. In Muller, R.L., Pottmyer, J.J. (eds.) Proceedings of the 1984 annual conference of the ACM on the Fifth Generation Challenge, pp. 50–54. ACM, New York (1984)
13. OECD.: Average length of stay in hospitals. In: Health at a Glance 2011: OECD Indicators. OECD Publishing (2011)
14. Salvador, C., Martins, M.R., Vicente, H., Neves, J., Arteiro, J.M., Caldeira, A.T.: Modelling molecular and inorganic data of amanita ponderosa mushrooms using artificial neural networks. Agrofor. Syst. **87**, 295–302 (2013)
15. Taheri, P.A., Butz, D.A., Greenfield, L.J.: Length of stay has minimal impact on the cost of hospital admission. J. Am. Coll. Surg. **191**, 123–130 (2000)
16. Vicente, H., Dias, S., Fernandes, A., Abelha, A., Machado, J., Neves, J.: Prediction of the quality of public water supply using artificial neural networks. J. Water Supply Res. Technol. AQUA **61**, 446–459 (2012)
17. World Health Organization: Primary Health Care. World Health Organization, Geneva (1978)

Bio-inspired Hybrid Intelligent Method for Detecting Android Malware

Konstantinos Demertzis and Lazaros Iliadis

Abstract Today's smartphones are capable of doing much more than the previous generation of mobile phones. However this extended range of capabilities is coming together with some new security risks. Also, mobile platforms often contain small, insecure and less well controlled applications from various single developers. Due to the open usage model of the Android market, malicious applications cannot be avoided completely. Especially pirated applications or multimedia content in popular demand, targeting user groups with typically low awareness levels are predestined to spread too many devices before being identified as malware. Generally malware applications utilizing root exploits to escalate their privileges can inject code and place binaries outside applications storage locations. This paper proposes a novel approach, which uses minimum computational power and resources, to indentify Android malware or malicious applications. It is a bio-inspired Hybrid Intelligent Method for Detecting Android Malware (HIMDAM). This approach performs classification by employing Extreme Learning Machines (ELM) in order to properly label malware applications. At the same time, Evolving Spiking Neural Networks (eSNNs) are used to increase the accuracy and generalization of the entire model.

Keywords Security • Android malware • Evolving spiking neural networks • Extreme learning machines • Radial basis function networks • Polynomial neural networks • Self-Organizing maps • Multilayer perceptron

---

K. Demertzis (✉) · L. Iliadis (✉)
Department of Forestry and Management of the Environment and Natural Resources, Democritus University of Thrace, 193 Pandazidou St, 68200 N. Orestiada, Greece
e-mail: kdemertz@fmenr.duth.gr

L. Iliadis
e-mail: liliadis@fmenr.duth.gr

S. Kunifuji et al. (eds.), *Knowledge, Information and Creativity Support Systems*, Advances in Intelligent Systems and Computing 416,
DOI 10.1007/978-3-319-27478-2_20

1 Introduction

Lately, the share of smartphones in the sales of handheld mobile communication devices has drastically increased. Among them, the number of Android based smartphones is growing rapidly. They are increasingly used for security critical private and business applications, such as online banking or to access corporate networks. This makes them a very valuable target for an adversary. Until recently, the Android Operating System's security model has succeeded in preventing any significant attacks by malware. This can be attributed to a lack of attack vectors which could be used for self-spreading infections and low sophistication of malicious applications. However, emerging malware deploys advanced attacks on operating system components to assume full device control [10]. Malware are the most common infection method because the malicious code can be packaged and redistributed with popular applications. In Android, each application has an associated *.apk* file which is the executable file format for this platform. Due to the open software installation nature of Android, users are allowed to install any executable file from any application store. This could be from the official Google Play store, or a third party site. This case of installing applications makes Android users vulnerable to malicious applications. Some of the most widely used solutions such as antivirus software are inadequate for use on smartphones as they consume too much CPU and memory and might result in rapid draining of the power source. In addition, most antivirus detection capabilities depend on the existence of an updated malware signature repository, therefore the antivirus users are not protected from zero-day malware.

This research effort aims in the development and application of an innovative, fast and accurate bio-inspired Hybrid Intelligent Method for Detecting Android Malware (HIMDAM). This is achieved by employing Extreme Learning Machines (ELMs) and Evolving Spiking Neural Networks (eSNNs). A RBF Kernel ELM has been employed for malware detection, which offers high learning speed, ease of implementation and minimal human intervention. Also, an eSNN model has been applied to increase the accuracy and generalization of the entire method. In fact, the bio-inspired model has shown better performance when compared to other ANN methods, such as Multilayer Perceptrons (MLP), Radial Basis Function ANN (RBF), Self-Organizing Maps (SOM), Group Methods of Data Handling (GMDH) and Polynomial ANN. A main advantage of HIMDAM is the fact that it reduces overhead and overall analysis time, by classifying malicious and benign applications with high accuracy.

1.1 Literature Review

Significant work has been done in applying machine learning (ML) techniques, using features derived from both static [7, 24, 29] and dynamic [4] analysis to identify malicious Android applications [13]. Amongst early efforts towards Android applications security was the "*install-time policy security system*" developed by Enck et al. which considered risks associated with combinations of the app permissions [9]. From another perspective, some works focused in the runtime analysis [20, 22] whereas others have tried a static analysis of apps [12]. For instance, Chin et al. [7] used a 2-means clustering [21] of apps' call activities, to detect Trojans. Fuchs et al. [11] used formal static analysis of byte codes [33] to form data flow-permission consistency as a constrained optimization problem. Barrera et al. [3] used app permissions in self-organizing maps (SOMs) to visualize app permission usage as a U-matrix [18]. Besides, their SOM component plane analysis allowed identification of the frequently jointly requested permissions. However, they did not relate categories and permissions. In [30], Tesauro et al. train ANN to detect boot sector viruses, based on byte string trigrams. Schultz et al. [27] compare three machine learning algorithms trained on three features: DLL and system calls made by the program, strings found in the program binary and a raw hexadecimal representation of the binary [23]. Kotler and Maloof [19] used a collection of 1971 benign and 1651 malicious executable files. N-grams were extracted and 500 features were selected using the Information Gain measure. The vector of n-gram features was binary, presenting the presence or absence of a feature in the file. In their experiment, they trained several classifiers: IBK k-Nearest Neighbors (k-NN), a similarity-based classifier called the TFIDF classifier, Naïve Bayes, Support Vector Machines (SVM) and Decision Trees under the algorithm J48 [28]. The last three of these were also boosted. In the experiments, the four best-performing classifiers were Boosted J48, SVM, Boosted SVM and IBK [28]. Also, Cheng et al. [6] proposed the use of ELM methods to classify binary and multi-class network traffic for intrusion detection. The performance of ELM in both binary-class and multi-class scenarios are investigated and compared to SVM based classifiers. Joseph et al., [16] developed an autonomous host-dependent Intrusion Detection System (IDS) for identifying malicious sinking behavior. This system increases the detection accuracy by using cross-layer features to describe a routing behavior. Two ML approaches were exploited towards learning and adjustment to new kind of attack circumstances and network surroundings. ELMs and Fisher Discriminant Analysis (FDA) are utilized collectively to develop better accuracy and quicker speed of method.

2 Methodologies Comprising the Proposed Hybrid Approach

2.1 *Extreme Learning Machines (ELM)*

The extreme learning machine (ELM) as an emerging learning technique provides efficient unified solutions to generalized feed-forward networks including but not limited to (both single- and multi-hidden-layer) neural networks, radial basis function (RBF) networks, and kernel learning [34]. ELM theories show that hidden neurons are important but can be randomly generated, independent from applications and that ELMs have both universal approximation and classification capabilities. They also build a direct link between multiple theories namely: ridge regression, optimization, ANN generalization performance, linear system stability and matrix theory. Thus, they have strong potential as a viable alternative technique for large-scale computing and ML. Also ELMs, are biologically inspired, because hidden neurons can be randomly generated independent of training data and application environments, which has recently been confirmed with concrete biological evidences. ELM theories and algorithms argue that "random hidden neurons" capture the essence of some brain learning mechanism as well as the intuitive sense that the efficiency of brain learning need not rely on computing power of neurons. This may somehow hint at possible reasons why brain is more intelligent and effective than computers [5].

ELM works for the "generalized" Single-hidden Layer feedforward Networks (SLFNs) but the hidden layer (or called feature mapping) in ELM need not be tuned.

Such SLFNs include but are not limited to SVMs, polynomial networks, RBFs and the conventional (both single-hidden-layer and multi-hidden-layer) feedforward ANN. Different from the tenet that all the hidden nodes in SLFNs need to be tuned, ELM learning theory shows that the hidden nodes/neurons of generalized feedforward networks needn't be tuned and these hidden nodes/neurons can be randomly generated [34]. All the hidden node parameters are independent from the target functions or the training datasets. ELMs conjecture that this randomness may be true to biological learning in animal brains. Although in theory, all the parameters of ELMs can be analytically determined instead of being tuned, for the sake of efficiency in real applications, the output weights of ELMs may be determined in different ways (with or without iterations, with or without incremental implementations) [34]. According to ELM theory the hidden node/neuron parameters are not only independent of the training data but also of each other. Unlike conventional learning methods which must see the training data before generating the hidden node/neuron parameters, ELMs could randomly generate the hidden node/neuron parameters before seeing the training data. In addition, ELMs can handle non-differentiable activation functions, and do not have issues such as finding a suitable stopping criterion, learning rate, and learning epochs. ELMs have several advantages, ease of use, faster learning speed, higher generalization performance, suitable for many nonlinear activation function and kernel functions [34].

2.2 *Evolving Spiking Neural Networks (eSNNs)*

The eSNNs are modular connectionist-based systems that evolve their structure and functionality in a continuous, self-organized, on-line, adaptive, interactive way from incoming information. These models use trains of spikes as internal information representation rather than continuous variables [25]. The eSNN developed and discussed herein is based in the "Thorpe" neural model [31]. This model intensifies the importance of the spikes taking place in an earlier moment, whereas the neural plasticity is used to monitor the learning algorithm by using one-pass learning. In order to classify real-valued data sets, each data sample, is mapped into a sequence of spikes using the Rank Order Population Encoding (ROPE) technique [8, 32]. The topology of the developed eSNN is strictly feed-forward, organized in several layers and weight modification occurs on the connections between the neurons of the existing layers.

The ROPE method is alternative to the conventional rate coding scheme (CRCS). It uses the order of firing neuron's inputs to encode information. This allows the mapping of vectors of real-valued elements into a sequence of spikes. Neurons are organized into neuronal maps which share the same synaptic weights. Whenever the synaptic weight of a neuron is modified, the same modification is applied to the entire population of neurons within the map. Inhibition is also present between each neuronal map. If a neuron spikes, it inhibits all the neurons in the other maps with neighboring positions. This prevents all the neurons from learning the same pattern. When propagating new information, neuronal activity is initially reset to zero. Then, as the propagation goes on, each time one of their inputs fire, neurons are progressively desensitized. This is making neuronal responses dependent upon the relative order of firing of the neuron's afferents [17, 37].

The aim of the one-pass learning method is to create a repository of trained output neurons during the presentation of training samples. After presenting a certain input sample to the network, the corresponding spike train is propagated through the eSNN which may result in the firing of certain output neurons. It is possible that no output neuron is activated and the network remains silent and the classification result is undetermined. If one or more output neurons have emitted a spike, the neuron with the shortest response time among all activated output neurons is determined. The label of this neuron is the classification result for the presented input [26].

3 Description of the Proposed HIMDAM Algorithm

The proposed herein, HIMDAM methodology uses an ELM classification approach to classify malware or benign applications with minimum computational power and time, combined with the eSNN method in order to detect and verify the malicious code. The general algorithm is described below:

Step 1:
Train and test *datasets* are determined and normalized to the interval [−1,1]. The *datasets* are divided in 4 main sectors with "*permission*" feature. *Permission* is a

security mechanism of mobile operating systems. For mobile phones any application executed under the device owner's user ID would be able to access any other application's data. The Android kernel assigns each application its own user ID on installation. To avoid the abuse of mobile phone functions, Android allows the user to effectively identify and manage mobile phone resources by setting permissions. If the application requires a certain function, the developer can announce permission. In the latest version of Android, there are a total of 130 permissions. To malware, some permissions are important and frequently needed, therefore they should be weighted. For example, the attacker needs permissions to transfer the stolen data to his account through the web, or to perform damaging behavior by sending out large number of SMS messages. The features involved can be divided in the sectors below:

1. *Battery + Permissions (5 features)*
2. *Binder + Permissions (18 features)*
3. *Memory + CPU + Permissions (10 features)*
4. *Network + Permissions (9 features)*

The *Hardware_Dataset* has been generated (16 features) including the most important variables from hardware related sectors (Battery, Memory, CPU, Network). On the other hand, the *All_Imp_Var_Dataset* (27 features) comprises of the most important variables from all of the sectors (Battery, Memory, CPU, Network, Binder). To calculate the importance of variables we replace them with their mean values one by one and we measure the root mean squared error (RMSE) of the "new" model. Original model error is considered to have a zero percent impact on the RMSE and 100 % impact is a case where all variables are replaced with their mean. The impact can easily exceed 100 % when the variable in a model is multiplied by another one or it is squared. A small negative percentage can also happen if a variable is merely useless for the model.

In order to create a very fast and accurate prediction model with minimum requirements of hardware resources, we randomly check two sectors (e.g. Battery and Binder or Memory and Binder) every time with the ELM classifier. According to the ELM theory [15], the Gaussian Radial Basis Function kernel $K(u,v) = exp(-\gamma\|u - v\|^2)$ is used. The hidden neurons are k = 20, w_i are the assigned random input weights and b_i the biases, where i = 1,…,N and H is the hidden layer output matrix.

$$H = \begin{bmatrix} h(x_1) \\ \vdots \\ h(x_N) \end{bmatrix} = \begin{bmatrix} h_1(x_1) & \cdots & h_L(x_1) \\ \vdots & & \vdots \\ h_1(x_N) & \cdots & h_L(x_N) \end{bmatrix} \tag{1}$$

$h(x) = [h_1(x), \ldots, h_L(x)]$ is the output (row) vector of the hidden layer with respect to the input x. Function $h(x)$ actually maps the data from the d-dimensional input space to the L-dimensional hidden-layer feature space (ELM feature space) H and thus,

$h(x)$ is indeed a feature mapping. ELM aims to minimize the training error as well as the norm of the output weights as shown in Eq. 2:

$$\text{Minimize: } \|H\beta - T\|^2 \text{ and } \|\beta\| \tag{2}$$

To minimize the norm of the output weights $||\beta||$ is actually to maximize the distance of the separating margins of the two different classes in the ELM feature space $2/||\beta||$.

The calculation of the output weights β is done according to Eq. (3):

$$\beta = \left(\frac{I}{C} + H^T H\right)^{-1} H^T T \tag{3}$$

where c is a positive constant and T is obtained from the *Function Approximation of SLFNs* with additive neurons with $t_i = [t_{i1}, t_{i2}, \ldots, t_{im}]^T R^m$ and $T = \begin{bmatrix} t_1^T \\ \vdots \\ t_N^T \end{bmatrix}$.

It has been shown numerically in ELM theory [15] that the above solution has better generalization performance. More specifically, the reasoning of the new Malware detection algorithm that has been developed in this research is as seen below:

1: ***If*** *both sectors' analysis with the ELM offers Negative results, no action is required and the next 2 sectors are examined.*
2: ***If*** *the ELM analysis results in a Negative result for the one sector and positive for the other then:*
3: ***If*** *both sectors belong to the general Hardware field (eg. Network and Battery) then the Hardware_Dataset is reexamined:*
4: ***If*** *the result is Negative then we go further.*
5: ***If*** *the result is Positive then the whole Original Dataset with all 40 features is checked.*
6: ***If*** *the ELM analysis of both sectors produces Positive results then the whole Original Dataset with all 40 features is checked.*
7: ***If*** *one of sectors belongs to the Binder field then the All_Imp_Var_Dataset is examined:*
8: ***If*** *the result is Negative then we go further.*
9: ***If*** *the result is Positive then the whole Original Dataset with all 40 features is checked with eSNN classification method.*
10: ***If*** *the ELM analysis of both sectors produces Positive result, then the whole Original Dataset is checked with eSNN classifier.*

Step 2:
The train and test datasets are determined and formed, related to n features. The required classes (malware and benign applications) that use the variable *Population Encoding* are imported. This variable controls the conversion of real-valued data

samples into the corresponding time spikes. The encoding is performed with 20 Gaussian receptive fields per variable (Gaussian width parameter beta = 1.5). The data are normalized to the interval [−1,1] and so the coverage of the Gaussians is determined by using i_min and i_max. For the normalization processing the following equation is used:

$$x_{1_{norm}} = 2 * \left(\frac{x_1 - x_{min}}{x_{max} - x_{min}} \right) - 1, \ x \in R \tag{4}$$

The data is classified in two classes namely: ***Class positive*** which contains the benign results and ***Class negative*** which comprises of the malware ones. The eSNN is using modulation factor m = 0.9, firing threshold ratio c = 0.7 and similarity threshold s = 0.6 in agreement with the vQEA algorithm [26, 37]. More precisely, let $A = \{a_1, a_2, a_3 \ldots a_{m-1,} a_m\}$ be the ensemble of afferent neurons of neuron i and $W = \{w_{1,i}, w_{2,i}, w_{3,i} \ldots w_{m-1,i}, w_{m,i}\}$ the weights of the m corresponding connections; let mod ∈ [0,1] be an arbitrary modulation factor. The activation level of neuron i at time t is given by Eq. 5:

$$\mathrm{Activation}(i,t) = \sum_{j \in [1,m]} \mathrm{mod}^{\mathrm{order}(a_j)} w_{j,i} \tag{5}$$

where order(a_j) is the firing rank of neuron a_j in the ensemble A. By convention, order(a_j) = +8 if a neuron a_j is not fired at time t, sets the corresponding term in the above sum to zero. This kind of desensitization function could correspond to a fast shunting inhibition mechanism. When a neuron reaches its threshold, it spikes and inhibits neurons at equivalent positions in the other maps so that only one neuron will respond at any location. Every spike triggers a time based Hebbian-like learning rule that adjusts the synaptic weights. Let t_e be the date of arrival of the Excitatory PostSynaptic Potential (EPSP) at synapse of weight W and t_a the date of discharge of the postsynaptic neuron.

$$\text{If } t_e < t_a \text{ then } dW = a(1 - W)e^{-|\Delta o|\tau \text{ else } dW} = -aW\,e^{-|\Delta o|\tau}. \tag{6}$$

Δo is the difference between the date of the EPSP and the date of the neuronal discharge (expressed in term of order of arrival instead of time), a is a constant that controls the amount of synaptic potentiation and depression [8]. ROPE technique with receptive fields, allow the encoding of continuous values. Each input variable is encoded independently by a group of one-dimensional receptive fields (Figs. 1 and 2). For a variable n, an interval $[I^n_{min,}\ I^n_{max}]$ is defined. The Gaussian receptive field of neuron i is given by its center μ_i and width σ by Eq. 8.

$$\mu i = I^n_{min} + \frac{2i-3}{2} \frac{I^n_{max} - I^n_{min}}{M-2} \tag{7}$$

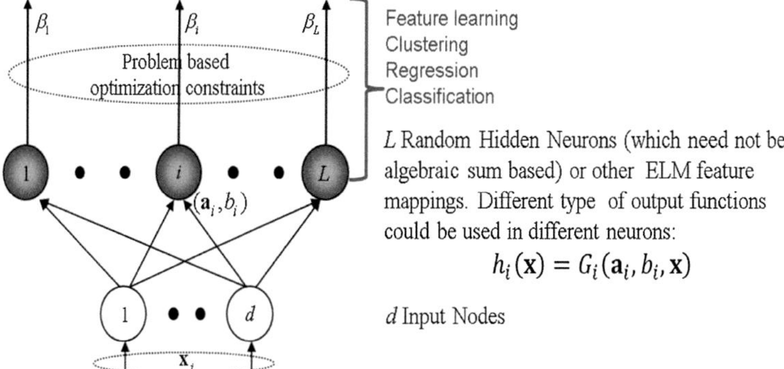

Fig. 1 Extreme learning machine (ELM) [34]

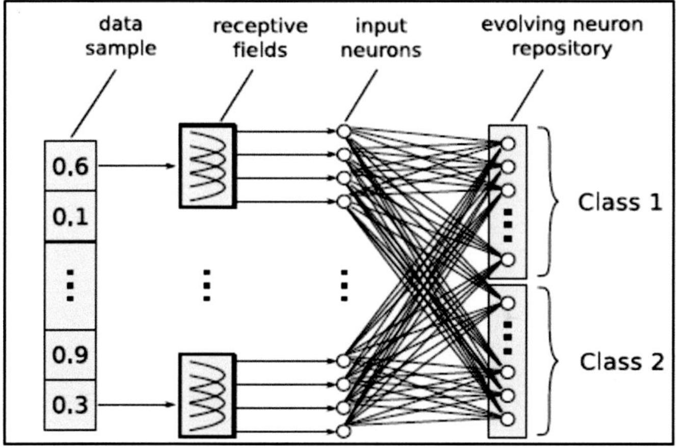

Fig. 2 The Evolving Spiking Neural Network (eSNN) architecture [17]

$$\sigma = \frac{1}{\beta}\frac{I^{n}_{max} - I^{n}_{min}}{M - 2} \tag{8}$$

where $1 \leq \beta \leq 2$ and the parameter β directly controls the width of each Gaussian receptive field. Figure 3 depicts an encoding example of a single variable.

For an input value v = 0.75 (thick straight line) the intersection points with each Gaussian is computed (triangles), which are in turn translated into spike time delays (right figure) [37].

Step 3:
The eSNN is trained with the *training* dataset vectors and the testing is performed with the *testing* vectors. The procedure of one pass learning is described in the following Algorithm 2 [17, 37].

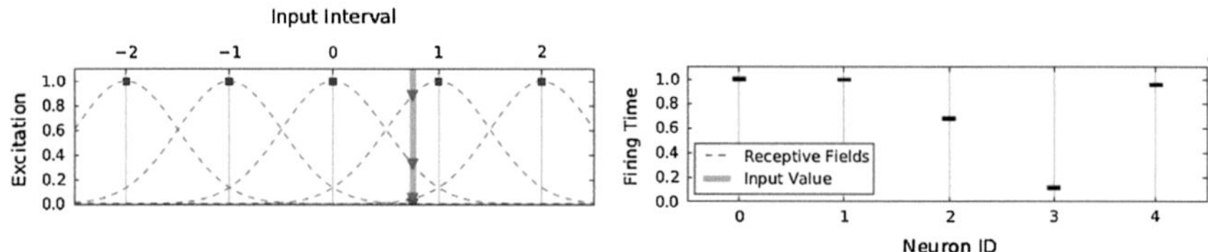

Fig. 3 Population encoding based on Gaussian receptive fields. *Left figure* Input interval—*right figure* neuron ID [17]

Algorithm 1: Training an evolving Spiking Neural Network (eSNN) [37]

Require: m_l, s_l, c_l **for a class label** $l \in L$

initialize neuron repository $R_l = \{\}$
for all samples $X^{(i)}$ belonging to class l **do**
$\quad w_j^{(i)} \leftarrow (m_l)^{order(j)}$, $\forall$ j | j pre-synaptic neuron of i
$\quad u_{max}^{(i)} \leftarrow \sum_j w_j^{(i)} (m_l)^{order(j)}$
$\quad \theta^{(i)} \leftarrow c_l u_{max}^{(i)}$
if $\min(d(w^{(i)}, w^{(k)})) < s_l$, $w^{(k)} \in R_l$ **then**
$\quad w^{(k)} \leftarrow$ merge $w^{(i)}$ and $w^{(k)}$ according to Equation 7
$\quad \theta^{(k)} \leftarrow$ merge $\theta^{(i)}$ and $\theta^{(k)}$ according to Equation 8
else
$\quad R_l \leftarrow R_l \cup \{w^{(i)}\}$
end if
end for

For each training sample i with class label **l** which represent a benign software, a new output neuron is created and fully connected to the previous layer of neurons, resulting in a real-valued weight vector $w^{(i)}$ *with* $w_j^{(i)} \in R$ denoting the connection between the pre-synaptic neuron j and the created neuron i. In the next step, the input spikes are propagated through the network and the value of weight $w_j^{(i)}$ is computed according to the order of spike transmission through a synapse j: $w_j^{(i)} = (m_l)^{order(j)}$, $\forall j \mid j$ pre-synaptic neuron of i. Parameter m_l is the modulation factor of the Thorpe neural model. Differently labeled output neurons may have different modulation factors m_l. Function order(j) represents the rank of the spike emitted by neuron j. The firing threshold $\theta^{(i)}$ of the created neuron I is defined as the fraction $c_l \in R$, $0 < c_l < 1$, of the maximal possible potential

$$u_{max}^{(i)} : \theta^{(i)} \leftarrow c_l u_{max}^{(i)} \tag{7}$$

$$u_{max}^{(i)} \leftarrow \sum_j w_j^{(i)} (m_l)^{order(j)} \tag{9}$$

The fraction c_l is a parameter of the model and for each class label $l \in L$ a different fraction can be specified. The weight vector of the trained neuron is compared to the weights corresponding to neurons already stored in the repository. Two neurons are considered too "similar" if the minimal *Euclidean* distance between their weight vectors is smaller than a specified similarity threshold s_l (the eSNN object uses optimal similarity threshold $s = 0.6$). All parameters of eSNN (modulation factor m_l, similarity threshold s_l, PSP fraction c_l, $l \in L$) included in this search space, were optimized according to the Versatile Quantum-inspired Evolutionary Algorithm (vQEA) [26]. Both the firing thresholds and the weight vectors were merged according to Eqs. 10 and 11:

$$w_j^{(k)} \leftarrow \frac{w_j^{(i)} + N w_j^{(k)}}{1 + N}, \forall j | j \text{ pre} - \text{synaptic neuron of } i \tag{10}$$

$$\theta^{(k)} \leftarrow \frac{\theta^{(i)} + N\theta^{(k)}}{1 + N} \tag{11}$$

Integer N denotes the number of samples previously used to update neuron k. The merging is implemented as the (running) average of the connection weights, and the (running) average of the two firing thresholds. After merging, the trained neuron i is discarded and the next sample processed. If no other neuron in the repository is similar to the trained neuron i, the neuron i is added to the repository as a new output.

4 Data and Results

For this experiment, we used the free dataset provided by B. Amos [2]. The author developed a shell script to automatically analyze *.apk* Android application files by running them in available Android emulators. For each *.apk* file, the emulator simulates user interaction by randomly interacting with the application interface. This is done using the Android "*adb-monkey*" tool [14]. Based on inspection of the source code, we can conclude that each feature vector of the dataset is collected at 5 s' intervals. The memory features were collected by observing the "*proc*" directory in the underlying Linux system of Android. The CPU information was collected by running the Linux "*top*" command. The Battery and Binder information was collected by using "*intent*" (Action listener) [1].

The original dataset has a total of 1153 data (feature vector) samples with 660 benign samples (classified as positive class) and 493 malicious samples (classified as negative class). It was divided randomly in two parts: 1) a training dataset containing 807 patterns (467 positive and 340 negative patterns) 2) a testing dataset containing 346 patterns (193 positive and 153 negative patterns). To identify the integrity of HIMDAM we have compared the ELM and eSNN classifiers with other neural network methods. The performance of both classifiers was evaluated on

Table 1 Accuracy (ACC) comparison between MLP, RBF, ELM, GMDH PNN, eSNN

	MLP		RBF		SOM		ELM		GMDH PNN		eSNN	
	Acc (%)	Time	Acc (%)	Time	Acc (%)	Time	Acc (%)	Time	Acc (%)	Time	Acc (%)	Time
All features	95.38	38.31	85.72	0.34	87.34	0.20	89.19	**0.17**	90.50	13.00	**97.10**	20.22
Battery + perm	73.58	1.69	71.96	0.19	64.28	0.23	72.49	**0.17**	71.30	2.00	**79.30**	1.22
Binder + perm	93.35	7.95	79.87	0.08	72.82	0.08	82.98	**0.05**	84.80	5.00	**93.60**	8.71
Memory CPU Perm	82.08	4.84	79.65	0.13	77.16	0.23	80.08	**0.12**	83.10	4.00	**84.90**	4.40
Network perm	**81.21**	3.02	69.83	**0.13**	70.10	0.19	71.27	0.14	73.70	3.00	80.10	3.11
Important var. from all features	97.69	14.59	83.76	**0.20**	80.34	**0.20**	89.47	**0.20**	91.00	8.00	**98.20**	11.02
Important var. from hardware	94.22	6.69	88.25	0.22	75.90	0.17	88.00	**0.14**	89.40	6.00	**94.80**	5.91

malware datasets. The results showed that the kernel based ELM has much faster learning speed (run thousands times faster than conventional methods) and the eSNN has much better generalization performance and more accurate and reliable classification results. The comparisons were performed on a dual boot PC with a P4 at 3.1 GHz CPU and 4 GB RAM. For the eSNN classification, the *Linux Ubuntu 12.04 LTS* OS with *PyLab* (*NumPy, SciPy, Matplotlib* and *IPython*) was employed. The MLP, RBF and SOM tests were performed with the *Weka 3.7* [35], ELM with *Matlab 2013* and GMDH PNN with GMDH shell software [36]. The performance comparisons of NN algorithms are shown in Table 1. The confusion matrices for ELM and eSNN can be seen in Table 2 (Fig. 4).

Table 2 Confusion matrices for ELM and eSNN algorithms

Confusion matrices for ELM				Confusion matrices for eSNN			
All features				All features			
	Benign (predicted)	Malware (predicted)	Accuracy (%)		Benign (predicted)	Malware (predicted)	Accuracy (%)
Benign (actual)	180	13	**93.26**	Benign (actual)	190	3	**98.40**
Malware (actual)	23	130	**85.12**	Malware (actual)	7	146	**95.80**
Overall accuracy			**89.19**	Overall accuracy			**97.10**
Battery permissions							
Benign (actual)	149	44	**77.20**	Benign (actual)	160	33	**82.90**
Malware (actual)	50	103	**67.78**	Malware (actual)	37	116	**75.70**
Overall accuracy			**72.49**	Overall accuracy			**79.30**
Binder permissions							
Benign (actual)	164	29	**84,97**	Benign (actual)	185	8	**95.80**
Malware (actual)	29	124	**80,99**	Malware (actual)	13	140	**91.40**
Overall accuracy			**82.98**	Overall accuracy			**93.60**
Memory CPU permissions							
Benign (actual)	159	34	**82.38**	Benign (actual)	168	25	**87.00**
Malware (actual)	34	119	**77.78**	Malware (actual)	26	127	**82.80**
Overall accuracy			**80.08**	Overall accuracy			**84.90**

(continued)

Table 2 (continued)

Confusion matrices for ELM				Confusion matrices for eSNN			
All features				All features			
	Benign (predicted)	Malware (predicted)	Accuracy (%)		Benign (predicted)	Malware (predicted)	Accuracy (%)
Network permissions							
Benign (actual)	142	51	**73.57**	Benign (actual)	161	32	**83.40**
Malware (actual)	48	105	**68.97**	Malware (actual)	36	117	**76.80**
Overall accuracy			**71.27**	Overall accuracy			**80.10**
Importance variables from all							
Benign (actual)	180	13	**93.26**	Benign (Actual)	191	2	**99.00**
Malware (actual)	22	131	**85.68**	Malware (Actual)	4	149	**97.40**
Overall accuracy			**89.47**	Overall Accuracy			**98.20**
Importance variables from hardware							
Benign (actual)	175	23	**90.67**	Benign (actual)	184	9	**95.30**
Malware (actual)	23	130	**85.33**	Malware (actual)	9	144	**94.30**
Overall accuracy			**88.00**	Overall accuracy			**94.80**

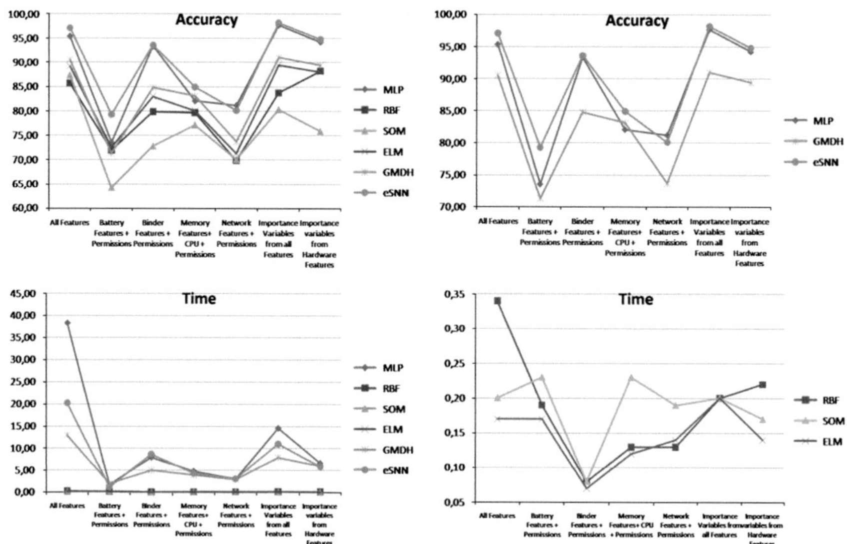

Fig. 4 Accuracy and time comparison of MLP, RBF, ELM, GMDH PNN and eSNN

5 Discussion—Conclusions

An innovative bio-inspired Hybrid Intelligent Method for Detecting Android Malware (HIMDAM) has been introduced in this paper. It performs classification by using ELM (a very fast approach to properly label malicious executables) and eSNN for the detection of malware with high accuracy and generalization. An effort was made to achieve minimum computational power and resources. The classification performance of the ELM and the accuracy of the eSNN model were experimentally explored based on different datasets and reported promising results. Moreover the hybrid model detects the patterns and classifies them with high accuracy. In this way it adds a higher degree of integrity to the rest of the security infrastructure of Android Operating System. As a future direction, aiming to improve the efficiency of biologically realistic ANN for pattern recognition, it would be important to evaluate the eSNN model with ROC analysis and to perform feature minimization in order to achieve minimum processing time. Other coding schemes could be explored and compared on the same security task. Also what is really interesting is a scalability of ELM with other kernels in parallel and distributed computing in a real-time system. Finally the HIMDAM could be improved towards a better online learning with self-modified parameter values.

References

1. Alam M.S., Vuong S.T.: Random forest classification for detecting android malware. In: IEEE IC on Green Computing and Communications and Internet of Things (2013)
2. Amos, B.: Antimalware. https://github.com/VT-Magnum-Research/antimalware (2013)
3. Barrera, D., Kayacik, H., Oorshot, P., Somayaji, A.: A Methodology for Empirical Analysis of Permission-Based Security Models and its Application to Android. ACM (2010)
4. Burguera, I., Zurutuza, U., Nadjm-Tehrani, S.: Crowdroid: behavior-based malware detection system for android. In: 1st ACM Workshop on on SPSM, pp. 15–26. ACM (2011)
5. Cambria E., Huang G.-B.: Extreme learning machines. IEEE Intell. Syst. (2013)
6. Cheng, C., Peng, W.T, Huang, G.-B.: Extreme learning machines for intrusion detection. In: WCCI IEEE World Congress on Computational Intelligence Brisbane, Australia (2012)
7. Chin E., Felt A., Greenwood K., Wagner D.: Analyzing inter-application communication in android. In: 9th Conference on Mobile Systems, Applications, and Services, pp. 239–252. ACM (2011)
8. Delorme, A., Perrinet, L., Thorpe, S.J.: Networks of Integrate-and-fire neurons using rank order coding b: spike timing dependant plasticity and emergence of orientation selectivity. Neurocomputing **38–40**(1–4), 539–545 (2000)
9. Enck, W., Ongtang, M., McDaniel, P.: On lightweight mobile phone application certification. In: Proceedings of the 16th ACM Conference on Computer Security, CSS (2009)
10. Fedler, R., Banse, C., Krauß, Ch., Fusenig, V.: Android OS security: risks and limitations a practical evaluation, AISEC Technical Reports, AISEC-TR-2012–001 (2012)
11. Fuchs, A., Chaudhuri, A., Foster, J.: ScanDroid: automated security certification of android applications, Technical report, University of Maryland (2009)
12. Ghorbanzadeh, M., Chen, Y., Zhongmin, M., Clancy, C.T., McGwier, R.: A neural network approach to category validation of android applications. In: International Conference on

Computing, Networking and Communications, Cognitive Computing and Networking Symposium (2013)
13. Glodek, W., Harang R.R.: Permissions-based detection and analysis of mobile malware using random decision forests. In: IEEE Military Communications Conference (2013)
14. Google, UI/Application Exerciser Monkey. http://developer.android.com/tools/help/monkey.html (2013)
15. Huang, G.-B.: An Insight into Extreme Learning Machines: Random Neurons, Random Features and Kernels. Springer (2014). doi:10.1007/s12559-014-9255-2
16. Joseph, J.F.C., Lee, B.-S., Das, A., Seet, B,-C.: Cross-layer detection of sinking behavior in wireless ad hoc networks using ELM and FDA. IEEE IJCA **54**(14) (2012)
17. Kasabov, N.: Evolving connectionist systems: Methods and Applications in Bioinformatics, Brain study and intelligent machines. Springer Verlag, NY (2002)
18. Kohonen, T.: Self-organizing networks. In: Proceedings of the IEEE (1990)
19. Kolter, J.Z., Maloof, M.A.: Learning to detect malicious executables in the wild. In: International Conference on Knowledge Discovery and Data Mining, pp. 470–478 (2006)
20. Lange, M., Liebergeld, S., Lackorzynski, A., Peter M.: L4Android: a generic operating system framework for secure smartphones. In: ACM Workshop on SPSM (2011)
21. MacQueen, J.: Some methods for classification and analysis of multivariate observations. In: Proceedings of the 5th Berkeley Symposium on Mathematical Statistics and Probability (1967)
22. Portokalidis, G., Homburg, P., Anagnostakis, K., Bos, H.: Paranoid Android: versatile protection for smartphones. In: 26th Annual Computer Security Applications Conference (2010)
23. Sahs, J., Khan, L.: A Machine learning approach to android malware detection. In: European Intelligence and Security Informatics Conference (2012)
24. Scandariato, R., Walden, J.: Predicting Vulnerable Classes in an Android Application (2012)
25. Schliebs, S., Kasabov, N.: Evolving spiking neural network—a survey. Evolving Systems **4** (2), 87–98 (2013)
26. Schliebs, S., Defoin-Platel, M., Kasabov, N.: Integrated Feature and Parameter Optimization for an Evolving Spiking Neural Network, 5506, pp. 1229–1236. Springer (2009)
27. Schultz, M.G., Eskin, E., Zadok, E., Stolfo, S. J.: Data mining methods for detection of new malicious executables. In: SP '01, pp. 38. IEEE Computer Society, Washington, DC (2001)
28. Shabtai, A., Fledel, Y., Elovici, Y.: Automated static code analysis for classifying android applications using machine learning. In: IC Computational Intelligence and Security (2010)
29. Shabtai, A., Fledel, Y., Elovici Y.: Automated static code analysis for classifying android applications using machine learning, in CIS. In: Conference on IEEE, pp. 329–333 (2010)
30. Tesauro, G.J., Kephart, J.O., Sorkin, G.B.: Neural networks for computer virus recognition. IEEE Expert **11**(4), 5–6 (1996)
31. Thorpe, S.J., Delorme, A.: Rufin van Rullen: Spike-based strategies for rapid processing. Neural Netw. **14**(6–7), 715–725 (2001)
32. Thorpe, S.J., Gautrais, J.: Rank order coding. In: CNS '97: 6th Conference on Computational Neuroscience: Trends in Research, pp. 113–118. Plenum Press (1998)
33. www.wala.sourceforge.net/wiki/index.php
34. www.extreme-learning-machines.org/
35. www.cs.waikato.ac.nz/ml/weka
36. www.gmdhshell.com/
37. Wysoski, S.G., Benuskova, L., Kasabov, N.K.: Adaptive learning procedure for a network of spiking neurons and visual pattern recognition. In: Advanced Concepts for Intelligent Vision Systems, pp. 1133–1142. Springer Berlin/Heidelberg (2006)

Predicting the Impact of Advertisements on Web Pages Aesthetic Impressions

Gianni Fenu, Gianfranco Fadda and Lucio Davide Spano

Abstract In this paper we study the advertisements impact on the users' aesthetic judgement of a web page layout. We obtain the prediction from a regression model, considering a set of image features calculated on the web page rendering. In particular, we analyse the effects on two different predictors: the colourfulness and the visual complexity. We compared the prediction against ground-truth values, obtained through a user study. We conclude that the prediction model behaves correctly for the perceived complexity, but it is not able to predict the increase on the colourfulness ratings. Finally, we discuss a browser extension that takes advantage of the predictive model for analysing the effects of both the advertisement layout and position on the user's aesthetic impression.

Keywords Online advertising · Aesthetics · Predictive models · Visual features · Design tools

1 Introduction

Users make their opinion on websites they visit for the first time after a few seconds [6]. This is a well known fact in the HCI community. Therefore, designers carefully select both the position and the layout of different contents in order to attract the users' attention. On the one hand, commercial websites profit including advertisements. On the other hand, their inclusion changes the users' first-look impression. Even if usually they skip the advertisements included in banners or presented

G. Fenu (✉) · G. Fadda · L.D. Spano
Department of Mathematics and Computer Science, University of Cagliari,
Via Ospedale 72, 09124 Cagliari, Italy
e-mail: fenu@unica.it

G. Fadda
e-mail: gianfranco_fadda@unica.it

L.D. Spano
e-mail: davide.spano@unica.it

S. Kunifuji et al. (eds.), *Knowledge, Information and Creativity Support Systems*,
Advances in Intelligent Systems and Computing 416,
DOI 10.1007/978-3-319-27478-2_21

as simple text, such distracting content decreases the performance while searching information or while interacting with the page, and increases the cognitive workload [2].

In this paper, we focus on predictive models for the users' aesthetic impression on web pages. We analyse the effects of advertisements on both the predicted and the perceived values. To achieve this goal, we exploited the prediction model proposed in [8]. We subsequently prepared a dataset including websites of different categories (e.g. news, e-commerce, business, etc.). For each website, we included in our experiment two different versions: one with and one without advertisements. In the experiment we compared the responses of the predictive model with the ground-truth results obtained through the user test.

After that, we describe a Chrome web browser extension that provides a support for the analysis of both the advertisements layout and position in the web page, exploiting the predictive models and the findings reported in this paper.

Finally, we summarize the results and we discuss the possible developments of this work.

2 Related Work

The user's opinion on an interface is influenced not only by its usability, but also by the aesthetic impression: the perceived usability is affected by the visual appeal of the interface [5]. This is particularly important on the web, where it is easy for users to switch between different sources. If a homepage does not impress a user, she will leave the page after a short time, without even evaluate the usability of the website [6].

User creates their first impression during the first 50–500 ms of a webpage visit [6], grounding the idea reported in different work that focused on predicting such impression through image-related metrics. We can recall here the work by Zheng et al. [10], which shows how different low-level image statistics correlate with the user's aesthetic ratings. In this paper, we exploit the work in [8] that has been validated with a large number of users. We provide a summary of their results in the next section.

In particular, we analyse the advertisements impact on the aesthetic perception of a web page. Owens et al. in [7] studied the so-called "text advertisement blindness", which is the ability of users to ignore the advertisement content inside web pages, even for textual information and not only for banners. However, even if users are able to skip such content, their presence increases the users' cognitive load [2].

The visual properties correlated with the effectiveness of an advertisement include the spatial position and its visual appearance [1]. All of them have an impact on with the overall aesthetic judgement of a website. In this work, we consider two different measures that may be affected by advertisements: the perceived colourfulness and the visual complexity.

3 Advertisement Impact on Rating Prediction

In this paper, we exploit the predictive model described in [8], which considers two different factors for predicting the visual appeal of a web page: the *colourfulness* and the *visual complexity*. In order to obtain the ground-truth data on the users' perception, the authors selected a set of 450 websites and they proposed a randomly selected sample of 30 website to 242 volunteers, which rated the colourfulness and the visual complexity with a 9-point Likert scale (1 is the lower and 9 is the higher value).

With these ground-truth values they created a regression model for predicting the ratings according to different image metrics related to colours (e.g. percentage of pixels close to the sixteen W3C colours, the average pixel value in the HSV space, and the two definitions of colourfulness in [9] and [4], quadtree decomposition [10] and space-based decomposition [3]). The two regression models are reported in Table 1.

Table 1 Regression model for the perceived colourfulness ($r = 88, p < 0.001, R^2 = 0.78, R^2_{adj} = 0.77$) and visual complexity ($r = 80, p < 0.001, R^2 = 0.65, R^2_{adj} = 0.64$) in [8]

Colourfulness	b	SE_b	$p >$
Constant	−0.68	0.73	0.05
Gray	2.45	0.56	0.001
White	2.74	0.71	0.001
Maroon	1.60	0.71	0.05
Green	1.85	0.57	0.01
Lime	−3.17	1.42	0.05
Blue	−3.91	1.06	0.001
Teal	1.01	0.56	0.05
Saturation	0.01	0.002	0.01
Colorfulness [4]	0.03	0.003	0.001
Image areas	0.06	0.008	0.001
Quadtree leaves	3.74E-4	0.000	0.001
Text area	−1.48E-6	0.000	0.05
Non text area	1.86E-6	0.000	0.001
Complexity	b	SE_b	$p >$
Constant	0.637	0.179	0.001
Text area	0.000	0.000	0.001
Non text areas	0.000	0.000	0.001
Number of leaves	0.005	0.002	0.05
Number of text groups	0.052	0.008	0.001
Number of image areas	0.056	0.012	0.001
Colorfulness [9]	0.011	0.005	0.05
Hue	0.005	.002	0.05

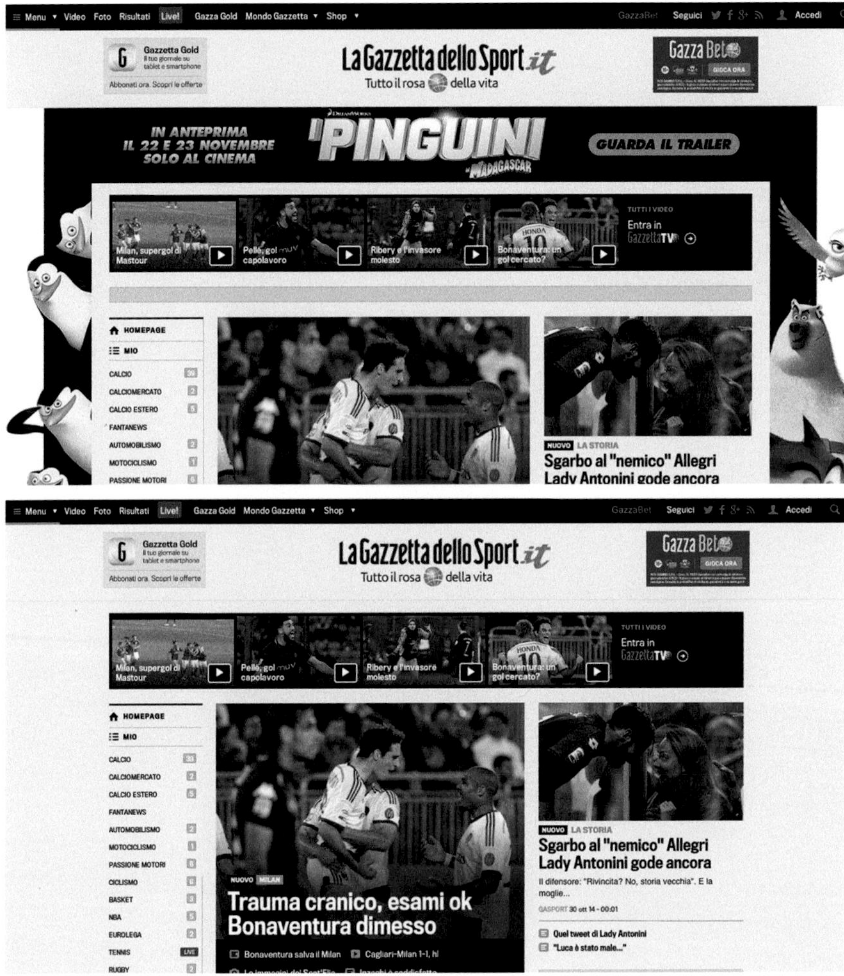

Fig. 1 Sample website with (*top part*) and without (*bottom part*) advertisements

Reinecke et al. demonstrated that both complexity and colourfulness can be used for predicting the mean aesthetic user rating for a website. Additional details on both the models and the metrics are available in [8].

In this study, we analyse the impact of the advertisements on the prediction of both colourfulness and visual complexity of websites. We are interested in understanding if there is a statistically relevant difference between the rating of the website without advertisements and the version including them.

In order to measure this difference, we created a dataset of 50 websites belonging to different categories (news, weather, sports etc.). We selected the websites considering a list of the top 500 accessed websites.[1] We did not consider websites without advertisements (e.g. institutional websites). We obtained a set of screenshots of the websites interfaces through a screen capture tool. After that, we manually removed the advertisements in the picture, replacing them with the website background color, pattern or image. Figure 1 shows two versions of the same website in the dataset with and without advertisements.

For each image in the dataset, we calculated the predicted values for the colourfulness and the visual complexity according to the regression models in Table 1. Then, we compared the ratings through a paired t-Test, in order to establish if the presence or the absence of advertisements introduced a significant difference on the predicted values.

For the colourfulness we registered a not significant difference between the values predicted for the websites with and without advertisements ($p = 0.77$). The 95 % confidence interval for the difference is $\bar{x} = 0.18 \pm 1.25$ and the rating is expressed in 1 to 9 Likert scale. Therefore, introducing the advertisements on websites does not produce effects on the predicted colourfulness.

Instead, for the visual complexity we registered a significant difference between the versions ($p = 10^{-7}$). Analysing the 95 % confidence interval around the difference, we found that removing the advertisements decreases the predicted complexity of about one (out of 9) point ($\bar{x} = 1.0 \pm 0.27$).

4 Advertisement Impact on User Ratings

In this section, we report on a user study we performed for establishing whether the effect of advertisements on the predicted ratings arises also for the perceived colourfulness and visual complexity.

In order to collect the ground-truth data for assessing the effects of the advertisements, we created a web application. The first presentation requires the user to enter some demographic data, such as age, education level and the experience with web browsing. After providing this information, the application shows 40 screenshots of 20 different websites, presenting both the version with and without advertisements in a random order. The user rates the perceived colourfulness and complexity in a 1 to 9 Likert scale.

Fifty users participated to the test, 35 male and 15 female. They belonged to different age ranges: 2 were less than 15 years old, 10 between 16 and 25, 13 between 26 and 35, 14 between 36 and 45 and 11 are older than 45. The education level varied from middle school (4), high school (11), bachelor degree (17), master degree (16) and Ph.D. (2).

We recruited the participants using social networks and emails. We performed the test using a remote setting. Once we collected the data, we performed a two way ANOVA analysis for repeated measures, in order to study the effect of the website

[1] http://www.alexa.com/topsites.

Table 2 ANOVA table for colourfulness ratings

Colorfulness	SS	df(SS)	Error	df(Error)	F	p-value
Version	273	1	207.81	49	64.3842	10^{-10}
Website	1131	6.46779	2626.73	316.92171	21.1018	$<10^{-16}$
Version * Website	121	11.77563	1057.22	577.00587	5.6271	10^{-9}

version (with or without advertisements) on the colourfulness and the visual complexity ratings. In the analysis we consider two different factors that influence the rating outcome: the website and the version.

4.1 Colourfulness

We start from the analysis of the colourfulness ratings, which is summarised in Table 2. In this analysis we considered as factors the evaluated website (*Website*) and its configuration with or without advertisements (*Version*). For each user, we repeated the measures for all the configurations of the two factors.

The Mauchly's test revealed that the sphericity assumption had been violated for the website factor ($\chi^2(189) = 6.33 \times 10^{-6}$, $p < 10^{-16}$) and for the interaction between the version and the website ($\chi^2(189) = 0.00136885$, $p = 3.2 \times 10^{-5}$). Therefore, we corrected the degrees of freedom applying the Greenhouse-Geisser estimation of sphericity, respectively with $\varepsilon_w = 0.3401$ and $\varepsilon_{v,w} = 0.61977$.

Table 2 shows an obvious significant effect of the website on the perceived colourfulness. It is more interesting to notice that both the version and the interaction between the version and the website have a significant impact on the users' ratings.

Even if the interaction is significant, the post-hoc analysis with the Tukey's HDS test revealed that the difference size is not meaningful for most values. The analysis is summarized in Fig. 2: the upper part contains the differences for the pairs having the version with advertisements, while the lower parts shows the pairs without advertisements. As it is possible to see, most of them contain 0 inside the confidence interval.

With respect to the version factor, the Tukey's HSD test revealed that the ratings of websites with and without advertisements differ significantly ($p < 10^{-16}$). On average, the difference between the version with and without advertisement is 0.739, and the 95 % confidence interval is [0.57339; 0.90461]. In a 1 to 9 Likert scale such difference is meaningful and, contrary to the predicted values, it demonstrates that users do perceive a colourfulness difference between the two versions.

This can be explained with the selective user's focus on a web page content: advertisements are perceived as an "added" content that does not really belong to the page, thus their presence always add colors to visual representation. The prediction instead is based on image features and, if the advertisements exploit roughly

Fig. 2 95 % family-wise confidence level for colorfulness means comparison

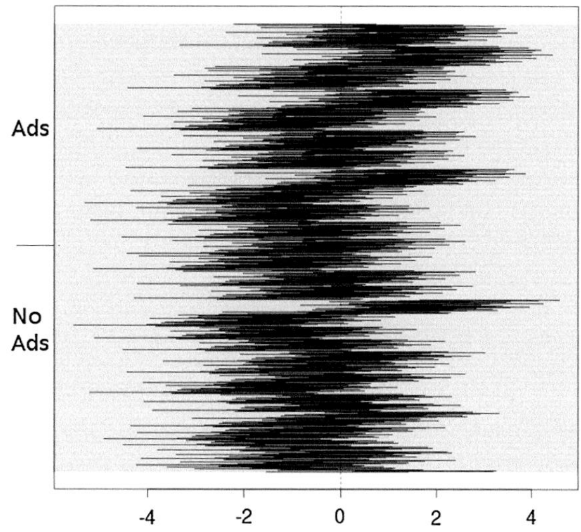

Table 3 ANOVA table for complexity ratings

Complexity	SS	df(SS)	Error	df(Error)	F	*p*-value
Version	99	1	166.4	49	29.4154	10^{-6}
Website	361.6	8.74304	2932.2	428.40896	6.0434	10^{-14}
Version * Website	41.4	11.43933	1382.7	560.52717	1.4680	0.08873

the same colours contained in the page, their presence may not affect the measurement. Therefore, we suggest to remove the advertisements from the website image for estimating the colourfulness perception and to increase the value by 0.5 points.

4.2 *Complexity*

The analysis of the complexity ratings is summarized in Table 3, considering the same factors and conditions with respect to the previous experiment. The Mauchly's test revealed again the sphericity assumption had been violated for the website factor ($\chi^2(189) = 1.38 \times 10^{-4}, p < 10^{-16}$) and for the interaction between the version and the website ($\chi^2(189) = 9.93 \times 10^{-4}, p = 2.9 \times 10^{-6}$) and we corrected the degrees of freedom with Greenhouse-Geisser ($\varepsilon_w = 0.46016$, $\varepsilon_{v,w} = 0.60207$).

Table 3 reveals again the obvious significant effect of the website factor for the complexity rating. In addition, it reveals that the interaction between the website and the version is not significant, therefore our data can be represented with an additive model.

The most interesting part of this analysis is the significance of the version factor ($p < 10^{-6}$). In order to evaluate the difference in the ratings between the version with and without advertisements, we performed a post hoc Tukey's HDS test. It revealed that, on average, websites with advertisements are almost half point (0.447 in a 1 to 9 Likert scale) more complex than those without advertisements. The difference is contained in a [0.26316; 0.63084] interval ($p = 2 \times 10^{-6}$).

Therefore, the study confirms that the predicted increase of complexity is perceived by users, even if it was overestimated. In this case, we suggest to use website images that contain the advertisements in order to predict an upper bound for the perceived complexity of the page.

5 Tool Support

We provide a tool for helping designers in both selecting the advertisements and their position in the page, controlling the perceived complexity. The tool exploits the regression model for the evaluation and allows designers to compare the difference between including or excluding the advertisements.

The tool consists of two components: a web service and its front-end. The web service calculates the following metrics applying the regression model described in [8]:

1. The prediction of the user's aesthetic judgement.
2. The evaluation of the user perceived colourfulness.
3. The evaluation of the user perceived complexity.

All operations require an image rendering of the web page layout and return a rating in a 1 to 10 interval. It is possible to invoke the web service through both HTTP and SOAP requests.

Designers are not supposed to invoke the web service directly, but they can use a simple Chrome browser extension that provides its front-end. We opted for a web browser extension rather than a javascript library for two reasons. Firstly, a browser extension allows to separate the analysis tool interface from the web application. Secondly, the javascript library script requires to be injected into existing web pages (e.g. adding a `script` tag), while an extension provides the analysis support without any change to the considered page. The drawback of our solution is that it works only on the Chrome browser, while a pure javascript implementation would support other browsers.

We summarize the steps for analysing the complexity of a web site in Fig. 3. First of all, the designer opens the analysis tool clicking on the extension icon, which has the shape of a magnifying lens, near to the browser address bar (Fig. 3, top part). Then, the designer takes a screenshot of the current a web page, selecting one of the open tabs and pressing the screenshot button (the blue button in Fig. 3, top-left part). After that, the tool shows a screenshot thumbnail in the extension user interface. The designer can repeat this step until she is satisfied with the displayed image and she can continue pressing the right arrow button (Fig. 3, top-left part).

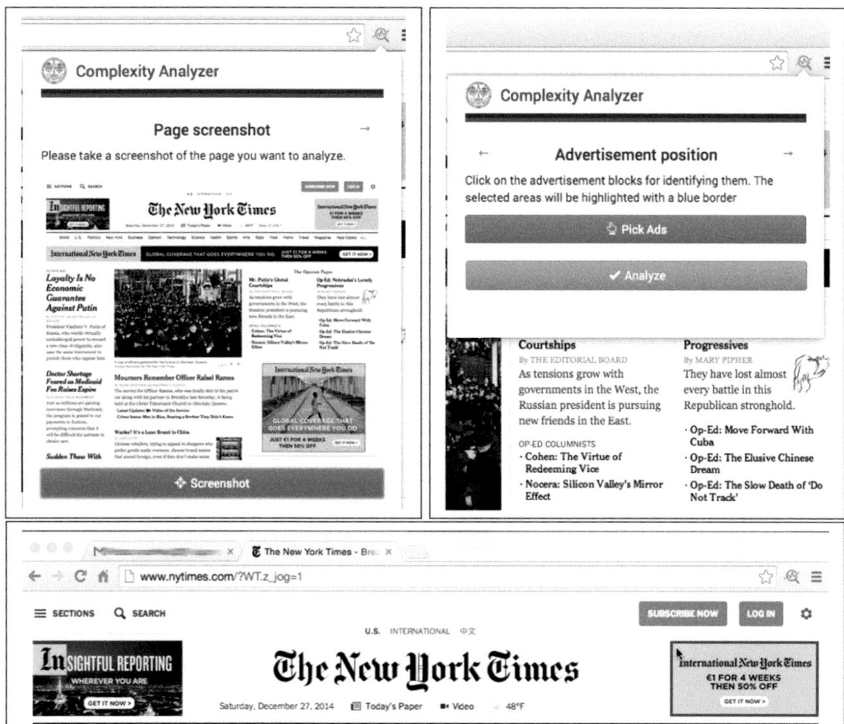

Fig. 3 Web page design evaluation process. The designer first creates a screenshot of the web page, then he identifies the advertisements blocks and finally completes the evaluation sending the input to the remote web service

In the second step, the tool gather information about the advertisements position. Pressing the "Pick Ads" button, visible in the top right part of figure Fig. 3, the tool allows the designer to click on the different advertisement blocks. The tool stores the top-left corner position (considering the page coordinate system), the width and height of the corresponding DOM box. Once selected, the tool highlight the advertisement box using a blue border (Fig. 3, bottom part).

Finally, the tool has gathered all the information for preforming the complexity analysis. The extension sends a set of requests to the web service (one for each operation) specifying as input the web page screenshot. Once the analysis has been performed at server side, the tool displays the results as shown in Fig. 4. In the upper part, the tool reports the overall prediction of the user perceived complexity, including both a visual representation (a star rating line) and the precise numeric value. In the middle part, the tool visualizes the web page thumbnail with the advertisement block highlighting. The bottom part includes the details the colourfulness and complexity evaluation. For both measures, the tool shows the corresponding values through a bar.

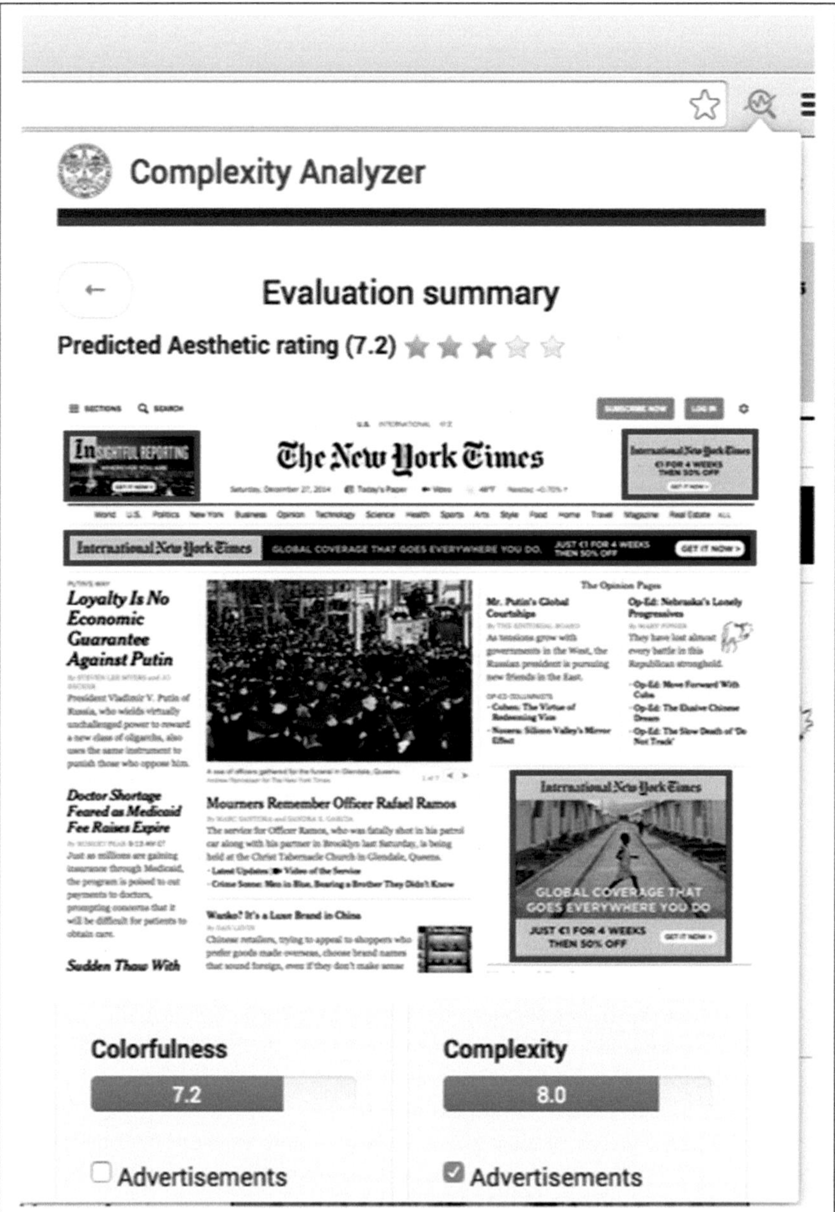

Fig. 4 Web page design evaluation summary. The *top part* shows the predicted aesthetic rating for the displayed design, while the *bottom part* shows the detail for both the colourfulness and complexity measures

It is possible for designers to evaluate the impact of advertisements on both measures including or excluding them through a dedicated checkbox. In order to exclude the advertisements, the tool replaces their boxes with the web page background colour before sending the image to the web service. The values of both the considered measure (colourfulness or complexity) and the overall aesthetic prediction are updated accordingly. Designers can save the analysis result for future comparisons in the browser local storage.

In general, considering the results discussed in this paper, we suggest to consider the advertisements for calculating the web page perceived complexity and to remove them for calculating the colourfulness.

6 Conclusions and Future Work

In this paper, we analysed the impact of advertisements on predicting the user's rating of both complexity and colourfulness of web pages, which are correlated with the first aesthetic impression of the user. In order to perform such analysis, we selected the model in [8], which is able to predict the user's rating according to a set of metrics calculated on the website image. We found that the presence of advertisements does not change significantly the colourfulness prediction in the discussed model. Instead, it does have an impact on the predicted complexity value: on average, websites with advertisements are one point more complex in a 1 to 9 Likert scale.

We conducted a user study for analysing the effects on real user ratings. We found a significant increase on the perception of both colourfulness and complexity ratings. While the latter is replicated also by the prediction model, the first one was not. Therefore, we suggest to evaluate the colourfulness of the web page contents rather than considering also the advertisements for predicting the user's aesthetic judgement.

Finally, we discussed a Chrome browser extension for evaluating a web page design through the regression model. It helps designers in selecting advertisements without increasing the perceived complexity. In addition, the tool helps them in comparing different positions and layouts for including advertisements in a web page.

In the future, we plan to further investigate the relationship between colourfulness and advertisements. Our goal is to obtain a prediction model that is able to correctly evaluate the inclusion impact of contents that are not directly related with the page topic.

Acknowledgments We gratefully acknowledge Sardinia Regional Government for the financial support (P.O.R. Sardegna F.S.E. Operational Programme of the Autonomous Region of Sardinia, European Social Fund 2007–2013—Axis IV Human Resources, Objective l.3, Line of Activity l.3.1 "Avviso di chiamata per il finanziamento di Assegni di Ricerca".

References

1. Azimi, J., Zhang, R., Zhou, Y., Navalpakkam, V., Mao, J., Fern, X.: The impact of visual appearance on user response in online display advertising. In: Proceedings of the 21st International Conference Companion on World Wide Web, pp. 457-458. WWW'12 Companion, ACM, New York, NY, USA. http://doi.acm.org/10.1145/2187980.2188075 (2012)
2. Burke, M., Hornof, A., Nilsen, E., Gorman, N.: High-cost banner blindness: Ads increase perceived workload, hinder visual search, and are forgotten. ACM Trans. Comput.-Human Interact. (TOCHI) **12**(4), 423–445 (2005)
3. Ha, J., Haralick, R., Phillips, I.: Recursive x-y cut using bounding boxes of connected components. In: Proceedings of the Third International Conference on Document Analysis and Recognition, 1995, vol. 2, pp. 952-955, Aug 1995
4. Hasler, D., Suesstrunk, S.E.: Measuring colorfulness in natural images. In: Proceedings of the SPIE Human Vision and Electronic Imaging, vol. 5007, pp. 87-95 (2003)
5. Lindgaard, G., Dudek, C., Sen, D., Sumegi, L., Noonan, P.: An exploration of relations between visual appeal, trustworthiness and perceived usability of homepages. ACM Trans. Comput.-Hum. Interact. 18(1), 1:1-1:30 (2011). http://doi.acm.org/10.1145/1959022.1959023
6. Lindgaard, G., Fernandes, G., Dudek, C., Brown, J.: Attention web designers: you have 50 milliseconds to make a good first impression! Behav. Inf. Technol. 25(2), 115-126 (2006). http://dx.doi.org/10.1080/01449290500330448
7. Owens, J.W., Chaparro, B.S., Palmer, E.M.: Text advertising blindness: The new banner blindness? J. Usability Stud. 6(3), 12:172-12:197 (2011). http://dl.acm.org/citation.cfm?id=2007456.2007460
8. Reinecke, K., Yeh, T., Miratrix, L., Mardiko, R., Zhao, Y., Liu, J., Gajos, K.Z.: Predicting users' first impressions of website aesthetics with a quantification of perceived visual complexity and colorfulness. In: Proceedings of the SIGCHI Conference on Human Factors in Computing Systems, pp. 2049-2058. CHI'13, ACM, New York, NY, USA. http://doi.acm.org/10.1145/2470654.2481281 (2013)
9. Yendrikhovskij, S., Blommaert, F., De Ridder, H.: Optimizing color reproduction of natural images. In: Color and Imaging Conference, vol. 1998, pp. 140-145. Society for Imaging Science and Technology (1998)
10. Zheng, X.S., Chakraborty, I., Lin, J.J.W., Rauschenberger, R.: Correlating low-level image statistics with users–rapid aesthetic and affective judgments of web pages. In: Proceedings of the SIGCHI Conference on Human Factors in Computing Systems, pp. 1-10. CHI'09, ACM, New York, NY, USA. http://doi.acm.org/10.1145/1518701.1518703 (2009)

OntOSN—An Integrated Ontology for the Business-Driven Analysis of Online Social Networks

Richard Braun and Werner Esswein

Abstract The analysis of Online Social Networks (OSN) like Facebook is an emerging field of research and of high relevance for businesses in order to derive information about current and prospective customers. Generally, OSN analysis requires the participation of different stakeholders; for example domain experts, data mining experts and IT engineers. Each of these groups operates with specific modeling approaches what can cause communication problems, redundancy or model discrepancies. We proclaim the design of an integrated ontology for OSN analysis in order to avoid these problems. Therefore, a set of requirements on the aimed ontology is evolved based on characteristic use cases coming from specific business purposes. In a second step, a range of existing ontologies are examined and compared with the elaborated requirements. Finally, the OntOSN ontology is designed.

Keywords Ontology Engineering · Domain-specific Ontology · Social User Ontology · Online Social Networks · OSN

1 Introduction and Motivation

Social media in general and Online Social Networks (OSN) in particular are substantial technologies of the daily life [24, 37], which provide new opportunities for businesses in the field of Customer Relationship Management (CRM), Social Business Intelligence and Social Media Analytics [22, 23, 26, 59]. On the other hand, OSN can cause several threats and risks for business like reputational damages, data

R. Braun (✉) · W. Esswein
Chair of Wirtschaftsinformatik, esp. Systems Development, Technische Universität Dresden, 01062 Dresden, Germany
e-mail: richard.braun@tu-dresden.de

W. Esswein
e-mail: werner.esswein@tu-dresden.de

S. Kunifuji et al. (eds.), *Knowledge, Information and Creativity Support Systems*, Advances in Intelligent Systems and Computing 416,
DOI 10.1007/978-3-319-27478-2_22

leakages or IT-related risks [13]. Due to a power shift in OSN, businesses are in a more reactive position and need to monitor OSN users and their interactions associated with the business. From an academic point of view, there are also several topics for investigation like the analysis of social influence structures [36]. It can be stated, that OSN are in the scope of both businesses and academia for a range of reasons [37, 52] and the multi-perspective analysis of OSN is an emerging field of research within the information systems discipline [6, 52].

1.1 Problem Statement and Research Gap

Multi-perspective analysis of OSN describes the situation where experts from different domains are interested in gaining insights from OSN for specific purposes. Therefore, purpose-specific analysis of OSN is necessary [59]. Stieglitz and Dang-Xuan [58] proclaims a framework in order to systematize analysis in Social Media: First, domain experts initially create a demand for specific data or information. Generally, analysis purposes are explicated as informal statements for some informational need. Occasionally, conceptual domain models outline these analysis purposes [11, 53], but these models are semi-formal, abstract and neither detailed or implementation-oriented.

In contrast to that, OSN data tracking describes in detail, what kind of data is available and accessible in OSN and what kind of technologies can be applied for tracking and parsing data. This area is usually managed by IT engineers in order to implement algorithms and queries on code level or API level. Ontologies are prevalent in this context for presenting common OSN concepts in specific applications and facilitate their adaptation in dedicated tracking tools. The theoretical background for the fulfillment of specific analysis purposes comes from analytical methods which constitute as formal models or graph models [9, 31, 36]. It become obvious, that each area works with distinct models for the description of its particularly relevant context of OSN. The integration of different areas is challenging due to the heterogeneity of description languages and different levels of abstraction, formalization and specific domain concepts (see Fig. 1).

Although, OSN analysis is an emerging research topic, this problem is not tackled so far. The following works at least touch the issue or state its relevance: [59] emphasizes the need for common technical frameworks, both [6] and [23] see the need of integrated data models, [52] states the relevance of OSN conceptualization in general, [59] addresses problems in business integration, [46] criticizes the missing integration of social graph and social text data and both [10] and [41] see the need for a multi-perspective social user model. We suppose, that the insufficient consideration in research is less an evidence for missing relevance, but rather caused by a lack of OSN analysis frameworks, which leads to missing awareness of the issue.

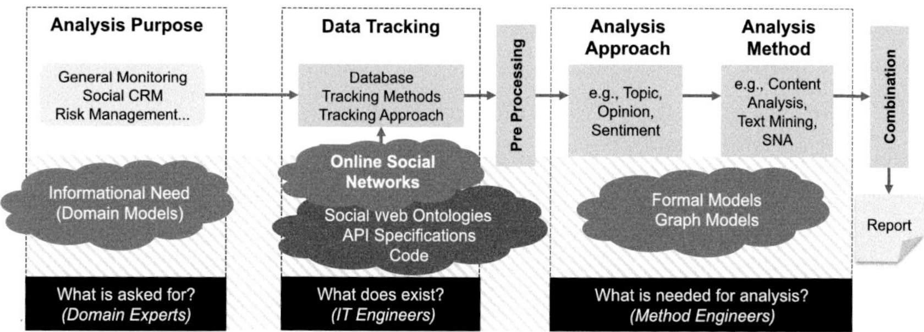

Fig. 1 Adapted and extended framework of [58] for the representation of OSN analysis areas operating with different modeling approaches and techniques. The proposed ontology addresses the *Analysis Purpose* area and the *Data Tracking* area

1.2 Research Objective and Research Method

Based on the stated research gap, we aim to provide an ontology that integrates concepts from business-driven analysis purposes and the data tracking area. We argue, that an appropriate ontology has the potential to close the gap between different modeling approaches and to facilitate the representation of knowledge regarding the analysis of OSN. Thus, both a better understanding and a better analysis of OSN are promised [14]. Two research questions (RQ) are related to the stated objective:

- **RQ 1**: Which requirements for such an ontology typically exist?
- **RQ 2**: Which ontologies, models or frameworks can be reused for ontology design and which concepts have to be designed from scratch?

We use a Design Science approach that aims to construct the artifact (ontology) based on a deep consideration of relevant use cases and the current state of the art [38]. Therefore, Sect. 2 provides fundamentals from the field of Ontology Engineering and evolves a set of requirements coming from characteristics of specific business use cases in OSN and previous work from literature. Consequently, techniques from the field of Requirements Engineering are conducted in order the build the requirements catalogue rigorously. Afterwards, in Sect. 3 existing approaches from the field of (social) user modeling are compared with the derived requirements in order to find evidence for reusable concepts and missing concepts. Finally, we design the OntOSN ontology based on existing concepts and a set of new concepts and properties in Sect. 4. The paper ends with a short summary and the consideration of further research aspects in Sect. 5. Due to the novelty of the topic and the scope of the ontology design, we focus solely on the analysis and design phase and plan to evaluate the artifact in a separate research article [5].

2 Requirements for Ontology Engineering

2.1 Adapted Ontology Engineering Approach

Ontologies can be defined as explicit specification, which conceptualize a delimited discourse for a specific group of actors having a defined aim [30]. Essential elements of ontologies are concepts, properties, relations and relation types [19]. In contrast to conceptual models, ontologies do not strictly distinguish type level and instance level. Thus, they can define specific rules and integrity constraints between particular concepts [56]. Generally, ontologies are independent of specific applications, but rather aim to provide sufficient abstract concepts for an appropriate description of the addressed problem domain [61]. Typically, these concepts are derived by a set of similar tasks or environments [56]. Ontologies are primarily used for knowledge management and semantic web issues, but have also relevance for engineering tasks [32]. Thus, we proclaim ontologies as suitable means for modeling business-relevant aspects in OSN. This is also motivated by the issue of integrating heterogeneous model understandings (following [62]).

The field of Ontology Engineering provides a range of methods for the design and development of ontologies [60]. A widespread Ontology Engineering method is the Methontology approach, that provides a procedure model for the stepwise development of ontologies [27]. Within this research article, we focus on ontology conceptualization and design and apply Methontology as follows: First, the aimed ontology is specified by the analysis and derivation of requirements coming from characteristic business use cases in the context of OSN analysis. According to Requirements Engineering, we therefore applied document-centric analysis with a particular "perspective reading" approach that reveals relevant use cases and analysis purposes from literature [48]. Therefore, especially literature from the field of Social CRM [26] and Risk Management [13] was examined as the ontology is intended to be a multi-perspective description of users and structures within OSN. Also, fundamental OSN analysis literature is considered.

Subsequent, required concepts for the aimed ontology were inferred from these use cases based on a rational discourse or specific parts from literature. All requirements are summarized within the final requirements catalogue, that responses to the first research question (RQ 1). Afterwards, existing approaches are examined in order to ensure an appropriate reuse and integration of existing approaches and facilitate the recognition of missing concepts [27, p. 24]. Thus, the second research question can be elaborated (RQ 2). Finally, the OntOSN ontology is designed and presented in Sect. 4.

2.2 Requirements from Social Customer Relationship Management

Social CRM (SCRM) is defined as the enrichment of Customer Relationship Management with external data from social applications [26]. Below, SCRM use cases and their relevant information objects are described in accordance to the SCRM phases *Pre Sale*, *Sale* and *After Sale* [4]. In the *Pre Sale* phase, OSNs can mainly be used for the identification of potential customers [9, 54]. It is also possible to identify customer wishes or feature requests based on specific statements, interests or demographic data from user profiles [4, 17, 54]. Analyzing the role of a user within the OSN or within a group of users can facilitate the identification of suitable lead users or opinion leaders for viral marketing campaigns [4, 9, 25]. In the *Sale* phase, OSNs can be used as additional sales channel, which supports individual customer communication and brand loyalty [55]. The analysis of user preferences and statements could enable personalized product offers, target advertising or custom product configurations [4, 54]. Therefore, it should be possible to map users from OSN to current customers from the CRM system in order to integrate both information spaces [53]. In the *After Sale* phase, OSNs are appropriate means for customer support and the management of questioning or even censorious statements [2, 37]. Due to the personalized relation to users, OSNs can also be used for customer recovery management [55]. Besides, OSN channels also facilitate including business or service partners [4, 54].

In order to satisfy the informational need coming from the stated use cases, the field of Social Media Analytics provides a range of analysis techniques such as opinion mining, sentiment mining, trend analysis [59] or at least the monitoring of users and their interactions [2, 9]. Consequently, it is necessary to provide modeling concepts, which cover basic user attributes, behavioral aspects, interrelations and content. Below, the derived requirements are categorized into the aspects *identity*, *structure*, *activity* and *knowledge* [49].

The *identity* aspect describes who a user is and which characteristics distinguish him.

- **Req. 1**: The ontology should provide concepts for personal attributes such as demographic data.
- **Req. 2**: Also, it seems to be promising to get insights into the particular context of a user (e.g., his current location).
- **Req. 3**: Additionally, it is necessary to provide concepts for the presentation of user preferences (e.g., interests or favorites).

The *structure* aspect describes the relation between users in OSN [49]. Additionally, the relation of a user to some kind of a resource (information object) is covered by the structural aspect.

- **Req. 4**: Basically, the ontology should provide concepts for representing relations, groups and sub network types within OSN. That enables structural analysis of the entire network by generating measures like density, path length, clustering coefficients or eigenvector centrality [36, 65]. Also, node-oriented centrality measures can be applied [28].
- **Req. 5**: Besides, the ontology needs to provide concepts for the detailed specification of relationship types between users and resources (e.g., information objects [42]). It should be possible, to assess some relationship regarding its distance [28], its strength [29] or its influence directions (positive, neutral, negative [50]). It should be also possible to depict the reasons of relationships [45].
- **Req. 6**: A single user can be associated with a specific role within OSN or within a specific group. A particular role always depends on the specific analytical view on the OSN. Examples for such roles are experts, opinion leaders, influencers or passive users [12, 49]. Thus, it is necessary to provide concepts for the specification of roles in OSN.

The *activity* aspect describes how active some user is [49]. Typical user activities are the exchange or publication of resources by posting any kind of statements (see examples in [8]).

- **Req. 7**: Generally, the ontology should support modeling of social interactions [36]. Similar to Req. 5, it should be possible to specify these interactions regarding to specific aspects such as electronic word-of-mouth-effects between users [18, 57].
- **Req. 8**: The ontology should also provide concepts for the specification of activity types [7, 8].

The *knowledge* aspect describes special know-how of a user regarding to analysis-specific topics (referring [49]).

- **Req. 9**: As stated in [49], it is promising to know competences of a user from a business point of view. Thus, the ontology should provide concepts for modeling skills and abilities of a user.
- **Req. 10**: The adaption of transmitted information objects is influenced by a certain level of trustworthiness, the credibility of the sender, the expertise of the sender, the quality of the message and contextual involvement of the participated users or similarity between users [43, 57]. Hence, the ontology needs to provide concepts for modeling certain levels of trust and user reputation.

It is also noteworthy to emphasize, that information objects are substantial for any kind of analysis. Comments, ratings, favorite lists, events, opinions and profiles are very important analysis objects [2, 20].

- **Req. 11**: The ontology needs to provide concepts for describing various information objects, its properties and its actual content.

- **Req. 12**: According to [2], information objects have three key properties: Topic type (e.g., private, business, public), intention (purpose of the statement) and sentiment. These properties need to be modeled by the aimed ontology. Further, some basic information such as timestamp data needs to be considered.
- **Req. 13**: As stated above, reputation and trust are of great importance within OSN. Hence, the ontology should provide concepts for the assessment of information object confidence and quality.
- **Req. 14**: From a business point of view, statements can have a priority and are associated to a specific business object [2]. Thus, the ontology should enable the explication of a particular relevance for a business.

As stated above, it is also necessary, to consider an appropriate integration between users in OSN and customers.

- **Req. 15**: The ontology should facilitate the integration of users and customers, as it is necessary to match user and customer information [53, 54]. Besides, an annotation of specific customer value attributes would be helpful [44].

2.3 Requirements from Risk Management

In contrast to the above mentioned benefits, businesses also face several technical, socio-technical and organizational threats and risks in OSN like data leakage; identify theft, social engineering attacks or industrial espionage [13]. The main risks are loss of control and reputational damage, which is mainly related to user interactions and transferred information objects [13]. Thus, it is necessary to monitor business-related conversations from a risk perspective. Data and information leakage is an other severe risk that is primarily caused by privacy problems [3, 39].

- **Req. 16**: A deep understanding and description of the accessibility of data objects within OSN is necessary to reduce uncertainty and complexity and to enhance the understanding of the personal data dissemination in OSN. Hence, the aimed ontology should provide concepts for describing visibility of information objects and respective privacy aspects.

3 State of the Art

Within this section, existing ontologies in the context of OSN are presented in order to get an overview of the state of the art and to prepare the later ontology integration. Independent from research on Social Media, user modeling approaches already came up in the 1990 s for the development of user adaptive applications. Below, relevant approaches are briefly presented in alphabetical order.

3.1 Existing Ontologies in the Context of OSN

Bozzon et al. [10] The approach of [10] addresses modeling crowd sourcing scenarios and adapts user ontologies and content description models. New concepts like *confidence level*, *network roles* and *topical affinities* are defined [10].

FOAF is an OWL based domain ontology for modeling both users and structures in the Semantic Web [16]. FOAF is primarily used for user-adaptive systems and for user aggregation issues. Core concepts are *user basics*, *personal information*, *accounts*, *projects*, *groups* and *documents*.

GenOUM is a generic, OWL based ontology for user modeling [21]. The focus of GenOUM lies on modeling user behavior and his knowledge regarding to particular concepts. Basic user attributes are not considered.

GUMO is an ontology for the generic description of users within their life cycle [34]. GUMO is mainly used for data exchange between user adaptive systems. Special concepts are *emotional state*, *knowledge* and *privacy preferences*.

HiddenU is less an ontology but rather a framework that aims to facilitate social nexus for the integration of scattered social data [40]. Thus, HiddenU has an aggregation and integration focus and provides more abstract concepts (e.g., *competences*, *social relationships* and *preferences*).

Kapsammer et al. [41]: In view of the variety of approaches, [41] proclaims a comprehensive and extensible reference model for the description of social user profiles. The reference model is divided into the packages *core*, *meta info* (privacy, provenance, quality, context), *resource* (structure and behavior) and *relation* (universal, social; [41]).

OntobUM addresses ontology based user modeling for knowledge management [51]. Therefore, OntobUM differentiates between explicit and implicit user information. Explicit user information stand for fundamental attributes like email addresses, skills or preferences. Also, behavior and user relations are understood as explicit information. On the contrary, implicit information address derived or deduced information like type classifications (e.g., writer or lurker), activity levels or initiatives for specific topics [51].

OpenSocial was intended to be an API standard for data exchange of OSN apps. Thus, it is very data-oriented and provides concepts like *activity*, *friends*, *meta data* and *messages* [33].

SIOC is a RDF based domain ontology for the connection of discussion media such as blogs, forums or OSN [15]. Core concepts are *posts*, *roles* and *items*.

SocIoS is an application-oriented ontology for the independent development of OSN apps that provides concepts like *activities*, *events* or *reputations*.

SWUM is a RDF based application-oriented ontology for the aggregation of user profiles in Social Media. Core concepts are *user role*, *relationships*, *locations* and *demographic data* [47].

U2M is a RDF based ontology for comprehensive user modeling and was designed as integration of GUMO and UserML [35]. Important concepts are *preferences*, *demographics* and *roles*.

URM is an application-oriented ontology for cross-system personalization and user modeling. Essential concepts are *preferences*, *relationships* and *tasks* [64].

U-Sem is a semantic web service for enriching and mining user data, which contains user modeling in U-Sem profiles [1]. Description groups within U-Sem are *observations* (activities, objects), *user characteristics* and *domain knowledge*. Social aspects are not considered.

Yan and Zha [63] propose an OWL based ontology for customer knowledge management [63]. Their approach allows defining different customer categories as well as statements regarding loyalty, knowledge contribution or feedback.

Reference Models: Further, literature provides some interesting data-oriented reference models in the context of Social Media. For instance, in the context of SCRM systems [53] and risk analysis of OSN [11]. Mukkamala et al. [46] evolves a formal model for social data analysis and formalize concepts of social graphs and social text.

3.2 *Requirements Catalogue*

Based on the analysis of the OSN domain, sixteen requirements were identified. First, identity related attributes like general personal data, context information and user preferences need to be addressed. Second, relations between single users and their respective type of relation need to be considered. Also network-specific measures are assigned to the structure group in order to describe a larger set of users. The role type enables the specification of user roles within a specific group or situation (e.g., opinion leader, product expert or critic). Third, the activity group encompasses the description of user activities and the description of social interaction. Fourth, the knowledge group stands for the ability of describing competences and for the representation of his reputation within the OSN, which influences his level of trust within the network. In the fifth place, the aimed ontology should describe information objects regarding their content, their basic data, their level of confidence or quality as well as their business relevance. The last group contains more analysis-specific features. Therefore, the ontology should provide concepts for the integration of users with existing customers of a business and for the annotation of relevant properties. For reasons of internal security and risk management, users should also be monitored in regard of their privacy settings and some kind of assessment of the business confidentiality of their statements.

As can be seen from Fig. 2, the majority of ontologies has a technical and data-oriented focus and aims to solve data exchange and integration issues. There are deficits regarding the representation of social attributes, interaction features, activity types and respective roles. Also, there is a lack regarding modeling topical affinities, reliabilities and trust [10, p. 17]. These aspects might be relevant for the identification of opinion leaders [9, 36]. Also, not all languages support the adequate description of information objects, especially regarding their confidence. Generally, a suitable integration of existing languages is missing and there is no universal data model or

Requirement:	Approach:	Bozzon et al. (2013)	FOAF	GenOUM	GUMO	HiddenU	Kapsammer et al. (2012)	OpenSocial	OntobUM	SIOC	Soclos	SWUM	U2M	U-Sem	Yan & Zha (2010)	Rosemann et al. (2012)	Braun & Esswein (2012)	**Consequence for Desigm**
	Identity																	
R1	Personal Data	●	●	○	●	●	○	○	●	○	●	●	●	⊙	●	●	●	reuse
R2	Context	●	○	○	○	○	⊙	●	○	○	⊙	●	○	○	○	○	⊙	reuse
R3	Preferences	○	○	●	⊙	●	⊙	⊙	●	○	○	⊙	○	⊙	⊙	⊙	●	reuse
	Structure																	
R4	Relation, Groups	●	●	○	○	●	⊙	●	●	○	●	○	○	⊙	○	○	●	reuse
R5	Relation Types	●	⊙	○	○	○	○	○	⊙	⊙	○	○	○	○	○	○	⊙	reuse
R6	Role Types	●	⊙	○	○	○	○	○	●	⊙	○	●	○	○	○	○	⊙	reuse
	Activity																	
R7	Social Interaction	○	⊙	○	○	●	⊙	⊙	⊙	○	●	○	⊙	○	○	○	○	reuse
R8	Activity Types	○	○	○	○	○	⊙	○	●	○	○	○	○	○	●	○	○	new
	Knowledge																	
R9	Competences	○	○	●	⊙	●	⊙	○	⊙	○	○	⊙	⊙	⊙	●	○	○	reuse
R10	Trust, Reputation	⊙	○	○	⊙	○	⊙	●	○	○	○	○	○	○	○	○	○	new
	Information Objects																	
R11	Content	●	⊙	○	○	○	○	●	○	●	●	○	⊙	○	●	○	○	reuse
R12	Basic Data	●	○	○	○	⊙	○	●	○	●	⊙	○	○	○	○	○	○	reuse
R13	Confidence, Quality	●	○	○	○	○	⊙	○	⊙	○	⊙	○	○	○	⊙	○	○	new
R14	Business relevance	○	○	○	○	○	○	○	○	○	○	○	○	○	●	⊙	⊙	new
	Others																	
R15	Customer Integration	○	○	○	○	○	○	○	○	○	○	○	○	○	⊙	⊙	⊙	new
R16	Privacy, Visibility	○	○	○	⊙	●	⊙	○	○	⊙	○	○	○	●	○	⊙	○	reuse
	Specification	●	●	●	○	○	⊙	●	●	●	⊙	●	○	○	●	●	●	

● (met) ⊙ (partially met) ○ (not met)

Fig. 2 Requirements of the aimed OSN ontology in comparison with existing ontologies, frameworks and conceptual models. A particular consequence for the OntOSN design is summarized for each requirement. Therefore, *reuse* stands for adapting or simply reusing concepts from existing approaches. *New* reflects the necessity for designing new concepts or properties

standard, that both contains structural, personal and social features! Consequently, there is also a lack of explicit considerations of business relevant features; for example in terms of a customer user integration.

4 Ontology Engineering

By reviewing the stated approaches, it became obvious that none of the ontologies or reference models meet all requirements. Thus, we decided to specify a new integrated ontology called **OntOSN**. With regard to Methontology approach, we tried to reuse as much as possible concepts from existing ontologies [27]. Therefore, each ontology was modeled graphically in order to identify relevant concepts, properties, relations and inheritance structures. Unfortunately, only a few ontologies are well-defined within OWL specifications or some kind of a meta-model (see the row "specification" in Fig. 2). The majority of the approaches does not provide an integrated specification (e.g., GUMO, U2M). This circumstance makes it difficult to identify definable concepts from several approaches. Thus, it was necessary to evolve relevant concepts based on informal descriptions or frameworks within the stated articles. Based on all explicated models, the OntOSN ontology was consolidated by integrating existing concepts and adding new concepts. The ontology was designed by using the OWL Lite ontology language and we decided to explicate the ontology in diagrams in order to provide an appropriate overview. The legend of the notation can be found at the lower right corner of Fig. 3. For reasons of readability, the *range* and *domain* specifications of property relations as well as specifications of inheritances (*subClassOf* and *subPropertyOf*) are hidden in the respective diagrams. Concepts that were reused and adapted from existing ontologies are marked by the respective namespace in front of the concept name (FOAF, SIOC, U2M, Ka+11 (Kapsammer et al. 2011), Bo+13 (Bozzon et al. 2013), SWUM, SocIoS, YZ10 (Yan and Zha (2010)). Thus, an integration or mapping of the new ontology with existing ontologies remain possible. Besides, owned properties of concepts (especially *Person*, *Event*, *Rating* and *User Role*) are not stated with respect to the limited space of this paper. Particular property values can be found in the original ontologies.

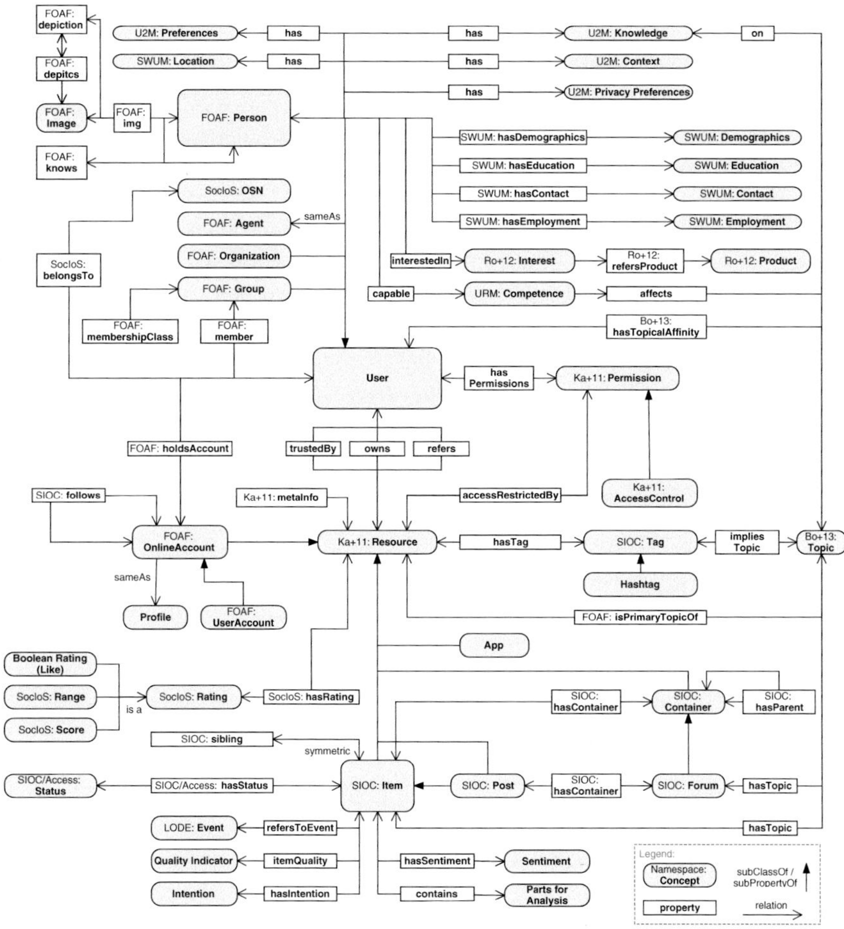

Fig. 3 Part of the OntOSN ontology representing essential user concepts and content description aspects

4.1 OntOSN: User and Information Objects

Figure 3 presents the part of OntOSN that depicts large parts of the user concepts and the information object concepts. Thus, the node elements of the abstract OSN structure are considered. The central concept of the OSN is the user concept covering all personal attributes and features of either a person (normal case), an organization or a group. Next to demographical attributes, a single user has a range of attributes like employment information, preferences, specific knowledge or competences, whereby the latter two affect a specific topic. A user always belongs to a specific OSN (e.g., Facebook) and holds an online account within this OSN. If a user maintains several accounts, the *holdsAccount* property enables an inference to the actual user.

The second very important concept within the analysis of OSN is the resource concept that stands for all OSN objects that can be created, altered, transferred, referenced or deleted by users. In particular, the resource types from SIOC (*item*, *post*, *forum* and *container* [15]) were adapted, whereby forums are understood both as structured discussions (e.g., within user groups) and as conversation between a set of users. Please note, that the *item* concept is the central point for analysis and stands for an atomic element that can be analyzed regarding its contents. The other concepts are more abstract concepts lacing single items. Each resource can be semantically enriched by tags (keywords) in order to emphasize its topic. Further, a resource has some meta information (e.g., creating timestamp, language) and could refer to an other resource (e.g., as a link). The relation between users and resources is considered in several aspects: In accordance to [41], every user has a specific permission for a resource (e.g., full access, no access or read-only) that limits the scope of elements he can access within the OSN. In particular, a user can refer to specific resources and can be referenced by a resource (e.g., as link to the profile). If the user is the creator or author of a resource, he owns it. Besides, the proclaimed ontology allows to specify a *trustedBy* relation between users and resources to give implicit statements on the trustworthiness of some content.

As stated above, the *item* concept is central for content-related analysis. With reference to the SIOC access module, an item may have a specific status (e.g., draft). Also, it can correspond to some kind of event. Events can be described by the LODE event description ontology. Further, an item generally has a specific sentiment, an intention and implies a certain level of quality (see [41]). Besides, the content of an item provides several elements (tokens) for analysis; such as product names, acronyms, negative words or locations (e.g., [2]).

4.2 OntOSN: Structural Aspects

Figure 4 depicts the part of OntOSN that focuses on the structural aspects of an OSN; namely social relations and interactions between users or users and resources. Thus, *edge* elements of the abstract OSN structure are considered. Generally, a user has several social relations to other users. Each relation can be specified by several properties like *realWorldEquivalent* (real relations), *distance* (number of intermediate nodes or users) or *strength* (e.g., [29]). Additionally, the level of trust between two users, the level of influence and the prevailing influence direction between two users can be modeled (see [43]). The set of social relations constitutes the social graph that can be measured by several socio-central measures [28, 65]. The similarity between two users (e.g., [57]) is covered by the *hasSimilarity* property.

Further, a user can perform social actions and interactions to or with other users. A social action (e.g., posting, liking, sharing) results in a resource (e.g., a public message), which could reference other users. Also, some social actions can refer to an earlier action. The set of social actions and interactions constitutes the activity graph that can be measured by activity-based measures like the page rank centrality index [36, 49]. Resources may also cause the social cognition of users by associating them with different information objects. The relation of a single user within a set of users is depicted in the middle of Fig. 4. Thereby, a *set of users* is understood as an abstract concept covering several group-related concepts like cliques, ego-centric

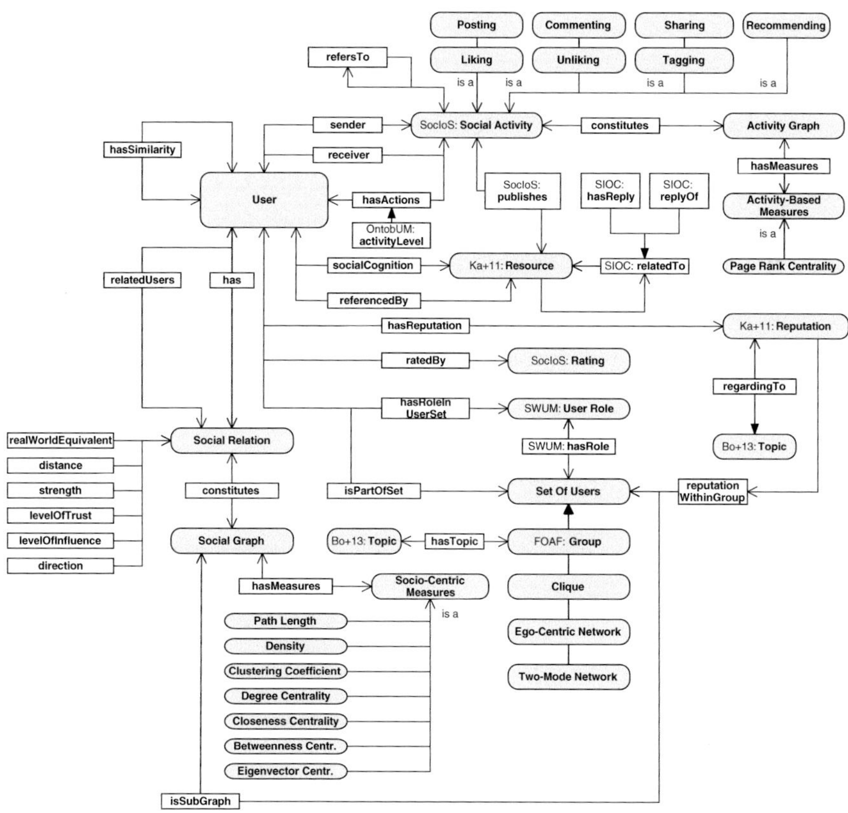

Fig. 4 Part of the OntOSN ontology representing structural aspects

networks or two-mode networks within the entire OSN. We used this construction for pointing out different roles a user can have within a OSN; depending on the related set of other users. Hence, users can also have different reputations; depending on the relevant group or topic.

4.3 OntOSN: Business Aspects

As stated in Sect. 3.2, any kind of business relevant concepts is missing in existing ontologies since the approaches mainly focus technical aspects. With respect to the increasing pertinence of OSN for businesses, it is necessary to integrate business relevant concepts. We therefore decided to both consider beneficial aspects (customer integration) and risky aspects (e.g., avoid internal data leakage). Figure 5 depicts the part of OntOSN that focuses on business aspects. It is possible to match some user to a customer. The customer concept can be extended with some kind of information coming from a CRM system (e.g., customer history). Also, users can be related to other stakeholders in the context of a company (e.g., competitors or business partners). It might be necessary to monitor these users in the context of competitive intelligence, for example. In contrast to that, a user can also represent an employee of a company. The identification of employees is crucial for recognizing and avoiding internal data leakage (see [13]). Such internal data might be disclosed by explicit statements resulting in any kind of content items. With regard to both positive and negative effects of OSN for businesses, items can be identified as business relevant for analysis-specific reasons (concept *business relevant content*). An item can imply a specific business relevance type like a complaint, a question or some product ratings. That type has a specific priority (for reaction) and can be specified whether it affects the own company or a competitor.

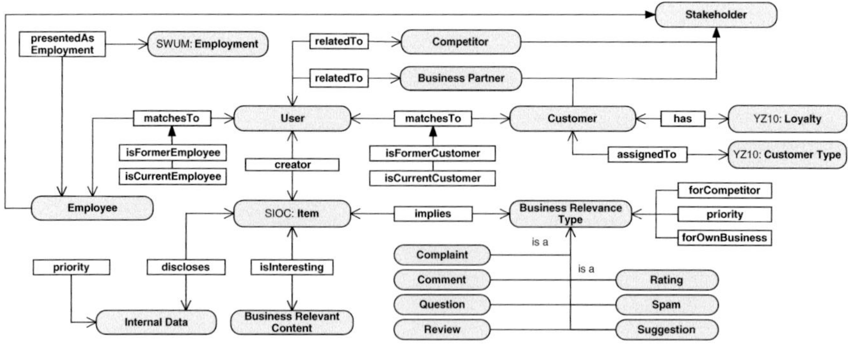

Fig. 5 Part of the OntOSN ontology representing concepts for business analysis

5 Conclusion and Further Research

This research paper designs an ontology for the analysis of OSN by integrating existing ontologies and adding new concepts and properties. The evolved ontology can be used for the analysis of OSN from different viewpoints and facilitates the analysis-specific design and development of OSN analysis tools due to the unified language base. Hence, it is possible to provide an integrated, comprehensive understanding of the emerging research field of OSN. Also, the ontology allows its extension due to the well-defined specification in OWL Lite. In order to enable an integration with specific business purposes, the ontology explicitly considers concepts for the integration of customers.

Thus, the contributions of this paper are as follows: First, development of a requirements set for a both business oriented and technical oriented OSN ontology. Second, examination of existing ontologies in regard to the requirements set. Third, proposal for an integrated OSN ontology under special consideration of business relevant aspects.

Nevertheless, there are some points for further research and improvement. First, the evolved ontology needs to be published as RDF (XML) file in order to facilitate its application. Therefore, also owned properties of concepts need to be explicated. Second, the consideration of further views on OSN needs to be discussed. Especially, an integration with concepts from data mining is promising. Third, existing concepts of the OntOSN ontology can be enriched by specific vocabularies or further ontologies in order to enable an even better expressiveness of single concepts.

References

1. Abel, F., Celik, I., Hauff, C., Hollink, L., Houben, G.J.: U-Sem: Semantic enrichment, user modeling and mining of usage data on the social web. arXiv:1104.0126 (2011)
2. Ajmera, J., Ahn, H.i., Nagarajan, M., Verma, A., Contractor, D., Dill, S., Denesuk, M.: A CRM system for social media: challenges and experiences. In: Proceedings of the 22nd WWW Conference, pp. 49–58 (2013)
3. Al Hasib, A.: Threats of online social networks. Int. J. Comput. Sci. Netw. Security **9**(11), 288–293 (2009)
4. Alt, R., Reinhold, O.: Social customer relationship management (social CRM)-application and technology. Bus. Inf. Syst. Eng. **4**(5), 287–291 (2013)
5. Alturki, A., Gable, G.G., Bandara, W.: A design science research roadmap. In: Service-Oriented Perspectives in Design Science Research, pp. 107–123. Springer (2011)
6. Appleford, S., Bottum, J.R., Thatcher, J.B.: Understanding the social web: towards defining an interdisciplinary research agenda for information systems. ACM SIGMIS Database **45**(1), 29–37 (2014)
7. Benevenuto, F., Rodrigues, T., Cha, M., Almeida, V.: Characterizing user navigation and interactions in online social networks. Inf. Sci. **195**, 1–24 (2012)
8. Berthon, P.R., Pitt, L.F., Plangger, K., Shapiro, D.: Marketing meets web 2.0, social media, and creative consumers: implications for international marketing strategy. Bus. Horiz. **55**(3), 261–271 (2012)

9. Bonchi, F., Castillo, C., Gionis, A., Jaimes, A.: Social network analysis and mining for business applications. ACM Trans. Intell. Syst. Technol. (TIST) **2**(3), 22 (2011)
10. Bozzon, A., Fraternali, P., Galli, L., Karam, R.: Modeling crowdsourcing scenarios in socially-enabled human computation applications. J. Data Semant. 1–20 (2013)
11. Braun, R., Esswein, W.: Corporate risks in social networks—towards a risk management framework. In: Proceedings of the 18th ACMIS (2012)
12. Braun, R., Esswein, W.: Towards a reference architecture model for virtual communities. In: Proceedings of of the IADIS International Conference on Web-Based Communities, pp. 48–56 (2011)
13. Braun, R., Esswein, W.: Towards a conceptualization of corporate risks in online social networks. In: Proceedings of the 17th EDOC, pp. 267–274. IEEE (2013)
14. Braun, R., Esswein, W.: OntOSN—towards an integrated ontology for the analysis of online social networks. In: 9th International Conference on Knowledge, Information and Creativity Support Systems, pp. 263–276 (2014)
15. Breslin, J.G., Decker, S., Harth, A., Bojars, U.: SIOC: an approach to connect web-based communities. Int. J. Web Based Communities **2**(2), 133–142 (2006)
16. Brickley, D., Miller, L.: FOAF vocabulary specification 0.99. http://xmlns.com/foaf/spec/ (2014)
17. Casteleyn, J., Mottart, A., Rutten, K.: How to use data from facebook in your market research. Int. J. Mark. Res. **51**(4), 439–447 (2009)
18. Cheung, C.M., Thadani, D.R.: The effectiveness of electronic word-of-mouth communication: a literature analysis. In: 23rd Bled eConference, Bled, Slovenia (2010)
19. Corcho, O., Fernández-López, M., Gómez-Pérez, A.: Methodologies, tools and languages for building ontologies. where is their meeting point? Data Knowl. Eng. **46**(1), 41–64 (2003)
20. Costa, P.R., Souza, F.F., Times, V.C., Benevenuto, F.: Towards integrating online social networks and business intelligence. In: IADIS International Conference on Web Based Communities and Social Media, vol. 2012 (2012)
21. Cretton, F., Calvé, A.: Generic ontology based user model: Genoum. Centre Universitaire D'Informatique, Université de Genève (2008)
22. Culnan, M.J., McHugh, P.J., Zubillaga, J.I.: How large U.S. companies can use Twitter and other social media to gain business value. MIS Q. Executive **9**(4), 243–259 (2010)
23. Dinter, B., Lorenz, A.: Social business intelligence: a literature review and research agenda. In: Proceedings of the ICIS (2012)
24. Ellison, N.B., et al.: Social network sites: definition, history, and scholarship. J. Comput.-Mediated Commun. **13**(1), 210–230 (2007)
25. Ermecke, R., Mayrhofer, P., Wagner, S.: Agents of diffusion-insights from a survey of facebook users. In: 42nd Hawaii International Conference on System Sciences, 2009. HICSS'09, pp. 1–10. IEEE (2009)
26. Faase, R., Helms, R., Spruit, M.: Web 2.0 in the CRM domain: defining social CRM. Int. J. Electron. Customer Relat. Manage. **5**(1), 1–22 (2011)
27. Fernández-López, M., Gómez-Pérez, A., Juristo, N.: Methontology: from ontological art towards ontological engineering (1997)
28. Freeman, L.C.: Centrality in social networks conceptual clarification. Soc. Netw. **1**(3), 215–239 (1979)
29. Granovetter, M.: Economic action and social structure: the problem of embeddedness. Am. J. Sociol. 481–510 (1985)
30. Gruber, T.R.: A translation approach to portable ontology specifications. Knowl. Acquisition **5**(2), 199–220 (1993)
31. Gundecha, P., Liu, H.: Mining social media: a brief introduction. Tutorials Oper. Res. **1**(4) (2012)
32. Happel, H.J., Seedorf, S.: Applications of ontologies in software engineering. In: Proceedings of Workshop on Sematic Web Enabled Software Engineering, pp. 5–9 (2006)
33. Häsel, M.: Opensocial: an enabler for social applications on the web. Commun. ACM **54**(1), 139–144 (2011)

34. Heckmann, D., Schwartz, T., Brandherm, B., Schmitz, M., von Wilamowitz-Moellendorff, M.: Gumo-the general user model ontology. In: User Modeling 2005, pp. 428–432. Springer (2005)
35. Heckmann, D., Schwarzkopf, E., Mori, J., Dengler, D., Kröner, A.: The user model and context ontology gumo revisited for future web 2.0 extensions. Contexts and Ontologies: Representation and Reasoning, pp. 37–46 (2007)
36. Heidemann, J., Klier, M., Probst, F.: Identifying key users in online social networks: a pagerank based approach. In: Proceedings of the ICIS (2010)
37. Heidemann, J., Klier, M., Probst, F.: Online social networks: a survey of a global phenomenon. Comput. Netw. **56**(18), 3866–3878 (2012)
38. Hevner, A.R., March, S.T., Park, J., Ram, S.: Design science in information systems research. MIS Q. **28**(1), 75–105 (2004)
39. Huber, M., Mulazzani, M., Leithner, M., Schrittwieser, S., Wondracek, G., Weippl, E.: Social snapshots: digital forensics for online social networks. In: Proceedings of the 27th Annual Computer Security Applications Conference, pp. 113–122. ACM (2011)
40. Kappel, G., Schönböck, J., Wimmer, M., Kotsis, G., Kusel, A., Pröll, B., Retschitzegger, W., Schwinger, W., Wagner, R., Lechner, S.: The hidden u—a social nexus for privacy-assured personalisation brokerage. In: Proceedings of the ICEIS, pp. 158–162 (2010)
41. Kapsammer, E., Mitsch, S., Pröll, B., Schwinger, W., Wimmer, M., Wischenbart, M.: A first step towards a conceptual reference model for comparing social user profiles. In: Proceedings of the International Workshop on User Profile Data on the Social Semantic Web (2011)
42. Le Malécot, E., Suzuki, M., Eto, M., Inoue, D., Nakao, K.: Online social network platforms: toward a model-backed security evaluation. In: Proceedings of the 1st Workshop on Privacy and Security in Online Social Media, p. 3. ACM (2012)
43. Lis, B.: In eWOM we trust. Bus. Inf. Syst. Eng. 1–12 (2013)
44. Malthouse, E.C., Haenlein, M., Skiera, B., Wege, E., Zhang, M.: Managing customer relationships in the social media era: introducing the social CRM house. J. Interact. Mark. **27**(4), 270–280 (2013)
45. Mislove, A., Marcon, M., Gummadi, K.P., Druschel, P., Bhattacharjee, B.: Measurement and analysis of online social networks. In: Proceedings of the 7th ACM SIGCOMM, pp. 29–42. ACM (2007)
46. Mukkamala, R.R., Vatrapu, R., Hussain, A.: Towards a formal model of social data. IT-Universitetet i København, Technical report (2013)
47. Plumbaum, T., Wu, S., De Luca, E.W., Albayrak, S.: User modeling for the social semantic web. In: SPIM, pp. 78–89 (2011)
48. Pohl, K., Rupp, C.: Requirements Engineering Fundamentals: A Study Guide for the Certified Professional for Requirements Engineering Exam-Foundation Level-IREB compliant. Rocky Nook (2011)
49. Probst, F., Grosswiele, L., Pfleger, R.: Who will lead and who will follow: identifying influential users in online social networks. Bus. Inf. Syst. Eng. 1–15 (2013)
50. Rafaeli, S., Raban, D.R.: Information sharing online: a research challenge. Int. J. Knowl. Learn. **1**(1), 62–79 (2005)
51. Razmerita, L., Angehrn, A., Maedche, A.: Ontology-based user modeling for knowledge management systems. In: User Modeling 2003, pp. 213–217. Springer (2003)
52. Richter, D., Riemer, K., vom Brocke, J.: Internet social networking—research state of the art and implications for enterprise 2.0. Bus. Inf. Syst. Eng. **3**(2), 89–101 (2011)
53. Rosemann, M., Eggert, M., Voigt, M., Beverungen, D.: Leveraging social network data for analytical crm strategies. In: Proceedings of the ECIS (2012)
54. Sigala, M.: ecrm 2.0 applications and trends: The use and perceptions of greek tourism firms of social networks and intelligence. Comput. Human Behav. **27**(2), 655–661 (2011)
55. Sinclaire, J.K., Vogus, C.E.: Adoption of social networking sites: an exploratory adaptive structuration perspective for global organizations. Inf. Technol. Manage. **12**(4), 293–314 (2011)
56. Spyns, P., Meersman, R., Jarrar, M.: Data modelling versus ontology engineering. ACM SIGMod Rec. **31**(4), 12–17 (2002)

57. Steffes, E.M., Burgee, L.E.: Social ties and online word of mouth. Internet Res. **19**(1), 42–59 (2009)
58. Stieglitz, S., Dang-Xuan, L.: Emotions and information diffusion in social media—sentiment of microblogs and sharing behavior. J. Manage. Inf. Syst. **29**(4), 217–248 (2013)
59. Stieglitz, S., Dang-Xuan, L., Bruns, A., Neuberger, C.: Social media analytics. Bus. Inf. Syst. Eng. **6**(2), 89–96 (2014)
60. Sure, Y., Tempich, C., Vrandecic, D.: Ontology engineering methodologies. Semantic Web Technologies: Trends and Research in Ontology-based Systems, pp. 171–190 (2006)
61. Uschold, M., King, M.: Towards a methodology for building ontologies. Citeseer (1995)
62. Wache, H., Voegele, T., Visser, U., Stuckenschmidt, H., Schuster, G., Neumann, H., Hübner, S.: Ontology-based integration of information-a survey of existing approaches. In: IJCAI-01 Workshop: Ontologies and Information Sharing, vol. 2001, pp. 108–117. Citeseer (2001)
63. Yan, Y., Zha, X.: Applying owl to build ontology for customer knowledge management. J. Comput. **5**(11), 1693–1699 (2010)
64. Zhang, F., Song, Z., Zhang, H.: Web service based architecture and ontology based user model for cross-system personalization. In: IEEE/WIC/ACM International Conference on Web Intelligence, pp. 849–852 (2006)
65. Zhang, M.: Social network analysis: history, concepts, and research. In: Handbook of Social Network Technologies and Applications, pp. 3–21. Springer (2010)

Educators as Game Developers—Model-Driven Visual Programming of Serious Games

Niroshan Thillainathan and Jan Marco Leimeister

Abstract Students use their own mobile devices like laptops, smartphones and tablets to work and study anywhere and anytime. Consequently, there is a high interest in using one's own mobile devices in the educational context. This presents a chance as well as a challenge for educators, who can now make use of mobile learning for supplying their students with tailored didactical content that is accessible anywhere and anytime. Therefore, in our work, we focus on presenting educational content using serious games. In order to allow educators without technical knowledge to create custom serious games, we propose a model-driven visual programming framework. In this paper, we present our framework, which consists of GLiSMo—a domain-specific modeling language—and the visual programming environment for serious games (VIPEr) that allows educators to use their didactical knowledge with a point and click graphical editor to create their own serious games.

Keywords Visual programming · Domain-specific modeling language · Model-driven development · Serious game

1 Introduction

In recent years, mobile computing has been one of the fastest growing areas of the technology industry worldwide. With innovative technologies inside small and portable devices a new generation of applications in various environments is rising. The fact that mobile devices, such as smartphones, tablets and notebooks, are more affordable, accessible and easier to use than desktop computers, makes their use

N. Thillainathan (✉) · J.M. Leimeister
Information Systems, Kassel University, Pfannkuchstr. 1, 34121 Kassel, Germany
e-mail: thillainathan@uni-kassel.de

J.M. Leimeister
e-mail: leimeister@uni-kassel.de

S. Kunifuji et al. (eds.), *Knowledge, Information and Creativity Support Systems*,
Advances in Intelligent Systems and Computing 416,
DOI 10.1007/978-3-319-27478-2_23

more appealing for the younger generation, who include them more and more into their everyday life. This has also been noted by the NMC Horizon Reports for Higher Education [6, p. 7].

Pearson and Harris Interactive have conducted a survey among 1206 college students in the U.S. Their results show that 91 % of the students use their laptops, 72 % their smartphones and 40 % their tablets regularly. Out of these students, 96 % have used a laptop, 66 % have used a smartphone and 83 % have used a tablet for educational work. From the learners' perspective, this shows that students are eager to use their mobile devices in the educational context [18]. From the educators' perspective, mobile computing enables educators to fulfill both the requirements noted by Johnson et al. [6] as well the learners' expectations and thus support the learning process [11].

This resulting educational paradigm of mobile learning is a special form of the established e-learning and has gained widespread attention as one way to enable students to learn anywhere and anytime. The specific type of mobile learning tool in our work are so-called serious games. Serious Games differ from traditional e-learning or mobile learning approaches as they combine education with game-play, by presenting learning objectives and contents in a game-like context. By providing a game-like structure, the learners stay motivated and continue using the game, and, at the same time, they have didactical content conveyed to them. So, the more fun the game is, the more the student is motivated to play and learn and thus has a higher learning rate [4].

While serious games have been shown to hold great potential as a didactical tool, developing them for specific learning objectives involves several challenges for the educators who wish to apply them. They have the didactical knowledge about what and how they want to teach. They typically do not have the knowledge needed for the technical realization, though and need support for developing the serious game for diverse mobile devices. In this paper, we present our model-driven visual programming framework, which enables non-technical educators to create simple serious games based on the "point and click" graphical adventure genre. Therefore, we first present some serious game development approaches and describe our research methodology. Next, the model-driven visual programming framework and its three components are introduced. The remainder of this paper depicts an experimental modeling and implementation of a serious game, an evaluation of the framework, by using the cognitive walkthrough method, followed by a conclusion and future work. This paper is based on our previously published paper, which can be found in [25].

2 Foundation: Serious Game Development Approaches

Creating a serious game is nearly impossible for most educators due to lack of programming skills and existing serious games generally do not allow the adaption of the content to their own learning objectives. Furthermore, serious game

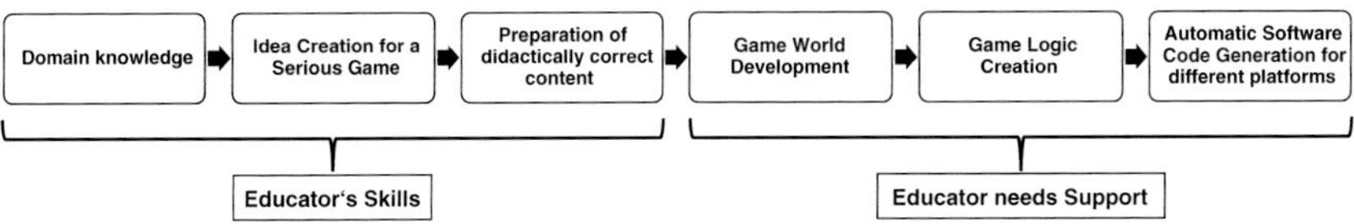

Fig. 1 Serious game development process

development not only consists of programming the software, but also designing the game world and game logic, which also is not the expertise of most educators. Figure 1 shows the process of serious game development by educators. As highlighted, the educator needs support during the game world development, game logic creation and software code generation. Therefore, educators have two options. The first option is to hand over the development process and providing the didactical content, to a professional game developer, which comes at high development cost [30]. We however argue for the second option: the educators develop serious games by themselves, supported during game development and the implementation process by a serious game development framework.

Since it is not feasible to turn non-technical educators into programmers—learning a programming language is highly complex [5]—we consider the so-called visual programming approach as the way to support educators. Visual programming languages (VPL) have been shown to be easy to understand and help non-programmers create software [13]. A VPL implementation consists of a visual programming environment (VPE), which represents the user interface to create application models, a syntactic and semantic validator for these models, as well as a compiler to generate software code from them [8]. To support a wide variety of diverse mobile devices, we employ a model-driven development process as the underlying principle of the VPL. Model-Driven Development (MDD) is a methodology in software engineering, which combines techniques for automated generation of software code from formal models. To formalize the application structure, the behavior and its requirements, formal languages called domain-specific modeling languages (DSML) are designed and implemented for a specific domain [21]. The VPE is used to create game models and the game world, which are then transformed into different programming languages for specific platforms using the model-driven development approach.

In current literature, only very limited research concerning development tools for serious games can be found. For our own research, we have identified two tools that allow the development of customized serious games, Sealund's Serious Games Engines and Worldweaver's DX Studio. *Sealund's Serious Games Engines* [22] is suited to developing 2D board games, with multiple choice questions. The educator has the possibility of selecting the type of game and integrating a predefined number of questions. The main disadvantage, however, is that this tool does not allow modeling the logic of a serious game. It is particularly desirable to develop games with individual behavior instead of predefined logic. *DX Studio* [29] provides a development environment for 3D- and learning applications. It allows creating a game environment consisting of objects and characters within the world. This tool lacks the possibility of creating the logic and behavior of the game easily and intuitively. To do so, the developer must have knowledge in JavaScript programming.

The presented tools have several weaknesses and disadvantages. Either the existing approaches are lacking in functionality or are too complex to be used by non-technical domain experts. Therefore, this induces the need for a solution for the development of serious games, which allows educators to develop their own serious games, without technical or game design knowledge. We want to enable educators to develop serious games adapted to their needs by presenting an environment that is intuitive and easy to handle.

3 Methodological Approach

To provide the framework for educators to create working serious games for their students without having to write software code, we apply a design-oriented research approach. In design-oriented research, artifacts, i.e. artificial, man-made things, are specially crafted in order to fulfill certain purposes and goals given their functions and adaptability. A design-oriented approach allows researchers to assess the impact of their artifacts and check whether they were able to achieve intended effects. Among the list of possible artifacts in IS research compiled by March and Smith [10] there are constructs, models, methods and instantiations. The framework we describe in this paper is comprised of constructs—the conceptualization used to describe the problem of creating mobile serious games by educators without technical background—as well as instantiations—the realization of artifacts in their target environment—in form of the concept and realization of our visual programming editor (VIPEr), the modeling language for serious games (GLiSMo) and the model-driven development Toolchain.

In contrast to approaches found e.g. in the social sciences, which are well-defined and broadly accepted as rigorous, a corresponding universal approach for design-oriented research does not exist. However, there are notable similarities between phases of design-oriented research found in the key literature on design science (e.g. March and Smith [10], Takeda et al. [23]) that allow deriving a basic consensus on a basic structure for design-oriented research.

The first step in design-oriented research is to identify the problem and its context, as such research follows a problem-solving paradigm. Taking this as input, a scenario for a possible solution is designed. This includes both the definition of the goals for the artifacts as well as the desired effects of the artifact. In a third step, the artifact or artifacts are realized and put to use in the context they were designed for. Artifact evaluation in the fourth step is based on the changes that can be observed concerning the problem and its context. From this evaluation, conclusions are drawn and documented, e.g. as new input for another iteration of the process or

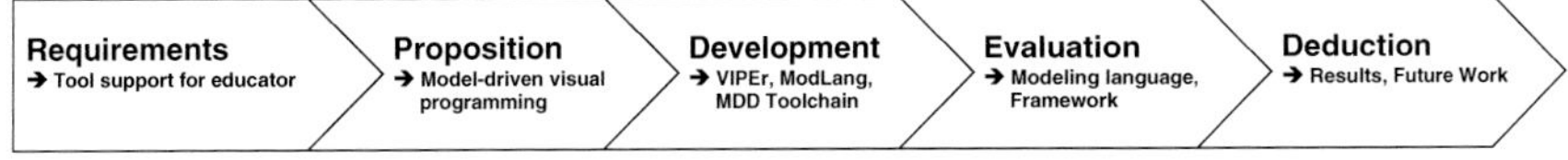

Fig. 2 Process steps followed in this research (adapted from Takeda et al. [23])

as the basis to theorize. Figure 2 shows the research process adapted form Takeda et al. [23], which we follow in this contribution.

4 A Framework for Model-Driven Visual Programming of Serious Games

As described in the introduction our aim is to present a solution for educators, which allows the development of mobile serious games without having knowledge in game development or game programming. Figure 3 shows an overview of our framework, which supports the educator during development by offering a visual programming interface and creates the game for multiple different mobile computing platforms by employing a model-driven development approach. Our framework can be divided into the three abstract components visual programming, domain-specific modeling language and model-driven development Toolchain (MDD Toolchain).

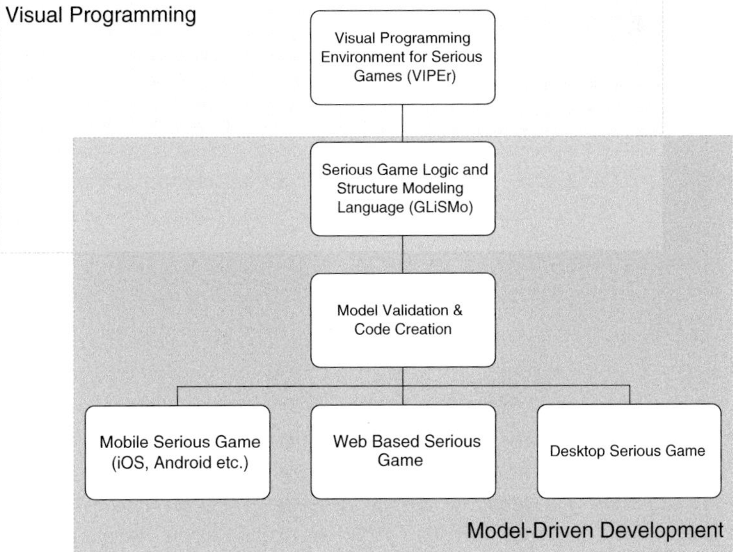

Fig. 3 Model-driven visual programing framework for serious games

In the visual programming component, we have the visual programming environment for serious games, which functions as the user interface for the educators. Here, the educators are able to design and develop the serious game world in an intuitive manner. To create the logic and behavior of the game, we use a domain-specific modeling language, which mediates as an interface to connect the visual programming and the MDD Toolchain component. Within the visual programming component, the DSML is used to model the logic of the game. In the following step, the DSML is used for validating the created models in the MDD Toolchain. After the validation process the software code of the serious game will be generated for different platforms, like mobile devices, web browsers or desktop computers.

In the subsequent sections we present the three major components of your framework, beginning with the core DSML, followed by the visual programming environment and MDD Toolchain.

4.1 Serious Game Modeling Language (GLiSMo)

In order to model a holistic serious game, a DSML tailored for serious games is needed. Therefore, we have created the Serious Game Logic and Structure Modeling Language (GLiSMo) as the underlying domain-specific modeling language, which allows designing both the structure as well as the logic of a serious game. The game structure describes the world the serious game is set in. It contains the layout of the game world, which characters and objects are included, and how the user/player interaction takes place. The logic characterizes the behavior of characters and objects in terms of how does the game react on specific actions performed by the player, or events occurring during game play. The logic also describes the assessment of player actions and the game adaption according to the adaption results. A detailed description of our domain-specific modeling language including the introduction of each language element can be found in our previous publication [24]. Examples of GLiSMo models are presented in Sect. 5.2.

4.2 Visual Programming Environment for Serious Games (VIPEr)

While the Serious Game Modeling Language enables developers to describe a serious game in abstract terms, i.e. in form of a model, it does not ease serious game creation for educators without a technical background. In order to make model creation feasible for this target group, we have designed the Visual Programming Environment for Serious Games (VIPEr) to provide educators with a point-and-click tool for model creation (see steps 4 and 5 in Fig. 1). As VIPEr can be regarded as a frontend for creating valid and meaningful models for the game's structure and logic, the development of VIPEr and GLiSMo are tightly linked.

As there already exists a plethora of development environments, we have chosen to extend an existing tool that is already in use rather than developing an integrated development environment (IDE) from scratch. For our purposes, we have chosen to create VIPEr as a plugin for the cross-platform game engine Unity, an established tool for game creation that already includes a built-in integrated development environment (IDE). VIPEr is especially made to develop point and click graphical adventure serious games in a simple manner. This game genre was chosen as the slow pace during game play, the possibility to study the environment and the fact that all tasks are based on problem solving are characteristics, which are relevant from a didactical point of view. These characteristics increase the learning success while playing [26]. We use GLiSMo as the underlying DSML with which the game logic and behavior can be created as intuitive as possible. To support this, we have taken good care while designing the clearly arranged andy user-friendly UI. Our focus is on displaying only the required elements within the user interface to make sure that educators can use VIPEr intuitively without frustration. At the same time, our intention is to relieve as much development work as possible from the educator. An example for this is that we have designed a room generator, which allows creating a given number of interconnected rooms automatically within the game world.

Our visual programming environment supports both game structure and logic development, as specified by GLiSMo. It is divided into game world design (room creation, placement of objects and characters) and game logic design (connection of individual tasks and actions in an editor), thus not only creating rooms and objects but also following the game logic happening directly in the player's view. By this, we want to make sure, only the most necessary information needs to be entered by educators.

As shown in Fig. 1, the educator needs support during the game world development, game model creation and software code generation. We have taken this as the basis to design the game development process in VIPEr. Developing a serious game with VIPEr needs only four simple steps. In the first step, the educator generates a basic structure of the level using the room generator. This is done by entering a level name, the desired number of rooms and by selecting a predefined setting (office, school, etc.). As an output of this process, a game world with the defined number of rooms will be automatically generated, having the associated doors in between the rooms. The second step consists of placing all needed objects, characters and decorations into the game world. The educator has to insert default characters and objects via drag and drop. Furthermore, it is crucial to set one of the selected characters as the main player character, which will be followed by the main camera during gameplay. At this point, the structure of the serious game, thus the game world with its objects and characters, is built. These two steps cover the game world development from Fig. 1. To support the game model creation in step 3 the logic of the serious game is designed. As provided by GLiSMo the logic describes the behavior of the game by simply using the following elements: (i) actions,

(ii) tasks and (iii) assessments. These elements are arranged by dragging and dropping into the visual programming interface. For example to create a conversation between two characters, the associated discussion element will be positioned into the visual programming interface, followed by entering the text of the conversation.

During this development process, the educators can test the game by using the play button. Finally, the game can be built by clicking on the build button, which starts the compiling process and ends with the final playable game. Both actions trigger the functions of the third component of our framework, the model-driven development Toolchain.

4.3 Model-Driven Development Toolchain

In the introduction, we have described model-driven software development as a methodology in software engineering, which consists of techniques for automated generation of software code from formal models. These techniques include domain-specific modeling languages, such as GLiSMo in our case, and model validators and code generators. The model validator analyzes and checks the developed models for correctness, whereas the code generator outputs the equivalent software code. Changes made to the software will be done by updating the model [21].

As presented in Fig. 3, our framework for serious games development contains a model-driven development Toolchain component, which consists of a model parser, model validator and a code generator. This component is connected with the visual programming component over GLiSMo as the interface. Figure 4 gives an insight into the process within this MDD Toolchain, which is responsible for the transformation of GLiSMo models to the final serious game software code. The starting point in this process are GLiSMo models of the structure or logic of a serious game, which will be designed by the educators visually with VIPEr. These models will internally be parsed and as an output of this parsing process, a semantic model will be generated, which describes the meaning of the GLiSMo models. These semantic models will subsequently be validated and checked for correctness, which results in validated models. Based on these models the associated software code will be generated for various platforms, i.e. mobile, web based and desktop devices.

The parsing algorithm as well as the validator are developed by us and are components of our serious game development framework. However, the code generator consists of two parts responsible for the two types of GLiSMo models.

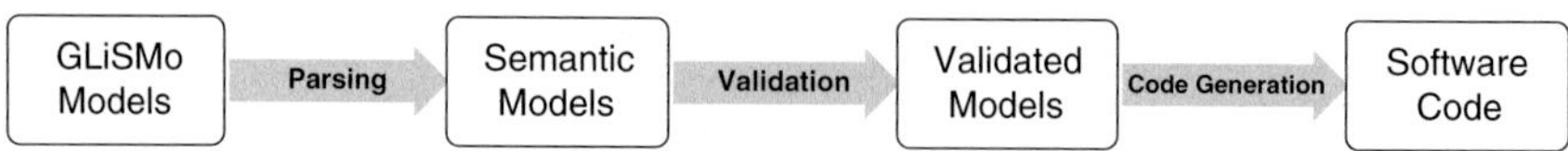

Fig. 4 Process from a GLiSMo model to the final software code

The first part generates the software code for the serious game world, which is mainly automatically done by Unity itself. The second part is responsible for the code generation of the game logic, which is programmed by us and is also part of our framework.

5 Evaluation of the Serious Game Development Framework

Our initial system evaluation of the proposed framework has been twofold. In an advanced modeling language evaluation, we conducted an expert review of the crucial modeling approach to ensure its feasibility. After this, we started to test the whole framework with the cognitive walkthrough method and a guided realization of a serious game together with instructors and domain experts. In the following sections, we will describe the results of our evaluation.

5.1 Advanced Modeling Language Evaluation

For a thorough evaluation focusing on the crucial serious game modeling language, we presented the current version of GLiSMo to four software engineering experts, from which two also are considered expert educators in higher education. During this step, we asked the evaluators to provide feedback on the modeling language, both from a technical standpoint, i.e. the technical feasibility of the modeling language as a basis for the visual programming environment as well as the model-driven development approach, as well as a game design standpoint, especially taking into account serious games for education. From the feedback we have received by observing their approach of analyzing the modeling language as well as the explicit comments from the experts, we could conclude that the current version of GLiSMo allows modeling the structure as well as the logic of the subset of serious games we aimed to support with the visual programming environment (VIPEr). While this would be a limitation for a comprehensive modeling language, this also indicates that no unnecessary features are included in the present form of GLiSMo.

5.2 Evaluation of the Visual Programming Environment for Serious Games

Usability Inspection with Cognitive Walkthrough

After the successful evaluation of the modeling languages feasibility for our purposes, we performed a usability inspection for VIPEr, which is a generic name for a collection of different usability inspection methods. Nielsen [17] distinguishes different types of usability inspection methods, which are heuristic evaluation [12, 14–16], cognitive walkthroughs [9, 20, 27], formal usability inspections [7], pluralistic walkthroughs [2, 3], feature inspection [1], consistency inspection [28] und standards inspection [28].

We have selected the cognitive walkthrough method as the usability inspection method, to evaluate the usability of our visual programming environment for serious games. The cognitive walkthrough method is used to identify solutions for a set of usability problems. Therefore, the software is given to a group of users, who test it based on specific, pre-defined criteria. The aim is to determine if the user is able to master some given tasks, with less or without any instructions. During the cognitive walkthrough process, the participants of the evaluation will try to solve the given tasks, by performing several actions. Meanwhile, it is crucial to reveal difficulties and confusions regarding the user interface. Further details about the cognitive walkthrough method are given in Polson, Lewis [20] or Wharton, Rieman [27].

In the following, we describe the application of the cognitive walkthrough method to VIPEr to identify usability problems in this visual programming environment. The main aim is to check, if educators without programming or game design skills are able to create their own serious games with VIPEr. Therefore, we have selected five evaluation participants, who should create a serious game scenario with VIPEr, which consists of game world and game logic development. The selected evaluation participants are teachers without any game design or software development skills, but with common affinity to computers. The game scenario, which the participants have to build is as follows: First, a building with four rooms should be created with the help of the room builder. This is followed by the selection of a non-player character, which has to be placed within the game world. After placing this character, two random objects have to be selected and also placed within the game world. In the next step, the evaluation participants have to open the logic editor and create the logic of this serious game scenario. Precisely, a conversation between the main character and the non-player character has to be created, followed by a multiple-choice task. After performing that, this small serious game scenario ends.

As a first step, during the cognitive walkthrough process, we have divided this serious game scenario into five tasks, which have a description of the task and an action sequence to fulfill the task. Table 1 shows one example for such a task, where the participant has to create a building with four different rooms in it. Besides this task, there were also the following tasks: *insert character, insert objects, insert conversation with logic editor, insert multiple-choice task with logic editor.* The evaluation participants have to solve these tasks, without having any clues about how to solve them.

After performing this cognitive walkthrough with all five participants, we have analyzed the evaluation results. Through this evaluation, we could identify several usability issues in VIPEr, which have to be fixed for a better user experience. We could find out that some function, which we thought of being intuitive, could not be easily seen by the participants. One example for this is that some buttons could not

Table 1 One sample task and action sequence for the cognitive walkthrough

Task CW-01: create room	
Task	Create a building, consisting of four rooms, with the room builder
Action sequence	To perform this task, the following four actions are necessary:
	1. Click on the button *Level Editor* • VIPEr: Show the level editor and the input mask
	2. Click into the input field named *Level Name* and enter a name for the building
	3. Click into the input field named *Number of Rooms* and enter the integer 4
	4. Click on the button *Build Rooms* • VIPEr: Show the newly created building with four rooms and the automatically created main player character

get the attention of the users. So, in those cases, the participants had to search for a long time to finally find the correct button to perform their intended action. Figure 5 shows a multiple-choice task created during the cognitive walkthrough process.

Guided Realization of a Serious Game

Besides the cognitive walkthrough, we have also performed a guided realization of a serious game, called Shack City 3D, for apprentices in the field of sanitation, heating and cooling (SHaC) together with instructors and domain experts from that field. In this guided realization, we provided an initial idea for a serious game based on previous interviews with instructors in SHaC indicating the need for educational content covering real-life scenarios for apprentices (Step 1). We subsequently asked

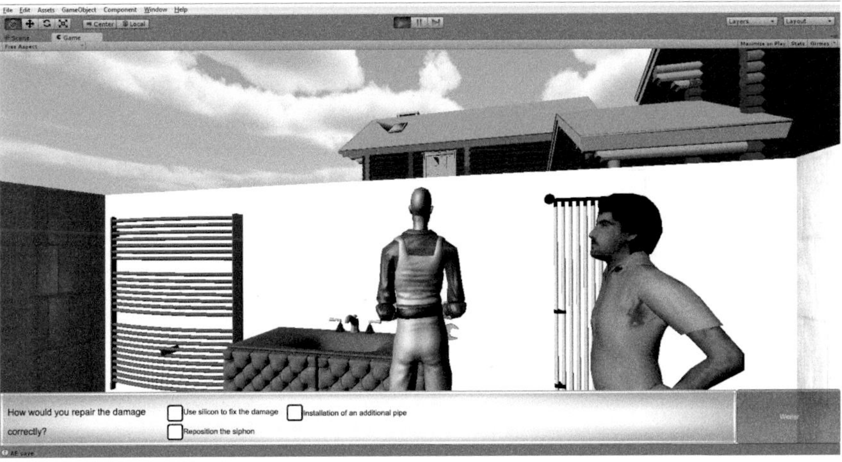

Fig. 5 Multiple-choice task created in the serious game scenario during cognitive walkthrough

the instructors and domain experts to provide their didactical content, describe their goals for the game world as well as its behavior and model it in GLiSMo using VIPEr (Step 2). The results where then compiled into a web-based game using the MDD Toolchain for instructors and domain experts to review (Step 3).

As the result of Step 1, the basic idea of the game is about the main character, a SHaC apprentice, who accompanies his foreman to different customers within a city to solve their problems with their SHaC. This is also supported by a story, which leads the player through the game. At each customer's place at first a dialogue between the apprentice and customer is shown, where the customer describes his problem in a novice manner, as in real life. Based on the information the apprentice has to find out solutions for the problem by solving different tasks. Possible types of tasks are multiple-choice questions, highlighting certain previously defined spots on objects and images, combination of certain objects or filling missing parts in images. After solving such subtasks there will be a conversation with the customer, which describes further problems. This will be alternated until all tasks are solved. At the end, the player gets detailed information and feedback about the solved tasks and learning progress. In Step 2, we derived model elements from the overall game idea, combined them with more detailed descriptions of instructors and their didactical content. This was then used to model both the structure and logic in GLiSMo using VIPEr. We performed multiple iterations of this modeling together with instructors, incorporating their expert feedback in each new iteration.

Figure 6 shows the excerpt of the logic model within VIPEr, which describes an interaction sequence with the customer "Maschberger". The structure of the logic model in this game consists of an alternating sequence of conversations, tasks and assignments. To be precise, the conversation deals in this example with the problem context, where the customer describes what has to be repaired by the apprentice. With the support of an interesting background story, the player gets further details

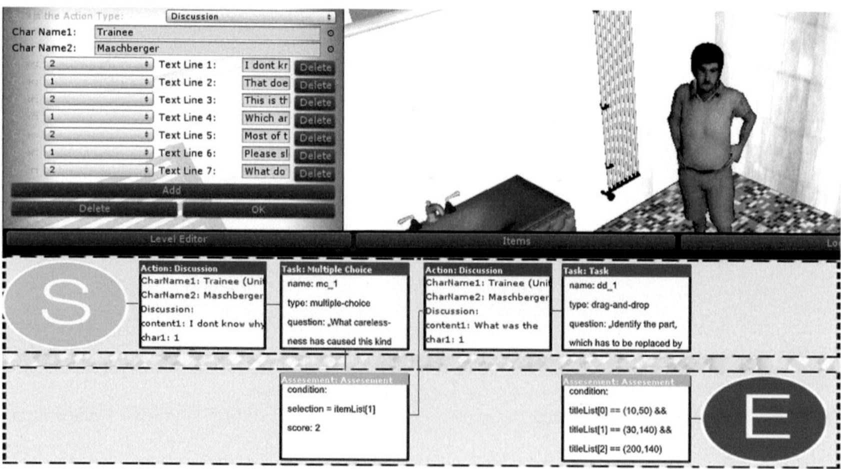

Fig. 6 Logic modeling interface in VIPEr

about the problem context. During these dialogues, the player will face tasks, which have to be solved by the player, followed by the assessment by giving feedback and scores. Among the world elements hence are the rooms, in which the conversations and tasks take place, as well as a player figure, representing the apprentice and non-player characters (NPC) for the customers providing the tasks. The resulting structure model in this case consists of following elements: (i) serious game root, (ii) different acts representing customer orders, (iii) scenes belonging to the acts and (iv) characters.

In Step 3 of the guided evaluation, the first prototype of Shack City 3D, based on the models created in the step before, can be tested using the MDD Toolchain. As described by the logic model, a typical situation at a customer alternates between discussions and tasks. A multiple-choice task is given, which follows a discussion. The player has to select the correct answer, which then will be evaluated by the game based on the content of the associated assessment element described in the logic model. First user tests with apprentices in the field of SHaC have been conducted together with their instructors, evaluating the didactical usefulness of the serious game. The apprentices have agreed that a serious game like this is a promising approach to motivate them for learning in casual settings. Furthermore, they have described the game as interesting and could imagine using such games during their studies as a supporting learning material. Their instructors and domain experts confirmed that the game built matches what they would like to provide for their apprentices when creating a serious game. They expect that the apprentices could benefit from using such serious games, by achieving better learning results and enjoying studying more at the same time.

6 Conclusion and Future Work

From the evaluation of our initial proof of concept, the idea of supporting educators to design and develop serious games through visual programming and model-driven development techniques appears to be a very promising approach. While educators have the domain knowledge and are able to create ideas for a serious game and prepare didactically correct learning content for them, they need support during the serious game design and development process, consisting of game world development, game model creation and software generation. We hence presented a framework that can be broken down into three abstract components: (i) visual programming, (ii) domain-specific modeling language and (iii) model-driven development Toolchain. Component (i) is realized in our visual programming environment VIPEr. VIPEr allows educators to apply their skills to come up with an idea and the content for a serious game and create a model of both via a drag-and-drop style interface and without the need for any technical knowledge. The modeling language we have designed for component (ii), GLiSMo, serves as interface and glues between the visual programming paradigm for the educator and the model-driven development to support a wide range of end devices.

Finally, component (iii) takes the model representation of the user input, validates it and generates code for multiple mobile platforms from it. We have evaluated the feasibility of this component-based framework and our instantiation of tools for each component by designing a serious game called Shack City 3D. During this evaluation, our main focus lies on GLiSMo as the integral interfacing part between the two worlds of visual programming and model-driven development. The results from our evaluation show the feasibility and applicability of GLiSMo for modeling Shack City 3D and the ability to generate software code from GLiSMo models for different mobile devices.

A first evaluation has shown that GLiSMo is suitable for modeling the structure and logic of serious games. Furthermore, the prototype of Shack City 3D, the outcome of our first serious game created using our framework, has been considered as useful by apprentices and educators from the field of SHaC. Having established the feasibility of GLiSMo, the next evaluation will focus on the detailed evaluation of VIPEr as a visual development environment for educators, i.e. performing non-guided development experiments with instructors. Finally, we would also like to evaluate the serious games further, which are the final product of the educator's development process. This has to be done both with educators, i.e. to determine whether the serious game resembles what they anticipated to create, as well as with potential users in order to determine whether they reach the potential outlined by Brennecke and Schumann [4] as well as Pivec [19].

References

1. Bell, B.R., Walkthroughs, Using Programming: Using Programming Walkthroughs to Design a Visual Language. University of Colorado at Boulder. Boulder, CO, USA (1992)
2. Bias, R.: Interface-Walkthroughs: efficient collaborative testing. Software **8**(5), 94–95 (1991)
3. Bias, R.G.: The pluralistic usability walkthrough: coordinated empathies. In: usability inspection methods. Wiley, New York (1994)
4. Brennecke, A., Schumann, H.: A general framework for digital game-based training systems. In: International Conference Game and Entertainment Technologies. Algarve (2009)
5. Gomes, A.J., Santos, A.N., Mendes, A.J.: A study on students' behaviours and attitudes towards learning to program. In: 17th ACM Conf. on Innovation and Technology in Computer Science Education, pp. 132–137. Haifa (2012)
6. Johnson, L., et al.: The NMC Horizon Report: 2013 Higher Education Edition (2013)
7. Kahn, M.J., Prail, A.: Formal usability inspections. In: Usability inspection methods, pp. 141–171. (1994)
8. Kang, Z., Zhang, D.Q., Jiannong, C.: Design, construction, and application of a generic visual language generation environment. IEEE Trans. Softw. Eng. **27**(4), 289–307 (2001)
9. Lewis, C., et al. Testing a walkthrough methodology for theory-based design of walk-up-and-use interfaces. In: Proceedings of the SIGCHI conference on Human factors in computing systems. ACM, New York (1990)
10. March, S.T., Smith, G.F.: Design and natural science research on information technology. Decision Support Syst. **15**(4), 251–266 (1995)
11. Martin, S., Peire, J., Castro, M.: M2Learn: Towards a homogeneous vision of advanced mobile learning development. IEEE (2010)

12. Molich, R., Nielsen, J.: Improving a human-computer dialogue. Commun. ACM **33**(3), 338–348 (1990)
13. Myers, B.A.: Taxonomies of visual programming and program visualization. J. Visual Lang. Comput. **1**(1), 97–123 (1990)
14. Nielsen, J.: Traditional dialogue design applied to modern user interfaces. Commun. ACM **33** (10), 109–118 (1990)
15. Nielsen, J. Enhancing the explanatory power of usability heuristics. In: Proceedings of the SIGCHI Conference on Human Factors in Computing Systems. ACM (1994)
16. Nielsen, J.: Usability Engineering. Elsevier (1994)
17. Nielsen, J.: Usability inspection methods, in Conference Companion on Human Factors in Computing Systems, pp. 377–378. Denver, Colorado, USA, ACM (1995)
18. Pearson: Pearson Student Mobile Device Survey 2013 National Report: College Students. Pearson (2013)
19. Pivec, M.: Play and learn: potentials of game-based learning. Br. J. Educ. Technol. **38**(3), 387–393 (2007)
20. Polson, P.G., et al.: Cognitive walkthroughs: a method for theory-based evaluation of user interfaces. Int. J. Man Mach. Stud. **36**(5), 741–773 (1992)
21. Schmidt, D.C.: Model-driven engineering. Computer **39**(2), 25 (2006)
22. Sealund. Sealund Serious Games—Serious Games Engines. Undated 16.08.2014 http://www.sealund.com/sgengines.php
23. Takeda, H., et al.: Modeling design processes. AI Mag. **11**(4), 37–48 (1990)
24. Thillainathan, N., Hoffmann, H., Leimeister, J.M.: Shack city—a serious game for apprentices in the field of sanitation, heating and cooling (SHaC). In: Informatik 2013. Koblenz (2013)
25. Thillainathan, N., Leimeister, J.M.: Educators as serious game designers—a model-driven visual programming framework. In: 9th International Conference on Knowledge, Information and Creativity Support Systems. Limassol, Cyprus (2014)
26. van Eck, R.: Building artificially intelligent learning games. In: Intelligent Information Technologies: Concepts, Methodologies, Tools, and Applications, pp. 793–825. IGI Global (2008)
27. Wharton, C., et al.: The cognitive walkthrough method: a practitioner's guide. In: Usability inspection methods, pp. 105–140. Wiley, New York (1994)
28. Wixon, D., et al.: Inspections and design reviews: framework, history and reflection. In: Usability inspection methods. Wiley, New York (1994)
29. Worldweaver. DX Studio 2003. http://www.dxstudio.com Accessed 19 Aug 2013
30. Younis, B., Loh, C.: Integrating serious games in higher education programs. In: Academic Colloquium. Ramallah (2010)

First Year Students' Algorithmic Skills in Tertiary Computer Science Education

Piroska Biró, Mária Csenoch, Kálmán Abari and János Máth

Abstract Faculties of Informatics are facing the problem in Hungary that students starting their tertiary education in Computer Sciences do not have a satisfactory level of algorithmic skills, their knowledge seems superficial, and the dropout rate is extremely high in these courses. We have launched a project to test how students' algorithmic skills have been developed in their primary and secondary education, how students evaluate their knowledge. The test proved that an extremely high percentage of the students arrive at the Faculty of Informatics with underdeveloped algorithmic skills, with unreliable knowledge, and they do not consider the recently emerged, non-traditional environments as programming tools and facilities for developing algorithmic skills.

Keywords Algorithmic skills · First year students of informatics · Programming tools

1 Introduction

Over recent years Computer Sciences and Informatics (CSI) education has become accepted; in basic curricula digital competency has appeared alongside traditional competencies, and beyond that CSI has emerged as a separate school subject.

P. Biró (✉) · M. Csenoch
University of Debrecen, Faculty of Informatics, Debrecen, Hungary
e-mail: biro.piroska@inf.unideb.hu

M. Csenoch
e-mail: csernoch.maria@inf.unideb.hu

K. Abari · J. Máth
University of Debrecen, Faculty of Arts and Humanities, Debrecen, Hungary
e-mail: abari.kalman@arts.unideb.hu

J. Máth
e-mail: math.janos@arts.unideb.hu

S. Kunifuji et al. (eds.), *Knowledge, Information and Creativity Support Systems*,
Advances in Intelligent Systems and Computing 416,
DOI 10.1007/978-3-319-27478-2_24

However, in spite of the great expectations regarding this new science, students arriving into tertiary CSI education do not show great confidence in algorithmic skills, and consequently in digital competency. It seems that they mostly navigate the GUI (Graphic User Interface) unplanned, unaware of their activities, trying to solve computer related problems without algorithms.

A large number of studies have been published in the last two decades focusing on the methods adopted to teach programming, to develop algorithmic skills, and to measure students' knowledge [1, 2, 4, 5, 10, 11, 13–15]. To extend these studies we launched the TAaAS project (Testing Algorithmic and Application Skills) in the 2011/2012 academic year at the University of Debrecen, Hungary [2, 3, 5, 7], in which 800 students were tested. The project focuses on the level of the students' algorithmic skills, their usage of terminology, and their problem solving abilities in traditional and non-traditional programming environments at the beginning of their tertiary studies.

2 Methods

Three programming tasks [2] compared in this study differ in terms of their environments, and consequently in the way they were presented to the students. Task 1 is designed to test the logical operators in a multilevel IF structure. Along with the source code a matrix is presented with nine pairs of inputs and the empty cells for the outputs, which are one of the four numbers from the algorithm.

In Task 2 three pseudo codes—Task 2.1, 2.2, and 2.3—are given with background information on the environment. The results of the task are three natural language sentences to describe what the codes do. This task, beyond the decoding, requires the ability to "translate" the code of an artificial language to a natural language in the presented environment [5, 10].

Task 3 is a multilevel spreadsheet function, an array formula. It tests how the students decode formulas, and understand vectors and logical operators in spreadsheet environment, in a functional language. To Task 2, the solution should be given in a natural language sentence. The task is included to see how students would be able to handle programming task presented in a non-traditional programming environment.

Beyond the evaluation of the students' results in the above mentioned three tasks, their self-assessment values in programming were recorded [7].

3 Hypotheses

The collected data and the statistical analyses of the TAaAS project can provide further information on the students' background studies in CSI. The following hypotheses were checked with the analyses.

H1: The order of difficulty of the pseudo-codes is: Task 2.1, 2.2, 2.3, from the least to the most difficult, respectively.

H2: The students arrive in tertiary education with a low and unsatisfactory level of algorithmic skills. Their knowledge is haphazard and non-viable.

H3: The students' uncertain knowledge of the subject can be differentiated: (1) a group of students with a very limited but quite convincing knowledge until they reach their limits, and (2) another group whose knowledge is somewhat arbitrary.

4 Results

4.1 Knowledge-Based Groups of Students

Task 1 was found to be the most successful among the three tasks, so it was used to define clusters of students for the further analyses. The regrouping of the students was based on nine variables—the nine questions of Task 1—and the four optional answers of this task—0, 1, 2, 3. These values served as categorical data for the Two-step Cluster Analysis in SPSS. Based on these premises, 3 + 1 clearly distinguishable clusters were found.

Those students are in Cluster 1 who solved the task with an excellent result (99.7 %, Table 1). In Cluster 2 there is an increase in the number of wrong answers, but this is still lower than the percentage of the right answers, which is 64.7 %. The students in Cluster 3 answered those questions correctly which do not have 0 input. These students answered 0 if one of the inputs was 0. It seems that they found that the code works for inputs A and B, but with 0 input the only output was 0. This can be explained by the false interpretation of the algorithm, which does not provide an explicit output for all the 0 inputs.

The detailed results of Task 1 with the three recognized clusters are mapped in Fig. 1. The titles of the graphs show the order of the questions and the input values, the X-Y pairs of the algorithms, to each question, while the columns the percentage of the students' correct answers in each cluster for the nine questions. The right answer to each question can easily be followed if we track the columns of Cluster 1 (black), since these students answered all the questions correctly. The gray and the

Table 1 The self-assessment values and the test results in the 3 + 1 clusters of Task 1

	Cluster 1	Cluster 2	Cluster 3	Cluster 0
N	213	134	122	331
self-assessment (%)	45.4	34.8	35.1	36.2
Task 1 (%)	**99.7**	**64.7**	**53.7**	**0.0**
Task 2 (%)	39.5	22.5	26.4	12.9
Task 3 (%)	30.7	24.6	30.6	19.8

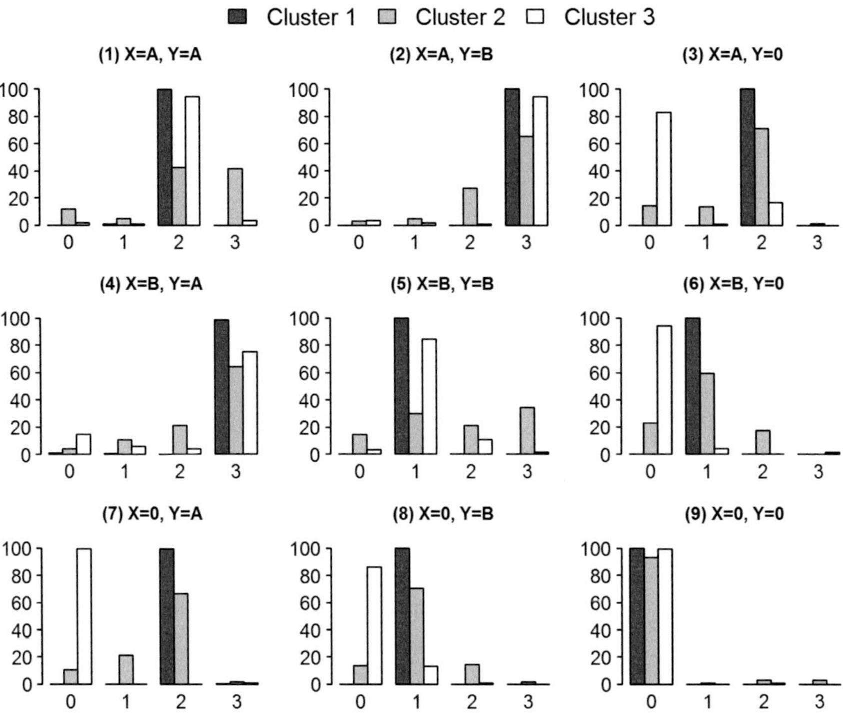

Fig. 1 The percentage of the answers of the 3 clusters in Task 1

white columns show the results of Clusters 2 and 3. As was mentioned in the previous section, the answers to Cluster 3 are 0 s if one of the inputs is 0, X = 0 and/or Y = 0.

Table 1 presents the results of the three tasks of the 3 + 1 clusters. In Cluster 0 are students who did not attempt or complete all the nine answers in Task 1. It is clear from the data that Cluster 1 has the highest scores in all the three tasks, followed by Cluster 2 and 3, while Cluster 0 scored the lowest.

As was mentioned at the beginning of this section, the regrouping of the students was based on the results of Task 1; however, it is noteworthy that the results of Task 1 and the self-assessment values do not follow the same pattern. This can be explained by the major characteristic of the task: the output of the task is neither a source code nor a description of code, but numbers of four choices, and as such, is different in nature from traditional programming tasks. It will be obvious from the further analyses (Sect. 4.3) that it is not Task 1 but 2 which fills the advisory role for the self-assessment values.

4.2 Connections Between the Three Tasks

Task 2 is divided into three subtasks, which are three pseudo-codes. Task 3 is one multilevel function consisting of an AND connection and a counting of those objects which fulfill both of the requirements. Using the 3 + 1 clusters of Task 1 the results of Tasks 2 and 3 are presented in Fig. 2. It is clear from the percentage of the correct answers mapped in the graphs that the difficulty of the three pseudo-codes increases from Tasks 2.1 to 2.3 (percentage correct: 33.3, 20.5, 17.1 %, respectively), which proves the H1 hypotheses, and Task 3 is as difficult as Task 2 in general (25.2, 23.6 %, respectively). Similarly to Task 1, the results of Cluster 1 (black) are the highest in these two tasks, Clusters 2 and 3 performed worse, and Cluster 0 has the lowest scores.

In the following the focus is on the connections between Task 1 and the other two tasks. In the comparison the differences in the results of the 3 + 1 clusters are examined. Four Kruskal-Wallis tests proved that there is a significant difference in the four clusters in terms of the percentage correct in Tasks 2.1, 2.2, 2.3 and 3 ($\chi^2_{2.1} = 68.1$, $df_{2.1} = 3$, $p_{2.1} < 0.001$; $\chi^2_{2.2} = 73.4$, $df_{2.2} = 3$, $p_{2.2} < 0.001$; $\chi^2_{2.3} = 90.7$, $df_{2.3} = 3$, $p_{2.3} < 0.001$; $\chi^2_3 = 14.3$, $df_3 = 3$, $p_3 = 0.002$) compared to Task 1, which proves the relationship between Task 1 and the other four tasks—Tasks 2.1, 2.2, 2.3 and 3.

Regarding the complexity of the tasks, two artificial tasks were created: simple and complex. Tasks 2.1 and 3 were grouped as simple, while Tasks 2.2 and 2.3 as complex. The averages of the two artificial tasks were calculated from the averages of the composing tasks in each cluster (Fig. 2).

The results of Cluster 3 in Task simple and complex are the most noticeable. As was found in Task 1, the result of Cluster 3 is highly dependent on the complexity of the task, i.e., up to a certain degree of complexity the students are able to achieve high results; however, when the complexity of the task reaches their limits their results fall back. Similar results were found in Tasks simple and complex. The graphs in Fig. 2 clearly show that in Task simple the results of Cluster 1 are about as high as of Cluster 3, while Cluster 2 scored lower. There is a significant difference between the results of Clusters 2 and 3 in Task simple (Mann-Whitney test: $p = 0.034$). However, this difference diminishes in Task complex ($p = 0.784$); the

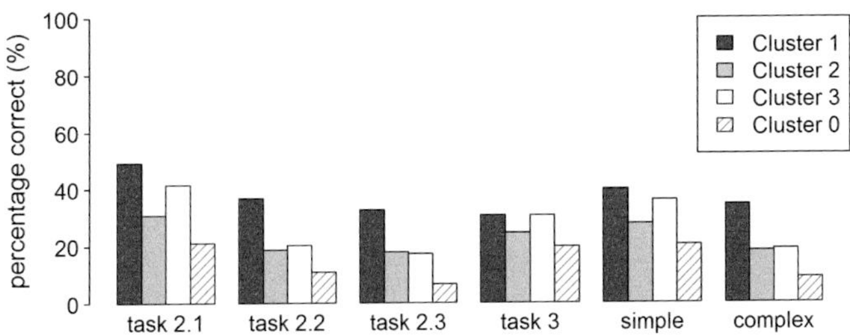

Fig. 2 The comparison of the results of Tasks 2 and 3 in the 3 + 1 clusters of Task 1

results of Cluster 3 fall back to the level of Cluster 2, and both are significantly lower than the result of Cluster 1. In summary, in the simple task the result of Cluster 3 is similar to the result of Cluster 1 and the difference is not significant ($p = 0.449$), while in the more complex task the results of Cluster 3 are significantly lower than of Cluster 1 ($p < 0.001$).

4.3 Self-Assessment

To decide which task(s) among the three would serve as guideline(s) for the self-evaluation in programming, the self-assessment values were compared to programming tasks. A strong positive correlation was only discovered between Tasks 2.1–2.3 and the self-assessment values ($r_{2.1} = 0.50$, df = 788, $p < 0.001$; $r_{2.2} = 0.42$, df = 788, $p < 0.001$; $r_{2.3} = 0.41$, df = 788, $p < 0.001$), while there was only a weak positive correlation with Task 3 ($r_3 = 0.16$, df = 789, $p < 0.001$). This result proves that students only consider Task 2 as a programming task, which involves traditional pseudo codes, neither Task 1 nor 3. The programming self-assessment values do not seem able to measure the non-traditional programming knowledge.

The self-assessment analysis of the 3 + 1 clusters shows that Cluster 1 is different from the other clusters (Kruskal-Wallis test: $\chi^2 = 16.7$, df = 3, $p < 0.001$), since other comparisons of pairs did not prove any significant difference ($\chi^2 = 0.05$, df = 2, $p = 0.977$). Consequently, those who had the highest results in Task 1 had the highest self-assessment values.

The self-evaluation overestimates the test results both in Tasks 2 and 3 in all the clusters (Table 1). The measure of the overestimation was calculated, and it was examined whether or not the overestimation was different in the 3 + 1 clusters. Based on the results of the Kruskal-Wallis probe, no significant difference was found between the clusters testing Task 3. However, there is a significant difference in the case of Task 2 ($\chi^2 = 38.9$, df = 3, $p < 0.001$). Following the Kruskal-Wallis probe, the multiple comparison test proved that Cluster 0 is significantly different from the other three clusters. In this group the self-assessment values are 23.3 % percent higher than the average results of Task 2, which is in accordance with the well-known Dunning–Kruger effect, and seems valid in connection with computer related activities, too [9]. These values in Clusters 1–3 are 5.9, 12.3 and 8.7 %, respectively. However, no significant differences were found between the other pairs.

5 Summary

The analysis revealed that the students' algorithmic skills have not developed in their previous education to a level which would serve as firm background knowledge for advanced level CSI studies (H2). The underachievement of the students cries for investigation since our education system [12] was among those which introduced formal CSI studies in primary and secondary schools, as early as the mid '90s.

In our analyses we were able to distinguish four knowledge-based groups of students. Between the best and the worst groups there are two groups with uncertain knowledge: a group of students were found who up to a certain point are quite convincing in their knowledge. However, they are not able to solve problems which require knowledge beyond their limits; instead they try to find escape routes, and they are not aware of their lack of knowledge [9]. The other group with uncertain knowledge proved rather uncertain both in their knowledge and their self-assessment (H3). Based on previous studies and reports, we can assume that the students of this latter group are socialized in an environment where the first level of mastery, the awareness of concept [6], in other words the construction of a mental model of a computer, is skipped and they only focus on the second level, which is usage with surface-approach metacognitive methods [2, 4, 8].

The results of the TAaAS project clearly present that formal CSI education without deep approach problem solving methods [2, 4, 8], by the pure existence, is not able to develop the students' algorithmic skills to solve computer related problems effectively. Being aware of these facts tertiary education has at least two things to consider. If students start their studies in Computer Sciences with underdeveloped algorithmic skills they have to be offered extracurricular activities to develop these skills before starting "serious" studies in programming. If colleges and universities ignore the students' actual knowledge, the dropout rate will not be reduced. On the other hand, and more importantly, teacher education in CSI has to be reconsidered and courses have to be offered which focus on the development of the algorithmic skills of students in primary and secondary education. Recent studies have proved that the computational thinking, the algorithmic skills of the students can be developed both in traditional and non-traditional environments—programming and document handling along with data retrieval, respectively. Consequently, CSI teachers have to be trained to accept and use the newly emerged tools also for the development of the students' computational thinking.

Acknowledgment The research was supported by the TÁMOP-4.2.2.C- 11/1/KONV-2012-0001 project. The project has been supported by the European Union, co-financed by the European Social Fund. The research was supported partly by the Hungarian Scientific Research Fund under Grant No. OTKA K-105262.

References

1. Biggs, J.B., Collis, K.E.: Evaluating the Quality of Learning: The SOLO Taxonomy. Academic Press, New York (1982)
2. Biró, P., Csernoch, M.: Deep and surface structural metacognitive abilities of the first year students of informatics. In: 4th IEEE International Conference on Cognitive Info-communications, Proceedings, pp. 521–526. Budapest (2013)
3. Biró, P., Csernoch, M.: Programming skills of the first year students of informatics, XXIII. In: International Conference on Computer Science, EMT, in Hungarian, pp. 154–159 (2013)
4. Biró, P., Csernoch, M.: Deep and surface metacognitive processes in non-traditional programming tasks. In: 5th IEEE International Conference on Cognitive Infocommunications CogInfoCom 2014 Proceedings. IEEE Catalog Number: CFP1426R-USB, pp. 49–54. Vietri sul Mare, Italy (2014). ISBN: 9781479972791
5. Clear, T., Whalley, J., Lister, R., Carbone, A., Hu, M., Sheard, J., Simon, B., Thompson, E.: Reliably classifying novice programmer exam responses using the SOLO taxonomy. In: Mann, S., Lopez, M. (eds.) 21st Annual Conference of the National Advisory Committee on Computing Qualifications 2008. Auckland, New Zealand (2008)
6. Computer Science Curricula 2013: Curriculum Guidelines for Undergraduate Degree Programs in Computer Science. ACM and IEEE Computer Society 20 Dec 2013
7. Csernoch, M., Biró, P.: Teachers' assessment and students' self-assessment on the students' spreadsheet knowledge. In: Gómez Chova, L., López Martínez, A., Candel Torres, I. (eds.) EDULEARN13, pp. 949–956. IATED, Barcelona, Spain (2013)
8. Csernoch, M., Biró, P.: Spreadsheet misconceptions, spreadsheet errors. Oktatáskutatás határon innen and túl. HERA Évkönyvek I., ed. Juhász Erika, Kozma Tamás, pp. 370–395. Belvedere Meridionale, Szeged (2014)
9. Kruger, J., Dunning, D.: Unskilled and unaware of it: how difficulties in recognizing one's own incompetence lead to inflated self-assessments. J. Pers. Soc. Psychol. **77**(6), 1121–1134 (1999)
10. Lister, R., Simon, B., Thompson, E., Whalley, J.L., Prasad, C.: Not seeing the forest for the trees: novice programmers and the SOLO taxonomy. In: Proceedings of the 11th annual SIGCSE, pp. 118–122. New York, NY, USA (2006)
11. Mayer, R.E.: The psychology of how novices learn computer programming. ACM Comput. Surv. **13**(1), 121–141 (1981)
12. NAT National Curriculum.: In Hungarian: Nemzeti Alaptanterv. Korona Kiadó, Budapest (1995)
13. Schraw, G., Horn, C., Thorndike-Christ, T., Bruning, R.: Academic goal orientations and student classroom achievement. Contemporary Edu. Psycho. **20**(3), 359–368 (1995)
14. Sheard, J., Carbone, A., Lister, R., Simon, B., Thompson, E., Whalley, J.L.: Going SOLO to assess novice programmers. SIGCSE Bull. **40**(3), 209–213 (2008)
15. Soloway, E.: Should we teach students to program? Commun. ACM. **36**(10), 21–24 (1993)

Implementing a Social Networking Educational System for Teachers' Training

Stavros Pitsikalis and Ilona-Elefteryja Lasica

Abstract The purpose of this paper is the development, documentation and best use of a social networking tool to create a Web-based Social Networking Educational System aimed at the implementation of a training program designed along the lines of instructional design and deploying diverse teaching models available in relevant literature. For trainees to experience a real-life problem, an educational scenario has been devised, constituting the central theme of the training program throughout which participants are encouraged to receive training and act with different roles. The results derived from the evaluation of the training program are, finally, presented and discussed.

Keywords Social networking · ICT use · Teacher training

1 Introduction

Rapid technological innovations affect several walks of people's everyday life, including education. From the traditional classroom to the digital one, learners' participation in the learning process has changed considerably. The development and integration of valid, reliable, user-friendly applications and tools is of vital importance to improving the speed of accessing information and enhancing learners' social skills.

In this paper, the utilization of Social Networking Applications in support of Teacher Training is proposed. The aim is the creation of a Web-based Social Networking Educational System (WSNES) that will serve as the basis of a learning

S. Pitsikalis (✉)
General Secretariat for Lifelong Learning & Youth, Marousi Athens, Greece
e-mail: pitsikalis@sch.gr

I.-E. Lasica
Foundation for Youth and Lifelong Learning, Athens, Greece
e-mail: e_ilona@outlook.com

S. Kunifuji et al. (eds.), *Knowledge, Information and Creativity Support Systems*,
Advances in Intelligent Systems and Computing 416,
DOI 10.1007/978-3-319-27478-2_25

community supported by teaching methods [2]. Through the WSNES in case (available at http://learn2learn.gr/social/), the sharing of digital educational content is encouraged and the promotion of constructive, collaborative and exploratory learning is attempted by providing new technological tools and by enabling better sharing and reuse of resources so that the opportunities offered by Information and Communication Technologies (ICT) can be exploited to the full. By means of creating a community that supports its members, the cultivation of trust and teamwork as well as the change of attitudes concerning issues that include self-awareness, self-esteem and self-control are developed via simple Educational Activities. The actual challenge is the implementation and application of the educational activities in a manner that will prove useful in real-life learning process scenarios. Specifically, for the trainees to gain experience in real-life problem-solving, an educational scenario, lying at the core of the training program, has been designed.

It is ultimately an approach towards creating attractive educational practices within the context of a social network. Trainees' development of creativity constitutes a central goal. Advertising and promotion set a solid ground for the development of creative thinking [7]. The trainees function as 'Executives' within the WSNES shouldering the responsibility of the Advertising and Promotion Divisions. The proposed scenario, addressing 2 classes of 16 trainees (part of a wider management training program on the use and exploitation of Digital Tools and Systems towards improving Promotional Services) is titled: "Training Executives on YouTube Optimization and Use to Promote their Services". Time allotted to the training program in question is 17 didactic hours (6 days–2 weeks).

The trainees acting as real executives are offered the opportunity to actively participate in interesting and intrinsically motivating activities. This results in trainees maintaining a positive attitude towards the content and getting actively engaged with it. The educational problem is related to the trainees' capacity for profoundly comprehending the technologies they observe through a specific process. The trainees develop skills relevant to the knowledge acquired, cultivate their critical and systemic thinking and collaboration skills so as to use them as stepping stones once creating appealing educational environments.

2 Design of the Educational Approach

The aim of the proposed pedagogical approach is to encourage trainees' active participation in the training program instead of merely attending it. The central objective is to focus on learning through setting problems, exploring possible solutions, developing projects, coming up with ideas and eventually presenting them. For the aforementioned reasons, the program highlights the raise of trainees' awareness, delineates the activities they participate in as well as the deliverable resulting from performing the activities. The trainer and the digital educational content hold a central role in the program's successful completion.

Trainees' prior knowledge on ICT-related issues is a prerequisite considering that the program has been designed to make the most of instructional strategies founded on new technologies and digital tools. Through the program in question trainees familiarize both with the tools they are prompted to use and the manner they can exploit the use of the latter to design their projects.

2.1 Trainees' Characteristics and Needs

The trainees assume the role of business/organizations executives (responsible for the promotion and advertising divisions). Furthermore, they display adequate knowledge and skills in Introductory IT Concepts (Computer Use, Office Software, Internet and Communications). Some of the adult trainees' characteristics that were taken into consideration involve [10, 11]: (1) not always disclosing their real intentions for joining the training program (may have joined to make new acquaintances, to become members of a team and, to a smaller extent, to gain knowledge), (2) wishing to make social relationships (feel at ease), (3) appreciating the trainer's genuine interest, (4) needing reassurance they can succeed in whatever they wish to do or learn, (5) needing reassurance and encouragement, (6) needing to feel the satisfaction deriving from achieving goals, (7) being restless, pressed for time and rushing into implementing what they have learned or into displaying the skills acquired, (8) having a wealth of experience to share with the group, (9) appreciating a concrete, well-designed learning experience, and (10) being quick to assess and evaluate good teaching. A key component of success for training programs is the implementation methodology and the emphasis placed on the participants' effective needs' analysis. The constituent parts of the methodology followed within this context (appropriately customized to fit the needs of the training program in question) comprise five stages composing the Systemic Approach to Training (SAT) (see Fig. 1).

As regards the trainees' learning needs, YouTube constitutes a digital tool to be used in the strategic promotion design process, due to its being massively acknowledged, usable and having the potential for social networking encouraged by its design. This innovative service may have an added value for the strategic design of promotional practices. The trainees feel the need to explore, discover and realize new technologies' actual dimension through their practical use. At the same time,

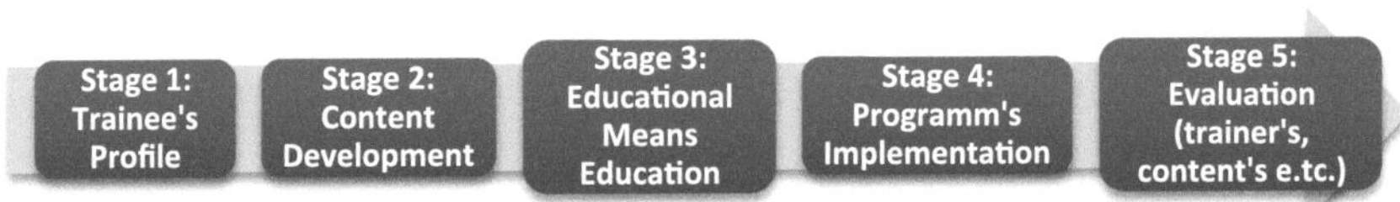

Fig. 1 Implementation methodology of the training program

they are in need of actively participating in the educational process and learning of how to act in a constructive manner as members of a team.

2.2 Task and Learning Objectives Analysis

With reference to the training program as a whole, these are the details concerning the implementation and nature of its activities: (1) Primary Information: Delimit 'accidental' correlations between two elements. Learning primarily takes place through memorization. Primary information, linked to YouTube and being the object of the program taught, refers to the following: History and website terminology, typologies, its structure, categories and video types, alternatives to video creation, qualitative evaluation indicators, search and creation (videos, channels, etc.) best practices, (2) Concepts: The organized knowledge a person possesses in terms of the critical features of persons, objects, events, ideas, statements and procedures. In this case the clarification and understanding of the following concepts are sought after: search, evaluation, creation, creativity, exploitation, quality, safety, motivation, social conditions, community, communication, functionality, progress, and shared experience, (3) Principles-Generalizations: Causal or regulatory relationships linking concepts and serving the purpose of mentally simplifying the complexity of reality while assisting the prediction of facts and situations. Within the context of YouTube use, generalizations allude to: early use, the role of the video creator-administrator, the role of the YouTube user, the principles of guiding and instructing the trainees, the creation and use of video and YouTube, (4) Procedures: While primary information, concepts and generalizations refer to the information and understanding an individual displays of the world, procedures allude to an individual's actions within society. Creating a successful video or YouTube channel, as well as the systematic practice of attracting large number of users (participation in success through acceptance), are prioritized within the training program, (5) Interpersonal Relationships: Interaction skills fall under this category. The design and administration practices of a YouTube video are largely dependent on communication skills and on building a digital online community where views/ideas can be exchanged, grounded in widely accepted social norms, (6) Attitudes—Values: Internal, well-established predispositions of an emotional and evaluative nature that affect a person's relationship with items and social surroundings. The main aim of the training program is the recognition of the videos' value, of the ease they have been created with and of their increased potential to be broadcasted.

Through the preceding needs analysis a range of knowledge and skills corresponding to the training program has been determined. By deploying the abovementioned activities the general purpose of the training program is identified as: increasing trainees' understanding of the YouTube functions and features, enhancing trainees' recognition of its added value within the context of promotion and enabling trainees to recognize and plan quality videos with a promotional orientation. A crucial axis for the achievement of the general aim is the WSNES displaying the necessary functions and educational content to support trainees. To clearly define the content of the program 'functionalization' of the general purpose is being conducted, formulating specific objectives. According to Bloom's taxonomy, such objectives fall into three domains: cognitive domain (identify, characterize, understand, comprehend, perceive, interpret, display etc.), skills (create, use, leverage, describe, characterize, make, become familiar etc.) and affective domain (will, wish, grasp, recognize, appreciate, realize etc.) [1].

2.3 Training Program Design and Teaching Strategies

The issue of the comparative evaluation of online systems and in particular of social networks' effectiveness as compared to traditional teaching systems is hotly debated. However, the majority seem to agree that no technological or communication system can enhance learning effectiveness in its own unless it is accompanied or is part of a pedagogic logic and serves a structured pedagogic process. The areas where attention has been paid to in terms of the quality of distance educational programs include: (a) the overall program design (serving general principles of education, taking into account particularities of distance teaching), (b) the design of teaching and learning and the fact that the objectives and expected outcomes of learning should be closely associated, (c) the program's management and implementation (ensuring that trainees will achieve their learning goals through active involvement in the learning process and by means of continuous communication and feedback among the trainers, trainees and the program's designers), (d) encouraging trainees to take control of their learning, (e) trainees' assessment.

The program takes place in four distinct phases containing individual modules as illustrated in Fig. 2. The modules have been configured in accordance with the learner-centered learning-related strategy [8] which advocates, among others, the following: (a) identification of prerequisites; required knowledge is specified so that participants have a clear view of what they should know before attending the program, (b) intimacy; relevant concepts are presented prior to unknown ones (for instance, the social phenomenon of social sharing is presented first accompanied by

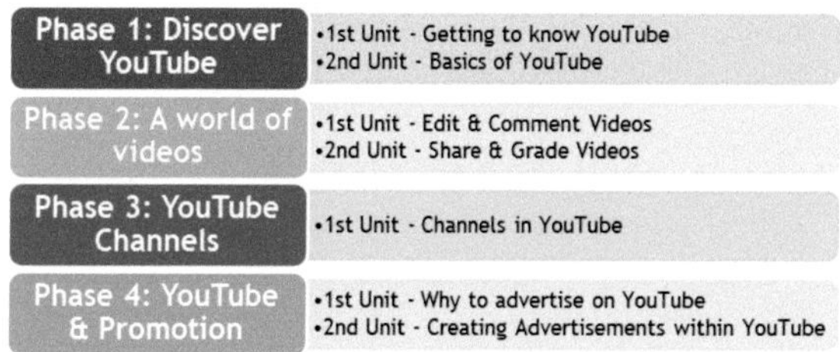

Fig. 2 Training program phases

the socio—psychological effects of using video in marketing strategies), (c) graded difficulty; the easy parts are taught first progressively moving to more complicated ones (based on this principle, the processes of account creation in YouTube, search and video presentation, precede the teaching of dynamic channels creation and ways to attract public attention), (d) motivation; at first, trainees are presented with highly motivating issues (such as presentation of popular videos and the reasons for employing YouTube to create promotional videos). For each of the program's modules a series of activities based on carefully selected instructional strategies has been provided.

The nature of the teaching content, the trainees' characteristics and the program's objectives necessitate the selective and combined use of four teaching strategies, namely: (a) Pre-instructional Strategy (can cater for the need of comprehending concepts and generalizations involved in each module while strengthening trainees' cognitive structures [6]), (b) Effective Teaching Strategy (it is aimed at formulating and automating processes that ensure quick and accurate promotion channels' creation by the trainees. Trainees receive individual practice in the new skill while the trainer provides guidance and assistance), (c) 5W1H Strategy (is designed to split a problem into individual sub-problems, addresses the development of problem-solving skills and consists of an organized series of questions inquiry focused (What, Where, When, Why, Who, How) [4]. It is employed within the training program to support trainees in developing original ideas as regards the use of YouTube in promotional practices) and (d) Webquests (series of guided inquiry activities in which learners undertake the responsibility of solving a problem using the internet as the primary source of information. Webquests' structure involves: (1) Introduction, (2) Task or Mission, (3) Process, (4) Assessment, (5) Conclusion, (6) Teacher's corner, (7) Resources.). Within this training program, the webquest strategy has been employed, as studies show that it has the potential to develop online search skills, comprehension of concepts but also to enhance trainees' motivation and collaboration [3].

Such strategies fall under the cooperative learning strategies spectrum on the grounds that they promote and improve individual and collective learning through collaboration, provision of mutual assistance among trainees, enhancement of attitudes/values and social skills. They are primarily founded on a theory of cognitive development [12] which emphasizes that cognitive processes can be modeled in the social world before internalized by the individual. The selected learning theory and teaching strategies are supported within the proposed WSNES, with Web 2.0 technologies that reinforce cooperation and active participation in learning [5].

3 Educational Content Development

The educational content supporting the program's activities makes allowances for both participants' cognitive and affective level. With reference to the WSNES' structure, an appealing environment has been created to enhance participants' positive attitude and mood; at the same time, being user friendly the WSNES allows novice users to easily navigate themselves throughout and actively participate in existing educational activities by means of merely possessing basic computer and internet use skills [5]. The tools employed for the development of the teaching materials or for reusing existing educational content can be hierarchically divided in the categories to follow: (1) Central Online Social Networking Training System, (2) Supportive Material for Phases/Modules, (3) Educational Activities' Material, and (4) Educational Content. Figures 3 and 4 represent parts of the educational activities/content available.

Oxwall has been employed as the Central Online Social Networking Educational System. The creation of an online community was considered important within the context of the training program as trainees can be offered the opportunity to actively participate, communicate, share and feel like being members of a team sharing common interests and thoughts. The Supportive Material and key material for the educational activities are briefly presented in Table 1.

Finally, regarding the educational content, the majority of Learning Objects has been either designed within the context of the educational program in case or has been the result of relevant Internet resources. To name a few, Learning Objects Libraries available in the World Wide Web, such as Cosmos, Merlot and OER Commons have been employed. Additionally, supportive tools, necessary for uploading files (.pdf) on the Web (scribd) have been deployed so that their integration within the WSNES could be rendered possible and for the sake of fast and easy image processing (pxlr).

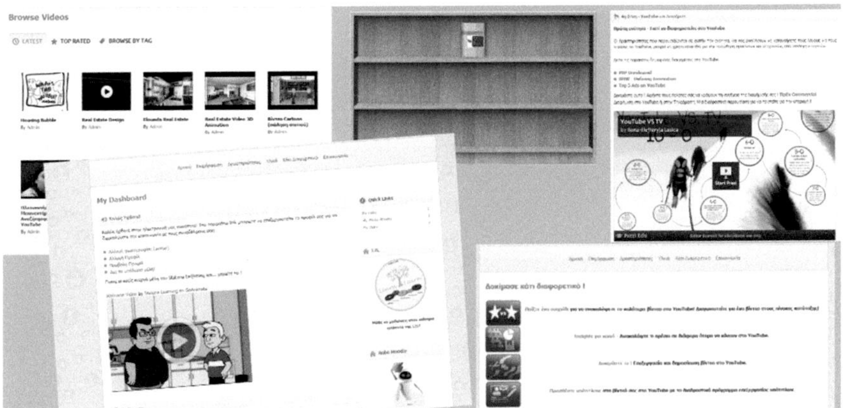

Fig. 3 Sample of educational activities

Fig. 4 Sample of educational content

Table 1 Supportive material for the training program's phases/modules and educational activities' key material

Tool/Software	Short description
FlipSnack	Online tool for interactive e-books creation. Employed for the creation of additional supportive material that informs trainees on theoretically presented issues. Allows the selection of a book from the library, "browsing" and transition to information through hyperlinks (*free version available*)
GoAnimate	Online tool for animation creation. Employed to capture trainees' interest. The animations created are brief, appealing; the dialogues are entertaining and have been placed at carefully selected WSNES spots to attract attention (*free version available*)
Onedrive	Online file sharing application. A closed group has been created, within the limits of which trainees can share any files (*free*)
Garfield & Marvel Comic Creator	Online tool for comics creation. Used to attract trainees' interest, motivate them to participate and capture trainees' interest through labeling inappropriate video content (*free version available*)
Prezi	Online tool for interactive presentations creation. Used to draw trainees' attention and provide them with useful information (*free version available*)
MyUdutu	Online eCourses builder. Used to aid the presentation of YouTube functions, e.g. the creation of a YouTube account. Enables serial attendance of the proposed modules employing digital media (image, video, hyperlinks) and text (*free version available*)
Zunal	Online tool for WebQuests creation. Webquests include guided inquiry activities in which trainees have to solve a problem deploying the internet as a primary source of information (*free*)

(continued)

Table 1 (continued)

Tool/Software	Short description
Jing	Snapshot and video recording tool. Used for short instructional videos creation (video tutorials), presenting steps to be followed by trainees to realize specific actions (*free version available*)
What2Learn	Online short game builder to keep trainees' interest alive. Was used to create hangman with key, YouTube-related, concepts (*free version available*)
CamStudio	Video editing tool, with an adding-comments potential, placing emphasis on key points and explanation input. Used for enriching videos created (*free*)

4 Training Program Evaluation Results

The trainees' (2 classes, 16 persons) evaluation is conducted by means of [9]: (a) the analysis of the trainees views added to the forum and to various supportive educational tools, (b) the functionality, aesthetics and content of the videos created by them, (c) the trainees' active participation in the activities of the training program, (d) specific online data collection forms, (e) quiz questions and online games, and (f) through uploading their deliverables to the StalonaLe@rning's OneDrive application (Fig. 5).

To facilitate the process of both qualitative and quantitative result extraction two (2) online questionnaires, one aimed at assessing the use of the Online Social Networking Educational System (StalonaLe@rning) and one for assessing the training program, have been designed and administered [9]. In addition, the participants' views added to the forum within the context of the tasks assigned and their face-to-face discussions were also taken into consideration. The online tool Survey Monkey (http://www.surveymonkey.com), which analyzes and

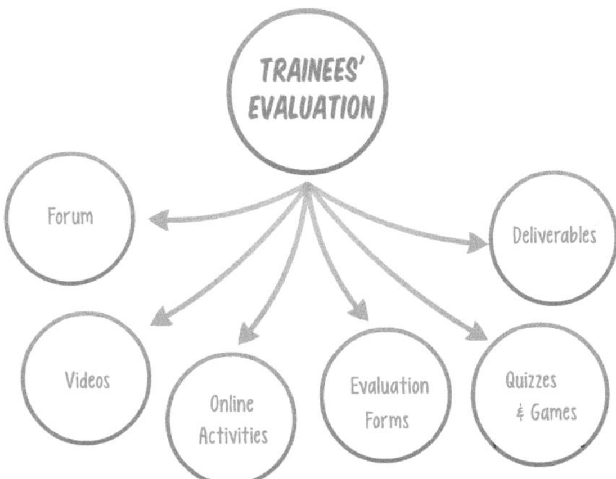

Fig. 5 Trainees' evaluation

automatically extracts real time results, has been employed for the creation of the questionnaires. This tool's potential for sharing, allows the online distribution of questionnaires through social networks, e-mail or through getting embedded in a website, thereby covering the individual preferences of the trainees. Besides, the potential for setting rules to the questions enhances the reliability of the responses and ensures everyone's involvement in the questionnaire. Finally, the potential for receiving up to 100 responses to a questionnaire (free edition of the tool), allows obtaining a satisfactory participant sample, depending on the training program implemented.

The overall number of the questionnaires' respondents was set to 29 persons, as 3 of the trainees did not participate in the process (90.6 % participation rate, 9.4 % abstention rate). It is worth mentioning that 55 % were trainees displaying a technical orientation (engineers, etc.), while 20 % were coming from IT disciplines and 25 % had a theoretical background (literature and science teachers, etc.). With regard to the first questionnaire, all trainees stated that their overall impression of the WSNES has been ranging from "very good" to "outstanding", while none of them had ever used a similar educational system. 80 % claimed that the educational content available in digitized audiovisual form is interesting and the diversity of content (combination of notes, tasks, examinations, discussions), as opposed to the static lecture notes' presentation is more intriguing. The ease of communication among the participants and the distribution of educational content and ideas that encouraged a fun atmosphere throughout the program, have been reported as the WSNES' strong point. Another feature of the WSNES that has been reported by the trainees involves information management and familiarization with diverse technologies (primarily reported by trainees displaying a theoretical background). Finally, 60 % of the trainees expressed their intention to integrate corresponding WSNESs into their own teaching programs.

In terms of the second questionnaire, all trainees felt that the training program was "very good" to "outstanding" and would definitely recommend it to their colleagues. 75 % of the trainees mentioned that their expectations of the program were fully met and that the teaching method employed was "very good" to "outstanding". Also, 80 % claimed that all activities met the program's objectives but also argued that the program's duration could be redefined.

5 Conclusions

The educational environment in which the training program and learning take place has been designed and implemented in compliance with the Instructional Design and Educational Approach principles presented in the above sections. Within this context: (a) the flexibility of applying various teaching methods depending on the learning conditions and the trainees' cognitive demands is ensured, (b) the trainees are dynamically involved in the teaching process having opportunities for communication with other trainees and with the instructor, enjoying a no longer passive role, (c) direct interaction and collaboration between the trainer and the trainees and the trainees themselves are ensured, (d) the trainees are getting engaged with the educational content and activities, taking over and completing various tasks, (e) they develop analytic, generalization, synthesis and application of knowledge skills, critical thinking and interest display in scientific matters.

The present educational program which has been implemented to assist trainees into acting as actual managers completing specific projects so as to evaluate and correspond to the demands of real life situations has been welcomed by the trainees who realized the need for developing good practices to be shared and the opportunity for getting involved in their own educational process.

Finally, the trainees were particularly interested in the interactive and digitized educational content, completed all activities and actively participated in the community created. It needs to be mentioned that the trainees got impressed by the design and the selection of a WSNES; concurrently, they reported that the balanced use of corresponding systems could have been exploited towards attracting trainees' interest and in support of traditional educational systems as well.

References

1. Anderson, L.W., Krathwohl, D.R.: A Taxonomy for Learning, Teaching, and Assessment. A Revision of Bloom´s Taxonomy of Educational Outcomes. Longman, New York (2001)
2. Ghislandi, P., Calidoni, P., Falcinelli, F., Scurati: C. e-university: α cross-case study in four Italian Universities. Br. J. Educ. Technol. **39**(3), 443–455 (2008)
3. Gulbahar, Y., Madran, R.O., Kalelioglu, F.: Development and evaluation of an interactive WebQuest environment:"Web Macerasi". J. Educ. Technol. Soc. **13**(3), 139–150 (2010)
4. Jang, S., Woo, W.: 5W1H: Unified user-centric context. In: 7th International Conference on Ubiquitous Computing. Springer-Verlag Berlin (2005)
5. Jothi, P.S., Neelamalar, M., Prasad, R.S.: Analysis of social networking sites: a study on effective communication strategy in developing brand communication. J. Med. Com. St. **3**(7), 234–242 (2011)
6. Joyce, R.B., Weil, M., Calhoun, E.: Models of Teaching, 7th edn. Prentice Hall, Upper Saddle River, N.J (2008)
7. Kilgour, M., Koslow, S.: Why and how do creative thinking techniques work?: Trading off originality and appropriateness to make more creative advertising. J. Acad. Mark. Sci. **37**(3), 298–309 (2009)
8. Morrison, R.G., Ross, M.S., Kemp, E.J.: Designing Effective Instruction, 3rd edn. Wiley, New York (2001)
9. Pitsikalis, S., Lasica, I.E.: StalonaLe@rning: a social networking educational system for teacher training. In: 9th International Conference on Knowledge, Information and Creativity Support Systems, Limassol, Cyprus (2014)
10. Smith, C., Hofer, J.: The characteristics and concerns of adult basic education teachers (NCSALL Rep. No. 26). National Center for the Study of Adult Learning and Literacy, Boston (2003)
11. Tight, M.: Education for adults, Volume 1: adult learning and education (2012)
12. Vygotsky, L.: Thought and Language. MIT Press, Cambridge, MA (1986)

Virtual Environment for Creative and Collaborative Learning

Anna Bilyatdinova, Andrey Karsakov, Alexey Bezgodov and Alexey Dukhanov

Abstract New guidelines for Russian technical universities and creative industries require competitive professionals and as a result we are urged to apply novel educational methods to make their training efficient. This tendency calls for new tools to be adopted and implemented in the learning purposes. Russian modern curriculum is based on three levels of desirable exit qualifications students need to develop—Hard Skills, Soft Skills and Professional competencies in a specialized field. Novel not only in content but also in form Elective courses in Scientific Visualization, and Virtual Reality of Double Degree Master's Program in Computational Science (ITMO University, Russian Federation and University of Amsterdam (UvA), The Netherlands) incorporate 3D anatomical atlas of human body, which became a prototype for the Virtual Learning Environment (VLE). In this paper, we describe VLE architecture and our experience of its introduction to the master program courses and master theses preparation. We use our flexible VLE to acquire knowledge, develop collaborative skills, and monitor knowledge acquisition within the same unified environment using varied methods.

Keywords Virtual reality · Virtual environment · Curriculum design · Learning competence

A. Bilyatdinova (✉) · A. Karsakov · A. Bezgodov · A. Dukhanov
ITMO University, Saint Petersburg, Russia
e-mail: a.bilyatdinova@gmail.com

A. Karsakov
e-mail: kapc3d@gmail.com

A. Bezgodov
e-mail: demiurghg@gmail.com

A. Dukhanov
e-mail: dukhanov@niuitmo.ru

S. Kunifuji et al. (eds.), *Knowledge, Information and Creativity Support Systems*,
Advances in Intelligent Systems and Computing 416,
DOI 10.1007/978-3-319-27478-2_26

1 Introduction and Related Works

Interconnection of the fields of study and application fields implies cultivation of multi-domain Hard and Professional Skills, which is reflected in growing popularity of interdisciplinary master's programs [11]. However, apart from audio, video and social media communication tools for learning and teaching purposes the lack of the multi- and interdisciplinary educational methods still exists. This situation is observed in most CIS countries due to rigid traditional approaches to training—remnants of past era.

Diverse virtual reality environments are gaining their place in education and learning processes for the last 15 years. An Educational Virtual Environment (EVE) or Virtual Learning Environment (VLE) could be defined as a virtual environment based on a certain pedagogical model, and incorporates or implies one or more didactic objectives, provides users with experiences they would otherwise not be able to experience in the physical world and redounds specific learning outcomes [22]. Dillenbourg's [10] and Mikropoulos' [21] papers clearly determine VLE common and unique features. Currently multiple research works discovering effectiveness and possibilities of VLEs are in the process [7, 8, 12, 19, 25–27].

Actual application fields of VLEs are wide enough and vary depending on the learning activities, i.e. virtual teams [4], and learning outcomes [18] to be attained using different environments. Quite a large number of existing works dedicated to the research of implementation of the existing solutions in the educational process, e.g. a game-like environment Second Life [6, 15, 17]. Others use their own solutions [1, 23, 24], that are also based on the gaming activities.

Separately it is possible to allocate the VLEs to train uniquely encyclopedic knowledge, such as anatomic atlases used to provide access to information in conjunction with a three-dimensional representation of the object of study. In this case, the object of study is a human body 3d model. Implementation architecture of anatomical atlases is rather diverse starting from web-based applications accessible through web browser [5, 16, 20] with the advantage that the developers can always monitor the relevance of knowledge and up to the applications that fully store and run directly on client's computer [9, 13, 14]. Key issue of using such types of environments is that all of them are mainly locked to datasets integrated by developers without abilities for users to amend these datasets or data itself.

Scientific Visualization Team of eScience Research Institute of ITMO University was organized in 2007, equipped with the state-of-the art facilities and during these years successfully fulfilled several ambitious projects in scientific visualization domain, e.g. Problem Solving Environment for Flood Protection Barrier of Saint Petersburg, original 3D engine with realistic sea waves, ship motion and sun system simulation, and a 3d anatomic atlas of normal human body [2, 28]. The latter project has become a prototype for the VLE implemented in the syllabi of several elective courses of the Master's Program in Computational Science "Supercomputer technologies in Interdisciplinary Research". Below we describe one of our projects of 2-days group work with the elements of competition.

Simultaneously students had a chance to test the key functions of *technical* (optimization of graphic engine to create complex models, introduction of Virtual Reality Tools such as dual-head stereo and 3D Vision, support of human-computer interaction—Microsoft Kinect and multi-touch support), and *functional* component (efficient structure of graphical user interface, content management system) of VLE developed by their lecturers.

2 Virtual Environment Infrastructure

Virtual Environment based on client-server architecture. Client side of the environment consists of two parts—Content Management System (CMS) and User Shell (US). A server part of the VE is an ordinary Apache or IIS web-server with PHP-interpreter (v.5.2 or higher) deployed on a remote PC. In addition, the system requires a database server with MySQL 5.1.7 or higher. CMS and the data storage are implemented using PHP language and MySQL.

CMS is a simple web application and is accessible through the web browser inter-face. Left side of the page has several drop-down sections with the lists of options to select. Right side contains editable fields, which purpose and content depend on selected option in the left part of the page. CMS has "Edit content", "Statistics and reports", "Users" and "System setup" expandable groups of options. "Edit content" section contains a list of the elements of a hierarchical tree of 3D models loaded into the VE for editing the content. "Statistics and reports" section contains statistical information about the use of the environment. "Users" section contains options to manage the user's rights. "System setup" contains global CMS and VLE parameters.

Differentiation system allows to us restrict access to the different CMS sections depending on the user types:

- "Students" have access only to US application (view-only rights).
- "Advanced students" in addition have access to the "Edit content" section of CMS.
- "Lecturers" have additional access to "Users" section, could create new users and define their type, except "System Administrator" type.
- "System administrator" has complete access to US and CMS.

We came up to a conclusion that the suggested above division of the groups of rights provides the most efficient use of our VE in its implementation cycle in compliance with educational purposes.

US is a main visualization application with graphical user interface. It is based on self-developed Fusion Engine platform, which is mainly implemented using C#. It runs under Windows 7 or higher and uses Direct3D 11 technology for visualization.

On start-screen it is required to login and then user gets to the model selection screen to select already loaded 3D model for exploration or to load his own 3D models. Upload of the model is an ordinary process and similar to all "Open File" actions on Windows platform. Currently VLE supports only widespread format for 3D models exchange—Autodesk FBX. Having selected of the desired model, user gets to the main interface of the application.

The interface is represented as a scrollable horizontal panel divided to the vertical sections. Main sections of the panel are Toolbar, Projection Window, Element Description and Media Gallery. Width of the two last sections depends on the amount of the uploaded through CMS accompanying materials. Toolbar contains icons of the main tools to work with environment and 3D model:

- Tools for navigation through hierarchy of the model and alphabetical list of the model's elements;
- Shading parameters—normal shading, x-ray shading and highlight color selection;
- Visibility parameters—show/hide selected element, show all elements, isolate selection;
- Supporting tools—extend/shrink projection window, open options and open manual.

Projection Window displays 3D model. It is possible to select desired elements of the model, orient camera around its focus point, place camera focus point to the center of selected element(s). Hovering the mouse over model's element allows to see its name in the bottom of the window. Element Description and Media Gallery serves to display all accompanying materials about selected element that stores on the server. Element Description shows a path to the element in model's hierarchical tree, element's name, text information and uploaded additional documents. Media Gallery is represented in a form of the tiles board, where thumbnails of uploaded images and videos are displayed. Each tile can be enlarged and reduced by clicking on it. After the increase of video file thumbnail, playback panel appears in the bottom of the screen.

Due to the Fusion Engine's support of the NVidia 3DVision and dual-head stereo technologies, it is possible to view all available content in 3d-stereo mode. In addition, engine's support of the modern human-computer interaction tools like touch-tables or Microsoft Kinect enable extra functionalities to the US.

With the help of presented infrastructure of VLE students can explore 3D models, and either learn hierarchical connection of its components in spatial representation, or study supplementary materials. Optionally they can edit it (both description of the objects and information about it). In fact, we have a tool for collaborative learning, tool for individual encyclopedic knowledge acquisition and knowledge control. More than that, our environment is not tied to a specific subject matter, which gives a wide range of opportunities to apply our learning environment in any field of knowledge, where the subject matter can be visualized in a spatial or temporal-spatial manner (animated objects, processes in or with the objects of study).

3 Implementation of VLE in the Learning Process

A global transition from post-industrial to knowledge economy raised new challenges for design and content of the courses in engineering education. So in 2012, ITMO University and University of Amsterdam (UvA) launched a double-degree Master's program in Computational Science. The program of 120 ECTS designed by the mutual effort of Russian and Dutch specialists lasts 2 years (60 ECTS in Russia and 60 ECTS in The Netherlands). The list of courses is based on three levels of desirable exit qualifications (competencies)—Soft Skills, Hard Skills and Professional Competencies. The goal to develop competencies, skills and abilities of students and not just to impose knowledge is a prerequisite for modernization and is supposed to bring the outcomes of educational process in line with international standards, in particular, Bologna Process.

At the level of design of the courses and following delivery of these courses, a lot of attention was given to their format and technical support to keep up with the standards of the UvA. Effective knowledge management and restructuring alongside with information share could help to stimulate the shift from traditional approach to teaching to student-centered and technology-enhanced learning to solve interdisciplinary tasks.

We conjecture that in our VLE researchers, teachers and students could work together aiming to acquire and create knowledge and enhance the ability to self-tuition in accordance with the principles of knowledge society. In general, inside this VLE students can plan, design, produce and apply new entities as a part of innovative practice forming research skills (Fig. 1).

Our test group project was carried out during 2 days (5 h during the first day, 5 h on the next day including group presentations) aiming mainly to run VLE in real

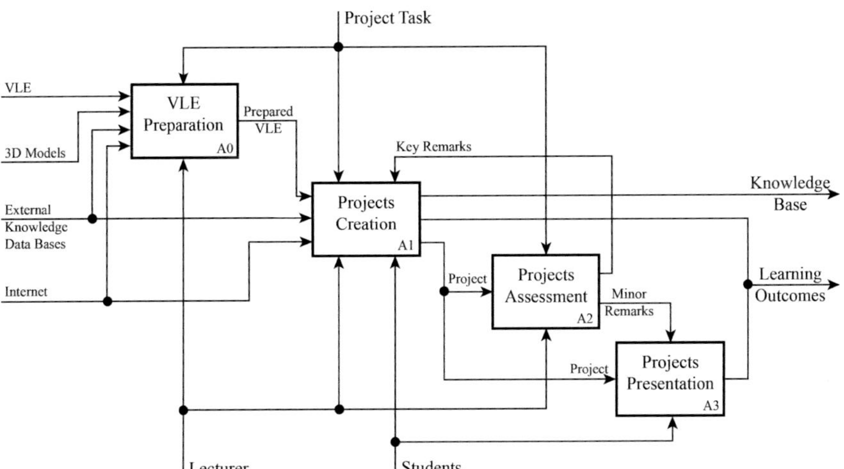

Fig. 1 Scheme of VLE implementation cycle

life and receive feedback on VLE operation or, to be more precise, how useful it would be to implement it in teaching courses (for example, Scientific Visualization and Virtual Reality, etc.) and group projects in the chosen specialization in the second semester of the master's program. We divided it into three principal tasks: to teach students not only to retrieve already available information (from lectures, www), but to generate knowledge, then define methods of research and assessment of learning outcomes and finally to identify the possible impact of the implementation of VLE.

Optionally, at the stage of opinion poll, we offered our students to reflect upon new fields of VLE implementation and influence of the proposed approach.

At the planning stage we implied that students will go from application of already received knowledge (discussion of the given task) to analysis (estimate resources and define the ways to solve the task) and synthesis (develop a working plan in a group) to evaluation (appraise, compare and assess their peers). To follow that logic we have chosen the following instructional techniques: group work (to improve collaborative and communicative skills), competition of the groups (to motivate students to win), presentations (students choose content to develop creativity), and self-assessment (to evaluate peers).

4 Implementation Method Description

While setting the task for the groups, we intentionally did not provide the detailed specification of the problem to solve. Only general criteria to assess the result of the group work were specified aiming at the development of creative thinking. 15 master students were divided into 5 groups of 3 people and 4 out of 5 groups had a female member. Each group was offered a system of a human body to explore (e.g. cerebrum, cardiovascular system, digestive system, muscular system, skeleton), find the information about it, define the most important from the point of view of the group parts of the system, to learn to use the functions of the VLE (such as different shading parameters, main constructive representation of the model).

For 2 days, students worked in a computer class equipped with all necessary software, databases provided by ITMO University library and internet access. The task was given in the first day and followed by 4 h of work, the next day 2 h of work were followed by final group presentations Each group had to prepare texts and pictures to illustrate their chosen hierarchy and to present their results during the session in computer class and in Scientific Visualization Center, which provides completed graphical view.

Participants were offered a questionnaire to assess the experience in general consisting of four parts: Program and Task Design, Competition of Group Presentations, Experience and Software feedback. The quality of presentations was estimated by both students themselves and four experts in computer simulation, physics, cloud computing and education. Total result of the group consisted of the average score of the parameters (80 %) and evaluation of the presentation skills of the group (20 %). All participants had to rate all five presentations by giving a rating number between 1 and 10 (1 = lowest, 10 = highest) and write down their opinion about such type of work, suggest possible implementation of given VLE, and moreover provide Software User's feedback (bug reports, functional improvements, global improvement ideas, etc.).

At the final presentation, each group had to convince the listener that their choice of hierarchy within the system is acceptable, which in turn requires proper reasoning (communicative skills) and carefully selected visual aids shown with the help of VLE (creativity). The presentations were estimated due to the following criteria: View, Shading, Adequacy of Display, General Visual Quality, Illustrations, Texts and Adequacy of Hierarchy of the given system.

5 Results and Conclusions

The final scores of the groups give a lot of food for thoughts in terms of development of Hard and Soft Skills of the students. As we can see from Table 1, two groups were very close to each other in the average number of parameter (9.08 and 8.98 out of 10), but the 20 % for presentational skills were crucial at the final stage of competition.

During the course of the group project, it is worth to mention that one of the challenges our master students face most often is the lack of collaborative skills since in this type of work do not resemble their day-to-day interaction. Members of a certain group with a strong female leader faced serious problems at the planning stage. However, a heated discussion of the plan of work, division of roles and time management issues, which even demanded the involvement of the experts, brought outstandingly good results. Element of competition also had a curious effect on those members of the group who did not successfully interact before the group work task. While working with VLE and learning to use it, members of one team built fruitful communication and performed well together. At the same time, some of the groups with sound technical background (e.g. high average marks) could not show the proper use of VLE functionality and their final product raised many questions and gave mixed impressions in general.

Figure 2 shows some visual the results of the students' group work. The first two (Fig. 2a and b) groups presented rather curious views of their systems, but the skeleton view (Fig. 2c) did not go further than plain demonstration of general view with a frontal projection and color division of general functional groups. Therefore, the new trend in Russian graduate education to put emphasis on Soft Skills (such as common culture, communicative skills, self-criticism and creativity) is not groundless. In the final analysis, we might conclude that VLE could be used as a tool to form creative and collaborative skills and students respond positively to the presented format. Furthermore, we assume that VLE could be used at best in combination with other forms of study, e.g. individual and group projects, division of roles by the teacher or by students themselves inside the groups.

Table 1 Results of the groups

	View	Shading	Constructivism	Visual quality	Illustrations	Texts	Chosen elements	Overall score
Cerebrum	9.29	9.07	**9.29**	**9.29**	8.93	8.36	8.64	8.98
Cardiovascular	**9.36**	**9.14**	8.79	9.21	**9.00**	8.50	**9.57**	**9.08**
Digestive	7.14	9.00	9.00	9.21	8.71	**9.21**	9.14	8.77
Muscular	7.00	8.14	8.43	8.43	7.43	8.79	8.29	8.07
Skeleton	7.50	7.43	7.86	7.57	8.57	9.00	8.93	8.12

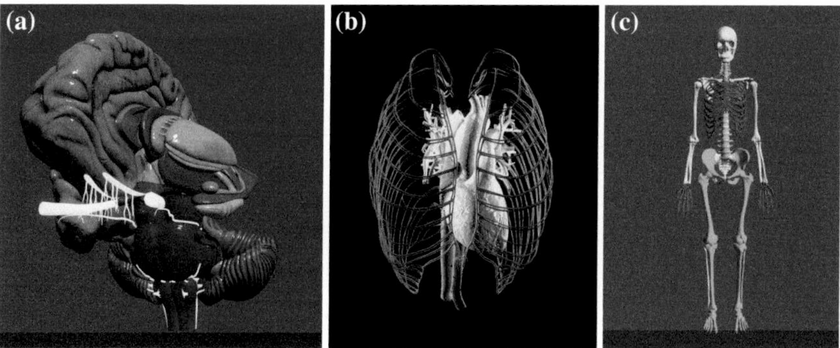

Fig. 2 Samples of group work's results in VE

One of the advantages of implemented VLE cycle, described in this paper is that it is not restricted to a subject field and could serve as a basis of multidisciplinary learning and research. In addition, it is could be deployed with different cycle duration (from a separate 'module style' group project lasting several day up to a part of a full course). Moreover, educators who took part in this group work project cannot but notice that students of our master's program with STEM background are more likely to adopt the new creative ways of tackling a problem when they are puts in interaction with both their peers and VLE.

6 Discussions and Future Work

Our experience of the introduction of VLE in the master research project reaffirms that the effectiveness of VLE use depends on the combination of objective (task design and description, implementation of various teaching and learning strategies, software precision) and subjective conditions (participants' feedback in different fields of knowledge, ability to extract and analyze information, organize and plan group and individual work). 35.7 % out of the students who participated in the test group project indicated in the questionnaire that they have not gained so much from group work, VLE and presentations of the groups. At the same time when offered to describe the experience by writing down a couple of words, we came up with rather positive feedback. Moreover, most of the students took seriously the bug report and gave detailed description of the errors they had encountered, but the students enjoyed the experience in general.

Implemented VLE cycle illustrates both acquisition of theoretical encyclopedic knowledge and group work to develop creativity. VLE could facilitate team-based learning and intensify formation of soft and professional skills. Possible application scenarios and development of existing VLE and/or development of a separate model of VLE corresponding to desired learning outcomes of the course could still be a subject of future research, since all VLE functionality is not fully discovered yet, and the only way to learn more is trying.

Due to a brief description of the technical implementation of VLE at the presentation and discussions at [3], some of the participants compared it with the packages for creating three-dimensional graphics. The two main differences of our development:

- VLE created on the basis of our own open source graphics engine Fusion;
- VLE is used to browse, explore and describe the 3d models, rather than to create them.

A variety of VLE implementation schemes in the educational process emerges based on the above-mentioned differences. First, the features of the VLE and Fusion engine can be modified and improved by the students while studying programming and computer graphics. Secondly, the VLE in its present form can already be used as demonstration tool in the domain of creation and preparation of 3d content for virtual reality systems. In the latter case, many questions arise concerning the differences from existing tools of the modern graphics engines. Nevertheless, we should not forget that the main purpose of these tools is the creation of final software, not for educational purposes. We were aiming to create a flexible tool for teacher to work in the same environment with their students that could be applied to multiple study areas. Currently, the presented VLE is being introduced as a part of elective courses in Visualization and Virtual Reality functioning as a tool for creative group projects, e.g. big 3d scene. These efforts are to be described in future works focusing on VLE solely as an element of multidisciplinary educational environments.

Acknowledgments This paper is supported by Russian Scientific Foundation, grant #14-11-00823.

References

1. Berns, A., Gonzalez-Pardo, A., Camacho, D.: Game-like language learning in 3-D virtual environments. Comput. Educ. **60**, 210–220 (2013)
2. Bezgodov, A., Esin, D.: Complex network modeling for maritime search and rescue operations. Procedia Comput. Sci. **29**, 2325–2335 (2014)
3. Bilyatdinova, A., Karsakov, A., Bezgodov, A., Dukhanov, A.: Virtual environment for creative and collaborative learning. KICSS'2014 Proceedings, pp. 313–320 (2014)
4. Bosch-Sijtsema, P.M., Haapamäki, J.: Perceived enablers of 3D virtual environments for virtual team learning and innovation. Comput. Human Behav. **37**, 395–401 (2014)
5. Brenton, H., Hernandez, J., Bello, F.: Using multimedia and Web3D to enhance anatomy teaching. Comput. Educ. **49**, 32–53 (2007)
6. Chau, M., Wong, A., Wang, M., Lai, S.: Using 3D virtual environments to facilitate students in constructivist learning. Decis. Support Syst. **56**, 115–121 (2013)
7. Dalgarno, B.: The potential of 3D virtual learning environments: a constructivist analysis. Electron. J. Instr. Sci. Technol. 1–19 (2002)

8. Dalgarno, B., Lee, M.: Exploring the relationship between afforded learning tasks and learning benefits in 3D virtual learning environments. In: Australian Society for Computers in Learning in Tertiary Education Annual Conference (2012)
9. DeLaurier, A., Burton, N.: The Mouse Limb Anatomy Atlas: an interactive 3D tool for studying embryonic limb patterning. BMC Dev. Biol. **8**, 83 (2008)
10. Dillenbourg, P., Schneider, D.K., Synteta, P.: Virtual learning environments. In: Proceedings of the 3rd Hellenic Conference "Information & Communication Technologies in Education", pp. 3–18 (2002)
11. Dukhanov, A.V., Krzhizhanovskaya, V.V., Bilyatdinova, A., Boukhanovsky, A.V., Sloot, P.M.: Double-degree Master's Program in computational science: experiences of ITMO University and University of Amsterdam. Procedia Comput. Sci. **29**, 1433–1445 (2014)
12. Grenfell, J.: Immersive interfaces for art education teaching and learning in virtual and real world learning environments. Procedia Soc. Behav. Sci. **93**, 1198–1211 (2013)
13. Hacker, S., Handels, H.: A framework for representation and visualization of 3D shape variability of organs in an interactive anatomical atlas. Methods Inf. Med. **48**, 272–281 (2009)
14. Hamrol, A., Górski, F., Grajewski, D., Zawadzki, P.: Virtual 3D Atlas of a human body-development of an educational medical software application. Procedia Comput. Sci. **25**, 302–314 (2013)
15. Hsu, L.: Web 3D simulation-based application in tourism education: a case study with Second Life. J. Hosp. Leis. Sport Tour. Educ. **11**, 113–124 (2012)
16. John, N.: The impact of Web3D technologies on medical education and training. Comput. Educ. **49**, 19–31 (2007)
17. Land, S. Van Der, Schouten, A.: Lost in space? Cognitive fit and cognitive load in 3D virtual environments. Comput. Hum. Behav. **29**, 1054–1064 (2013)
18. Lorenzo, G., Pomares, J., Lledó, A.: Inclusion of immersive virtual learning environments and visual control systems to support the learning of students with Asperger syndrome. Comput. Educ. **62**, 88–101 (2013)
19. Loureiro, A., Bettencourt, T.: The use of virtual environments as an extended classroom—a case study with adult learners in tertiary education. Procedia Technol. **13**, 97–106 (2014)
20. Lu, J., Pan, Z., Lin, H., Zhang, M., Shi, J.: Virtual learning environment for medical education based on VRML and VTK. Comput. Graph. **29**, 283–288 (2005)
21. Mikropoulos, T., Bellou, J.: The unique features of educational virtual environments. Proc. e-society 122–128 (2006)
22. Mikropoulos, T., Natsis, A.: Educational virtual environments: a ten-year review of empirical research (1999–2009). Comput. Educ. **56**, 769–780 (2011)
23. Parsons, S., Leonard, A., Mitchell, P.: Virtual environments for social skills training: comments from two adolescents with autistic spectrum disorder. Comput. Educ. **47**, 186–206 (2006)
24. Schmidt, M., Laffey, J., Schmidt, C.: Developing methods for understanding social behavior in a 3D virtual learning environment. Comput. Human Behav. **28**, 405–413 (2012)
25. Semradova, I., Hubackova, S.: Learning strategies and the possibilities of virtual learning environment. Procedia Soc. Behav. Sci. **83**, 313–317 (2013)
26. Semradova, I., Hubackova, S.: Virtual learning environment and the development of communicative competences. Procedia Soc. Behav. Sci. **89**, 450–453 (2013)
27. Simkova, M., Stepanek, J.: Effective use of virtual learning environment and LMS. Procedia Soc. Behav. Sci. **83**, 497–500 (2013)
28. Zagarskikh, A., Karsakov, A., Tchurov, T.: The framework for problem solving environments in urban science. Procedia Comput. Sci. **29**, 2483–2495 (2014)

A Semiotic and Cognitive Approach to the Semantic Indexation of Digital Learning Resources

Françoise Greffier and Federico Tajariol

Abstract This paper presents the design of a semantic system to index digital learning resources according to the metadata describing their cognitive features. By "cognitive feature" we mean the cognitive activities (e.g., reading, listening, body interactions, etc.) associated with the form of presentation (e.g., text, audio, image, etc.). The semantic system includes a parser, which detects the semiotic components of a resource, and two ontologies that formally describe the cognitive features and the semiotic descriptors, as well their association.

Keywords Differentiated instruction • Cognitive style • Semiotic component • Semantic indexation • Digital learning resources

1 Introduction

Information and Communication Technologies for Education enable teachers to carry out different types of Digital Learning Resources (DLR). A DLR is a file available on websites and on repositories of learning objects[1]: a text explaining a concept, a practical work to train students, an interactive simulation, an exercise to evaluate their level of knowledge, etc. Each DLR is the creative outcome of a teacher, who acts as a pedagogical designer. In fact, even if a teacher does not create learning resources *ex novo*, (s)he can reuse and adapt pre-existing DLR in order to fit learners' pedagogical goals, cognitive abilities and learning preferences.

[1]For instance, AMSER https://amser.org/.

F. Greffier · F. Tajariol (✉)
Objets et Usages Numériques Research Unit, ELLIADD Laboratory, FR-EDUC, University of Franche-Comté, Montbéliard, France
e-mail: federico.tajariol@univ-fcomte.fr

F. Greffier
e-mail: francoise.greffier@univ-fcomte.fr

S. Kunifuji et al. (eds.), *Knowledge, Information and Creativity Support Systems*, Advances in Intelligent Systems and Computing 416,
DOI 10.1007/978-3-319-27478-2_27

This activity of re-editing can be very important because the teacher-designer must read, select and detect specific chunks of existing digital resources: a text, a picture, a sound, etc. We call a chunk a "semiotic component".

To accomplish the task of creating personalized learning resources, a teacher lacks information about the *cognitive features* of each chunk that a DLR contains. By "cognitive feature" we mean the association between cognitive abilities specific to accomplish learning tasks (e.g., reading, listening, etc.) and the forms of presentation of a DLR (e.g., audio, image, text, etc.). So, for example, a learner who is asked to read a text uses different cognitive abilities and skills than when (s)he observes a diagram illustrating the same content, and her(his) understanding of the content could be not the same. To tackle this problem, we propose an automatic semantic system to index the semiotic components of a DLR according to their cognitive features.

The indexation needs of descriptors, which are metadata describing the semiotic components, their spatial relations (e.g. a chunk of text placed at the bottom of a picture) and the required cognitive abilities to process these chunks. The main goal of our work is to build these descriptors and to design a set of tools to help teachers to create differentiated DLR fitting student's learning abilities.

This paper is built as follows. In the next section we explain the scientific problem, we suggest our approach to index the cognitive features of a DLR and we define our set of descriptors concerning the cognitive features of a DLR. In the third section, we present the architecture of our system we plan to develop.

2 Design Differentiated Digital Learning Resources: A Creativity Problem

Personalized learning means to adapt the pedagogical contents and activities to the learner's cognitive styles [4, 14], as well as to design differentiated forms of a same pedagogical content [24]. Thus, when a teacher is in charge of the design of a DLR, (s)he also has to reason on which structure and form of presentation could be in adequacy with (her)his learners' specific cognitive styles [7]. For example, a philosophy teacher must prepare a lesson on modal logic intended for students in visual arts: (her)his first intention could only be to present logic symbols and formalisms, whereas those students would be more effective when they processed images and narrative forms, as if they owned a particular way of information processing. This research tackles the issue of design differentiated DLR: how could we help teachers to create differentiates DLR that would enable students to be more effective in their learning process? Amongst works on cognitive and learning styles [19, 20], and the complex links between these two constructs [26], we suggest a different theoretical approach, the Multiple Intelligence Theory [7].

According to the MI theory, the intelligence of a person is a set of seven sets of abilities and skills (s)he develops to solve problems and process information. These

Table 1 Grid analysis linking Gardner's theory to the semiotic components of a DLR

Semiotics components	Gardner's multiple intelligences
Sentences describing and explaining a concept	Linguistic
Keywords, short sentences, diagrams	Logical-mathematical
Sounds, music	Musical
Fixed images, animated images, videos	Visuo-spatial
Interactive animations in which the learner must perform some gestures (e.g. click and drop)	Kinaesthetic
Linguistic expressions showing the personal feeling of the author of the learning resource	Interpersonal
Linguistic expressions engaging the learner to think	Intrapersonal

sets would be relatively autonomous to one another: intrapersonal, interpersonal, kinaesthetic, linguistic, logical-mathematical, musical, visuo-spatial. For example, a person who has a visuo-spatial intelligence is able to better think through visual elements (e.g., images, graphs, cards, colours, etc.) and structures (e.g., patterns, diagrams, etc.) rather than speeches (linguistic intelligence), symbols (logical-mathematical), etc. Two main reasons trigger our theoretical choice. First, previous works have successfully applied the MI theory to the design of Technology Enhanced Learning environments [8, 23]. Secondly, as we showed in our previous works [9], the modularity of the MI theory let us formalize the association between a form of presentation (textual, audio, etc.), its semiotic components (words, sentence, images, etc.) and the cognitive activities required (reading, manipulating, etc.). In fact, we built a grid of analysis linking the semiotic components of a DLR (sentences, keywords, etc.) to the multiple intelligences (Table 1). For instance, a DLR that contains keywords, short sentences and diagrams, generally involves the logical-mathematical intelligence, whereas only images (fixed or animated) and videos would just require the visuo-spatial intelligence.

3 A Semantic System to Index the Digital Learning Resources

Our Semantic System to Index Digital learning resources (SSID) enables the automatic indexation of DLR. It analyses the semiotic components of a digital resource and produces the metadata describing the cognitive features of a DLR. The SSID consists of three components: a morphological analyser and two ontologies.

(i) *Morphological analyser* A parser detects the semiotic components of a digital resource, it calculates their number and their presence (page number). We developed the parser under Java (1.7.0_11), we used the Apache POI library

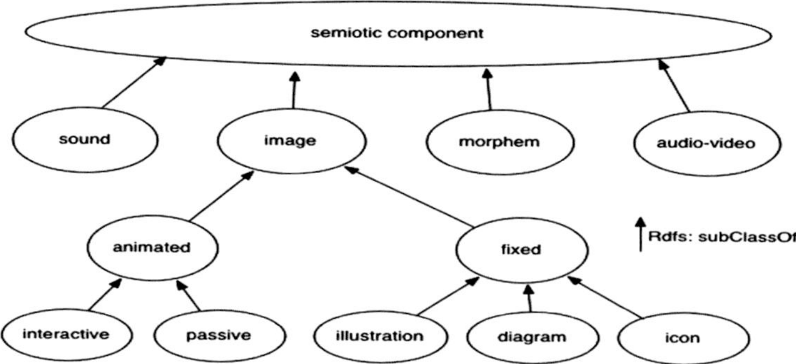

Fig. 1 SD: Domain ontology of the semiotic components in a digital resource

(http://poi.apache.org/) to retrieve the semiotic components in each file (e.g. .*ppt*, .*pptx*, .*docx*) as well the itext library (http://itextpdf.com/) for the .*pdf* files. For each slide of a .*ppt* file, the Apache POI creates a set of shapes. Then, by means of a set of algorithms we developed, the parser distinguishes between animated and fixed images. All output is recorded in a database and in an external spreadsheet file.

(ii) *Domain Ontologies* Considering the connections between semantic web and ontologies [2, 22], we created two domain ontologies [10]. In the *Semiotic Descriptors* ontology (Fig. 1) (SD) we formally specified the semiotic components of a DLR and their relations. The SD ontology formally describes that, for instance, an icon is a fixed image, whereas a .swf file is a passive image.

In the *Cognitive Style* ontology (Fig. 2) (CS), we formally described seven ways of presenting the learning contents of a DLR, according to the Multiple Intelligence Theory [7].

(iii) *Application Ontology* We designed two application ontologies to build both ontological classes and the relations between them. First, we built the *Linking Ontology* (LO) to associate formally the semiotic components and the seven intelligences of the Multiple Intelligence Theory (§ Table 1). For instance, when the parser detects and counts the semiotic descriptors "morphemes", the

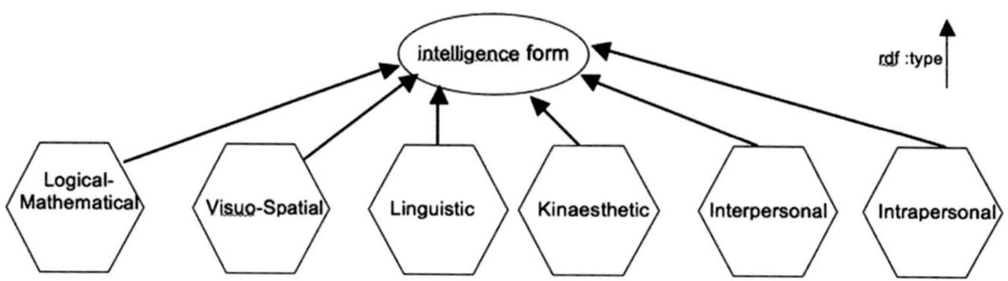

Fig. 2 CS: ontology of multiple intelligence forms

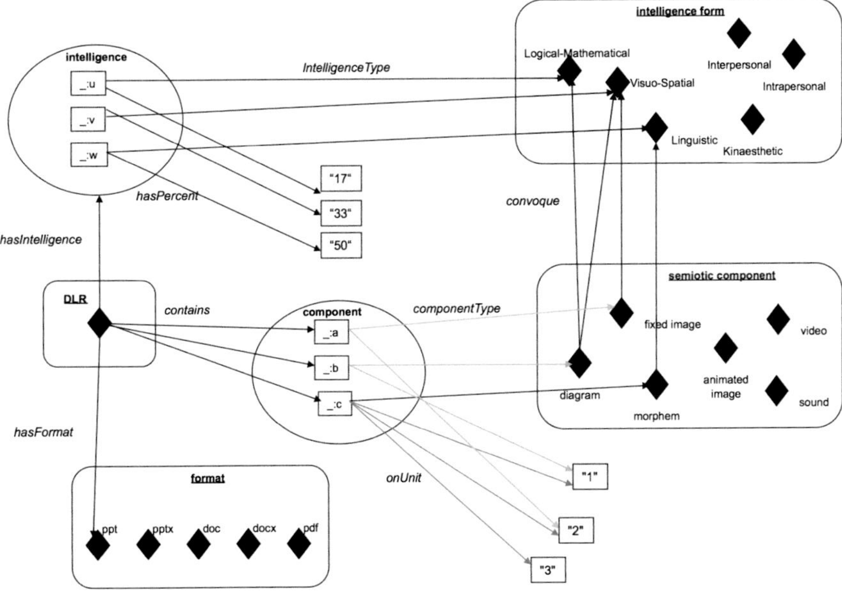

Fig. 3 Counting ontology

> Linking Ontology enables to associate the morpheme to the Linguistic form of intelligence. Second, we created the *Counting Ontology* (CO) to compute each form of intelligence required to process the semiotic components of digital resource (Fig. 3).

To explain how the CO works with other components of the SSID, we show and discuss an example. The parser analyses a DLR in the *.ppt* format (link: *hasFormat*) that consists of three slides (link: *onUnit*). The parser counts three slides (Table 2) and for each slide detects the number of semiotic components (link: *componentType* in Fig. 3).

At the next step, the SSID just detects the presence of each semiotic component, without comparing, for instance, the number of morphemes to the number of images. By means of the *Linking Ontology*, the SSID identifies the cognitive features required to process each semiotic component. Following our example, the SSID detects three different forms of intelligence in the digital resource (Table 3).

Following the example, there are six occurrences of three form of intelligence required to process the information of this specific digital resource: logical-

Table 2 Outcomes of the parser

	Diagram	Morpheme	Fixed image
Slide 1	1	10	
Slide 2		31	1
Slide 3		15	

Table 3 Forms of intelligence detected in each slide

	Logical-mathematical	Linguistic	Visuo-spatial
Slide 1	X	X	X
Slide 2		X	X
Slide 3		X	

mathematical (one occurrence), linguistic (three) and visuo-spatial (two). The SSID computes the ratio between the occurrences of a form of intelligence and the total amount of the forms of intelligence required by the digital resource. We obtain LM 17, L 50 and VS 33 % (link: *hasPercent* in Fig. 3).

To obtain this outcome, it is necessary to link both the results of the parser to the counting ontology and to express structured queries by means of Sparql.

4 Discussion

The function of a parser is necessary. In fact, in spite of some ontology-based image system [11], we observed that DLRs are seldom indexed in general, because pedagogical designers have to fill several and high complex descriptors, such as Learning Object Metadata (LOM). The parser uses some API (Java) such as itext or doc.office, which includes some *getters* (ex: getTitle, getSlides). The *file header* of each file gives us also useful information to distinguish a fixed from an animated image. First, an ontology enables the description of complex interrelationships between the concepts [1, 5] and remains independent by the processing of the inference engine [9]. So, for instance, the Cognitive Style ontology could be reusable in other software, such as an application monitoring when the user (teacher or student) tries to select the DLRs in adequacy with (his) her cognitive style [3, 21]. Our purpose is to establish the inter-ontology relations between the SSID metadata and other existing ontologies and standards [27], as the Dublin Core [6, 25], the Learning Object Metadata (LOM) [12], the Standard Content Sharable Object Model Reference (SCORM). So, we aim to establish the links between our CS ontology and the Dublin Core metadata, which index digital resources from the title, the author, the date of edition, etc.

Concerning the LOM [17] and MLR (Metadata for Learning Resources) [15, 16], we are going to study how integrating the metadata of the SSIID in LOM number of category 5 and 8 [18]. Category 5, entitled "Educational", includes the pedagogical characteristics of a resource (type of resource, role of the user, context of use), whereas category 8, called "Annotation", lets user add annotations on the pedagogical use of the DLR. About SCORM, in the fourth release we estimate to ingrate the metadata of SSID within the section "system of content description".

The design of the SSID is also an ergonomic challenge. How to show to the teacher-designer the details about the DLR (s)he choose? Which form of visualizing details is the most efficient for teachers? And what is the best time to show these details? During the teacher's creative process or just at the end? Or both? In order to answer to these research questions, at this stage of our work we are working very closely with a group of teacher-designer to define the user interface of the SSID following a user-centred design process [13]. Moreover, we are designing specific software to browse a DLR, to look through its set of semiotic components without opening it and to help teachers to select quickly the semiotic components according their cognitive features.

5 Conclusions

In this paper, we presented a system called SSID that can index DLRs according to their cognitive features. The SSID detects the cognitive abilities required to learn the content of a DLR. Moreover, it classifies the semiotic components of a DLR according to an established set of cognitive skills, from the Multiple Intelligences Theory, by means of a parser and two ontologies. The SSID assists teacher in creating differentiated learning resources in order to fit learners' specific cognitive abilities. The SSID enables teacher to select specific components of a DLR, to combine them without losing sight of the learners' cognitive skills in processing information.

References

1. Bennacer, N., Bourda, Y., Doan, B.L.: Formalizing for querying learning objects using OWL. In: Proceedings of International Conference on Advanced Learning Technologies, pp. 321–325 (2004)
2. Berners-Lee, T., Hendler, J., Lassila, O.: The semantic web, scientific american. http://www.sciam.com/2001/0501issue/0501berners-lee.html (2001)
3. Boyle, T.: Layered learning design: towards an integration of learning design and learning object perspectives. Comp. Educ. **54**(3), 661–668 (2010)
4. Brusilovsky, P., Kobsa, A., Nejdl, W. (eds.): The adaptive web: methods and strategies of web personalization. Springer, Heidelberg (2007)
5. Doan, B.L., Bourda, Y., Bennacer, N.: Using OWL to describe pedagogical resources. In: Proceedings of International Conference on Advanced Learning Technologies, pp. 916–917 (2004)
6. Dublin core metadata initiative. http://www.dublincore.org/
7. Gardner, H.: Frames of mind: The theory of multiple. Basic Books, New York (2004)
8. Greffier, F., Tajariol, F.: A semantic system to support teachers to create differentiated digital learning resources. In: 9th International Conference on Knowledge, Information and Creativity Support Systems, Limassol (2014)
9. Gruber, T.R.: The role of common ontology in achieving sharable, reusable knowledge bases. In: 2nd Proceedings of the International Conference on Principles of Knowledge Representation and Reasoning, pp. 601–602. Morgan Kaufmann, CA (1991)

10. Guarino, N.: Formal ontology and information systems. In: Guarino, N. (ed.) 1st International Conference Formal Ontology in Information Systems, pp. 3–15. IOS PRESS, Amsterdam (1998)
11. Hyvönen, E., Styrman, A., Saarela, S.: Ontology-based image retrieval. In: XML Finland Conference, Towards the Semantic Web and Web Services, pp. 15–27 (2002)
12. IEEE: LOM. Draft standard for learning object metadata. http://ltsc.ieee.org/wg12/files/LOM_1484_12_1_v1_Final_Draft.pdf (2002).
13. Kaptelinin, V., Nardi, B.: Acting with Technology. MIT Press, Boston (2009)
14. Lefevre, M., Cordier, A., Jean-Daubias, S., Guin, N.: A Teacher-dedicated tool supporting personalization of activities. In: ED-MEDIA 2009-World Conference on Educational Multimedia, Hypermedia and Telecommunications, pp. 1136–1141 (2009)
15. LOM to MLR element mapping, ISO/IEC JTC1 SC36 WG4 N0188 (2007)
16. MLR—Metadata for learning resources, ISO/IEC JTC1 SC36 19788
17. Neven, F., Duval, E.: Reusable Learning Objects: a survey of LOM-based Repositories. In: 10th ACM Conference on Multimedia, pp. 291–294 (2002)
18. Nilsson, M., Palmer, M., Brase, J.: The LOM RDF binding—principles and implementation. In 3rd ARIADNE Conference, Belgium (2003)
19. Petty, G., Haltman, E.: Learning style and brain hemisphericity of technical institute students. J. Stud. Tech. Careers **13**, 79–91 (1991)
20. Riding, R.: On the nature of cognitive style. Educ. Psychol. **17**(1 and 2), 29–49 (1997)
21. Sicilia, M.A., Lytras, M., Sánchez-Alonso, S., García-Barriocanal, E.: Modeling instructional-design theories with ontologies: Using methods to check, generate and search learning designs. Comp. Human Beh. **27**(4), 1389–1398 (2011)
22. Sowa, J.F.: Knowledge Representation. Logical, Philosophical and Computational Foundations. Brooks/Cole, CA (2000)
23. Tracey, M.W.: Design and development research: a model validation case. Ed. Tech. Res. Dev **57**, 553–571 (2009). doi:10.1007/s11423-007-9075-0
24. Türker, M.A., Görgün, İ., Conlan, O.: The challenge of content creation to facilitate personalized elearning experiences. Int. J. E-Lear. **5**(1), 11–17 (2006)
25. Weibel, S.: The Dublin core: a simple content description model for electronic resources. Bull. Am. Soc. Infor. Sci. Tech. **24**(1), 9–11 (1997)
26. Zhang, L., Sternberg, R.J.: A threefold model of intellectual styles. Ed. Psy. Rev. **17**(1) (2005). doi: 10.1007/s10648-005-1635-4
27. Zheng, L., Yintao, L., Wang, J., Yang, F.: Multiple standards compatible learning resource management. In: Proceedings of International Conference on Advanced Learning Technologies, pp. 657–661 (2008)

The Significance of 'Ba' for the Successful Formation of Autonomous Personal Knowledge Management Systems

Ulrich Schmitt

Abstract Just as computer science underwent a revolution in the 1980s with the wide-spread use of personal computers, it is possible that Knowledge Management (KM) will in the twenty-first century experience a decentralizing revolution that gives more power and autonomy to individuals and self-organized groups." Levy's scenario—nearly seven decades after Vannevar Bush's still unfulfilled vision of the Memex—stresses the dire need to provide overdue support tools for knowledge workers in the rising Creative Class and Knowledge Societies. With a prototype system addressing the issues about to be converted into a viable Personal Knowledge Management System (PKMS), the author consolidates recent activities and publications and visualizes and presents a comprehensive Meta-PKMS-Concept covering the main environmental, design, development, deployment, and capacity building concerns. The paper also portrays its substantial interrelationship with Nonaka's concept of 'Ba' and its impact on the novel PKMS approach aiming to aid team-work, life-long-learning, resourcefulness, and creativity of knowledge workers throughout their academic and professional life and as contributors and beneficiaries of organizational performance.

Keywords Personal knowledge management systems · Knowledge worker · Memes · Memex · SECI model · Ba · Meta-concept · Knowcations

1 Pioneering Concepts for Autonomous Personal KM Systems

A recent article [1] contributes to a theme where the original purposes of established Knowledge Management (KM) Models and Theories are re-examined to determine their value for the next KM generation. After qualifying the Personal KM

U. Schmitt (✉)
University of Stellenbosch Business School, P O Box 610, Bellville 7535, South Africa
e-mail: schmitt@knowcations.org

S. Kunifuji et al. (eds.), *Knowledge, Information and Creativity Support Systems*, Advances in Intelligent Systems and Computing 416,
DOI 10.1007/978-3-319-27478-2_28

(PKM) concept and prototype to be introduced as a 'Next KM Generation' system, their functionalities are reverse engineered in order to crystallize over forty KM Models and Theories which constitute the ingredients for the ongoing integrating, adapting, broadening, deepening, repurposing, or innovating activities of the PKM System (PKMS) design and which—in terms of the PKMS—have passed this test of time.

To provide a comparative visual overview of the models most instrumental to the PKMS concept, Fig. 1 adopts an expanded 4-level version of Boisot's Information-Space and Social Learning Cycles [2] and maps Kolb's Learning Model [3], Wierzbicki's and Nakamori's Creative Space and I^5-System [4, 5] with its incorporation of Nonaka's SECI Spiral [6], and Pirolli's and Card's Foraging and Sensemaking Loop for Intelligence Analysis [7]. While Kolb concentrates on the individual, Boisot considers the multiple dimensions of abstraction, codification, and diffusion, which Wierzbicki and Nakamori widen by adding the accumulated human record to the individual and group perspective and by further differentiating the tacit knowledge into intuition and emotions. To integrate the different knowledge types, guidance is provided by the I^5-System as well as the 18-step Sensemaking Process depicted. Over the past two years, a series of papers and presentations [1, 8–27] have accompanied the progressive transformation of these models into the PKM's system design, again transparently mapped within the Information-Space [13, 20, 23, 25].

However, the comprehensive novel approach also needs to be forward engineered. The overarching meta-concept adopted (Fig. 3) forms the focus of this paper. Its role is to integrate the established and re-purposed with the new and innovative KM constructs for meeting the multiple needs encountered and for planning ahead for the productive system deployment. The result exemplifies the successful fusion of Bush's 70 year old vision of the Memex (1945), Simon's 45 year old call for Attention Management (1971), Nonaka's (2000) 15 year old notion of Personal Autonomy as well as the scenarios recently put forward by Levy (2011).

1.1 *Vannevar Bush's Vision of the 'Memex' (1945)*

In 1945, convinced that "the world has arrived at an age of cheap complex devices of great reliability", President Truman's Director of Scientific Research imagined the 'Memex'. It resembles a hypothetical sort of mechanized private file/desk/library-device, acting as an enlarged intimate supplement to one's memory, and enables an individual to store and recall, study and share the "inherited knowledge of the ages".

It facilitates the addition of personal records, communications, annotations, and contributions and the building of non-fading trails of one's individual interest

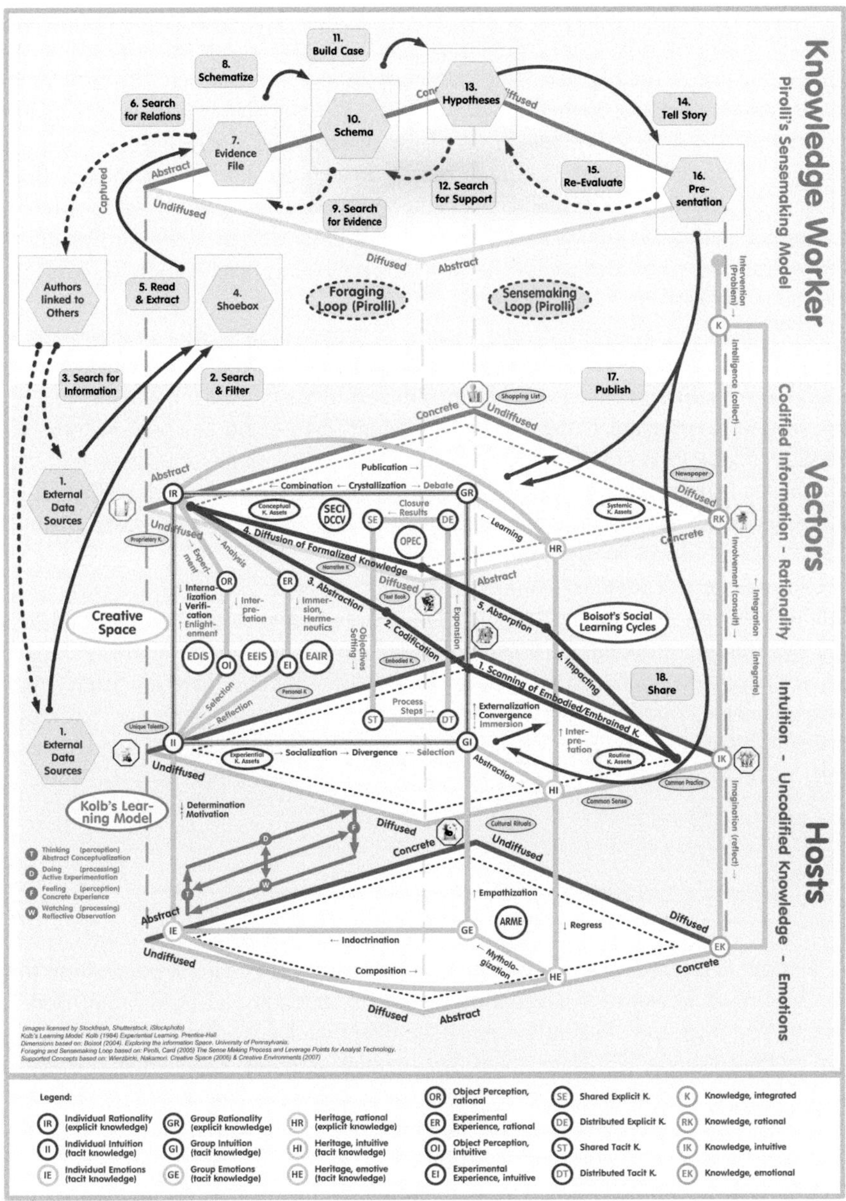

Fig. 1 Comparative overview of KM models instrumental to the PKMS concept

through the maze of materials available; all of which are easily accessible and sharable with the Memexes of acquaintances. Bush based his vision on the observations of a steadily "growing mountain of research" and an "increased evidence that we are being bogged down" as specialization extends further in the name of progress. He regarded our methods of transmitting and reviewing the results of research "to be generations old" and "totally inadequate for their purpose". Bush foresaw the appearance of new forms of encyclopedias with an extensive mesh of associative multi-disciplinary trails already built-in as well as a new profession of trail blazers who find delight in the task of establishing useful trails through the enormous mass of the common record.

A 'Memex' would provide the means to make intellectual excursions more enjoyable and an individual would "reacquire the privilege of forgetting the manifold things he does not need to have immediately at hand, with some assurance that he can find them again if they prove important" and could enjoy more time for learning or knowledge development. As an added benefit of the trails captured, "the inheritance from the master becomes, not only his additions to the world's record, but includes for his disciples the entire scaffolding by which they were erected" [28].

Unfortunately, a Memex-like device has not yet emerged; in 'Still building the Memex', Davies concludes: "Personal knowledge management [PKM] is [still] a real and pressing problem. [...] Yet it does not appear that Vannevar Bush's dream has yet been fully realized on a wide scale" [29]. Although there are many PKM tools available, "they are not integrated with each other", and the "currently available PKM systems can provide only a partial support to knowledge workers" [30].

1.2 Simon's Notion of 'Attention Management' (1971)

In 'Designing Organizations in an Information-rich World', Simon pointed out that the "wealth of information creates a poverty of attention and [with it] a need to allocate that attention efficiently among the overabundance of information sources that might consume it". Thus, "it is not enough to know how much it costs to produce and transmit information; we must also know how much it costs, in terms of scarce attention, to receive it. [...] In a knowledge-rich world, progress does not lie in the direction of reading information faster, writing it faster, and storing more of it. Progress lies in the direction of extracting and exploiting the patterns of the world—its redundancy—so that far less information needs to be read, written, or stored" [31].

Since then, accelerating technological progress and economic pressures are bringing about organizational, commercial, social and legal innovations with profound impacts on the way we work and live, including the change from information scarcity (few sources/channels, high associated costs) to a never before experienced ever-increasing attention-consuming information abundance. Accordingly, Simon's notion is more relevant than ever. As Kahle observed: "While today we have many powerful applications for locating vast amounts of digital information, we lack effective tools for selecting, structuring, personalizing, and making sense of the digital resources available to us" [32].

1.3 Nonaka's Theory of Organizational Dynamic Knowledge Creation (2000)

Nonaka's and Takeuchi's theory of organizational dynamic knowledge creation represents one of the most widely cited theories in knowledge management. Its SECI Model (socialization, externalization, combination, and internalization) addresses a major objective of Organizational KM (OKM) by aiming to make the tacit knowledge (gained only experientially and difficult to articulate, explain, share —as opposed to formal or explicit knowledge) of knowledge workers explicit, so it can be measured, captured, stored, protected, shared and further utilized in a 'spiral' of knowledge creation for the benefit of the organization and its stakeholders and independent of the availability of the persons concerned [6].

In the further development of his model, Nonaka defines Knowledge Creation as "a continuous, self-transcending process through which one transcends the boundary of the old self into a new self by acquiring a new context, a new view of the world, and new knowledge" and—in order to guide this journey—introduces the concept of 'ba' (Fig. 2) as a shared context or place (physical or virtual) in which knowledge is shared, created, interpreted and utilized [33].

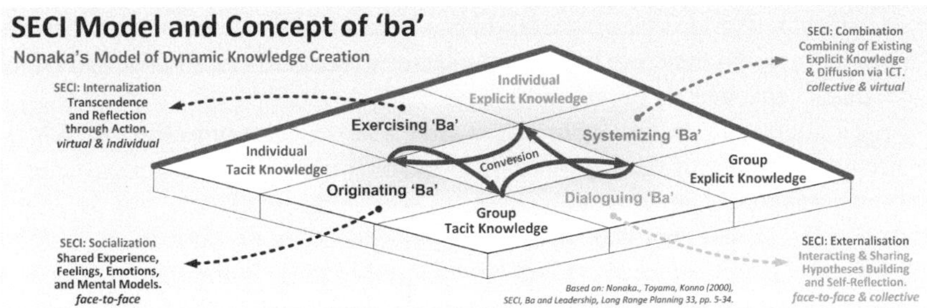

Fig. 2 Nonaka's SECI model and the space concept of 'ba' [33]

To provide adequate organizational leadership for building, connecting, and energizing such dynamic knowledge-creating environments and their expansion (plurality of 'ba'), Nonaka furnishes management with recommendations [33] which share a common ground with the PKMS concepts and systems to be introduced, such as to:

- Foster and utilize one's personal knowledge-related proficiencies to keep track of one's individual knowledge assets, and utilize them when they are needed.
- Continuously 'map' personal knowledge assets by taking account of any new stock dynamically created from one's existing knowledge assets.
- Intentionally stimulate 'ba' by employing adequate physical or virtual space, or mental space such as distinctive goals or a knowledge vision.
- Choose the right mix of collaborators and promote their interaction.
- Preserve the conditions for an enabling environment, such as autonomy,[1] creative chaos, redundancy, requisite variety, and love, care, trust and commitment.

1.4 Levy's Autonomous PKM Capacities and Creative Conversations (2011)

In 2011, Levy envisaged: "just as computer science underwent a revolution in the 1980s with the widespread use of personal computers, it is possible that Knowledge Management (KM) will in the twenty-first century experience a decentralizing revolution that gives more power and autonomy to individuals and self-organized groups".

His scenario is based on a future of decentralized autonomous PKM capacities, networked in continuous feedback loops. Nourished by the creative conversation of many individuals' PKM devices, the systems are expected to assume an elementary role that enables "the emergence of the distributed processes of collective intelligence, which in turn feed them". Accordingly, "one of the most important functions of teaching […] will therefore be to encourage in students the sustainable growth of autonomous capacities in PKM" [34].

Levy's notions are in line with Wiig's view that the "root objective of PKM is the desire to make citizens highly knowledgeable" in order to "function competently and effectively in their daily lives, as part of the workforce and as public citizens" [35]. These thoughts mark a departure from the—until then—narrow individualistic confinement PKM had been placed in [36]. In limiting its scope, it has been labelled as sophisticated career and life management with a core focus on

[1]"Autonomous individuals and groups in knowledge-creating organizations set their task boundaries for themselves in pursuit of the ultimate goal expressed by the organization" [33].

personal enquiry [37] or as a means to improve some skills or capabilities of individuals, negating its importance relating to group member performance, new technologies or business processes [38].

2 The Needs for Autonomous Personal KM Systems

Due to technological and economic forces, the world of work is changing rapidly[2] and the "responsibility for self-development and lifelong learning is [said to be] now in the hands of the individual, who increasingly controls the development of his/her career and destiny. […] In the world of the modern knowledge worker, it has become necessary for individuals to maintain, develop and market their skills to give them any chance of competitive advantage in the job market in both the short and long term" [37] and of participating productively as professionals in the SECI activities of the organizational world realities.

Hence, autonomy is a key virtue in both the world of work as well as the world of education, it increases the initiative and commitment of an individual for finding valuable information, for utilizing existing knowledge and generating new knowledge. But, as sovereign professionals move from one project or responsibility to another, they will want to take their version of a KM system with them. Accordingly, an effective PKMS has to support the notion that "knowledge and skills of a knowledge worker are portable and mobile. Unlike manual workers, they have numerous options on where, how, and for whom they will put their knowledge to work" [43].

In following the pioneering concepts presented, the PKM System-in-progress addresses these needs and aims to provide knowledge workers with the overdue support for aiding life-long-learning, resourcefulness, creativity, and team-work throughout their social, academic, and professional life and as contributors and beneficiaries of organizational and societal performance.

[2]Work has suffered from a process of fragmentation which will continue to accelerate. Gratton analyzes its implications and highlights the slipping control over constant interruptions, the loss of time for real concentration, and less learning by observation and reflection [39]. With specializations and domain-specific knowledge on the rise, the identification of people has also shifted from their company to their occupation and profession, and "the vertical hierarchy and traditional career ladder have been replaced by sideways career moves between companies, and a horizontal labor market" [40]. Moreover, the 2013 study 'Future of Employment' analyzed 702 occupations and estimated about 47 % of total US employment to be still at risk due to recent technological breakthroughs able to turn previously non-routine tasks into well-defined problems susceptible to computerization [41]. Similar predictions have been put forward for Europe by Bowles [42].

3 The Benefits of a Comprehensive System Concept and Design

The complexities involved in realizing *Bush's* functionalities and in addressing *Simon's* attention-related concerns as well as the challenges[3] posed by *Nonaka's* plea for autonomy and *Levy's* push for creative conversations and capacity development call for an all-embracing high-level system concept and design.

Its aims are to mobilize and motivate the relevant audience and to instill in them a sustainable commitment to interact with and/or promote the PKMS technologies for keeping their personal human capitals à-jour (see Mostert's Six Levels of Appreciation [44] presented in [14, 16]). To facilitate the respective development process, it:

- Points out the main functionalities of the Personal Knowledge Management System,
- Guides the publication of papers to be pitched at particular chapters of a later book, to be utilized for supporting face-to-face coursework as well as e-learning modules, to provide content for system help and for demonstrating the meme-based repositories,
- Proposes alternative pathways of how learners can access any content to learn about the PKM system and/or Personal/Organization Knowledge Management in general.

Fortuitously, what might have appeared initially as difficult to reconcile or conflicting (e.g. objectives, philosophies, methodologies) has found its way into an overarching meta-concept which even provided the system's name 'Knowcations[4]'

[3]The novel PKMS approach entails a departure from the current heavyweight, prohibitive, centralized, top-down, institutional developments in order to strengthen individual sovereignty by employing grass roots, bottoms-up, lightweight, affordable, personal applications. The needs have been summarized in the form of Five Vital Provisions [14]: a. Digital personal and personalized knowledge is always in possession and at the personal disposal of its owner or eligible co-worker, residing on personal hardware and/or personalized cloud-databases. b. Contents are kept in a standardized, consistent, transparent, flexible, and secure format for easy retrieval, expansion, sharing, pooling, re-use and authoring, or migration. c. Information and functionalities can continually be used without disruption independent of changing one's social, educational, professional, or technological environment. d. Collaboration capabilities have to be mutually beneficial to facilitate consolidated team and enterprise actions that convert individual into organizational performances. e. The PKMS design and its complex operations are based on a concept, functionalities, and interventions which are clearly understood and are painlessly applied in practice.

[4]The artificial name 'Knowcations' crafted for the system is made up of KNOW as a reference to knowledge and know-how and CATIONS as an intended association to the locations or spaces as well as to the vocations or abilities which are so vital to further our expertise and careers. The 'Knowcations' idea originated during the author's Ph.D. studies in the early 90s. Since then, the resulting prototype has been continuously expanded and used personally for career support as a management consultant, scholar, professor, and academic manager. With the development platforms and cloud-based services available now, an innovation opportunity has presented itself for converting and advancing the prototype into a viable PKM system across platforms.

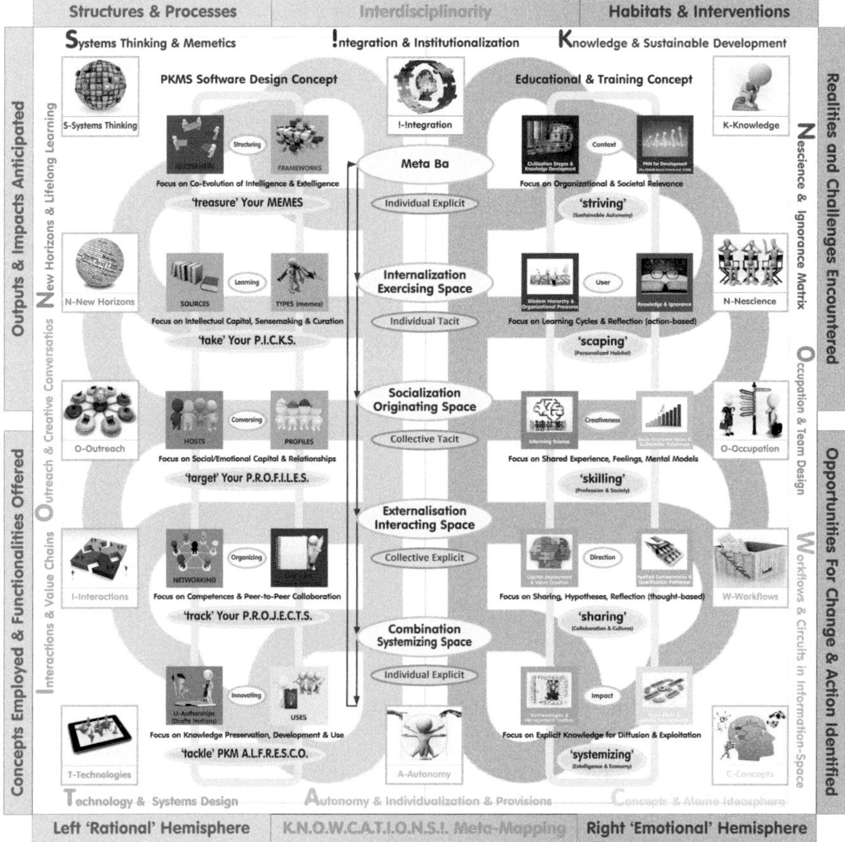

Fig. 3 Visualizing of the 'KNOWCATIONS!' overarching meta-concept

with a further lease of life as an acronym as well as a heuristic method for similar conceptual design challenges. The present state of this meta-concept has been visualized in a clock-like fashion in Fig. 3 (numbers 1 to 12 used refer to position on a watch dial).

The 'Knowcations!' Meta-Concept (Fig. 3) visualizes closely associated aspects:

- The 12 topics (clock-wise from 1 to 12 o'clock) deal—in the appropriate sequence—with the realities and challenges encountered, the opportunities for change and action identified, the concepts employed and functionalities offered, and the outputs and impacts anticipated (legends right and left of Fig. 3).
- Thus, the right-hand side (1–5) represents the external system-independent environment (habitats and interventions) and the left-hand side (7–11) the central features and objectives of the PKM system technicalities (structures and processes). Furthermore, barriers preventing and provisions enabling the transition from the current inadequate state (right) to the recommended scenario

Table 1 Legend for right-hand (1–6) and left-hand (7–12) side of Fig. 3

Topic		Legend	With particular focus on	Paper
1	K	Knowledge and sustainable development	Four PKM challenges	[18, 26]
2	N	Nescience and ignorance matrix	Acquisition/preservation	[17, 22]
3	O	Occupation and team design	Capacity development	[19, 21]
4	W	Workflows and circuits in information-space	Collaborative level	[8, 20, 24]
5	C	Concepts and meme ideosphere	Conceptual level	[9, 23]
6	A	Autonomy, individualization and provisions	Transformation	[14, 25]
7	T	Technology and systems design	Combination	[12][a]
8	I	Interactions and value chains	Externalization	[15][a]
9	O	Outreach and creative conversations	Socialization	[11, 27]
10	N	New horizons and lifelong learning	Internalization	[16][a]
11	S	Systems thinking and memetics	Meta-level, Meta-'ba'	[10, this paper]
12	!	!ntegration and institutionalization	Integrating PKM and OKM	[1, 13]

[a]In-work

(left) are given (6) and the importance of integrating personal and organizational KM is emphasized (12).

- Due to the closely related issues addressed by the topics (challenges on the right versus solutions on the left), topics can also be paired (1&11, 2&10, 3&9, 4&8, and 5&7) as indicated by the bridging pathways between the right and the left hemisphere.
- The system analysis and design objectives covering these topics have resulted in the 'PKMS Software Design Concept' (depicted as a separate yellow rounded rectangle in the middle of the left hemisphere) whose main entities are shown as icons placed appropriately on the bridges connecting them to their most relevant counterparts.
- Similarly, key content to be addressed in the accompanying roll-out of an 'Educational & Training Concept' have been exemplified and positioned (depicted as a further separate yellow rounded rectangle in the middle of the right hemisphere).
- Accordingly and in a reversal of the initial left-right analysis-design direction, a right-left development-deployment path is shown from the system to the environment.
- The five bridges symbolizing the distinctive environment-design-development-deployment-links are passing through the mid-section (five larger ellipses in the center section) which themselves are interconnected and

represent Nonaka's SECI model and 'Ba' concept, individually and in their plurality.

The Meta-Concept depicted provides a good example of how ongoing analytical and design activities can inform the setting of a blueprint which then, in turn, informs the further integration and advancement of all system components. As a consequence:

- Terminology, colors, acronyms, and icons are used consistently across system layers and dimensions resulting in a unified and coherent user familiarization/experience.
- Published papers have been individually pitched at each of the 12 topics (as indicated in Table 1). Their meme-based[5] representations have been integrated in the PKMS knowledge bases and form part of the test data for system validation. They are providing not only assistance and help functionalities for the user in the deployment phase but also case study examples and knowledge structures to enable further authorship.
- This closely integrated content is also setting the stage for a PKM/OKM lecturing concept serving higher education and professional training; it enables a delivery based on a cohesive, tried and tested framework rather than on the fragmented collections of KM notions which typify too many Knowledge Management course outlines. To paraphrase Bush: "As an added benefit of the trails captured, the [content provided by the developer] becomes not only his additions to the world's record, but entails for his [customers] the entire scaffolding by which they [have been] erected" [28].
- The visual representations and mappings in the Information-Space (exemplified in Fig. 1) offer not only a concise system overview for users but also provide a road map together with—as Bush put it—"an extensive mesh of associative multi-disciplinary trails already built-in of alternative pathways" [28]. These structures will conveniently accommodate the establishment and navigation of eLearning courses, planned after the completion of the face-to-face course design.

[5]The novel distributive facet of the PKMS has a dual dimension. The first, from centralized organizational OKM systems to decentralized personalized PKM devices, has already been alluded to. The second, from document-centric repositories to meme-based knowledge bases is portrayed in a second KICCS conference paper [9]. Rather than whole documents, what is referred to and what is stored at the same time, is smaller and more distinct, a basic building block of knowledge in the eyes of the beholder (a meme). The information-structure captured should be perfectly understandable alone by itself without piggybacking irrelevant or potentially redundant information. In this quasi-atomic state it can be repeatedly utilized (and traced back) at any later stage in combination with other memes to form memeplexes or knowledge assets, defined as "nonphysical claims to future value or benefits" [45].

4 The Concept of 'Ba' as the Underpinning of 'Knowcations'

At the heart of the Meta-Concept in Fig. 3 is the SECI-Ba-model which also has been depicted in the I^5-System [5] and the 'Information-Space' [2] (see Fig. 1) alluded to. Although the analysis and design activities have not been based initially on Nonaka's concepts, this center stage has become apparent due to the closeness and correlation with the emerging findings and design choices.

The four 'ba' and their plurality emphasize the shared contexts or places in which the PKMS philosophy is implemented, understood, followed, shared, utilized, and further developed as well as the meme-based R^8-practices of repositing, referencing, relating, recalling, reflecting, retrieving, reusing and repurposing. In the context of the software application, the 'Ba' translate into a T^5-Approach (treasure, take,[6] target,[7] track,[8] and tackle[9]) and in in terms of the real-world deployment in a S^5-Approach (striving, scaping, skilling, sharing, and systemizing).

While the 'Exercising Ba' addresses the development and curation of the PKMS user's intellectual capital via study and desktop research, the 'Originating Ba' is present in the social and emotional interaction during field research and teamwork activities. The 'Interacting Ba' surfaces as Human-Computer Interface of the PKMS and its cloud-based networking endeavors with own and others' memes, and the 'Systemizing Ba' is represented during the crafting and dissemination of the system outputs and related feedbacks and, of course, by the Creative Conversations referred to by Levy.

The 'Ba' spaces are equally relevant at on-line and off-line PKMS usage times as well as during user training and education. They provide an important linkage between personal and organizational KM philosophies (as referred to in Sect. 1.3), but also apply at the grass-roots level of repurposing and combining relevant memes to form knowledge assets:

Memes "can either be encoded in durable vectors (e.g. storage devices, books, great art, major myths, or artefacts) spreading almost unchanged for millennia, or they succeed in competing for a host's limited attention span to be memorized (internalization) until they are forgotten, codified (externalization), or spread by the spoken word to other hosts' brains (socialization) with the potential to mutate into new variants. To gain an advantage in competing for attention and survival, it pays

[6]P.I.C.K.S. refers to the **P**ersonal **I**gnorance, **C**ompetencies, & **K**nowledge **S**cape [1].

[7]P.R.O.F.I.L.E.S. is an acronym for the relationships between hosts and vectors (social capital) which define **P**rofessional Experiences, **R**esearch Activities, **O**utcomes/Results, **F**ormal Education, **I**ntellectual Capital, **L**iteracies/Skills, **E**motional Capital, and **S**ocial Networks.

[8]P.R.O.J.E.C.T.S. is an acronym for the internal logics and logistics online activities: **P**rojecting, **R**eferencing, **O**riginating, **J**ustifying, **E**videncing, **C**ontexting, **T**opicing, and **S**cripting.

[9]A.L.F.R.E.S.C.O. depicts external off-line tasks: **A**dministering, **L**ocating, **F**eed-backing, **Re**flecting, **E**ngaging, **S**ourcing, **C**rafting, and **O**bserving. The system, however, also provides functionalities to plan, manage, document, track, and evaluate these external tasks.

to form symbiotic relationships (combination) with other memes (memeplexes) to mutually support each other's fitness and to replicate together" [9].

Hence, in fostering the 'individualization versus institutionalization' integration, Nonaka's notion of Experiential, Routine, Conceptual, and Systemic Knowledge Assets (depicted in Fig. 1) as "firm-specific resources that are indispensable to create value" [33] is equally important in the PKMS context and its authorship capabilities. To add a hands-on perspective to the conceptual meme-based knowledge creation cycles, a recent 'demo' paper utilized the prototype for the paper's development which, in turn, described the iterative process steps involved in its creation [16].

5 The Road Travelled and the Road Ahead

The PKMS concept and prototype-in-progress add further detail to the Creative Spaces [4] and the JAIST Nanatsudaki Model with its sequence of seven waterfalls or creative spirals which describes academic/individually-oriented as well as practice/group-oriented knowledge creation processes [5] and which has been aligned to the PKMS's information-space cycles [8 and Fig. 1].

After finalizing the prototype test phase, an estimated 12 months migration to a viable PKM application is envisaged based on a production environment for cross-platform mobile, web and cloud applications. In parallel, the series of recent conference sessions[10] and publications will be concluded in line with the meta-concept:

- *Meta Ba* The Meta-Concept described is firmly rooted in two prior publications. The very first paper identifies four challenges for individuals managing their knowledge at acquisition/preservation, collaborative, capacity development, and conceptual level and looks at approaches to address them [26] and another which positions the PKMS in the context of human development [18]. This paper, an extended paper [10], describes the Meta-Concept in detail and will be followed by an exploration of PKM's impact on social, cultural, economic, and environmental sustainability.
- *Exercising Space* Two papers introduce an extended ignorance matrix and discuss the role of PKM systems in making citizens highly knowledgeable [17, 22], followed by the 'demonstration' paper mentioned [16]. It is planned to publish suggestions for a KM course structure and e-learning system in support

[10]In acknowledging the trans-disciplinarity of the PKM notion, the papers and presentations received feedback from and addressed a wide scope of conference themes, including Knowledge Management and Knowledge Technologies, Informing Science, Management and Social Sciences, Higher Education and Human Resource Development, Innovation and Creativity Support Systems, e-Skills and e-Learning, as well as Organizational Learning, Future Studies, and Sustainability.

of the educational needs Levy highlighted as well as to offer respective conference presentations and tutorials.

- *Originating Space* Two papers and a presentation look at the workplace and careers of knowledge workers in the business, educational, and developmental context [19, 21, 27]. They just have been concluded by presenting a framework of twelve PKM for Development (PKM4D) criteria (aligned to Maslow's Extended Hierarchy of Needs) which explored the close relationship between PKM and e-Skills [11].
- *Interacting Space* Three papers and a presentation demonstrate the close proximity of the PKM concept and functionalities with renowned KM models and organizational KM practices [8, 15, 20, 24]. A follow-up will explore the benefits of PKMS as a 'Structural Intellectual Capital Asset' for creating 'Corporate Knowledge Assets'.
- *Systemizing Space* The integration of the models and workflows in the context of the meme-based PKMS concept and its information-space have been presented [9, 23]. Papers-in-progress add further perspectives by contemplating how the combination's innovative potential is able to invigorate digital scholarship, individual and institutional curation [12], and the traceability of knowledge in a new configuration of citation systems. A more technical paper will compare the PKMS prototype to other related recent system development activities as well as semantic technologies.
- *Plus Autonomy versus the Integration of PKM and OKM* The PKMS concept advocates that the calls for 'Individualization' and 'Institutionalization' in the KM context have to be regarded as two sides of the same coin; four papers have taken this notion from initial contemplations to the identification of barriers and remedies [14, 25], and from a summarized proposal to a detailed account of what the 'next KM generation' might look like [1, 13].

References

1. Schmitt, U.: Quo Vadis, knowledge management: a regeneration or a revolution in the making? Forthcoming: Journal of Information & Knowledge Management (JIKM), Special Issue on KM Models and Theories. (2016)
2. Boisot, M.: Exploring the information space. University of Pennsylvania (2004)
3. Kolb, D.A.: Experiential learning: experience as the source of learning and development, vol. 1. Englewood Cliffs, Prentice-Hall (1948)
4. Wierzbicki, A.P., Nakamori, Y.: Creative space: models of creative processes for the knowledge civilization age. Series: Studies in Computational Intelligence, vol. 10. Springer Publishing Company (2006)
5. Wierzbicki, A.P., Nakamori, Y.: Creative Environments: Issues of Creativity Support for the Knowledge Civilization Age. Springer Publishing Company (2007)
6. Nonaka, I., Takeuchi, H.: The Knowledge-Creating Company. Oxford University Press, Oxford (1995)
7. Pirolli, P., Card, S.: The sensemaking process and leverage points for analyst technology. In: Proceedings of International Conference on Intelligence Analysis, vol. 5. (2005)

8. Schmitt, U.: Knowcations—positioning a meme and cloud-based 2nd generation personal knowledge management system. Forthcoming A. M. J. Skulimowski and J. Kacprzyk (Eds), Knowledge, Information and Creativity Support Systems: Recent Trends, Advances and Solutions (Selected Papers from KICSS'2013 - 8th International Conference on Knowledge, Information, and Creativity Support Systems, Nov 7-9, 2013, Kraków, Poland), 978-3-319-19089-1, Springer Series: Advances in Intelligent Systems and Computing (AISC) (2016)
9. Schmitt, U.: Significance of memes for the successful formation of autonomous PKMS. In: 9th International Conference on Knowledge, Information and Creativity Support Systems (KICSS), 06–08 Nov 2014, Limassol, Cyprus, pp. 339–345 (2014). 978-9963-700-84-4 Extended Version Accepted In: Knowledge, Information and Creativity Support Systems (Selected Papers from KICSS'2014, 9th International Conference held in Limassol, Cyprus, on November 6-8, 2014). Kunifuji, S., Papadopoulos, G.A., Skulimowski, A.M.J. (eds.), 978-3-319-27477-5, Springer Series: Advances in Systems and Computing (AISC) 2194-5357, Vol. 416. (2016)
10. Schmitt, U.: The significance of 'Ba' for the successful formation of autonomous personal knowledge management systems. In: 9th International Conference on Knowledge, Information and Creativity Support Systems (KICSS), 06–08 Nov 2014, Limassol, Cyprus, pp. 327–338 (2014). 978-9963-700-84-4
11. Schmitt, U.: Making sense of e-Skills at the dawn of a new personal knowledge management paradigm. In: Proceedings of the e-Skills for Knowledge Production and Innovation Conference of the South African iKamva National e-Skills Institute (iNeSI) and the Informing Science Institute (ISI), 17–21 Nov 2014, Cape Town, South Africa, pp. 417–447 (2014) 2375-0634
12. Schmitt, U.: Supporting digital scholarship and individual curation based on a meme-and-cloud-based PKM concept. In: Presentation at the Conference of the International Journal of Arts and Sciences (IJAS), Rome, 28–31 Oct 2014. Published in: Supporting digital scholarship and individual curation based on a meme-and-cloud-based personal knowledge management concept. Academic Journal of Science (AJS), Vol. 4/1, pp. 220-237, 2165-6282. (2015)
13. Schmitt, U.: Proposing a next generation of knowledge management systems for creative collaborations in support of individuals and institutions. In: Proceedings of the 6th International Joint Conference on Knowledge Discovery, Knowledge Engineering and Knowledge Management (IC3 K), 21–24 Oct 2014, Rome, Italy, pp. 346–353 (2014). 978-989-758-050-5
14. Schmitt, U.: Overcoming the seven barriers to innovating personal knowledge management systems. In: Proceedings of the International Forum on Knowledge Asset Dynamics IFKAD, 11–13 June 2014, Matera, Italy, pp. 3662–3681 (2014). 978-88-96687-04-8
15. Schmitt, U.: From circuits of knowledge to circuits of personal knowledge management concepts (Presentation). Paper presented at the conference on organizational learning, knowledge and capabilities OLKC, 22–24 April 2014, Oslo, Norway (2014)
16. Schmitt, U.: How this paper has been created by leveraging a personal knowledge management system. In: Proceedings of the 8th International Conference on Higher Education ICHE, 16–18 March 2014, Tel Aviv, Israel, pp. 22–40 (2014)
17. Schmitt, U.: The role of personal knowledge management systems in making citizens highly knowledgeable. In: Proceedings of the 8th International Technology, Education and Development Conference INTED, 10–12 March 2014, Valencia, Spain, pp. 2005–2014 (2014). 978-84-616-8412-0
18. Schmitt, U.: Personal knowledge management devices—the next co-evolutionary driver of human development?! In: Proceedings of the International Conference on Education and Social Sciences INTCESS14, 03–05 Feb 2014, Istanbul, Turkey. CD 12:4 (2014). 978-605-64453-0-9

19. Schmitt, U.: Leveraging personal knowledge management systems for business and development. Paper presented at the 2nd conference africa academy of management (AFAM), 9–11 Jan 2014, Gaborone, Botswana (2014)
20. Schmitt, U.: Knowcations—conceptualizing a personal second generation knowledge management system. In: Skulimowski, A.M.J. (ed.) Looking into the Future of Creativity and Decision Support Systems, Proceedings of the 8th Conference on Knowledge, Information and Creativity Support Systems KICSS, 07–09 Nov 2013, Krakow, Poland, pp. 587–598 (2013)
21. Schmitt, U.: Furnishing knowledge workers with the career tools they so badly need. In: Proceedings of the International Human Resource Development Conference HRDC, 17–18 Oct 2013, Mauritius (2013)
22. Schmitt, U.: Managing personal knowledge to make a difference. In: Proceedings of the 27th British Academy of Management Conference BAM, Liverpool, 10–12 Sep 2013 (2013). 978-0-9549608-6-5
23. Schmitt, U.: Managing personal knowledge for the creative class economy (poster). WorldFuture 2013—Exploring the Next Horizon, 19–21 July 2013, Chicago, U.S.A. (2013)
24. Schmitt, U.: Innovating personal knowledge creation and exploitation. In: Proceedings of the 2nd Global Innovation & Knowledge Academy GIKA, 9–11 July 2013, Valencia, Spain (2013)
25. Schmitt, U.: Knowcations—a meme-based personal knowledge management system-in-progress. In: Proceedings of the 8th International Conference on e-Learning ICEL, ACPI, 27–28 June 2013, Cape Town, South Africa, pp. 523–527 (2013). 978-1-909507-26-5
26. Schmitt, U.: Knowcations—The quest for a personal knowledge management solution. In: Proceedings of the 12th International Conference on Knowledge Management and Knowledge Technologies i-Know, 05–07 Sep 2012, Graz, Austria. Copyright 2012 ACM 978-1-4503-1242-4/12/09. Graz: ACM. http://dl.acm.org/citation.cfm?id=2362469 (2012)
27. Schmitt, U., Butchart, B.: Making personal knowledge management part and parcel of higher education programme and services portfolios. J. World Univ. Forum **6**(4), 87–103 (2014). 1835-2030
28. Bush, V.: As we may think. The Atlantic Monthly, Issue 176.1, pp. 101–108 (1945)
29. Davies, S.: Still building the memex. Commun. ACM **53**(2), 80–88 (2011)
30. Osis, K., Gaindspenkis, J.: Modular personal knowledge management system and mobile technology cross-platform solution towards learning environment support. Annual International Conference on Virtual and Augmented Reality in Education (VARE), pp. 114–124 (2011)
31. Simon, H.A.: Designing organizations for an information-rich world. In: Greenberger, M. (ed.) Computers, Communication & Public Interest. Johns Hopkins Press (1971)
32. Kahle, D.: Designing open educational technology. In: Iiyoshi, T., Vijay Kumar, M.S. (eds.) Opening up Education. MIT Press, Cambridge (2009)
33. Nonaka, I., Toyama, R., Konno, N.: SECI, Ba and leadership: a unified model of dynamic knowledge creation. Long Range Plan. **33**, 5–34 (2000)
34. Levy, P.: The Semantic Sphere 1. Wiley (2011)
35. Wiig, K.M.: The importance of personal knowledge management in the knowledge society. In: Pauleen, D.J., Gorman, G.E. (eds.) Personal Knowledge Management, pp. 229–262. Gower (2011)
36. Cheong, R.K., Tsui, E.: From skills and competencies to outcome-based collaborative work. Knowl. Process Manage. **18**(3), 175–193 (2011)
37. Pauleen, D.J., Gorman, G.E.: The nature and value of personal knowledge management. In: Pauleen, D.J., Gorman, G.E. (eds.) Personal Knowledge Management. Gower, pp. 1–16 (2011)
38. Davenport, T.H.: Personal knowledge management and knowledge worker capabilities. In: Pauleen, D.J., Gorman, G.E. (eds.) Personal Knowledge Management, pp. 167–188. Gower, England (2011)

39. Gratton, L.: The Shift—The Future of Work is Already Here. HarperCollins, UK (2011)
40. Florida, R.: The Rise of the Creative Class—Revisited. Basic Books (2012)
41. Frey, C.B., Osborne, M.A.: The Future of Employment. Oxford Martin (2013)
42. Bowles, J.: The computerisation of European jobs, Bruegel. www.bruegel.org/nc/blog/detail/article/1394-the-computerisation-of-european-jobs/ (2014)
43. Rosenstein, B.: Living in More Than One World. Berrett-Koehler Publishers (2009)
44. Mostert, M.: Systemic Leadership Learning. Knowres Publishing (2013)
45. Dalkir, K.: Knowledge Management in Theory and Practice, Butterworth-Heinemann (2005)

The Significance of Memes for the Successful Formation of Autonomous Personal Knowledge Management Systems

Ulrich Schmitt

Abstract Recent papers based on a prototype-in-progress focused on advancing Personal Knowledge Management (PKM) as the next generation of Knowledge Management Systems (KMS). Based on assumptions of autonomous capacities engaged in creative conversation, the personal devices are supposed to enable the emergence of the distributed processes of collective extelligence and intelligence, which in turn feed them. This shift from centralized organizational KMS to decentralized personalized KMS devices is expected to give more power and autonomy to individuals and self-organized groups. A second novel feature proposed is the substitution of document-centric repositories with meme-based knowledge bases. While a parallel conference paper has considered the significance of Nonaka's concept of 'Ba' for the overarching meta-concept of the PKM system-in-progress, this contribution concentrates on the particular aspect of memes and introduces their relevant features and potential allowing for information-rich, multi-dimensional information structures and trails as well as for more elaborate dissemination concepts and citation systems.

Keywords Personal knowledge management systems • Knowledge worker • Memes • Memetics • Digital scholarship • Citation systems • Knowcations

1 The Inadequacy of Today's Reference and Citation Systems

Traceability forms the backbone of quality management in modern manufacturing and stands for the ability to trace the history, application or location of an entity. "For discrete manufacturers, this means you can track and trace each component

---

U. Schmitt (✉)
University of Stellenbosch Business School, PO Box 802, Bellville 7535, South Africa
e-mail: schmitt@knowcations.org

S. Kunifuji et al. (eds.), *Knowledge, Information and Creativity Support Systems*, Advances in Intelligent Systems and Computing 416,
DOI 10.1007/978-3-319-27478-2_29

that comprises your product—from suppliers and manufacturers through assembly and final delivery to customers by creating an as-built genealogy. And for food and other process manufacturers, it's the ability to trace each ingredient of a product from 'farm to fork' through the creation of a batch genealogy" [15].

Unfortunately, for knowledge seekers and producers such an enabling state of management did not emerge with the advances and widespread affordability of Information and Communication Technologies (ICT). Instead, we are faced with a never before experienced ever-increasing attention-consuming information abundance.

It is still the academic paper-based citation system which dominates today's knowledge tracking and it has been—since the seventeenth century—the basis for the reputation economy in science. It "allows scientists to build on the earlier work without having to repeat that work [and] the citation both credits the original discoverer, and provides a link in a chain of evidence" [16].

In motivating his vision of the 'Memex'[1] seven decades ago, Bush already called these methods of transmitting and reviewing our research results "to be generations old" and "totally inadequate for their purpose" [4]. In order to take advantage of today's online realities, Nielsen calls for a "New Era of Networked Science" and urges removing barriers that prevent potential contributors from engaging in a wider sharing and faster diffusion of their ideas, sources, data, work-in-progress, preprints, and/or code for the benefit of more rapid iterative improvement [16]. Several of these counterproductive barriers have been identified, including incompatible formats and structures, fragmentations and inconsistencies, redundancies and discouraging services as well as the prohibitive institutional approaches and centralized developments controlled by large corporations interested in captured audiences [27] and in "knowledge being shut up in silos and balkanized within small closed communities" [14]. Missing, broken, or pretentious web links or references add further to the sorry state of inefficiencies, but so do the current reference and citation systems in general.

"Any correct reference indicates a discrete source (e.g. book, article, web site) with a page number or access date added only occasionally. It represents a granularity which might have been adequate in the stable context of paper-based worlds, but is far from sufficient in a volatile digital world where contents referenced are constantly altered or erased. It also is far too unrefined for the potential capabilities of digital Personal Knowledge Management Systems (PKMS).

[1]In 1945, Vannevar Bush (then President Truman's Director of Scientific Research) imagined the 'Memex', a hypothetical sort of mechanized private file/desk/library-device. It is supposed to act as an enlarged intimate supplement to one's memory, and enables an individual to store, recall, study, and share the "inherited knowledge of the ages". The 'Memex' would have facilitated the addition of personal records, communications, annotations, and contributions, but, above all, the recording of non-fading trails of one's individual interests through the maze of materials available —all easily accessible and sharable with the 'Memexes' of acquaintances. In the 70 years since Bush's famous article [4], a 'Memex' has never been realized [6, 12, 18] but it is one of the inspirations for the PKM system proposed [24].

A PKMS is expected to serve its master over a lifetime of educational and professional careers and is supposed to constantly evolve in the process. What we have to refer to and what we need to store at the same time, has to be smaller and more distinct, a basic building block of knowledge in the eyes of the beholder. Captured best in a quasi-atomic state, the information-structure should be perfectly understandable alone by itself but be able to be used at any later time in combination with other building blocks stored without piggybacking irrelevant or potentially redundant information" [27]. This basic building block of knowledge is also referred to as a meme.

2 It Is a Meme's World

Memes were originally described by Dawkins as units of cultural transmission or imitation [7]. They are (cognitive) information-structures [1] that evolve over time through a Darwinian process of variation, selection and transmission [5]. Able to self-replicate by utilizing mental storage in human hosts, they influence their hosts' behavior to promote further replication [1]. From the meme's-eye view, every human is a machine for making more memes, a vehicle for propagation, an opportunity for replication and a resource to compete for Blackmore [2]. But, memes exist only virtually and have no intentions of their own; they are merely information pieces in a feedback loop with their longevity being determined by their environment [5].

To live on, a meme has to be able to survive in the medium it occupies and the medium itself has to survive [1]. They can either be encoded in durable vectors (e.g. storage devices, books, great art, major myths, or artefacts) spreading almost unchanged for millennia [1], or they succeed in competing for a host's limited attention span to be memorized (internalization*) until they are forgotten, codified (externalization*) or spread by the spoken word to other hosts' brains (socialization*) with the potential to mutate into new variants (Fig. 1). To gain an advantage in competing for attention and survival, it pays to form symbiotic relationships (combination*) with other memes (memeplexes) [9] to mutually support each other's fitness and to replicate together (*-markings refer to comparable SECI Model stages [17, 24].

Hughes adds the point that human evolution and memes' endurances have not only thrived on big brain memory and communication technology with a high degree of accuracy but also on peoples' insatiable urge to use these technologies for the purposes intended [11]. Schmitt magnifies this perspective and argues that the progress of human development is primarily due to successive technological co-evolutions. At each stage, civilization runs into constraints which could only be overcome with the emergence of a further powerful co-evolution [30].

Today's constraint in need to be dealt with is clearly the over-abundance of information. Instead, the 'breakthrough' technology under way, termed the 'Industrial Internet' or 'Internet of Things', will exploit the opportunities of cloud-based

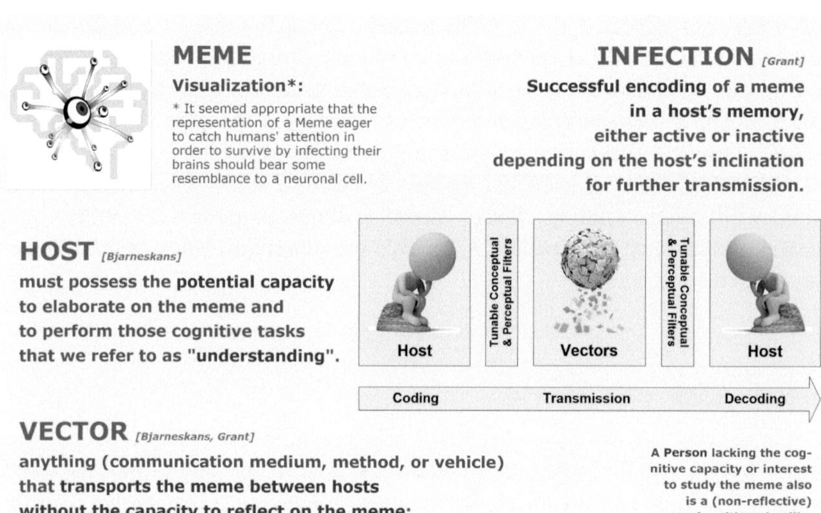

Fig. 1 The transmissions of memes between hosts via vectors [1, 9, 28]

platforms and applications by incorporating networked sensors and software into goods and machines resulting in the self-organizational capability of complex value chains. As a consequence, ever more information will be generated—further defeating the very attention[2] our cognitive capabilities are able to master [30].

3 The Ideosphere as the Memes' Habitat of Operation

Memetics studies ideas and concepts viewed as 'living' organisms, capable of reproduction and evolution in an 'Ideosphere' [19], an "invisible but intelligible, metaphysical sphere of ideas and ideation" where we engage in the creation of our world. "This means that the substance of the world is idea, which forms, reforms, and transforms itself via the conversations of humankind, synergetically organizing itself as an evolutionary, multidimensional network (with technology just an artefact of idea). The problem, however, is that the majority of 'humanity remains the consumer of ideas without being the producer" [13]. Hence, Kimura calls for an ideospheric transformation set off by a synergetic phenomenon that emerges "when individuals in sufficient numbers become authentic, independent thinkers, that is,

[2]Simon already noted over 40 years ago, "the wealth of information is creating a poverty of attention and with it a need to allocate that attention efficiently among the overabundance of information sources that might consume it" [34].

originators of ideas, producers of dialogues, and contributors to the network of conversations that comprises the world" [13]. Fittingly, Levy stresses the "need for a personal discipline for collection, filtering and creative connection (among data, among people, and between people and data flows)" and the need for education "to encourage in students the sustainable growth of autonomous capacities in Personal Knowledge Management" [14].

But, if memes and their inbuilt ideas are able to flourish in a virtual 'Ideosphere' as their habitat of operation, PKM Systems aiming at supporting individual capacity and repertoire for innovation, sharing and collaboration are well advised to utilize the very same space and resources and to form a digital counterpart of this 'Ideospere'.

In this respect, the three-dimensional Information-Space Model [3] which has been used to visualize the PKMS's entities, workflows, and its integrated KM models [21, 26, 31] also represents this 'ideosphere', and its memes and memeplexes are displayed in their amalgamated states as the PKMS user's knowledge assets and his/her intellectual, social, and emotional capital [22].

4 Traditional Document-Focus Versus Meme-Based PKMS

What this means for creative authorship has been exemplified in a recent 'demonstration' paper which added a hands-on perspective to the conceptual Information-Space cycles by utilizing the PKMS prototype for the paper's development; it describes the iterative process steps involved in its very own creation [28]. However, while the published paper version represents an ordered sequence of memes resulting in a book-like linear format, its internal virtual structure in the knowledge base remains multi-dimensional, information-rich, and its memes are ready to be re-used, rephrased, re-purposed, and shared—with their trails in place to be visited and revised. As envisaged in the 'Memex', "the inheritances from the master become, not only his additions to the world's record, but includes for his disciples the entire scaffolding by which they were erected" [4]. In this extended version of the initial paper [23], this 'scaffolding' is exemplified as a 'snapshot' (Fig. 2) and further explicated. The snapshot (from a meme's point of view) maps the growing relationships of one meme-scape from a document-centric (grey area) to a meme-based (red area) approach supported by networked PKM repositories (green area).

In a document (grey area), a meme has a preceding and a succeeding meme, can contain references (internal to figures, tables, footnotes, etc.; external to other authors' sources with occasional page numbers) or can be referenced by keywords or in an index. A meme-based authorship approach (grey and red area) retains the links to the referenced memes and conserves them with their relevant frames of references (e.g. origins, titles, formats, licenses). Referenced memes can be relevant

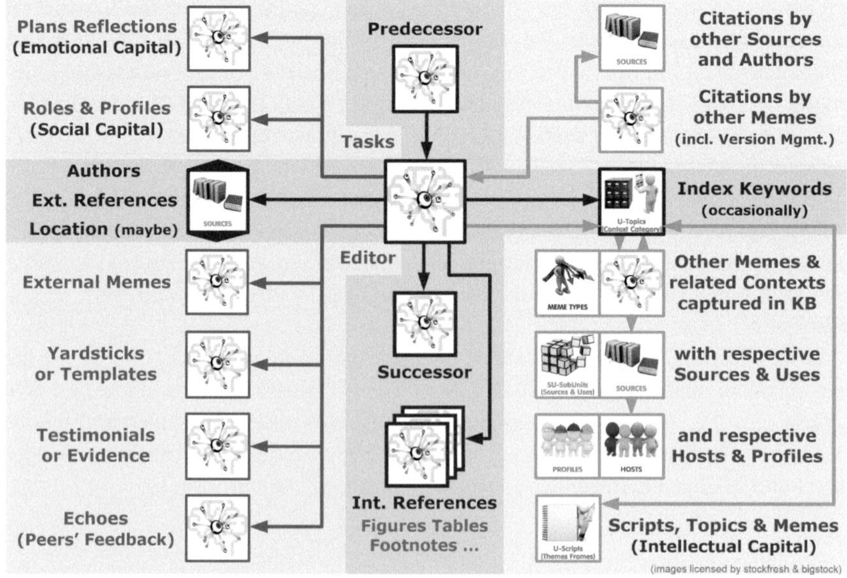

Fig. 2 Traditional document-centric versus meme-based PKM repositories

prior work ('standing on the shoulders of giants'), questionnaires or recommended practices (e.g. accreditation standards or business plan templates), evidence and testimonials (to back up any claims made in the meme), or meme-specific feedbacks in peer-review or mentorship processes. Additionally, links to one's own or other's plans and reflections are possible (including progress reporting and to do's) as well as one's own or others' roles and/or profiles).

Supporting networked PKM repositories (adding the green area) allow feeding-backward and enable the integration of subsequently created memes and documents which are utilizing the meme-under-consideration. Since it is encouraged to embed memes in a more-dimensional classification system for subsequent easy retrieval, and as a pure, pre-edited, re-purposed, and/or already re-combined version according to individuals' preferences and objectives (topics and script structures), potentially relevant other memes (own and others') can be accessed and examined.

5 The Benefits of a Meme-Based PKMS Approach

A PKMS is expected to serve its user unceasingly over life-long-learning periods of educational, professional, social and private activity and experience. The concept and prototype introduced focuses on the relevant textual, visual, audio, or video memes any document or message contains. By capturing and referencing these memes as quasi-atomic basic information building blocks (in the eyes of the beholder), the system allows recalling, sequencing and combining already stored memes with one's own new meme creations for integration in any type of authoring and sharing activity one would like to pursue. As a result, the user obtains the means to retain and build upon knowledge acquired in order to sustain personal growth and facilitate productive collaborations between fellow learners and/or professional acquaintances.

Any meme captured is able to further evolve during learning processes and to form symbiotic relationships with known or newly imagined memes during authorship or sensemaking sessions to morph into increasingly complex memeplexes or knowledge assets (e.g. articles, presentations, or scripts). As Distin points out: "In recombination, existing memes are appropriately recombined in new situations, creating new ways of thought and novel effects, perhaps as the result of previously recessive memes' 'effects' being revealed in the reshuffle". As a result of mutations, "copies will not always be exact, and the idea or skill in question may change in some way en route" [8].

Dawkins [7] points out three qualities of a meme to enhance its fitness in order to maintain a continued presence in future generations[3]: Fecundity, Longevity, and Copying Fidelity. All three features are profiting extensively from the secure, convenient, and standardized storage in the PKMS's knowledge bases and from the creative conversations between networked autonomous PKMS devices [14].

As the 'Memex' envisioned seven decades ago, the Knowledge Bases of a meme-based PKM System act as enlarged intimate supplements to a user's memory and enable us to store, recall, study, and share the 'inherited knowledge of the ages' relevant to us. They facilitate the addition of personal records, communications, annotations, contributions as well as non-fading trails of one's individual interest through the maze of materials and memes available [4]; all to be easily accessible and shareable with the PKMSs of acquaintances. In contrast to its organizational counterpart, a PKMS objective "is to enable self-reflecting monologues of its user over life-long-learning periods of educational, professional, social and private activity and experience. In these conversations with self, the knowledge under review is biographically self-determined and presents itself as a former state of personal extelligence captured" and new memes created by the user [27].

[3]Fecundity asserts that the more copies of itself a meme can encode in vectors, the higher its fitness. Longevity refers to a meme's replication rate which has to be appropriate to its environment and depends on cost-benefit ratios. Copying Fidelity rests on the accuracy with which a meme is copied resulting in either the integrity of information, sensible variations, or erroneous mutations [7].

Table 1 Main entity types kept in the meme-based PKMS knowledge bases

Hosts	Sources (sub-sources)	Uses (sub-uses)
Actors, agents, users	Testimonials (proof evidence)	Memes (information units)
Communities, networks	Yardsticks (rules standards)	Authorship (drafts notions)
Teams, groups	Periodicals (articles papers)	Nemes (takes Ideas)
Organizations, institutions	Events (reports papers)	Intentions (tasks diaries)
Research areas (classifications)	Books (chapters papers)	Forethoughts (plans)
Industry sectors (classifications)	Artefacts (assets components)	Evaluations (reflections)
Spaces (location classifications)	Repositories (sites files items)	Scripts (themes frames)
Meta-topics (e.g. RFCD, TOA)	Shoe boxes (records notes)	Topics (context category)

The PKMS workflows[4] capture a variety of entities (Table 1) and their diverse relationships[5] and support individuals in keeping their accumulated intellectual, social and emotional capitals [10] in compatible recorded formats and in an enabling state of maintenance and learning [22]. Stacks of time and attention currently lost due to redundant findings, mundane tasks, and rework can be mobilized for concentrating—instead—on the creative or innovative targets set and for facilitating consolidated actions that convert individual into organizational performances [31].

Accordingly, meme-based PKMS devices empower individuals to be autonomous in the development of their expertise as well as to determine how it will be used or exchanged with people, communities, or organizations close to them. This personal capacity also includes the freedom of choices in regard to the structuring of any contents to be disseminated.

However, while converting a PKMS's authorship project into a published paper or pdf version results in a book-like format distinguished by one-dimensionality and finality (as alluded to), its internal virtual structure in the knowledge base remains information-rich and multi-dimensional, and its memes are ready to be re-used, rephrased, or re-purposed—with their trails in place to be visited and revised.

As a result, the snapshot (Fig. 2) also represents a map of a particular meme with its current relations to its virtual meme neighborhood (lines depict non-fading trails). This map is recursive: any neighboring meme encountered (while following

[4]A recent paper [20] identified forty-one established knowledge management constructs which form the ingredients for the ongoing integrating, adapting, broadening, deepening, repurposing, or innovating activities of the PKM concept and system design.

[5]P.R.O.F.I.L.E.S. is an acronym for the relationships between hosts and vectors which define Professional Experiences, Research Activities, Outcomes and Results, Formal Education, Intellectual Capital, Literacies and Skills, Emotional Capital, and Social Networks. P.R.O.J.E.C.T.S. is an acronym for the internal logics and logistics activities: Projecting, Referencing, Originating, Justifying, Evidencing, Contexting, Topicing, and Scripting, while A.L.F.R.E.S.C.O. depicts external tasks to be monitored by the system: Administering, Locating, Feed-backing, Reflecting, Engaging, Sourcing, Crafting, and Observing.

the trails via the user interface) equips this meme with the maps' center of attention and details its own particular virtual meme neighborhood with an updated set of trails (being defined up to that moment by the accessible knowledge bases) to pursue. These dynamically created up-to-date trail structures created by autonomous meme-based PKMS devices linked for creative conversations open up a range of opportunities for e-learning support concepts, for more elaborate dissemination and collaboration models as well as for speedier and pointier citation-based reputation systems.

6 Conclusions for the Road Ahead

The novel meme-based Personal KMS approach incorporates a number of appealing outcomes which will be further explored in upcoming and planned papers:

- It provides solutions to the challenges faced by knowledge workers identified in the very first paper [32]. They have been reviewed against a newly devised framework of twelve PKM for Development (PKM4D) criteria in the context of e-Skills [22].
- Over more than a decade, KM thinkers have been pondering over the need for a new chapter in the evolutionary history of Knowledge Management. An upcoming paper paints a 'Next KM Generation' scenario with the PKMS prototype as a point of departure and reverse engineers its functionalities in order to crystallize those KM Models and Theories which—in terms of the PKMS—have passed the test of time [20].
- With all the author's papers captured in their meme-based representations and information-rich references, a KM course structure for face-to-face and e-learning as well as a respective conference tutorial are being established.
- Further opportunities in the educational context [33] are being elaborated on in the context of digital scholarship, mentorship, individual and institutional curation [25]. A paper-in-progress follows up on knowledge traceability and contemplates the potential impact of shared meme-based repositories for transforming citation systems to contribute to the "New Era of Networked Science" as advocated by Nielsen [16].
- Although the novel PKM concept and system aim at departing from the centralized institutional KM developments and at strengthening personal autonomy, it is not meant to be at the expense of Organizational KM Systems, but rather as the means to foster a mutually beneficial co-evolution. Accordingly, a paper-in-work details how individuals' intellectual, social, and emotional capital is able to interact with a firm's Human, Relationship, Strategic, and Structural Capital for converting individual into organizational performances and for promoting Corporate Knowledge Assets.

- A parallel 'KICSS 2014' paper [24] presents a Meta-PKMS-Concept to consolidate this portfolio of opportunities covering the main environmental, design, development, deployment, and capacity building concerns. It also portrays its substantial interrelationship with Nonaka's concept of 'Ba' and its impact on the PKMS's objectives.
- With the PKMS's envisaged role of making citizens highly knowledgeable [29], a paper-in-progress utilizes the PKM4D framework for an inquiry of the PKM's impact on professions and society (social), autonomy and collaboration (cultural), extelligence and leadership (economic), and personalized habitats (environmental).

References

1. Bjarneskans, H., Grønnevik, B., Sandberg, A.: The Lifecycle of Memes. http://www.aleph.se/Trans/Cultural/Memetics/memecycle.html (1999)
2. Blackmore, S.J.: The power of memes. Sci. Am. **283**(4), 52–61 (2000)
3. Boisot, M.: Exploring the information space: a strategic perspective on information systems. In: Working Paper Series WP04-003. University of Pennsylvania (2004)
4. Bush, V.: As we may think. Atlantic **176**(1), 101–108 (1945)
5. Collis, J.: Introducing Memetics. meme.sourceforge.net/docs/memetics.php (2003)
6. Davies, S.: Still building the memex. Commun. ACM **53**(2), 80–88 (2011)
7. Dawkins, R.: The Selfish Gene. Paw Prints, Chennai (1976)
8. Distin, K.: The Selfish Meme. Cambridge University Press, Cambridge (2005)
9. Grant, G., Sandberg, A., McFadzean, D.: Memetic. http://www.lucifer.com/virus/memlex.html (1999)
10. Gratton, L.: The Shift—The Future of Work is Already Here. HarperCollins, London (2011)
11. Hughes, J.: On the Origin of Tepees. Free Press, New York (2011)
12. Kahle, D.: Designing open educational technology. In: Iiyoshi, T., Vijay Kumar, M.S. (eds.) Opening Up Education. MIT Press, Cambridge (2009)
13. Kimura, Y.G.: Kosmic alignment—a principle of global unity. Kosmos J. http://www.via-visioninaction.org/via-li/articles/Kosmic_Alignment.pdf (2005)
14. Levy, P.: The Semantic Sphere 1. Wiley, New York (2011)
15. Motorola: Quick Reference Guide: Traceability for Manufacturing. Motorola Inc. (2011)
16. Nielsen, M.: Reinventing Discovery—The New Era of Networked Science. Princeton University Press, Princeton (2011)
17. Nonaka, I., Takeuchi, H.: The Knowledge-Creating Company. Oxford University Press, London (1995)
18. Osis, K., Gaindspenkis, J.: Modular personal knowledge management system and mobile technology cross-platform solution towards learning environment support. In: Proceedings of the Annual International Conference on Virtual and Augmented Reality in Education (VARE), pp. 114–124 (2011)
19. Sandberg, A.: Memetics. http://www.aleph.se/Trans/Cultural/Memetics (2000)
20. Schmitt, U.: Quo Vadis, knowledge management: a regeneration or a revolution in the making? Forthcoming: Journal of Information & Knowledge Management (JIKM), Special Issue on KM Models and Theories. (2016)
21. Schmitt, U.: Knowcations—positioning a meme and cloud-based 2nd generation personal knowledge management system. In: A. M. J. Skulimowski and J. Kacprzyk (Eds), Knowledge, Information and Creativity Support Systems: Recent Trends, Advances and Solutions (Selected

Papers from KICSS'2013 - 8th International Conference on Knowledge, Information, and Creativity Support Systems, Nov 7–9, 2013, Kraków, Poland), 978-3-319-19089-1, Springer Series: Advances in Intelligent Systems and Computing (AISC) (2016)
22. Schmitt, U.: Making sense of e-skills at the Dawn of a new personal knowledge management paradigm. In: Proceedings of the e-Skills for Knowledge Production and Innovation Conference of the South African iKamva National e-Skills Institute (iNeSI) and the Informing Science Institute (ISI), pp. 417–447. Cape Town, 17–21 Nov 2014. 2375-0634
23. Schmitt, U.: Significance of memes for the successful formation of autonomous PKMS. In: 9th International Conference on Knowledge, Information and Creativity Support Systems (KICSS), Limassol, Cyprus, pp. 339–345. 06–08 Nov 2014. 978-9963-700-84-4
24. Schmitt, U.: The significance of 'Ba' for the successful formation of autonomous personal knowledge management systems. In: 9th International Conference on Knowledge, Information and Creativity Support Systems (KICSS), Limassol, Cyprus, pp. 327–338. 06-08 Nov 2014, 978-9963-700-84-4. Extended Version Accepted In: Knowledge, Information and Creativity Support Systems (Selected Papers from KICSS'2014, 9th International Conference held in Limassol, Cyprus, on November 6-8, 2014). Kunifuji, S., Papadopoulos, G.A., Skulimowski, A.M.J. (eds.), 978-3-319-27477-5, Springer Series: Advances in Systems and Computing (AISC) 2194–5357, Vol. 416. (2016)
25. Schmitt, U.: Supporting digital scholarship and individual curation based on a meme-and-cloud-based PKM concept. In: Presentation at the Conference of the International Journal of Arts and Sciences (IJAS), Rome, 28–31 Oct 2014. Published in: Supporting digital scholarship and individual curation based on a meme-and-cloud-based personal knowledge management concept. Academic Journal of Science (AJS), Vol. 4/1, pp. 220–237, 2165–6282. (2015)
26. Schmitt, U.: Proposing a next generation of knowledge management systems for creative collaborations in support of individuals and institutions. In: Proceedings of the 6th International Joint Conference on Knowledge Discovery, Knowledge Engineering and Knowledge Management (IC3K), pp. 346–353. Rome, 21–24 Oct 2014. 978-989-758-050-5
27. Schmitt, U.: Overcoming the seven barriers to innovating personal knowledge management systems. In: Proceedings of the International Forum on Knowledge Asset Dynamics IFKAD, Matera, pp. 3662–3681. 11–13 June 2014. 978-88-96687-04-8
28. Schmitt, U.: How this paper has been created by leveraging a personal knowledge management system. In: Proceedings of the 8th International Conference on Higher Education ICHE, Tel Aviv, pp. 22–40. 16–18 Mar 2014
29. Schmitt, U.: The role of personal knowledge management systems in making citizens highly knowledgeable. In: Proceedings of the 8th International Technology, Education and Development Conference INTED, Valencia, 10–12 Mar 2014. 978-84-616-8412-0
30. Schmitt, U.: Personal knowledge management devices—the next co-evolutionary driver of Human development?! In: Proceedings of the International Conference on Education and Social Sciences INTCESS14, Istanbul, 03–05 Feb 2014, CD 12:4. 978-605-64453-0-9
31. Schmitt, U.: Knowcations—conceptualizing a personal second generation knowledge management system. In: Skulimowski, A.M.J. (ed.) Looking into the Future of Creativity and Decision Support Systems. Proceedings of the 8th Conference on Knowledge, Information and Creativity Support Systems KICSS, Krakow, pp. 587–598. 07–09 Nov 2013
32. Schmitt, U.: Knowcations—the quest for a personal knowledge management solution. In: Proceedings of the 12th International Conference on Knowledge Management and Knowledge Technologies i-Know, Graz. Copyright 2012 ACM 978-1-4503-1242-4/12/09. Graz: ACM, 05–07 Sep 2012. http://dl.acm.org/citation.cfm?id=2362469
33. Schmitt, U., Butchart, B.: Making personal knowledge management part and parcel of higher education programme and services portfolios. J. World Univ. Forum. **6**(4) (87–103): 1835–2030 (2014)
34. Simon, H.A.: Designing organizations for an information-rich world. In: Greenberger, M. (ed.) Computers, Communication and Public Interest. Johns Hopkins Press, Baltimore (1971)

Computer Creativity in Games—How Much Knowledge Is Needed?

David C. Moffat and Paul Hanson

Abstract In seeking to make computer systems that can be creative, there are several problems, including the definition of the concept of creativity, understanding how to design creative algorithms, encoding the domain knowledge needed for them, and evaluating the creative products of those algorithms. We make a case that games are in some ways a good domain in which to research and develop creative algorithms, in particular the design of computer programs that can play games in ways that human players would consider to be creative. Although the playing of games is not in itself a very important research aim, we argue that it is a good arena in which to test out attempts to make creative programs. We initially consider how the field of computer play may be said to have contributed to computational creativity so far, with games like chess that require a specialised knowledge base in order to perform well. As an example, we show the play of our program for a falling-blocks puzzle game, that uses simple search to play to a fairly good standard, and which a human player may well consider to be creative at times. In that case, we conclude that at least in some non-trivial domains, computational creativity might be achieved without the typically heavy requirement for a large or sophisticated knowledge-base or even machine learning.

1 Creativity—Concept and Requirements

1.1 Definition of Creativity

In the literature for creativity (and computational creativity), if a product or idea is generally said to be the result of a creative thought process then it must be both *novel*, and *valuable* (e.g. [1, 5]). What counts as valuable will of course depend on the domain; and the novelty of any idea should somehow be calculated by comparison with previous ideas in the domain. In many or most domains these are vague,

D.C. Moffat (✉) · P. Hanson
Department of Computing, Glasgow Caledonian University, Glasgow, UK
e-mail: D.C.Moffat@gcu.ac.uk

S. Kunifuji et al. (eds.), *Knowledge, Information and Creativity Support Systems*, Advances in Intelligent Systems and Computing 416,
DOI 10.1007/978-3-319-27478-2_30

informal and dependent on the intuition of domain experts. For example, this is the case for all the arts, clearly; and for practically all the so-called "creative industries." Indeed, we always start with human intuition as our only guide to what is creative, until we can develop more formal measures.

Some authors include further conditions alongside novelty and value. Notably, Boden says that a creative idea should generally be in some sense *surprising* [2]. This could mean that the idea is not obvious even to domain experts, and that they would not easily have thought of it themselves. It could also mean that the creative idea might seem elegant, once it is heard. In the more striking cases, it might then even seem obvious; but that is only with the benefit of hindsight.

While there remains debate about the definition of creativity, we shall take it to mean those three conditions, of *novelty*, *value* and *surprise*, with *elegance* an optional fourth. As well as the matter of definition, in the field of computational creativity (CC), there is the difficult matter of designing *algorithms* that could be capable of showing something like a human level of creativity; and the vexatious need to *evaluate* the creative products, even without any clear criteria that can be formulated.

1.2 Requirement for KA (Knowledge Acquisition)

As well as designing algorithms for CC, a prior problem is to provide them with any necessary knowledge to work with. When we aim to make programs that independently contribute to a field of human endeavour, they also need to have all or most of the knowledge that humans already discovered, in that field. This is a formidable requirement, in the general case requiring us to formalise a suitable knowledge representation scheme and then encode in it all relevant human knowledge.

It is essentially the AI version of the philosophical knowledge problem, whose best known and most ambitious work was the *Cyc* project, started in 1984, which has taken an open-source form as *OpenCyc*. The difficulties such projects experience in creating their large knowledgebases, and trying to ensure that they do not permit dubious inferences to be drawn, are a reflection of the inherent difficulty of knowledge. It was a surprise to AI researchers, for example, that common-sense reasoning turned out to be intractable so far, on realistic domain sizes.

This venerable old problem of AI and expert systems is still with us today in most domains of real-world expertise. Even if we had arrived at a fully competent knowledge representation language for all domains, the task of encoding any single domain into it would be extremely intimidating. It is often referred to as the "knowledge acquisition (KA) bottleneck."

Instead of encoding knowledge into a system (or as well as doing that), it is possible to make the system learn for itself, from its own experience in the domain. This approach of machine learning, however, is no nearer a general solution than the more direct approach of encoding. There being as yet no general solution to the KA-bottleneck, we have to choose our domain for creativity carefully, so that we can feasibly solve the KA issue for the chosen domain at least.

1.3 Requirement for Evaluation

Another reason to choose the creative domain carefully is to enable us to evaluate the ideas or other creative products as reliably and as uncontroversially as possible. The evaluation should also be quick, and formalisable, especially in CC fields, to incorporate them into the creative generation algorithms.

In general, however, it is not easy to find a consensus on what ideas in any given domain are truly *novel*, *valuable*, and *surprising* (not to mention *elegant*). Nor can we usually agree on a suitable formalisation of these concepts, which are still an active research area.

Therefore we generally fall back on naive human judgement, possibly with several judges, at least to begin with, in order to conduct at least some kind of evaluation of creative programs. This is the approach taken in this paper.

2 Gameplay as a Domain for Computational Creativity

Historically, games like chess have been important in the development of AI, leading to today's sophisticated algorithms that can beat the strongest human players. The way that these algorithms work is by a highly specialised kind of search process, that searches the very widely branching game tree guided by heuristic evaluations of the anticipated board positions. There is a lot of human knowledge and gameplay intuitions encoded into these heuristic functions. There is also a large component of machine learning, as the algorithms can play against themselves at high speed to better tune their functions. Despite the heavy reliance of the best chess programs on both expert knowledge and machine learning, however, we consider that such a reliance is not actually necessary for games, and so we select a game that can be tractable without demanding these functions.

Here we intend for game playing programs to be creative, just in the way they play their games. There are other ways to apply creative programs to the domain of games, such as by helping to design them (so called "procedural content generation"); but that is not the focus of this paper.

The attractions of gameplay as a testbed for CC (computational creativity) research are that it offers at least partial solutions to most of the above problems.

Firstly, games are typically easy to formalise briefly and clearly, because they are designed to have simple and exact rules. As games are often highly abstracted from their original worldly metaphor (e.g. real armies at war, in the case of chess), they do not require mountains of common sense or other domain knowledge to be encoded. Many games thus avoid or alleviate the *KA-bottleneck problem*.

Secondly, the playing of games has been a testbed application of AI since the beginning, so we have a depth of knowledge about how to write algorithms for this task. This largely solves the *algorithm problem*.

Thirdly, each game situation (or position) is potentially very similar to many that have occurred before in other games; and yet the smallest difference might have big consequences, so that there is rich novelty available, even against any database of games played in the past. Given such a database, it should be feasible to program a variety of possible estimates of *novelty*. Games typically award points or have other ways to quantify play performance, which can conveniently be used to assess *value*. Taken together, these two criteria of creativity are thus fairly measurable, which goes a long way to solving the *evaluation problem*.

Finally, the criterion of *surprise* or even *elegance* are less easy to formalise. We shall therefore use the game score as a basis for value, and rely on human intuitive judgement for the criteria of novelty, surprise or elegance.

3 PuyoPuyo as a Game for CC Research

The subject of our study is a simple arcade game called PuyoPuyo. First we implement a competent player for it. We then illustrate its performance that could be said to be creative gameplay, and end with some of its potential advantages as a testbed for CC research.

3.1 The Game Puyo-Puyo

PuyoPuyo is played at the same time by two opposing players, where each player has a game board of 6×13 spaces. However, our program only plays a simpler, single-player version, so we here leave out the features of the game which are relevant when there are two players.

Pairs of different coloured puyo are dropped onto the player's board, which must then be positioned to avoid filling up the game board. As in *Tetris*, if and when the board fills up entirely to the top, the player loses [8]. Spaces on the board can be cleared by popping four or more puyo of the same colour. When this happens, the player's score increases. After a chain of Puyo is cleared in this way, any other puyo above them fall into their spaces (again, similar to *Tetris*); and if that makes a new chain (of four or more puyo of another colour), then it is also cleared away, and so on until no more puyo can be cleared. Then the next pair of puyo enter the top of the board for the player to drop.

In the example a blue and green puyo pair is dropped. The blue one completes a chain, which then disappears to let the higher puyos fall. A green chain is thereby completed, which also disappears. This causes a blue chain and then a red chain to go as well, leaving only five puyos at the end of the move. Because four chains have gone in a single move, it has got a high score (much more than the four chains would have got separately). To build up the potential for such "cascades" of chains is thus the strategic aim of the game (Fig. 1).

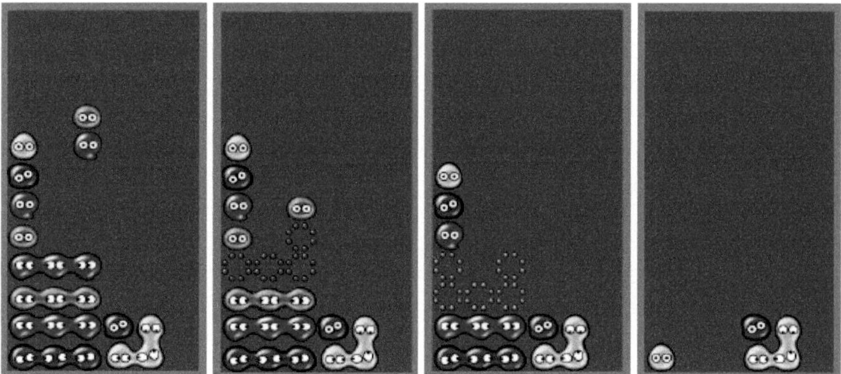

Fig. 1 Screenshots of *PuyoPuyo*. A *green/blue* pair falls, to make a chain of four *blues*, which clear away; then the five *greens* go; then the next four *blues*; and finally four *reds*

While the aim may be clear, it is unclear how to achieve it. In other games like *Tetris*, one can postulate general rules or heuristics about good ways to play the game. Players soon learn to fit blocks in patterns and wait or hope for blocks to fall that exactly fit into the gaps, so that they can clear lines. But in *PuyoPuyo* there is not such an obvious strategic method. We thus fall back on brute search methods, which are not guided by expert heuristics. It is for these reasons that MC algorithms are interesting to investigate.

3.2 Monte Carlo Search (MCS) for PuyoPuyo

Monte Carlo (MC) simulations are predominantly used as a substitute to heuristic evaluation functions in game AI—the advantage being that a value in a search tree can be determined without the need for complex, time consuming expert knowledge based functions. In order to capitalize on the advantages of MC simulations and to improve the search capabilities of it, the concept of MC evolved into what is known as the Monte Carlo Tree Search (MCTS).

The MCTS is a form of best-first search which uses the results of multiple MC simulations to determine which path to take [4] This form of search was used by Bouzy [3] who combined a standard MC search, with a shallow and selective global tree search. His ComputerGo programs Indigo and Olga play the game pretty well [3].

In this study we investigate the possible value of a simpler MC algorithm, as follows, which shows how even a small and simple algorithm can play with some apparent creativity.

Because the game *PuyoPuyo* has a large branching factor it becomes computationally difficult to exhaustively search the tree more than a few layers deep. Each new pair of *puyo* can be rotated four ways (similar to *Tetris*), and dropped in any of the six columns. Less the two positions where one of the *puyo* would be off the edge of the board, this means that the top level of the search tree branches 22 ways (i.e. $4 \times 6 - 2$). In this study the simple MCS algorithm only searches the first layer exhaustively, by visiting each of the 22 children in turn, in round-robin fashion. None of the nodes are favoured either; in contrast to what happens with many MCTS algorithms, where more attractive looking nodes are visited more often under the assumption that the best solutions may be found under those ones. These two simplifications of the more complex and advanced MCTS family of algorithms make this MCS algorithm much shorter to implement in code.

Below the first layer, the MCS algorithm switches into random search mode in which it simulates further plays of the game, by visiting and expanding only one child at random in each layer. However, it is not possible in *PuyoPuyo* to complete each so-called *playout* to the end of the game, until a terminal node is reached with a win-lose result. This is because the game can only terminate in a loss for the player, whose aim is to survive as long as possible, just as in other falling-block games like *Tetris*; and that loss should be far into the future, and well beyond the depth that could guarantee to be reasonably searched in a fixed time limit.

In this implementation of MCS therefore, we impose a depth limit to the search and take the increment in score when the limit is reached as our indication of how good the partial playout (the simulated random play to that depth) was. This score is then used to update the estimated value of the initial move at the first layer that preceded the random playout stage. Each of the 22 first-level nodes (children of the root node) is annotated with its own "best so far" score, from all the playouts that have yet originated from it. In this way the algorithm may be interrupted at any time, and the best of all those first-level nodes can be selected as the best move to play. As the game is to be played in real-time, the MCS algorithm is allowed to run for as long as it can, until the blocks have fallen so far that the AI player must now commit itself to one of the 22 possibile moves, and play it.

After some experimentation, we found that a depth-limit of only three was enough to give a good performance with real-time play on a modest PC (with Intel's Celeron two-core processor). This happens to be the most authentic depth-limit, as it corresponds to being aware of the next two blocks that will come after the current one. As a human player of PuyoPuyo can see the next two blocks queueing up at the side of the board (similar to *Tetris* again), this means the AI player is playing under the same conditions as the human player. More detail about the algorithm itself can be found in our earlier paper [6].

4 Performance of the PuyoPuyo Player

The MCS algorithm described plays PuyoPuyo pretty well. An example of one of its more surprising and possibly creative moves, see Fig. 2.

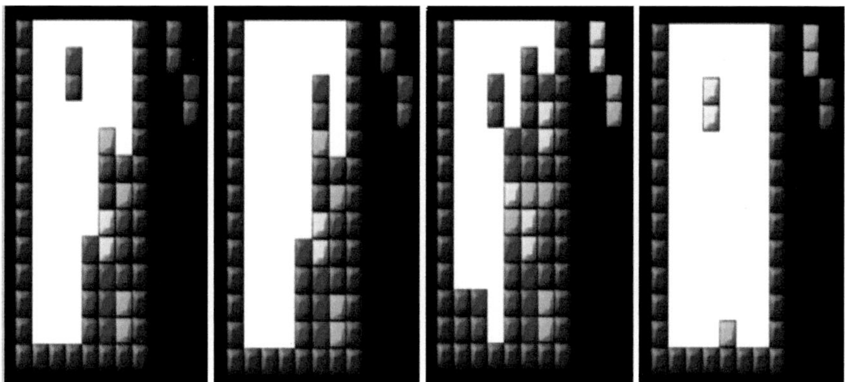

Fig. 2 Program playing *PuyoPuyo*. The program chooses a sequence of less obvious moves that seem to risk losing, but builds a good position on the *left* of the board, and finally exploits it to clear the whole board

In the figure we see that a good move would be to drop the red/blue pair from where it is, as that would clear first the reds, and then more chains after that, to get a good score. But the program instead moves the pair to the side, and then moves the next few pairs to the sides as well; but putting more blues and reds to the left. This endangers the play as the blocks nearly reach the top; but then the last pair shown (the blue/red) is dropped right into the hole, and nearly the entire board is cleared in one go.

This piece of play seems *elegant*, even to the two authors who are novice PuyoPuyo players; as well as apparently risky, which makes the final clearance more satisfying. As a high score is achieved by these moves, and a more obvious and still good move was foregone to make this play, it seems to be a good candidate for a creative play.

Although it may appear that the program is planning a long way ahead, it is in fact only searching to depth three. The reason that it appears to be conceiving of the game in terms of strategies like "building a strong position to the left of the board" and tactics like "aim to plug the gap with same coloured puyo" is probably due to our own human tendency to conceive the game in those terms. The program itself has no such complexities within it, however. It only has knowledge of the rules of the game, and it only looks three moves ahead. Anything else it appears to be doing must therefore be an artefact, or a result of the way we perceive the game.

5 Discussion

An earlier version of this paper was presented at the KICSS-14 conference [7], where it generated some interesting reactions that are worth reviewing here. This included both questions and discussion after the presentation, and comments by reviewers previously.

Questions were raised regarding the claim that little or no knowledge is required for creativity, at least in the restricted domain chosen of games, and in particular the game of *PuyoPuyo*. Some people wished that the amount of knowledge had been quantified more formally.

Because of the lack of *formal* evaluation, there was also some feeling by a reviewer that the case made by the paper was unconvincing, in that the creativity was not demonstrated. However, at the presentation there was no such doubt raised, perhaps because the video recording of the game as it was played was more convincing to the audience than merely reading about it as described in the paper.

In fact, the performance of the AI player seemed to impress all who saw it, just as we had hoped it might. After all, we ourselves as the authors found the play to be strikingly suggestive of creativity, and that it would probably have been deemed creative if played by a human. So it was good to see that the audience who saw the presentation at least seemed to agree thus far. But they began to doubt the creativity in the play once they knew how the AI player was made. One reaction was that the AI player was not properly creative as it had very little knowledge of the game it was playing, or how to play it. This reaction was gratifying to us, as it was in fact the main point of the paper. A related objection was that the AI player could not be truly creative because it was based on a blind search process. Again however, that was actually the point of the paper: to show that little knowledge was needed, and only a simple search strategy, to give the appearance of creative play.

Regarding the quantification of knowledge, we see no need for a formal treatment in this case. It is clear that the AI player has hardly any knowledge, since it is based on a Monte-Carlo search algorithm. It is the conflict between this lack of knowledge and the expectations based on theory (as well as the intuitions of the audience reactions), that was our purpose to establish.

Another reaction expressed, on hearing how the algorithm worked, was to assert that it could not be genuinely creative because it was determined in its behaviour, and would play exactly the same way in all future games, with identical initial circumstances. It is curious to note that this objection is nearly the opposite to the other objection raised above, which found the algorithm uncreative because it was too random in its progress.

However, these objections are all beside the point, it seems to us. The message of the paper is that the algorithm achieves its performance by means of a guided random procedure and no expert knowledge; and that the performance is judged to be creative simply enough by human intuition; first our own, and then the audience at the presentation. All objections that doubt this creative performance, only on the basis of knowledge about how the program is coded, are judging the matter on privileged information, which should strictly speaking be disregarded. If a program is to be thought uncreative simply because it is just a program, whether random or not, then no program will every be judged creative. The field of computational creativity would be doomed to fail from the outset in that case, so it cannot be a fair way to judge the matter.

Although it may seem counterintuitive to many that a program can be creative, and even more so that it can be so without any expert knowledge, and with only the help of a simple search algorithm; nevertheless that is the conclusion of the argument in this paper. The evidence is the AI player described and demonstrated, and its play that appears to be creative as judged by human observers.

6 Conclusion

Having identified some key ways in which KA problems are presenting some major road-blocks to CC research, we argued for the role of games (and in particular game-play) as a suitable domain.

We presented a simple game, *PuyoPuyo*, that can be understood by anybody who knows *Tetris*, but that does not yet have strategies formulated for how to play it well. Therefore a correspondingly simple Monte-Carlo search algorithm was encoded to play it, which had no special game strategy or other heuristic knowledge put into it. The algorithm did not learn either; and yet it was able to achieve a creditable performance. The example play of the game showed that the program is capable of some apparently novel, strong and even surprising play. In other words its play appears to be creative to some extent.

The main significance of the work as we see it, is that for the chosen game and algorithm, a modestly creative performance was achieved without any special knowledge, and without any machine learning.

Furthermore, the level of game expertise required by the CC researchers, who work with such a domain, is low; as is the complexity of the gameplay algorithm. This should allow more focused attention on the key research issues of how best to assess and account for the roots of any creativity found, which could help the research field to progress faster.

Common reactions to this kind of work sometimes appear to be based on precon-cieved notions about what creativity is and how it is achieved. However, they can be misleading and might lead to unfair judgements. The outputs of a creative program should be judged without reference to how they were achieved, or even as if they were produced by a human. In that case, the AI player shown here would be seen as creative, even though it was produced with little knowledge, and no learning. This goes against the prevailing views in the field, and is thus the contribution of this work.

References

1. Boden, M.A.: The creative mind. Abacus, London (1998)
2. Boden, M.A.: Creativity and artificial intelligence. Artif. Intell. **103**, 347–356 (1998)
3. Bouzy, B.: Associating shallow and global tree search with Monte Carlo for 9 × 9 go. In: International Conference on Computers and Games, LNCS 3846, pp. 67–80. Ramat-Gan, Israel (2004)

4. Chaslot, G.M.J.B., Winands, M.H.M., van den Herik, H.J., Uiterwijk, J.W.H.M., Bouzy, B.: Progressive strategies for Monte-Carlo tree search. New Math. Nat. Comput. **4**(3), 343–357 (2008)
5. Csikszentmihalyi, M.: Creativity: Flow and the Psychology of Discovery and Invention. Harper-Perennial, New York (1997)
6. Hanson, P., Moffat, D.C.: Monte Carlo search for a real-time arcade game, puyo-puyo. In: Proceedings of the Symposium on AI Games. AISB 2014 Convention, at Goldsmiths, University of London (2014)
7. Moffat, D.C., Hanson, P.: Computer creativity in games—how much knowledge is needed? In: Papadopoulos, G.A. (ed.) Proceedings of the 9th International Conference on Knowledge, Information and Creativity Support Systems, pp. 346–351. Limassol, Cyprus, 6–8 Nov (2014). http://kicss2014.cs.ucy.ac.cy
8. Sonic Team: PuyoPuyo. Sega (2009). Cartridge; Nintendo DS

When Engineering and Design Students Collaborate: The Case of a Website Development Workshop

Meira Levy, Yaron Shlomi and Yuval Etzioni

Abstract Designers are often characterized by their creativity, whereas engineers are often characterized by their analytical thinking style. Both capabilities are vital in the new economic era, particularly for innovation. Thus, the collaboration between designers and engineers is crucial and should be fostered by academia and industry. Our preliminary research examined perceptions and interactions among 14 students in engineering, art and design programs during a website development workshop. The research indicates that students enjoyed the experience and found it educational and socially engaging. The data collected through questionnaires and observations were qualitatively analyzed, revealing that students learned each other's jargon, conducted discussions that addressed both technological and design concerns and considered various opinions leading to joint decision making. Importantly, one of the teams was invited to continue development of its project in an entrepreneurship accelerator program. This study sheds light on the potential of multidisciplinary teams in development processes in general and in website development processes in particular.

Keywords Multidisciplinary collaboration · Website development · Qualitative research

M. Levy (✉) · Y. Shlomi
Department of Industrial Engineering and Management, Shenkar College of Engineering and Design, 12 Anna Frank Street, Ramat-Gan 52526, Israel
e-mail: lmeira@shenkar.ac.il

Y. Shlomi
e-mail: yshlomi@shenkar.ac.il

Y. Etzioni
Department of Textile Design, Shenkar College of Engineering and Design, 12 Anna Frank Street, Ramat-Gan 52526, Israel
e-mail: yetzioni@shenkar.ac.il

S. Kunifuji et al. (eds.), *Knowledge, Information and Creativity Support Systems*, Advances in Intelligent Systems and Computing 416,
DOI 10.1007/978-3-319-27478-2_31

1 Introduction

In the new economic era multidisciplinary knowledge and capabilities are required for gaining competitive advantages [1, 2]. Previous research addressed the need to enable effective collaboration among people from different disciplines, and found that successful multidisciplinary encounters depend on tailoring the selection of a theme, participants, and location to the encounter's particular objectives [3]. In particular, universities are preparing their students for a vital need of industry and society; i.e., to bridge the gap between producers and consumers of technology. To this end, universities are enriching their curricula with relevant themes from social sciences, humanities, cultural and management studies. Moreover, students from various disciplines need to understand the financial, business, environmental economic and social constraints in which businesses operate [4]. For this aim, multidisciplinary workshops are organized where participants are familiarized with one another's profession and learn to appreciate dissimilar viewpoints [5]. Although the literature suggests that multidisciplinary programs are beneficial for broadening the students' perspectives, there are scant reports that describe the interdisciplinary educational perspectives and the interactions that occur among their participants.

In this paper we report on a multidisciplinary workshop aimed at website development. Workshop participants were students from engineering, design and art departments. During the workshop they gained technical knowledge related to a platform for web design, as well as knowledge related to digital culture and images. Understanding digital culture is vital to the development of a website as it encompasses the social implications of the knowledge and images presented in the site. Our preliminary research aimed at understanding the students' perceptions and interactions during the workshop for drawing inferences on how multidisciplinary teams might work collaboratively. An earlier report on this study can be found in [6]. Our paper is organized as follows: we present the importance and challenge of collaboration in the design process and elaborate on digital culture. We then present our empirical study and suggest some conclusions and directions for future research.

2 The Importance and Challenge of Collaborative Design

Designers are often characterized as creators, empathizers, pattern recognizers and meaning makers [1]. They are trained to deal with fuzzy and unpredictable situations and projects that require personal perspective, visual presentations, emotional involvement and almost no quantification [7]. Conversely, engineers are often characterized as more logical and linear thinkers with computer-related capabilities. They study science, accounting, finance, and related analytical subjects. There is a growing awareness of the need to create multifunctional teams that include designers and engineers in new product development process. The multifunctional teams are viewed as a strategic tool aimed at fostering innovation and gaining competitive advantage [2, 8].

Working together, designers and engineers can change the way organizations are operating. Their collaboration will transform the organizations from the Information Age economy to the Conceptual Age economy characterized by inventive, empathic, and big picture capabilities [1]. Moreover, integrating the capabilities of sequential, logical and analytical thinking associated with engineers with the nonlinear, intuitive, and holistic thinking associated with designers [1] is essential to innovation. However, enabling the collaboration of engineers and designers is challenging. People from different professional backgrounds often have different value systems leading to misunderstanding and conflict [2]. Some of these differences can be traced to the training (i.e., the academic curriculum) that these professionals received (ibid).

3 Digital Culture

Digital culture exhibits the intersection of technology, knowledge, and culture in the digital age. The digital art object or image is a realization of human experience and hence considered an expression of human life in the field of aesthetics. Manovich [9] claims that the "New Media" refers to a new approach to culture representations including an art object or an image as means of communication. The new media technology does not create new images but it brings new representations and manipulations capabilities with enormous presentation opportunities in private and social spaces (ibid). Specifically, manipulated digital images yield less detailed representations of reality than those obtained with the naked eye. Thus, researchers predict that watching digital images desensitizes our awareness to missing information [10]. At the same time new digital media tools and software can produce or edit digital images that may replace reality with virtual visibility, and can reduce human ability to fulfill human dialogue and relationships at the reality sphere [11].

As we turn to digital culture our visual experience is both expressing and challenging every aspect of our life. We are facing changes in our relationships with human body, objects and materials, our attitudes to written text and toward local and global time and space (ibid).

4 Empirical Study

4.1 Method and Settings

The main objective of our study was to explore the students' perceptions, expectations and interactions that occur during a design-engineering workshop. The students were third year students at Shenkar College of Engineering, Design and

Art in Israel. Lecturers from the engineering, design and art departments formed teams for Shenkar's annual "Jam Week". Each team of lecturers prepared a 4-day workshop in which students from the different departments worked together on projects guided by the workshop organizers. The first and last authors led the workshop discussed in this article, entitled "The Internet Sphere and Our Lives Sphere, Reflection and Alteration" with the aim of studying the digital culture and web technology principles and developing a website according to the students' interests. During the workshop students learnt, besides the web design platform, what digital culture is, and in particular what digital image means for people, and how artists develop it [12]. Understanding the digital culture concept was fundamental and allowed design discussions beyond technological issues.

The students learnt the Wix[1] Platform, which is a commercial free platform for creating websites. The 4 days of the workshop were mostly devoted to the teams' independent work on their projects, with few breaks for short lectures given by a representative from WIX and by the two workshop organizers. The WIX representative also served as a tutor during 2 days of the workshop, while the lecturers advised the teams as needed. The workshop participants were 14 students (7 from design and art and 7 from engineering) who were divided to four multidisciplinary teams.

The workshop started with a social game in which students reported on three subjects of interest to them. Then we used a string to create a physical network among students who reported on related topics. Finally, students were assigned to teams based on common interests, with the constraint that each team include at least one student from engineering and one from design or art.

Data were collected through open-ended questionnaires (see Appendices 1 and 2) and through observations of the teams during their work. Students completed one questionnaire at the beginning of the workshop, and a second one at the workshop's conclusion. The gathered data were qualitatively analyzed [13].

4.2 Analysis and Findings

The analysis of the pre-workshop questionnaire revealed that all of the students were enthusiastic to learn a new web building platform that can serve them in the future. Eight students used one of the words "innovation"/"creativity"/"thinking out of the box"/"inspiring" in their list of associations. The words "expectations"/ "excitement"/"experience"/"curiosity"/"enjoyment"/"pleasant" were found in 11 questionnaires. Ten students mentioned that they want to get to know other workshop participants, especially those from a different discipline, for gaining other perspectives.

[1]http://www.wix.com/.

These findings suggest that students are eager to meet colleagues from various disciplines, for extending their perceptions and learning experience.

The data that were gathered during the workshop focused on conversations among the teams' members. Qualitative analysis of these exchanges revealed that students, at the beginning of the workshop, tended to comment on issues that were more related to their discipline. For example, a design student talked about the site's logo, and an engineering student commented "color is your field", while pointing at the design student. Engineering students focused on the functionality of the site or handling the work: "we need to reduce clicks"/"lets upload the content, we won't be able to finish", while design students focused on the design or the general site theme: "we need to add affects, like something is burning"/"the site should be more clean"/"The main page is the most important one, we must invest in the design".

As the workshop continued, we heard engineering students adopt the design role and terminology: "the site gives the feeling of wholeness"/"What is your design concept?"/"we need a digital image that expresses the hard work of the artist therefore we need a close-up picture". At the same time, design students adopted the more engineering perspectives: "we need to present a work process"/"we need to add a button to stop the music".

These findings suggest that although the workshop lasted only 4 days students got involved with issues that initially belonged to their colleagues from the counter discipline. The general atmosphere among the team members was that while the engineering or design student was considered the professional in her discipline, there was mutual respect to hear opinions from everyone in the team.

The analysis of the post-workshop questionnaires revealed that the students found the workshop pleasant, instructive, and socially beneficial. Twelve students mentioned "collaboration"/"team work" and "mutual work" as terms describing their thoughts at the end of the workshop. They mentioned the collaborative atmosphere and decision making processes where each student could express thoughts and preferences.

4.3 Project's Development in the Entrepreneurship Accelerator Program

After the Jam week ended, Shenkar Center for Innovation invited the workshops' participants to compete for enrollment in Shenkar's ACT (Accelerating Creative Talents) Accelerator program. The program was aimed athelping innovators in their early stage to promote their innovative ideas and make them economical feasible. One of our workshop's teams won the competition, and was invited to take part in an entrepreneurship accelerator program. We regard this invitation as an additional indicator of the workshop's success. The winning team developed an online marketplace that would allow students in Shenkar and other academic institutions to sell their design products. During the program, the students were familiarized with building a business plan, protecting intellectual property, creating presentations for funds raising and team building. The program also included business and technological guidance, initial funding and a business mentor. In addition, the program participants join a community of innovators that gain experience and insights from each other.

In a recent interview with the students in this team, we learned that the program allowed the team to enhance their initial idea, and crystallized their business plan. Moreover, they continue to develop the proposed website independently, i.e., even after completing the entrepreneurship program. One of their shared decisions is to add a designer to their team, who is more familiar with web development, for implementing the site.

5 Conclusion and Future Research

We report on a multidisciplinary workshop in which students from the engineering, design and art departments collaboratively developed websites according to their interests while being engaged in technological and cultural enrichment activities. Our view is that the interactions among the workshop participants exemplify collaboration among engineering and design students. Moreover, the study revealed that students were enthused to participate in such experience and found it educational and enjoyable. The interactions enabled mutual learning of each discipline's terminology while respecting various opinions when arriving at joint decisions regarding the characteristics of their websites. The multidisciplinary discussions, encompassing both technological and design perspectives, enriched the class and created a creative atmosphere that fostered creation of the websites. The atmosphere in the workshop which allowed social interactions as well as educational opportunities, developed trust among the workshop participants and provided the ground for creating innovative ideas and getting feedbacks from other groups. These dynamics eventually enabled one of the teams to earn the prestigious participation in the ACT Acceleration program. Our on-going research will follow this group and study their collaboration and future development.

Our findings concur with other proponents of collaboration between the art, design and innovation communities [14]. Supporters of these collaborative efforts realize the value of design in innovation processes which enrich the technological perspective with new user-centered viewpoints. The design-inspired innovation brings a new terminology to the industrial design process, leading to new products and services.

While this case study was successful, the generalization of this preliminary research may be limited. However, it sheds light on the potential of multidisciplinary teams in development processes in general and in web design process in particular.

Engineers and designers rely on a variety of design terminologies to define and model their problems and alternative solutions. Further, we assume that these professionals may not use the same design modeling tools (e.g., UML, wireframe). More generally, the two disciplines might be unfamiliar with each other's goals, perspectives, and tools. This is important because each of the disciplines relies on these elements to present its design solution (e.g., website).

Future research will continue to study the evolution of the innovative project that came out of the workshop. In particular, we will explore differences between the goals, perspectives and modeling tools of the two disciplines. In the long term, this exploration will guide us in creating a modeling tool that is accessible and beneficial to both disciplines. This tool will facilitate collaboration within multidisciplinary teams.

Appendix 1: Workshop Entry Questionnaire

Welcome to the workshop "The Internet Sphere and Our Lives Sphere, Reflection and Alteration". The workshop staff is interested in documenting your attitude to the workshop and to topics related to it. We would like to ask you to answer some questions at the beginning of the workshop. Thank you for your cooperation.

1. Name (optional): ___________
2. Email: ____________
3. Academic program: _______________________
4. Why did you choose this workshop?
5. What are your expectations and goals from the workshop?
6. List 10 words describing associations, perceptions, emotions, etc. that you can think of at the beginning of the workshop.
7. Describe your knowledge and experience as it relates to the workshop agenda as presented so far.
8. During the workshop you will meet and work with students from different academic programs. What are your expectations and goals from meeting and working together?
9. Do you have any professional work experience? If so, please elaborate.

Appendix 2: Workshop Summary Questionnaire

Towards the end of the workshop, you are requested to give the workshop organizers a feedback about the workshop activities and personal experiences you went through as individual and as a member of a joint team of engineering and design. Thank you for your cooperation.

1. Name (optional): ___________
2. Email: ____________
3. Academic program: ________________________
4. Did the activities in the workshop match your expectations and goals from the workshop?
5. Describe your experience working with colleagues from other academic programs
6. List 10 words describing associations, perceptions, emotions, etc, relating to the end of the workshop
7. What new subjects have you learned in the workshop?
8. Describe how decisions were made in the team, and whether there was a priority in decision-making related to the academic program (design/engineering). (in issues of choosing the site idea, site design, site content, technological issues, work documentation, team handling issues)
9. Do you have any recommendations for a similar workshop in the future?

References

1. Pink, D.H.: A whole new mind: why right-brainers will rule the future. Riverhead Books, New York, NY (2005)
2. Stamm, V.B.: Managing Innovation, Design and Creativity, 2nd edn. Wiley (2008)
3. Bridle, H., Vrieling, A., Cardillo, M.: Preparing for an interdisciplinary future: a perspective from early-career researchers. Futures **53**, 22–32 (2013)
4. Parashar, K.M., Parashar, R.: Innovations and curriculum development for engineering education and research in India. Procedia—Social and Behavioral Sciences, **56**, 685–690 (2012)
5. Taneli, Y., Yurtkuran, S., Kirli, G.: A Multidisciplinary Design Exercise: Myndos Excavation Site. Procedia—Social Behav. Sci. **106**, 120–129 (2013)
6. Levy, M., Shlomi, Y., Etzioni, Y.: When engineering and design students collaborate: the case of website development workshop. In: KICSS 2014 the 9th International Conference on Knowledge, Information and Creativity Support Systems, Limassol, Cyprus (2014)
7. Managers and Designers: Two Tribes at War? In: Oakley, M. (ed.) Design Management: A Handbook of Issues and Methods. Blackwell, Oxford (1990)
8. Rigby, D.K., Gruver, K., Allen, J.: Innovation in turbulent times. Harvard business review. (2009). http://hbr.org/2009/06/innovation-in-turbulent-times/ar/1. Accessed 6 March 2014
9. Manovich, L.: Avant-garde as Software. http://www.manovich.net/ (1999)
10. Rushkoff, D.: Program or be programmed: ten commands for a digital age. Berkeley: Soft Skull Press. http://share.pdfonline.com/6d37ec8261714ac18920d77697ad83b3/Program_or_Be_Programmed.htm
11. Verilio, P.: Speed and information: cyberspace alarm. http://www.ctheory.net/articles.aspx?id=72. Accessed 14 Dec 2014
12. Sagmeister, S.: Everything I do always comes back to me. http://www.sagmeisterwalsh.com/. Accessed 14 Dec 2014
13. Strauss, A., Corbin, J.: Basics of qualitative research grounded theory procedures and techniques. Sage Publications, Newbury Park (1990)
14. Utterback, J., Vedin, B.A., Alvarez, E., Ekman, S., Walsh Sanderson, S.W., Tether, B., Verganti, R.: Design-inspired Innovation. World Scientific Publishing, New Jersey (2006)

Knowledge-Managing Organizational Change Effectors, Affectors and Modulators in an Oil and Gas Sector Environment

Anthony Ayoola

Abstract Knowledge Management techniques (KMT) have been used effectively in a number of business domains, for dealing with complex management issues. A key area of interest is the use of KMT for managing organizational change in an effective manner. Developmental, transformational and transitional change types are generally influenced by a range of inter-dependent factors. These factors can be broadly categorized as change effectors, affectors and modulators. This paper investigates the use of Knowledge Management techniques for modulating salient change effectors, affectors and modulators influential in the oil and gas sector, within a Middle-Eastern context. Research findings from a recent multi-company case study of oil and gas sector organizations in the Gulf region are presented based on a conceptual framework predicated on crucial change effectors, affectors and modulators. Knowledge Management modulated change factors investigated in the study include worker empowerment, autonomy, participation, communication, culture, normative responsibility, training/re-skilling and reward systems.

Keywords Knowledge management · Interaction circles · Knowledge diffusion · Knowledge externalization · Sustainable change

1 Introduction and Background

Companies in the oil and gas sector have undergone significant change in recent years, driven by technological growth, business competition, market-place issues and public expectations. Technology is continually advancing and non-progressive companies find their organizational set-up outdated and inefficient. To counter this,

A. Ayoola (✉)
School of Business & Management, American University of Ras al Khaimah, Ras al Khaimah, United Arab Emirates
e-mail: Anthony.Ayoola@aurak.ae

S. Kunifuji et al. (eds.), *Knowledge, Information and Creativity Support Systems*, Advances in Intelligent Systems and Computing 416,
DOI 10.1007/978-3-319-27478-2_32

oil and gas sector firms embark on frequent bouts of change with attendant developmental, and management issues.

This paper provides a systematic framework for characterizing developmental, transformational and transitional change types and determining how these are influenced by a range of HRM-dependent factors. The causal HRM-related factors are systematically categorized as change effectors, affectors and modulators. This paper will establish which effectors, affectors and modulators exert the most significant levels of influence in the oil and gas sector; within a Middle Eastern context.

Studies by Stace and Dunphy [1] map the change driven formats to various change types. So, for example, they recommend the mapping of unilaterally driven change to the transformational change type, primarily for reasons of convenience and the much shorter timescales.

For transitional or developmental transitions, multilaterally driven (participative) change is recommended, since there is greater scope for worker involvement, and longer timescales. Knowledge-Managed HRM (KM-HRM) spans several people-management activities and is applied at multiple levels within the organization. When properly practiced, KM-HRM provides a framework for holistic people management, serving to select, train, motivate and performance manage the company's workforce. KM-HRM has significant strategic and operational implications. The work we have carried out systematically clarifies specific KM-HRM practices as being effectors, affectors or modulators.

2 Knowledge-Managing Change

Standard change management schemes [2] are typically linear, with feedback loops often used to regulate and fine-tune the process. It is possible to integrate Knowledge-Management formalism within standard change management. We term this approach KMCM (Knowledge-Managed Change Management). KMCM manages tacit and explicit change-related knowledge sources within the organization. KMCM is predicated on key Knowledge Management (KM) phases which are in turn based on Socialization, Externalization, Combination (Group-learning and Dissemination) and Internalization KM activities. Approaches of this nature are termed SECI-based Systemic Models [3–5] and [6]. We have used SECI-based models to formulate a more relevant and robust framework to underpin the proposed KMCM technique. The derived scheme is shown in Fig. 1.

We can elaborate further on the KMCM phases.

Phase 1: Initiation of Change Management Processes

Phase 1 entails identification of change stakeholders, change management assets and change effectors, affectors or modulators (to be defined in the next section) within the organization. It also requires the specification of usable change-defining

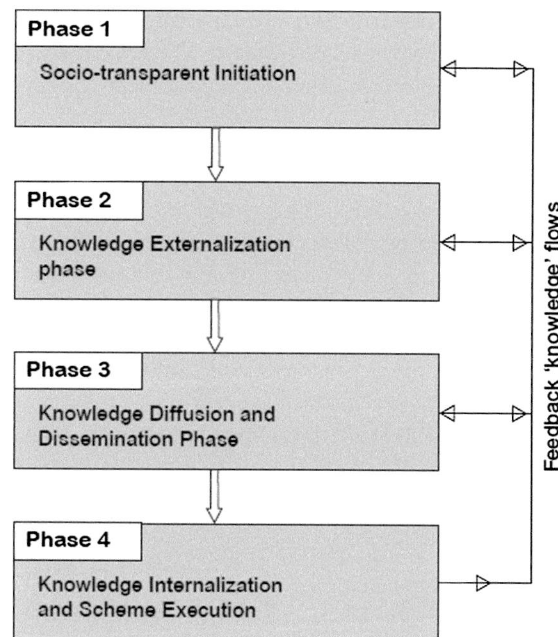

Fig. 1 Multi-phased KMCM workflow

metrics that conform to 'SMART' criteria. That is, they have to be specific, measurable, achievable, relevant and time-bound.

Phase 2: Knowledge Externalization Phase

The previous phase served to collate primary change management knowledge. A significant component of this is *tacit* or *implicit* change management knowledge. That is, many strands of the change management know-how are informal and known primarily to specific individuals and/or areas of the organization. In phase 2 we finalize the collation of underpinning tacit or implicit change management knowledge. Within this phase we formalize and render explicit all the main tacit knowledge strands pertaining to change management. A key requirement within this phase is the creation of *Change Knowledge Interaction Circles (CKICs)*—Small groupings of key personnel involved with change management within the organization (termed the Socialization Sub-phase).

Phase 3: Knowledge Diffusion and Dissemination

Phase 3 starts by expanding the *Change Knowledge Interaction Circles (CKICs)* into larger learning groups. The CKICs serve as *nuclear* or *seed* groupings holding externalized and formalized change management knowledge. Members of the CKICs are intra-organizational change management experts or super-users. By having them interact formally (and informally) with larger employee groupings, change management knowledge is *diffused* across the organization. The CKIC—employee interaction generates a group-learning scenario, serving to raise awareness of change management issues and practices.

Phase 4: Knowledge Internalization and Execution

Our discussion of Phase 3 suggested that change knowledge diffusion would, over a period, start to modulate change management behavior with the organization. This is essentially the early phases of change knowledge internalization. In Phase 4, all change stakeholders within the organization will have started to internalize more optimal change management processes and best practices. Additionally, at this point, the organization embarks on a full-scale execution of its change management program(s).

Within the KMCM workflow, a key issue is the detailed characterization of change-related HRM effectors, affectors or modulators. We will discuss how this is carried out in subsequent sections.

3 HRM Effectors, Affectors and Modulators

The KMCM work carried out systematically clarifies specific HRM practices as being effectors, affectors or modulators. Effector HRM practices seek to implement people management policies and processes. This includes recruitment activities, disciplinary processes, job design and performance appraisal. Effector HRM is driven by targets and performance levels. Affector HRM practices are concerned with building commitment, loyalty and trust within the workforce. Affectors seek to enhance the psychological contract between the company and its employees. Affector HRM provides the caring and supportive face to HRM. It also strives to elicit participation where this is in alignment with the organizational ethos or culture. Affector HRM roles include providing mechanisms and policies for benefits, dispute resolution, grievance processes etc. Affector HRM seeks to minimize worker discontent, and match employee expectations to what the company is able to offer.

Modulator HRM practices provide the necessary control mechanisms for people management. This could include organizing training for re-skilling or up-skilling, and the restructuring of worker reward systems and procedures. Modulator HRM is able to modulate or elevate states induced by effectors or affectors e.g. changing skills levels of newly recruitment employees through training or improving levels of worker motivation through well designed reward systems. In other words, Modulators can sometimes be viewed as secondary transitioning agents, whilst Effector and Affector HRM practices are primary influences. Having specified concepts pertaining to KM, change management and HRM, we can proceed to integrate the three strands within relevant parts of the KMCM workflow.

Change state transitions are thus affected by the quality and expediency of effector, affector and modulator HRM practice. The transitions are also subject to promoter and inhibitor factors.

Change promoters include

- Favorable (change-embracing) strong organizational culture.
- Management support and commitment to the change.
- Strong worker involvement and clarity about purpose of change.
- Effective communication
- Favorable external environment—e.g. sector with minimal levels of competition.
- Strong leadership.

Change inhibitors include

- Unfavorable, change–averse or change resistant organizational culture
- Insufficient or no management support
- Low worker involvement, and lack of clarity or transparency of change objectives
- Poor communications (resulting in greater uncertainty)
- Unfavorable external environment
- Short timescales or insufficient time
- Weak leadership

By judicious and effective use of modulator practices, the effects of the various inhibitor factors can be mitigated, or the promoter factors enhanced/strengthened. As an example, a grid can be used to map effectors, affectors and modulators to some of the relevant change parameters.

4 Methodology of the Study

A multi-company comparative case study of companies in the Middle Eastern Oil and Gas sector was carried out over several months [7]. The study was based on a critical realism research epistemology. A mixed inductive- deductive research approach was used. Data collection utilized a mix of primary and secondary data collection instruments. The primary data was collected using self-completed questionnaires (SCQ) and semi-structured interviews (SSI). Secondary data was used for additional triangulation. The time-frame was essentially longitudinal, with cross-sectional elements. The selected companies had been involved in significant change processes, and stratified samples of HR managers, line managers, training managers and regular employees were used. The work identifies the key HRM effectors, affectors and modulators that influence change management within the organizations, as a prelude to adopting and implementing a formal KMCM scheme for managing future change.

5 Study Findings

See Figs. 2, 3, 4, 5, 6 and 7.

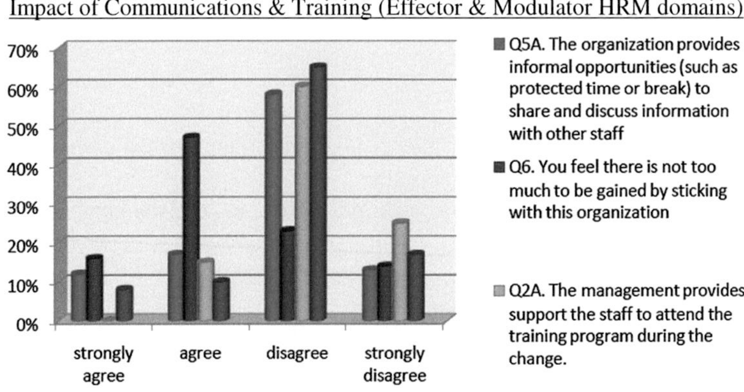

Fig. 2 Impact of communications and training

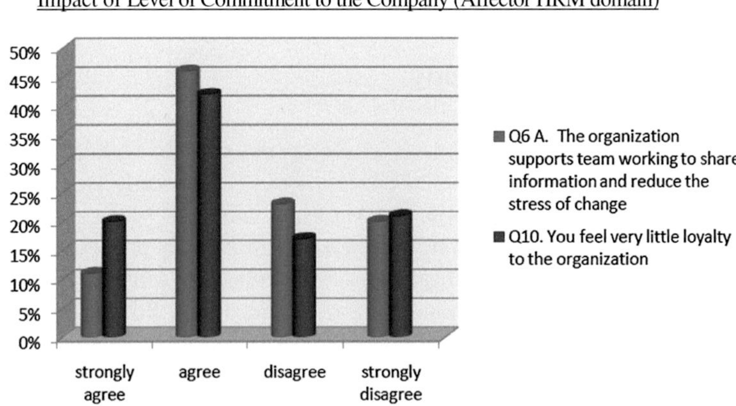

Fig. 3 Impact of level of commitment to the company

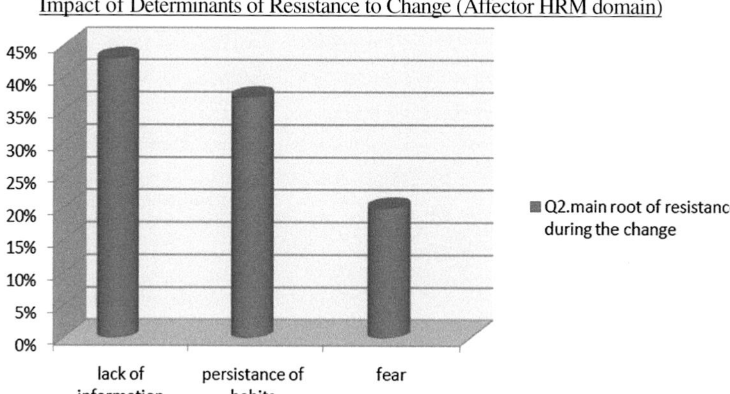

Fig. 4 Impact of determinants of resistance to change

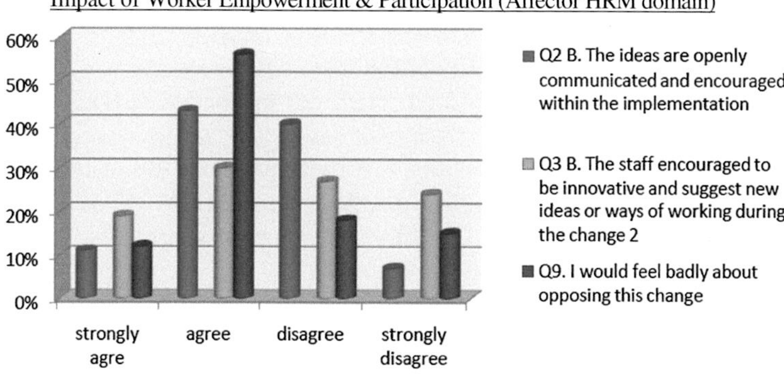

Fig. 5 Impact of worker empowerment and participation

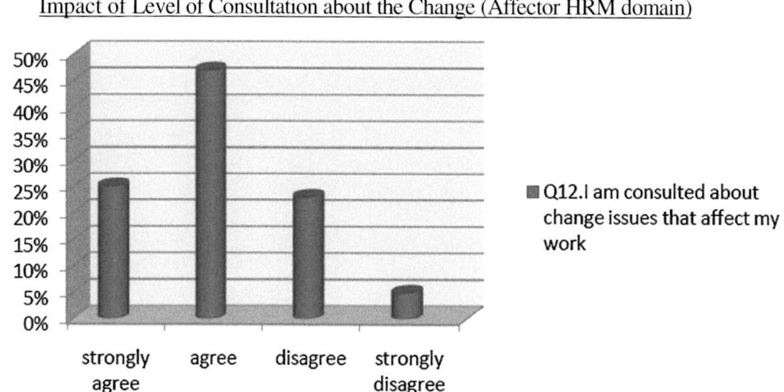

Fig. 6 Impact of level of consultation about the change

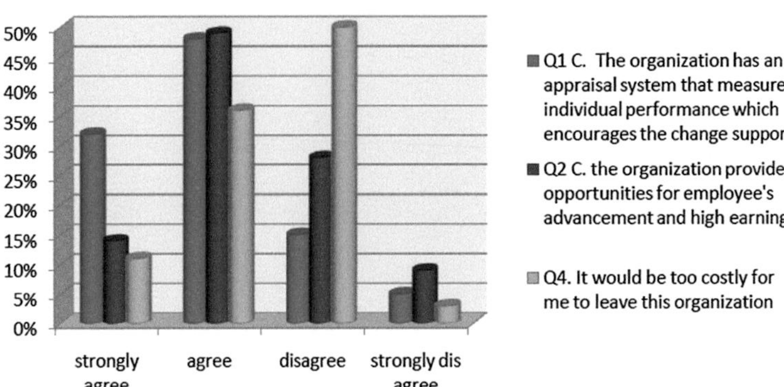

Fig. 7 Impact of employee rewards and performance appraisal

6 Comments and Conclusions

Varying levels of Effector, Affector and Modulator HRM practices were used by the companies during the period of change, with the qualitative responses and perceptions shown in the charts. In a majority of cases, the changes were judged to be difficult, and the overall conclusion was that more judicious use of the prevailing HRM Effector, Affector and Modulator practices, within a KMCM scheme, would have resulted in more efficient and effective change processes.

Acknowledgments The author gratefully acknowledges the support provided by the American University of Ras Al Khaimah, and is appreciative of the input provided by the consultancy team at Hanta Associates, as well as valuable discussions with relevant academic Faculty of the University of Bolton, Bolton-UK and Ras Al Khaimah-UAE Campuses.

References

1. Dunphy, D., Stace, D.: The strategic management of corporate change. Hum. Relat. **46**(8), 905–920 (1993)
2. Carlsson, S.A.: Knowledge managing and knowledge management systems in inter-organisational networks. Knowl. Process Manag. **10**(3), 194–206 (2003)
3. Knight, T., Howes, T.: Knowledge Management: A Blueprint for Delivery. Butterworth Heinemann, Oxford (2003)
4. Lane, P.J., Lubatkin, M.: Relative absorptive capacity and inter-organizational learning. Strateg. Manag. J. **19**(5), 461–477 (1998)
5. Newell, S., Robertson, M., Scarbrough, H., Swan, J.: Managing Knowledge Work and Innovation. Palgrave Macmillan, London (2009)
6. Scarbrough, H.: Knowledge Management: A Literature Review. Issues in People Management Series, Institute of Personnel and Development, London (1999)
7. Rezaeian, H.: MBA Dissertation, University of Bolton (2010)

From Computational Creativity Metrics to the Principal Components of Human Creativity

Pythagoras Karampiperis, Antonis Koukourikos and George Panagopoulos

Abstract Within the field of Computational Creativity, significant effort has been devoted towards identifying variegating aspects of the creative process and constructing appropriate metrics for determining the degree that an artefact exhibits creativity with respect to these aspects. However, in the effort to determine if an artefact is creative by human standards, it is also important to examine the perception of creativity by humans and to which extend this perception can be formalized and applied on the evaluation of creative works. In this paper, we investigate how the human perception for creativity exhibited in text artefacts can be correlated by the usage of appropriate formulations of computational creativity metrics. To this end, we propose a model for transitioning from traditional metrics to a space that adheres to the principal components of human creativity and reflects the way that human approach the assessment of the creative process.

Keywords Computational creativity · Creativity measurement · Creativity in texts · Cognitive models

1 Introduction

Human creativity is a multifaceted, vague concept, combining undisclosed or paradoxical characteristics. As a general notion, creativity adheres to the ability to move beyond traditional and established patterns and associations, by transforming them to new ideas and concepts or using them in innovative, unprecedented contexts and settings [6]. In general, the creativity of a person can be divided qualitatively by taking into account its origin in psychometric or cognitive aspects of

---

P. Karampiperis · A. Koukourikos (✉) · G. Panagopoulos
Computational Systems & Human Mind Research Unit, Institute of Informatics & Telecommunications, National Center for Scientific Research "Demokritos", Agia Paraskevi, Greece
e-mail: kukurik@iit.demokritos.gr

S. Kunifuji et al. (eds.), *Knowledge, Information and Creativity Support Systems*, Advances in Intelligent Systems and Computing 416,
DOI 10.1007/978-3-319-27478-2_33

their thinking process [1]. On the other hand, machines can mimic human creativity, or provide the necessary stimuli for encouraging and promoting the production of creative ideas and artefacts, it is not straightforward to assess the exhibited creativity by using automated techniques. Rather, most efforts have been focused on analyzing creativity on different aspects and producing different metrics, based on the nature of the examined artefacts.

In [4], we introduced our proposed approach for formulating text-based metrics for the core aspects of creativity as the latter are determined in the relevant literature and examined their conformance with the human perception of what constitutes a creative artefact. In this extended version we present the detailed experiments and elaborate on the deviations between these two perspectives (computational metrics and human judgment) and present in detail a model for transferring from automatics measures to a space that more accurately reflects the human opinion.

2 Understanding the Human Perception of Creativity

2.1 Metrics for Computational Creativity over Textual Content

The association of creativity metrics with quantifiable results derived from the textual data is the critical step for automating the evaluation process. The formalization of creativity metrics for textual content is a complex task, and the related work is focuses on very specific characteristics of the examined content in order to model creativity.

Zhu et al. [6] propose a machine-learning approach based on features derived from computer science and psychology perspectives. Other works focus on concrete linguistic and morphological characteristics of the text, e.g. analogies [2].

In our previous work [3], we presented a formulization of a set of Computational Creativity Metrics for Novelty, Surprise, Rarity, and Recreational Effort, over textual artefacts. In this paper, we extend this work in order to correlate these Computational Creativity Metrics with the perception of creativity by humans.

2.2 Correlation of Computational Creativity Metrics with the Human Perception of Creativity

In order to assess the adherence of the proposed metric formulization with the human perception for creativity, we organized and conducted an experimental session based on storytelling activities. For the execution of the experiment, we employed forty (40) human participants, split in ten (10) teams of four (4) members each. All teams were asked to construct a story, on a specified premise, the survival of a village's habitants under a ravaging snow storm. The stories were created incrementally, with twenty (20) fragments produced for each story.

Following the completion of the stories, the teams were organized in two groups, each consisting of five teams. Without any interaction between the groups, each team was called to rate the stories of the remaining four teams belonging to their group, using a rank-based 4-star scale (i.e. the best story received 4 stars, the second-best story received 3 stars etc.). In this way, we obtained a ranked list of the five stories in each group. The goal of our experiment was to determine if, using the ranked lists of one of the test groups and a formalized representation of the computational creativity metrics, we can identify their correlation and examine if the distribution of values for the metrics follow the pattern of human judgment. To this end, we define a constrained optimization problem over functions of the aforementioned metrics, which is described below.

2.3 *Extracting a Model for the Human Perception of Creativity*

Each artefact (story) S_n is characterized (via the execution of the computational creativity metrics presented in the previous section) by a set of 4 independent properties $g^{S_n} = \left(g_1^{S_n}, g_2^{S_n}, g_3^{S_n}, g_4^{S_n}\right)$ where g_1 stands for "Novelty", g_2 for "Surprise", g_3 for "Rarity" and g_4 for "Recreational Effort". We define as partial creativity function (PCF) related to artefact property g_k a function that indicates how important is a specific value of the property g_k when calculating the creativity of an artefact S_n. This function is defined by the following formula:

$$PCF_{g_k}\left(g_k^{S_n}\right) = w_{g_k} * \left(\frac{c_{g_k} * \left(1 - d_{g_k}\right)}{e^{\left(a_{g_k} * g_k^{S_n} + b_{g_k}\right)^2}} + \frac{d_{g_k}}{2}\right), \quad (1)$$

where $g_k^{S_n} \in [0, 2]$ is the value of property g_k for the artefact S_n, and $0 \le a_{g_k} \le 5$, $-4 \le b_{g_k} \le 4$, $0 \le c_{g_k} \le 1$, $0 \le d_{g_k} \le 2$ are parameters that define the form of the partial creativity function, whereas $0 \le w_{g_k} \le 1$ represents the weight of property g_k in the calculation of the overall creativity. The calculation of the above parameters for all g_k properties lead to the calculation of the complete creativity function (CCF), as the aggregation of the partial creativity functions, as follows: $CCF\left(g^{S_n}\right) = \frac{1}{4} * \sum_{k=1}^{4} PCF_{g_k}\left(g_k^{S_n}\right)$.

If CCF_{S_1} is the complete creativity of an artefact S_1 and CCF_{S_2} is the complete creativity of an artefact S_2, then the following properties generally hold for the complete creativity function:

$$CCF_{S_1} > CCF_{S_1} \Leftrightarrow (S_1)P(S_2),$$

$$CCF_{S_1} = CCF_{S_1} \Leftrightarrow (S_1)I(S_2),$$

where P is a strict preference relation and I is an indifference relation, as perceived by humans when evaluating the creativity of these artefacts.

Given a preference ranking of a reference set of artefacts, we define the creativity differences $\Delta = (\Delta_1, \Delta_2, \ldots, \Delta_{q-1})$, where q is the number of artefacts in the reference set and $\Delta_i = CCF_{S_i} - CCF_{S_{i+1}} \geq 0$ is the creativity difference between two subsequent artefacts in the ranked set.

We then define an error parameter E for each creativity difference:

$$\Delta_i = CCF_{S_i} - CCF_{S_{i+1}} + E_i \geq 0.$$

We can then solve the following constrained optimisation problem:

$$Minimise \sum_{i=1}^{q-1} (E_i)^2 \text{ s.t.} \begin{cases} \Delta_i \geq 0, & if (S_i)P(S_{i+1}) \\ \Delta_i = 0, & if\ (S_i)I(S_{i+1}) \end{cases}$$

This optimisation problem leads to the calculation of the partial creativity function parameters (a, b, c, d and w) for each property g_k.

Table 1 presents the values on the four creativity metrics calculated for the stories produced by each team in the two groups. Based on these values and the human assessment of the story rankings, the results of the constrained optimization problem defined in the previous section resolves in the calculation of the partial creativity parameters (a, b, c, d and w). The solution of the optimization problem is presented in Tables 2 and 3 for Group A and Group B respectively.

Regarding the impact of the various metrics in the computation of the overall creativity, we observed that Novelty is generally considered a particularly positive attribute creativity-wise for the stories, its partial creativity (PC) increasing as its

Table 1 Creativity metric values of the produced stories

	Story ranks	Novelty	Surprise	Rarity	R. Effort
Group A	#1	1.440	0.900	1.575	0.854
	#2	0.325	1.700	0.800	1.629
	#3	1.530	0.125	1.700	0.557
	#4	1.405	0.575	1.800	0.211
	#5	0.055	1.600	1.275	1.309
Group B	#1	1.480	0.675	1.650	0.720
	#2	0.575	1.125	0.950	0.969
	#3	1.735	0.350	1.750	0.743
	#4	1.690	0.175	0.875	0.014
	#5	0.405	1.950	0.175	1.786

Table 2 Optimal parameter values for Group A

	Group A NLP solution				
	a	b	c	d	w
Novelty	4.9932	−3.9748	1.0000	0.0000	1.0000
Surprise	3.0045	−2.5302	1.0000	0.0000	1.0000
Rarity	3.3335	−2.7751	0.9860	0.1611	0.9808
R. Effort	2.1063	−4.0000	0.9812	0.0214	0.9860

Table 3 Optimal parameter values for Group B

	Group B NLP solution				
	a	b	c	d	w
Novelty	5.0000	−3.9452	0.9600	0.0000	1.0000
Surprise	2.8049	−2.5186	1.0000	0.0001	0.9955
Rarity	3.4441	−2.8270	0.9938	0.1605	0.9618
R. Effort	2.0258	−3.8250	0.9986	0.0214	0.9753

value increases. In contrast, the remaining metrics reached their maximum partial creativity at a certain value, after which their partial creativity started to decrease, indicating that e.g. recreational effort greater than a certain point is not perceived as a direct indication of creativity. Hence, the obtained results indicate that, while the proposed computational creativity metrics are correlated with the perception of humans for creativity, this correlation is not direct for all metrics. The following section discusses on the implications of these observations and details our approach for using the proposed metrics towards building a dimensional plane that more accurately reflects the human perspective for creativity (Table 4).

Table 4 Computational creativity metrics values for Europarl and storybook datasets

	Computational creativity metrics				
	Doc no.	Novelty	Surprise	Rarity	R. Effort
Formal verbal transcriptions	1	0.05090	0.15521	0.16667	0.77820
	2	0.11686	0.84821	0.25000	0.01014
	3	0.04792	0.21635	0.14394	0.56020
	4	0.07355	0.13729	0.05000	0.50697
	5	0.01267	0.12373	0.25000	0.19011
Literary work	1	0.05138	0.10716	2.00000	1.78925
	2	0.05097	0.10142	0.26667	1.68172
	3	0.03030	0.16625	0.26667	1.60000
	4	0.06409	0.08024	2.00000	1.41075
	5	0.04940	0.14300	2.00000	1.69892

3 Transferring Computational Creativity Metrics to the Human Perspective

As stated, the four computational creativity metrics discussed, provide a good estimation for the respective creativity aspects exhibited by textual artefacts. However, in the process of using such metrics to approach the human notion for creativity, the derived results indicated some deviations between this formalization and the way humans think. In broad terms, we observe the following two characteristics that should be taken into account when trying to model the human perspective:

- Humans prefer to think *monotonically*, perceiving the value of a metric/dimension as analogous to the "quality" of the examined artefact in that dimension;
- Humans prefer to think *orthogonally*, perceiving each of the features as a dimension independent from the rest.

The first step towards identifying the adherence of our metrics with the human perspective is to examine the orthogonality of the proposed metrics formulation. To this end, we ran an experiment for calculating the four basic computational creativity metrics on two datasets derived from distinct and distant domains, and determined whether the four metrics are orthogonal.

The first dataset comprised transcriptions of European Parliament Proceedings [5]. Given the described formulation of computational creativity metrics, we consider as a "story" the proceedings of a distinct Parliament session and as a fragment the speech of an individual MP within the examined session. The second dataset was derived from a literary work, Stories from Northern Myths, by E.K. Baker, available via the Project Gutenberg collection. In this case, the story is a book chapter and the story fragment is a paragraph within the chapter. In total, we examined 50 distinct parliament sessions from the Europarl dataset and 40 chapters from the storybook.

Based on the obtained results, we calculated the correlation between the four computational creativity metrics. Tables 5 and 6 provide the correlation values between the four metrics. It is evident that the computational creativity metrics by themselves are not orthogonal.

Table 5 Computational metrics correlation: formal verbal transcriptions

	Novelty	Surprise	Rarity	R. Effort
Novelty	1.00000	0.13393	0.12329	−0.40681
Surprise	0.13393	1.00000	0.26453	−0.43151
Rarity	0.12329	0.26453	1.00000	−0.33499
R. Effort	−0.40681	−0.43151	−0.33499	1.00000

Table 6 Computational metrics correlation: literary work

	Novelty	Surprise	Rarity	R. Effort
Novelty	1.00000	−0.64243	0.10392	−0.10762
Surprise	−0.64243	1.00000	0.07376	−0.02538
Rarity	0.10392	0.07376	1.00000	−0.03882
R. Effort	−0.10762	−0.02538	−0.03882	1.00000

Hence, in order to better approximate the human perception for creativity, we propose the following abstraction for modelling the examined aspects of creativity to a space more closely resembling human thinking:

- *Innovation*, that is, the tendency to produce ideas and artefacts that are disassociated with the other elements on a given context.
- *Atypicality*, that is, the tendency to deviate from the norm without actually breaking through.

Innovation, by its nature, is a perspective which closely associated with the Novelty computational metric. To this end, the magnitude of Innovation for a textual artefact is equal with its value for the novelty metric (Tables 7, 8 and 9).

Table 7 Creativity dimensions values for Europarl and storybook datasets

	Creativity dimensions		
	Doc no.	Innovation	Atypicality
Formal verbal transcriptions	1	0.05090	0.29769
	2	0.11686	0.46572
	3	0.04792	0.26921
	4	0.07355	0.19103
	5	0.01267	0.17986
Literary work	1	0.05138	1.10149
	2	0.05097	0.50742
	3	0.03030	0.51839
	4	0.06409	1.00510
	5	0.04940	1.09751

Table 8 Correlation of creativity dimensions: formal verbal transcription

	Innovation	Atypicality
Innovation	1.00000	2.986E-07
Atypicality	2.986E-07	1.00000

Table 9 Correlation of creativity dimensions: literary work

	Innovation	Atypicality
Innovation	1.00000	1.436E-07
Atypicality	1.436E-07	1.00000

On the other hand, we consider Atypicality as a combination of the Surprise, Rarity and Recreational Effort metrics, each bearing a different weight towards determining Atypicality.

These two axes also provide a rough conceptualization of the two major qualitative aspects of creative work: whether the said work is *visionary*, i.e. it provides a groundbreaking approach on a given field; and whether it is *constructive*, i.e. it uses in a novel way established techniques and ideas in order to produce a high-quality artefact.

As indicated by the experiment described in Sect. 3, Innovation has an analogous and close to monotonic association with the human judgment for creativity. Therefore, and in order to satisfy our second requirement (orthogonality), we consider Innovation as the strictly defined dimension of our space and seek for the formulation of Atypicality that results to a dimension orthogonal to Innovation.

More specifically, let Atypicality of a text t be the normalized weighted sum of its Surprise, Rarity, and Recreational Effort:

$$A(t) = \frac{w_s Sur(t) + w_r Rar(t) + w_e Eff(t)}{w_s + w_r + w_e}, \text{ with the weights } w_s, w_r, w_e \in [-1, 1].$$

We aim to find the weight values that constitute Atypicality orthogonal to Innovation, i.e. those weight values for which *Correl(Innovation, Atypicality)* = 0. We thus define the following optimization problem:

$$Minimise \sum_{i=1}^{n} (Correl(Innovation_i, Atypicality_i))^2 \text{ s.t. } w_s, w_r, w_e \in [-1, 1],$$

where n is the number of the combined datasets (Fig. 1).

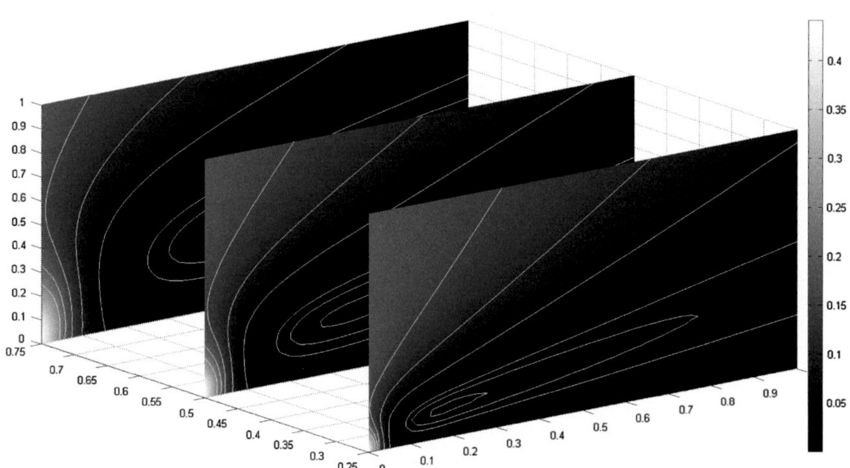

Fig. 1 Non-linear search space: surprise plane

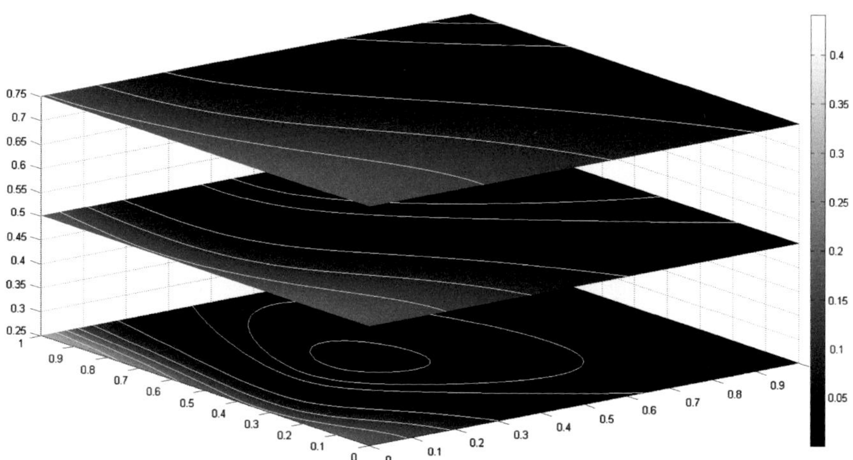

Fig. 2 Non-linear search space: R. Effort plane

Although the search space of the optimization problem above is highly non-linear, as it is demonstrated in the non-linear contours in Fig. 2, solving this problem is straightforward. The optimum weight values in our case are:

$$(w_s, w_r, w_e) = (0.13951, 0.10154, 0.06905)$$

The following tables present the innovation and atypicality in the two datasets, as well as, the correlation between these two dimensions for the found optimum weight values.

The resulting model defines two orthogonal axes, Innovation and Atypicality, which define the space for measuring and characterizing the observed creativity, as an Euclidean vector, the length of which indicates the quantitative aspect of the creativity exhibited by the artefact, while its direction indicates the tendency for either Innovation (visionary creativity) or Atypicality (constructive creativity).

Using this model for the evaluation of the story set of the initial experiment described in Sect. 3, and taking into account the vector length for each story, we obtained the same ranking as the one produced by the human evaluation. This is a strong indication that the proposed model accurately reflects human judgement, while also pertaining to core principles of the human perception of creativity.

4 Conclusions

Understanding the elements of the Creative Process, is a challenging research issue, which combines research results from several research fields, like neuroscience, psychology and philosophy. Within the field of Computational Creativity, there have been significant results for the identification of various aspects of the creative process and the construction of appropriate metrics for these aspects.

However, in the effort to determine if an artefact is creative by human standards, it is also important to examine the perception of creativity by humans and to which extend this perception can be formalized and applied on the evaluation of creative works.

The work described in the present manuscript showcases our findings towards transitioning from computational creativity metrics associating specific attributes of text artefacts with creativity aspects to a creativity calculation model that better reflects the human perception of creativity.

For the next steps of our research we aim to extend the proposed model for the evaluation of artefacts of different modalities towards designing a generic framework for measuring creativity.

References

1. Boden, M.: The Creative mind: Myths and Mechanisms, 2nd edn. Routledge, London (2004)
2. Chiru, C.-G.: Creativity detection in texts. In: Proceedings of the 8th International Conference on Internet and Web Applications and Services (ICIW2013), pp. 174–180. Rome, Italy (2013)
3. Karampiperis, P., Koukourikos, A., Koliopoulou, E.: Towards machines for measuring creativity: the use of computational tools in storytelling activities. In: Proceedings of the 14th IEEE International Conference on Advanced Learning Technologies (ICALT2014) Advanced Technologies for Supporting Open Access to Formal and Informal Learning, Athens, Greece (2014)
4. Karampiperis, P., Koukourikos, A., Panagopoulos, G.: From computational creativity metrics to the principal components of human creativity. In: Proceedings of the 9th International Conference on Knowledge, Information and Creativity Support Systems (KICSS 2014), Limassol, Cyprus, 6–8 November 2014
5. Koehn, P.: Europarl: A Parallel Corpus for Statistical Machine Translation. MT Summit (2005)
6. Zhu, X., Xu, Z., Khot, T.: How creative is your writing? A linguistic creativity measure from computer science and cognitive psychology perspectives. In: Workshop on Computational Approaches to Linguistic Creativity, pp. 87–93. ACL (2009)

A Social Creativity Support Tool Enhanced by Recommendation Algorithms: The Case of Software Architecture Design

George A. Sielis, Aimilia Tzanavari and George A. Papadopoulos

Abstract Reusability of existing knowledge for the design and development of new ideas is a key principle of the definition of creativity as a process. Professionals to create the Architecture design of innovative software tools use software Architecture Design as such process. This work proposes a framework/tool by which the design of SAD can be directed based on several types of recommendations and in particular of well-known design patterns [1]. This work approaches the recommendations the use of contextual factors in combination with recommendation methods and known recommendation algorithms.

1 Introduction

A creativity outcome is a sequence of thoughts and actions that lead to a novel adaptive production [5]. Plucker and Beghetto [8] define creativity as the interplay between ability and process by which an individual or a group produces an outcome or product that is both novel and useful as defined within some social context. Having assumed that creativity is an attribute that we all have, we reach the conclusion that it is necessary to find ways and means to assist in outsourcing this property. Creativity is a characteristic that can be cultivated and at the level of its development is different for each person. Cultivating and expressing the creative

G.A. Sielis (✉) · G.A. Papadopoulos
Department of Computer Science, University of Cyprus,
P.O. Box 20537, 1678 Nicosia, Cyprus
e-mail: sielis@cs.ucy.ac.cy

G.A. Papadopoulos
e-mail: george@cs.ucy.ac.cy

A. Tzanavari
Department of Design and Multimedia, University of Nicosia,
P.O. Box 24005, 1700 Nicosia, Cyprus
e-mail: tzanavari.a@unic.ac.cy

S. Kunifuji et al. (eds.), *Knowledge, Information and Creativity Support Systems*,
Advances in Intelligent Systems and Computing 416,
DOI 10.1007/978-3-319-27478-2_34

ability can be achieved through help and guidance. One approach to achieve that is through the use of Creativity Support Tools (CSTs). Creativity Support Tools are software systems that can emulate a realistic creative process by offering the users the ability to record thoughts and ideas with the use of ICT means. The utilization of the existing CSTs is bounded to the replacement of the conventional means of collecting ideas, such as the paper or the white board, with corresponding virtual environments, but they also offer the users the possibility to collaborate with other people from a distance.

This research aims to reinforce the importance of these tools with the addition of more advanced functionality. Generally, CSTs are the basis for strengthening the creative capacity and they can also be used for the cultivation of the creative ability. At this stage, these tools lack the mechanisms that can create the necessary stimulus to the user, which would make the user more productive in the process of creativity [10]. Such mechanisms are the recommendation systems for the support of the creativity process like for example the recommendations of people to collaborate with, related problems or solutions, related ideas and finally related resources such as articles, images or videos. Recommendation systems are systems that belong to the information filtering systems family. In general Recommendation systems seek to predict the rating or the preference that a user would give to an item. Thus the common methodology that the recommendation systems follow is to find the correlation between three types of modeling, user, rating and item modeling in order to produce recommendations. The use of recommendation systems to support a creativity process is a challenging task due to the multi-dimensionality of factors that influence the process and the complexity in identifying the type of recommendation that can foster creativity at each phase of the creativity process within a Creativity Support Tool.

The motivation for this work is driven from the notion that Creativity and Innovation are keys to success for any business. The development of a Social Creativity Support Tool (SCST) can become the means of bringing professionals and experts together for the exchange of ideas and collaborate together through creativity support spaces supported by user aiding functionality.

For example at the existing CSTs noticed that they lack of methods for finding people while formulating a collaboration team. The functionality they support regarding the group composition for collaboration is only the invitation of person to join a creativity session without analyzing her knowledge background and the level of competence related to the project. The use of a social network designed for creativity will give the opportunity to a practitioner to get recommendations of experts in a particular topic who are (or not) in the same company or organization and can add value to a creative project that is processed. A social creativity network can be beneficial for businesses in two ways. On the one hand it will be used as means for collaborations between companies (public social web network). On the other hand it will be used as a tool that connects employees of large scale international companies (intranet application). The supporting recommendations can vary according to the context of the creative project and the type of actions needed for the successful completion of a creative project. In that case the interfaces, the tools and the supporting functionality can be different based on the context of the project.

2 Problem Statement

In this research it is stated that social creativity tools integrated with context aware recommender systems can positively influence the creativity process. Context aware recommendations such as recommendations of users for collaboration or resources related to the context of a creative project but also other recommendations related to particular creativity phases can facilitate the creativity process. Therefore the problem approached has two parts: first the development of a Social Creativity Support Tool for the development of professional networks targeted in solving problems and producing innovative ideas and second the development of a context aware recommendation algorithm, which will have the flexibility to analyze the most important contextual factors for each phase of the creativity process and produce useful recommendations to foster the professionals' creative ability.

3 Approach

As mentioned above, the creativity process is not a straight-forward process. On the contrary, it is characterized by dimensions that may vary depending on the domain it is applied to. According to the problem statement, the expected outcomes of this work are related to creativity as the main research area. The wideness of the creativity topic though makes the design of a research plan very complex. For example a creativity process may be an art session, a book conversation or the development of a web site. So to study the research hypothesis a particular topic was selected in order to model creativity and to extract valid conclusions. Therefore we framed the problem into a particular creativity-related problem, that of the Software Architecture Design (SAD) process. SAD is by default a creativity process since it is a process that is executed in phases by individuals or groups of experts. The participants of such a process must have an important knowledge background in the topic and be able to exchange expert opinions/ideas for each phase of the SAD lifecycle process. For the study of SAD as a creativity process and following the problem statement, a Creativity Support Tool has been developed for the support of SAD. The SCST will be the tool on which the users will be able to create professional networks for collaborative creativity sessions. The development of the creativity tool and the integration with the proposed recommendation algorithms will be used as prototype for the examination of the current work's research questions. In particular, the research goals that were set for the current research work are listed in the rest of this section.

3.1 Research Goals and Methods

To assess the impact that Context Aware Recommendations have in creativity process within a Social Creativity network, the following set of objectives have been determined:

- Define the creativity context elements and design of a creativity model that will be applied in a social creativity network.
- Define the process of SAD through the designed creativity tool and identify the types of recommendations that must be applied at each step.
- Define the recommendation algorithms and in particular specific recommendation methods for recommending SAD oriented types of recommendations such as, recommendation of Design Patterns, for the examination of how such recommendations are influencing the SAD as a creative process.
- Design the evaluation plan to lead to evaluation results concerning:
  - The usability of the SCST
  - The accuracy and validity of the proposed recommendations
  - User satisfaction and experience by using types of recommendations such as recommendation of users to collaborate, resources and most importantly the recommendation of Design Patterns in order to facilitate users in designing Software Architecture independently to the knowledge level they currently have.
  - Impact of Context Aware Recommendations in creative ability of the users

The outcomes of this research are expected to be taken by applying the prototype in a large scale international IT company whose employees will be asked to create profiles in the prototype application. Then a group of Software Architect Analysts will be asked to design new projects with it. They will be asked to formulate groups of users to collaborate with, from branches of the company that are in other countries and use the recommender system to collect resources related to a given project. With the design of a prototype for a domain specific group of users will lead to the collection of results regarding the creativity impact as well as the overall users' satisfaction and the prototype usefulness. The results of the test will be used for further improvements of the prototype the improvement of the proposed recommendation algorithm and also lead to the design of new methods and variations of the recommendation algorithms that will be applied.

4 SAD Recommendations—Related Work

The last years Design Patterns for Software Architecture design are increasing. New patterns are appearing and applied widely. Gueheneuc and Rabih [4] proposed a methodology of recommending design patterns through the textual analysis of each pattern into the most important words and computing the similarity distance between those words and the words of the query given by the user. Gomes et al. [3] proposed a Case Based Reasoning Recommendation system for the recommendation of Design Patterns based on previous experiences using a Design Patterns Knowledge Base and related taxonomies. A similar system developed by [11] that recommends patterns using the Implicit Culture Framework (ICF). The recommendations are produced based on the users previous actions, based on conventional Information Retrieval and CBR methods. Palma et al. in [6] propose a DPR framework which recommends patterns based on predefined questions that the designers have to answer, and based on the given answer the framework has a weighting mechanism for the selection of the appropriate pattern. The initial identification of patterns that can be used through the DPR framework [6] are selected through LUCENE indexing and TF-IDF filtering of the query given and the intent description of each pattern.

5 Current Status

The research work described through the objectives definition in Sect. 3 is a work which is still in progress. Most of the objectives are currently met in the current status of the work, but some are still in progress. In particular, at the time being we defined the contextual model of creativity which is designed and represented in the form of ontology. The Contextual model is an extension of the proposed ontology proposed in previous work of ours in [9]. The new ontology is taking into account additional conceptual elements related to the phases of the creative process as a SAD process. With the use of the designed ontology it became possible to bring data from different data sources such as existing software design repositories like GitHub.com and Sourceforge.net and combine them with the new data that is created in the local database of the SCST. Based on this model we managed to define the most important relations that can be used as input data for the recommendation algorithms that we use.

Having the Creativity Contextual Model defined in the form of ontology, and taking into account the semantic data and relations of the model, we developed a Knowledge Base tool by which a user is able to browse through the existing SAD projects, the participants of the projects and the metadata related to each particular entity (Fig. 1).

As shown in Fig. 1 the ontology contains the concepts of a creativity project and the project phases that a project may have during its execution. Based on these concepts/classes several triples are defined such as for example (*Project, projectCreatedBy, projectCreator*), (*ProjectGroupMember, isMemberOf, ProjectGroup*), etc. Using the triples defined in the ontology and with the use of SPARQL queries it is feasible to collect data from several domain specific data sources and create an interactive Knowledge Base. The importance of ontology is not only tracked to the concentration of data from other sources into one

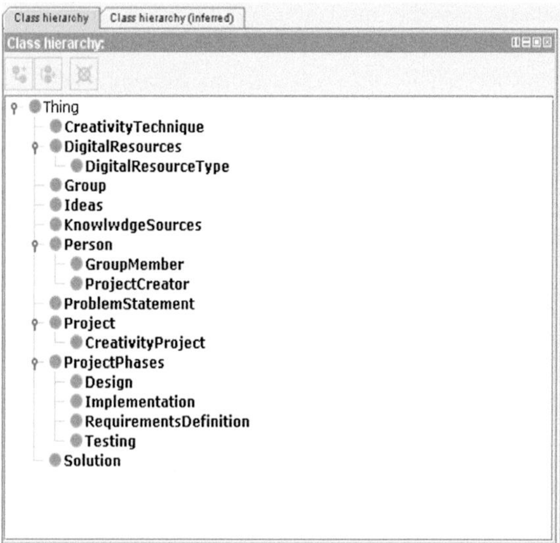

Fig. 1 Creativity ontology model

Knowledge Base, but also into the formalization of data into a structured form. By structuring the data following the defined concepts and their relations facilitates the filtering of data to be used in the computation of recommendations related to the design and development of new Software Architecture Design projects. The proposed creativity model consists of a structural representation of the ingredients of a project (abstract form, not particularly for SAD) such as the type of project, the Participants of project Groups; the Phases of executing a Project or the Creativity Techniques that are used for the execution of a Project.

In terms of the general concept of using recommendations for the enhancement of creativity we suggest types of recommendations like recommendation of relevant projects while a new project is created, recommendation of related resources, recommendation of peers to collaborate with and recommendation of Design Patterns to facilitate the Architecture Design process. The main objective of this work is the context aware recommendations of Design Patterns and the influence of the recommendations in the creative outcome of the process.

The knowledge Base tool is part of the overall SCST which in its whole contains a second sub-tool which is used for the design of Software Architecture Diagrams. The flow that can be followed by end users in order to open en existing Design project or create a new one defines the proposed workflow process for the SAD as creative process. Thus at each phase of the process we defined specific Context Aware Recommendation types that vary based on the process phase and the available context information. In addition to that, social networking functionality is also added in particular phases such as professional network connections between designers; *Follow* a professional's activities or view the progression of selected SAD projects, which are in progress by other professionals.

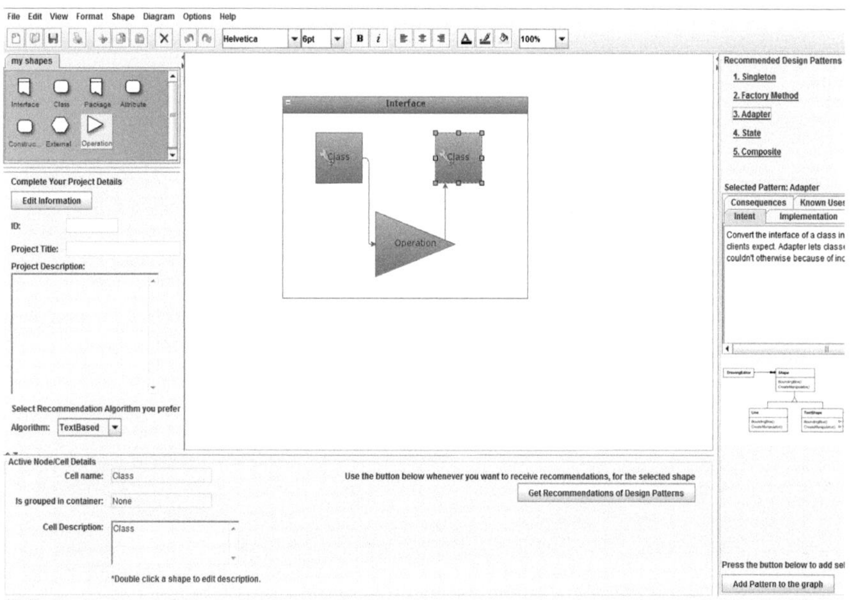

Fig. 2 Software architecture design editor enhanced by context aware recommendations of design patterns

A lot of effort at the current state of the research is given in producing recommendations on Design Patterns during the SAD in the form of diagrammatic design (Fig. 2).

The recommendation of Design Patterns consists one of the most important tasks of this work and to meet that objective we developed a mechanism by which, we use three different recommendation methodologies. The methodologies used are the following: 1. *Lucene indexing, TF-IDF* [7] *and Cosine Similarity,* 2. *Graph based recommendations using Design Patterns rules* and 3. *Recommendation of Design Patterns Based Context Elements and Utility Function.*

Lucene indexing: Assuming that during the definition of a new SAD project the designer defines a list of metadata information for the project and during the design of the diagram, designer is able to define metadata for each particular node of the diagram, we can construct queries, which can be used to filter patterns using text comparisons. The text comparisons can be made between the selected metadata (considered as context elements) and the attributes of each pattern. The patterns attributes are indexed with the use of Lucene Framework and the text a comparison between metadata and pattern attributes is done with the use of the TF-IDF [7] algorithm and the cosine similarity for the computation of the similarity distance between the matching pattern attributes.

Table 1 Sample of patterns list and the decision rules for each pattern used in method 2

Pattern name	Pattern category	Definition	Design purpose
Factory	Creational	Number of classes linked to a constructor element	Create individual objects in situations where constructor is inadequate
Singleton	Creational	If the class does not belong to a container or is member of a container which contains two and more containers	Ensure that there is exactly one instance of a class S. Be able to obtain the instance from anywhere in the application
Facade	Structural	Set of classes in package Packages all together in one interface	Provide an interface to a package of classes
Adapter	Structural	If existing functionality—external element is used	Allow an Application to use external functionality in a retargetable manner
Interpreter	Behavioral	If notes elements are linked to the classes	Interpret expressions written in a formal grammar
Iterator	Behavioral	If operation node exists between to classes	Provide a way to access the elements of an aggregate object sequentially without exposing the underlying representation

Graph based recommendations using Design Patterns rules: The definition of each pattern is based on some rules that must be satisfied in order to make the decision that a specific pattern is suitable enough to be applied for the solution of a problem. For the better understanding the following example is given: In literature the factory pattern is defined by the following rule: "*when a container contains functions or subclasses (nodes) that are objects of the same class then Factory pattern can be applied*". For example the class automobile can contain the subclass Ford, Toyota etc. In that case automobile is the container and each subclass is a node connected to the constructor of the container (*Sample of rules for particular patterns in* Table 1).

Recommendation of Design Patterns Based Context Elements and Utility Function: With this method the recommendations of Design Patterns are produced based on the users' preferences. The System provides to the user a set of predefined contextual elements for which the user is able to define the importance weight value for each particular element. The context aware recommendation method takes into account the user's input and produces its results based on the output of a utility function. Thus the context aware recommendation result vary according to which context element is considered as important by an individual user and to what is the importance level in relation to the received recommendations.

Each methodology is used based on the phase of the design and the available contextual information but in that case the selection of the methodology is determined only within the design editor sub-tool of the prototype. The Design Patterns were selected from the GoF [2] list of patterns and modelled in a separate ontology that is queried according to needed recommendation methodology input.

6 Future Work

The remaining work for the completion of this research work is mostly related to the Evaluation Objectives. We currently organize the evaluation sessions with professionals Software Architecture Designers from corporate organizations as well as academic ones. The evaluation will be done in a controlled environment, using specific use case scenarios and specific problem statements. That way we aim to collect feedback from the experts regarding the experience and usability, the accuracy of recommendations, the impact of recommendation in the creativity design, the usefulness of the tool and finally whether the social attributes of the prototype facilitated overall process.

By the completion of this work we will have the following accomplished tasks: 1. The creativity ontology model, which will be available to be used for other CSTs, The Design Patterns Ontology, a Semantic Web Application having a SAD knowledge Base and a graph editor supported by recommendation types that facilitate the creative process of SAD and finally three new methodologies for the recommendation of Design Patterns.

Finally, the analysis of the results regarding the users' satisfaction ant the prototype usefulness, as well as the validity of the context aware recommendations of Design Patterns will lead into important conclusions regarding the main objective of the current work, which is whether the provision of context aware recommendations during a creative process such as the Software Architecture Design improves the creativity of the designers.

References

1. Alexander, C., Sara, I., Murray, S.: A Pattern Language: Towns, Buildings, Construction. The Oxford University Press, New York (1977)
2. Gamma, E., Helm, R., Johnson, R., Vlissides, J.: Design Patterns: Elements of Reusable Object-Oriented Software. Addison-Wesley Longman Publishing Co. Inc., Boston, MA, USA (1995)
3. Gomes, P., Pereira, F.C., Paiva, P., Seco, N., Carreiro, P., Ferreira, J.L., Bento, C.: Using CBR for automation of software design patterns. In: Proceedings of the 6th European Conference on Advances in Case-Based Reasoning, pp. 534–548 (2002)
4. Gueheneuc, Y.G., Rabih, M.: A Simple Recommender System for Design Patterns. In: Proceedings of the 1st EuroPLoP Focus Group on Pattern Repositories (2007)
5. Lubart, T.: Models of the creative process: past, present and future. Creativity Res. J. **13**(3–4), 295–308 (2000)

6. Palma, F., Frazin, H., Gueheneuc, Y., Moha, N.: Recommendation system for design patterns in software development: an DPR overview. In: Proceeding of 2012 Third International Workshop on Recommendation Systems for Software Engineering (RSSE), Switzerland, pp. 1–5 (2012)
7. Pasquale, L., Gemmis, M., Semeraro, G.: Content-based recommender systems: state of the art and trends. In: Ricci, F., Rokach, L., Shapira, B., Kantor, P.B. Recommender Systems Handbook, Springer, US, pp. 73–15 (2011). 978-0-387-85819-7, http://dx.doi.org/10.1007/978-0-387-85820-3_3
8. Pluker, J.A., Begheto, R.A.: Why creativity is domain general, why it looks domain specific, and why the distinction does not matter. Creativity From Potential to Realization, 153–167 (2004)
9. Sielis, A.G, Mettouris, C., Papadopoulos, G.A., Tzanavari, A., Dols, R.M.G., Siebers, Q.: A context aware recommender system for creativity support tools. JJUCS **17**(12), pp. 1743–1763 (2011)
10. Sielis, A.G., Tzanavari, A., Papadopoulos, A.G.: Enhancing the Creativity Process By Adding Context Awareness in Creativity Support Tools. HCI 09 International, San Diego (2009)
11. Weiss, M., Birukou, A.: Building a pattern repository: benefitting from the open, lightweight, and participative nature of wikis. In: Workshop on Wikis for Software Engineering at ACM WikiSym, 2007 International Symposium on Wikis (WikiSym), Montre'al, Que'bec, Canada, Oct 21–23 (2007)

The Approach to the Extension of the CLAVIRE Cloud Platform for Researchers' Collaboration

A.V. Dukhanov, E.V. Bolgova, A.A. Bezgodov, L.A. Bezborodov and A.V. Boukhanovsky

Abstract This paper describes an approach to extend the CLAVIRE platform for sharing traditional scientific documents, data sources, executable services and 3D-visualization means for collaboration and peer review. This approach includes the development of an intellectual editor for scientific package integration into CLAVIRE, a C++ based graphical library and graphical engine "Fusion"; composite application with interaction between package and visualization tools. This means facilitating the process of sharing and presenting scientific results in a vivid and interactive manner. The considered approach was developed within the principles of research object ontologies, and with the use of the workflow-centric research object approach and method of interactive workflow. The examination of the approach was performed by evolving young scientists from cities located across Russia. At the end of the paper, the approach to rapid learning courses and resource design was developed.

Keywords Scientific collaboration · Scientific result sharing · Scientific package · Cloud service · CLAVIRE cloud platform · Research object

A.V. Dukhanov (✉) · E.V. Bolgova · A.A. Bezgodov · L.A. Bezborodov
A.V. Boukhanovsky
ITMO University, Saint Petersburg, Russian Federation
e-mail: dukhanov@niuimo.ru

E.V. Bolgova
e-mail: katerina.bolgova@gmail.com

A.A. Bezgodov
e-mail: demiurghg@gmail.com

L.A. Bezborodov
e-mail: lev.bezborodov@gmail.com

A.V. Boukhanovsky
e-mail: avb_mail@mail.ru

S. Kunifuji et al. (eds.), *Knowledge, Information and Creativity Support Systems*,
Advances in Intelligent Systems and Computing 416,
DOI 10.1007/978-3-319-27478-2_35

1 Introduction

Modern scientific society is characterized by many changes in methods, approaches, and technologies. Modern researchers should respond quickly to these changes and offer rapid, novel scientific results. Contemporary scientific society should not only be provided with traditional scientific databases complemented by 3D pictures, data sources, and social networks like ResearchGate, but should also be provided with platforms that have easy-to-master interfaces for sharing the executable scientific results (which require original computational resources) in the form of cloud service. They should also be provided with instruments for the easy development of visualization tools that support desktop graphical adapters, and top systems of 3D-Stereo visualization.

Contemporary researchers have a wide choice of ways to present their results in an interactive and vivid manner. However, some of these methods have limited functionality, while others require skills and abilities in programming, which may distract scientists from the main study. We have the following questions under consideration:

Q1. How to present (share) scientists' results in a vivid and interactive manner for collaboration and peer review via the use of cloud computing technologies and modern concepts, including Research Object?
Q2. How to simplify scientists' efforts (including reducing programming costs) to share their results in the manner mentioned in the first question?
Q3. How to create a scientific collaboration environment in keeping with the trend of modern technological achievements in cloud computing and result presentation, including 3D Visualization?
Q4. How to use the latest scientific achievements for training scientists and professionals, including within the framework of Master's and postgraduate courses?

This paper offers an approach to extend the CLAVIRE platform for sharing traditional scientific documents, data sources, executable services and 3D-visualzation methods for collaboration and peer review. This approach includes the development of an intellectual editor for scientific package integration into CLAVIRE, a C++ based graphical library, graphical engine "Fusion", and a composite application with interaction between package and visualization tools.

2 Related Works

Modern researchers (or research teams) have a wide choice of options to present (share) their scientific results. They can disseminate results in the traditional form of a scientific paper in journals, or conference proceedings to use special instruments, plug-ins and other computer services for sharing data arrays, visualizations, scientific workflows (eScience applications), multimedia content, and other digital materials (Q1).

2.1 Collaboration Approaches Based on Scientific Papers

Today, scientists can collaborate with each other on scientific papers via the use of various tools, platforms and social networks, such as Google Docs for Researchers,[1] Mendeley,[2] Research Gate,[3] and Life Science Network.[4] These resources allow not only the sharing of published papers, but also the provision of drafts and manuscripts for collaborative works and peer review, significantly increasing the number of qualified reviewers [1]. The peer review within these resources may be carried out much faster than it is during the traditional peer review procedure before acceptance of a paper in a journal or conference proceedings [2]. The "Content innovation" built in the "Article of the Future" project [3] provides authors with digital tools "…to disseminate their research in its full digital richness in Elsevier articles".[5] These tools allow a scientific paper to be published in Elsevier's journals with interactive 3D visualizations, data sources, interactive plots, and the like.

2.2 Using Traditional Models of Cloud Services

The services mentioned above allow authors to make theirs papers more vivid and interactive. However, researchers are limited by the feasibility of these services and sometimes do not have the ability to present all the features of their scientific products (application packages, script or project of a scientific environment, and so on). Using cloud computing approaches within the models of SaaS, PaaS and IaaS [4] allows researchers to present the functions of their original tools. They can provide a remote desktop for a launch application package, make a remote interface, or create another original decision for sharing the functionality of their scientific tools and middleware, such as by using DataTurbine, or OPeN Dap [5]. The first case is limited by the level of confidence among partners. Other cases may require a lot of effort and be time consuming. These efforts may be needed in cases where a distributed computational infrastructure is required. The growing amount of interdisciplinary problems, which require the use of two of more regular and special application packages working in original computational resources, was the crucial factor for implementing the Scientific Workflow Paradigm.

---

[1]http://gigaom.com/2014/07/23/macmillan-invests-in-google-docs-for-researchers-firm-writelatex/.

[2]http://www.mendeley.com/.

[3]http://www.researchgate.net/.

[4]http://www.lifescience.net/.

[5]http://www.elsevier.com/about/content-innovation.

2.3 Scientific Workflow Paradigm

Currently, the Scientific Workflow Paradigm (SWfP) is the fundamental instrument for scientific collaboration [6, 7]. SWfP enables the composition and execution of one, two or more application packages and tools in a distributed computational environment. On one hand, workflow (WF) is "… a high-level specification of a set of tasks and the dependencies between them that must be satisfied in order to accomplish a specific goal" [8]. On the other hand, the WF can be presented as a computer program written in high-level programming language. At present, scientists have a wide choice of ways and tools to represent workflows (including WF languages and graphical instruments—GUI [9] to develop WF as a Directed Acyclic Graph (DAG)), as well as a wide choice of Scientific Workflow Management Systems (SWfMS), including Taverna, Kepler, and Pegasus. In particular, Taverna has instruments to share WFs realized during the "MyExperiment" project [8]. One of the key properties of SWfMS is WF execution scheduling, including task to resource mapping. There are many algorithms for task scheduling, which covers most of the properties and features of WFs (a set of WFs), and distributed computational resources. For example, the set of scheduling algorithms for multiple deadline-constrained workflows [10] was developed for heterogeneous computing systems with time windows. These algorithms may be useful for WFs not only for shared scientific collaboration, but also for educational purposes, as the student has the ability to run one or more real packages when required computational resources are idle.

The concept of collaborative scientific workflows [11] considers the required features of collaborative WFs for the "…enablement of scientific collaboration", including the life-cycle of scientific collaboration. The research workflow platform approach described in [12] offers to unite most of the useful features mentioned above in one digital solution. This work has one successful result—the eSciDoc platform developed for the aims of the Max Planck Society. The authors of the paper note, "Building and maintaining completely independent platforms for each domains… may not be sustainable" [12]. The concept of the Research Object is expected to overcome this problem.

2.4 The Research Object Concept

The Research Object (RO) may be defined as a container of scientific resources [13] that are received during all the stages of the research life cycle. This container may include executable packages and WFs, data sources, program documentation, short descriptions, scientific papers, tools for collaborative work, and so on. In this regard, the principles of RO ontologies are presented in [14]:

- P1. Preserving Data and Methods
- P2. Overcoming Obfuscation through Annotation
- P3. Treating the Research Object as a Container
- P4. Citations and Credit in Research Objects
- P5. Treating Research Objects as Software

These five principles unite all the features of the previous achievements mentioned above, standardize descriptions of the RO components, and present the RO as a complete structure. The paper [14] considers the Research Object Digital Library (RODL)—a set of tools that "…collects, manages and preserves aggregations of scientific workflows and related objects and annotations…". The RO-enabled "MyExperiment"[6] is an extension of the virtual research environment mentioned above, and was made on the basis of workflow-centric RO.

Summarizing the above, we can see that modern researchers have a large selection of tools, environments and platforms to facilitate their research and to share related results in different forms for collaboration and peer review. These tools are based on widely recognized standards of computer-aided object descriptions; hence, they are easy for scientists to master, as they have a basic knowledge of ontologies, domain-specific languages and contiguous areas of IT. Nevertheless, researchers without the required knowledge and skills in appropriate areas of IT need to appeal to colleagues to prepare an RO, or must learn the required technologies, thus distracting them from the main study.

2.5 The CLAVIRE Platform as a Means to Facilitate Sharing of Workflow-Centric Scientific Results

The CLAVIRE platform, developed by ITMO University, is based on the concept of the iPSE (Intelligent Problem Solving Environment) [15, 16]. The iPSE concept, implemented in the CLAVIRE, allows us to describe the computational process of problem solving in the form of *abstract* WF, which is translated into instructions for executing application packages presented as cloud (remote) services (concrete WF). An abstract WF is described in the simple domain-specific language "EasyFlow" [16] (Q2). These steps, defined with the use of EasyFlow statements, contain definitions of input data and execution modes (*standard*, *urgent*, etc.) of application packages integrated into CLAVIRE. In order to present an installed application package as a cloud service (integrated into CLAVIRE), it is sufficient to describe it in the special language "Easy package" [16].

[6]http://alpha.myexperiment.org/.

The CLAVIRE platform contains several user interfaces (UI) to develop and execute workflows and present them as a cloud service (P5) based on the AaaS model. Four main UIs in the CLAVIRE (only a web-browser is required) were described in [16–18]. The WF-based UI and VSO UI are closest to the principles of the RO ontologies [15].

2.6 The Use of Scientific Results for Educational Purposes

Scientific results, presented in the form of collaboration and peer review, may be used as a tools in the educational process, especially for training Masters and PhD students. In this case, students have direct access to the latest achievements of scientific communities. Teachers can simply translate scientific results in the form of reusable learning objects (RLO) [19], which can be a component of the learning module or course.

The RLO contains a number of properties [20] that partly correspond to the five RO principles mentioned above. Among these properties, we can distinguish following:

- *Reusable*—RLOs can be used in multiple contexts, for multiple purposes, and at multiple times;
- *Self-contained*—each RLO focuses on a specific topic/learning objective;
- *Standardized*—RLOs follow the same organizational structure;
- *Searchable*—RLOs are tagged with metadata (information that describes the RLO);
- *Flexible*—RLOs are easy to update, and provide access to quality teaching and learning resources for a wide range of learners.

Contemporary educational societies have a large number of standards, methodologies, technologies, and practical solutions to build educational resources and learning environments [21, 22], including learning tool interoperability (LTI) [23].[7]

Since scientific results are often updated regularly, especially in interdisciplinary areas, the directed usage of scientific results for educational purposes is not easy. In the event of the emergence of a new scientific result or the revision of an existing one, teachers must take a number of steps to update the relevant courses.

[7]Developed within the IMS Global Learning Consortium—http://www.imsglobal.org/lti/.

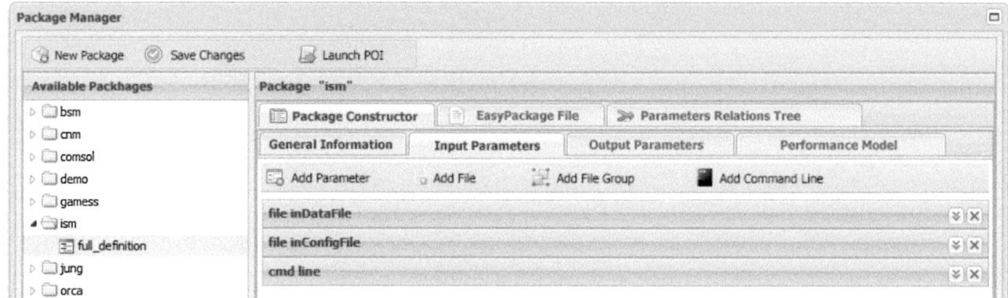

Fig. 1 Example of the package manager interface

3 Web-Based Intellectual Editor for Integrating Application Packages into the CLAVIRE

The Web-based Intellectual Editor "Package Manager" (PM) was developed to facilitate the process of integrating application packages into the CLAVIRE platform (Q2). The JavaScript framework, HTML and CSS-tables were used for the PM implementation. For data transfer and requests, the JSON text interchange format is used [15].

The PM provides the package owners with an interface, which allows users to input general information about packages, such as the package's authors, the organization, and license type, as well as its inputs and outputs (Fig. 1).

In addition, the PM allows users to insert parameters that are not input parameters of the application package. For each parameter, users can specify the expression in Ruby notation, which is dependent on existing parameters defined in the package description [15]. This option allows owners to present their packages in the desired format. If necessary, the PM has the ability to define a validation expression for any input parameter, such as to make it foolproof or to protect against legitimate user.

The integrated CLAVIRE package becomes a cloud service, and researchers can share it with all or with selected users. If the source package needs to be modified the owner can make the required changes in the package description in such a way that the input parameters of the relevant cloud service will be unaffected.

4 C++ Based Library for the Rapid Development of Visualization Tools and the FUSION Engine

The C++ based library for the rapid development visualization tools offers a template called "base_demo," which can be changed to build custom domain-specific visualization tools. This template contains the four abstract methods [15] that provide users with the ability to build prepared, dynamic 3D graphical scenes (Q3).

The template "base_demo" allows developers to connect the library Kinect for Windows SDK и Kinect for the Windows Developer Toolkit, hence providing

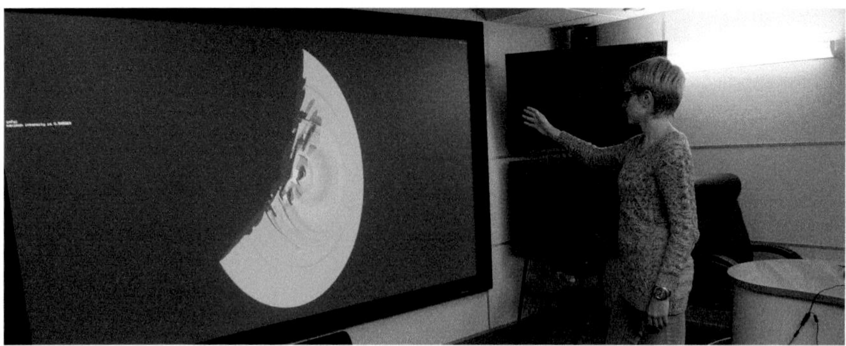

Fig. 2 The interaction between an operator and the visualization tool using MS kinect

operators with the ability to control scenes through hand gestures (Q1). This library supports the detection of the hand movements [15].

The C++ base library discussed above supports the mode of stereoscope visualization of various types, such as anaglyphs and polarization. Special glasses are required for immersion in a virtual environment. Figure 2 shows an operator interacting with the one of the visualization tools in the Center of Scientific Visualization and Virtual Reality of the ITMO University.

The library mentioned above was extended to the graphical engine called "Fusion", based on DirectX 11 Technology. This engine contains a set of C# classes that simplify the building of 3D scenes from standard graphical primitives (lines, bars, and so on), and typical 3D models (imported from files in FBX format) with contemporary effects support, including tessellation.[8] Also, the Fusion engine supports the NVidia 3DVision and dual-head stereo technologies, as well as modern human-computer interaction tools, such as touch-tables or Microsoft Kinect, thus enabling additional functionalities to be added to the user shell [22].

The Fusion engine was successfully applied in the design of a highly realistic virtual reality environment, such as the "Anatomical Atlas", which presents a highly detailed model of a human body, including its systems, organs, and supplementary materials. The use of graphical effects, including normal and x-ray shading, and highlighting color selections, users can build a required 3D picture of the complete human body or any part thereof, such as the central nervous system or the heart, in different environments and lightning modes [22]. This Atlas was used to create a 3D virtual learning environment for educational purposes (Q4) [24].

In order to present 3D images on the screens of devices that have weak graphics (without DirectX support), we developed the prototype of a tool that captures 3D pictures from Fusion engine's application and translates the related video stream to the web browser. Figure 3 shows examples of video streams presented in web browsers (Q3).

[8]http://www.nvidia.com/object/tessellation.html.

Fig. 3 Examples of web browser windows with video streams captured from the fusion engine's applications

5 Developing the Composite Application for Interactive Visualization

The composite application (WF) was developed through the WF-based UI or VSO interfaces, and contains at least two steps, run_package and run_vis, which summon integrated application packages, namely scientific packages (simulation package) and visualization tools. This composite application is implemented with the use of the IWF method: Interaction between one of the scientific packages and a visualization tool (Q1). The Example of this interaction was show in [15].

The IWF method allows researchers to present scientific results in a modern, visual format by using top-quality virtual reality systems, even scientific package and visualization tool work in different places (separate systems).

6 The Examination of the Developed Approach and Tools by Young, Evolving Russian Scientists

The developed approach, the CLAVIRE platform, PM, and C++ based visualization library, were examined by an evolving group of 20 young scientists (aged 23–33 years old), who are in the different cities in the Russian Federation, including Nizhniy Novgorod, Moscow, Ufa, Rostov-on-Don, and Novosibirsk. All the young scientists were either participants in or winners of a competition in computational science [25], but not more half of them were skilled in the area of programming.

We split this group into two equal subgroups (10 participants). Members of the first subgroup had to develop scientific packages relevant to their field and that supported the mode of the command line and a visualization tool, and to then integrate both solutions into CLAVIRE. Members of the second subgroup only had to integrate their scientific packages into CLAVIRE. The first subgroup had 6 weeks to solve all the tasks, and the second subgroup had 3 weeks. In addition, all the participants in the experiments were required to prepare a report with a description of the scientific area and the application package, the visualization concept, and a description of the package integration and of the visualization tool. The process of integrating application packages and using the visualization tool was shown in [15].

The participants then described their packages via the PM, and checked those descriptions using POI (the CLAVIRE's interface to run the integrated package or exists WF [15]). Almost all of the members of the first group then developed visualization tools and sent them to the administrator, who installed them on our CSVVR. Moreover, they developed the composite application, IWF, which included calls for the scientific package and the visualization tool. The members of the second group made a regular composite application with calls for only one or more of the scientific application packages.

All the participants prepared reports and included them in their WF project. We then checked the composite applications and the visualization tools.

All the members of both groups successfully integrated their scientific packages into CLAVIRE. Nine of the 10 members in the first group developed their visualization tools. Three of these tools (33, 3 %) showed an excellent picture, four (44, 5 %) showed a good picture, and two of them (22, 2 %) presented a picture of satisfactory quality. Therefore, almost all the participants mastered the C++ based library. One member did not develop his visualization tool since he was over-crowded during the period of the experiment. The participants took from one to 3 weeks to integrate their packages via PM. Only two of them asked many questions; the other members mastered the PM without serious problems. The broad variation in the length of time taken can be explained by the different levels of the members' workloads.

Figure 4 shows the results of four visualization tools, which represent calculations of packages built within the following research.

1. Active data storage processes visualization for the load-balancing dynamical distribution in computational nodes of the cluster [26] (Fig. 4a);
2. The paraxial distribution simulation of ultrashort laser pulses [27] (Fig. 4b);
3. Subsurface hydromechanics: Simulation of straining action during oil field development [28] (Fig. 4c); and
4. Electromagnetic field simulation with the use of iteration solvers for Maxwell's equations [29] (Fig. 4d).

Figure 4 shows that all pictures presented in a simple, flat forma are informatively and visually poor. This negative effect reveals that the studied phenomena have powerful 3D features in stereo mode, which are overlapped and interfused in the 2D picture. This feature implicitly confirms the necessity of applying methods

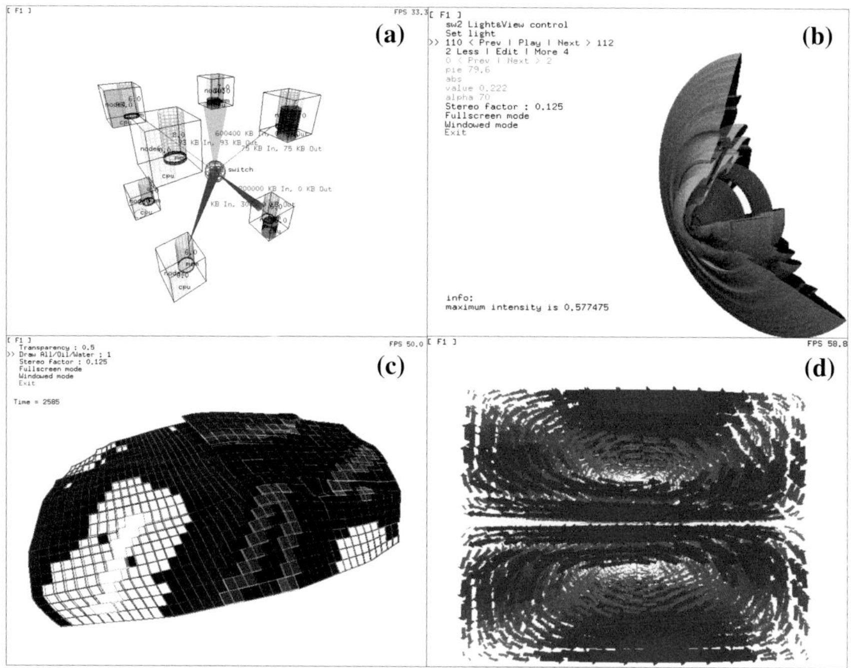

Fig. 4 Examples of pictures generated with the use of developed visualization tools

of interactive 3D (stereo) visualization for the study of complex processes and systems.

All the participants used PM, in which they described (including general information, authors and organizations, inputs and outputs—P2 and partly P4) and prepared executable cloud services. They used POI for running the prepared cloud service (P5), and WF-based UI for the development of the composite application (WF), adding data for the experiment and demonstration (including results presented in different formats), which are stored in the WF-project (P1). The participants also added reports and papers related theirs research areas (partly P3) that were prepared during the experiment into the WF project. We do not claim full compliance with the RO ontologies, but we showed that these principles are useful for creating scientific collaboration environments based on the CLAVIRE platform and the developed tools. Using these, scientists can easily and rapidly develop and share their scientific results in forms that are related to the modern achievements of eScience, including executable cloud services and 3D visualization tools.

The participants in the experiment did not use the VSO interface because they were solving a problem in one scientific area, and had only one or two scientific packages. The VSO concept and relevant interfaces are useful for those heading on scientific groups, or for those who lead inter- and multidisciplinary projects. The application of the VSO interface is one of the topics for future work.

7 Approaches to the Rapid Implementation of Scientific Results in the Educational Process

The approach presented above was applied to 20 young scientists and specialists in different scientific areas, who learnt to use the services of the CLAVIRE platform quickly, and who completed their projects successfully. As a result of this experience, we can offer the presentation of scientific results in various formats for the design and development of eLearning resources, including virtual learning labs (VLL) that are based on real, scientific application packages.

Figure 5 shows the possible collaboration between scientific and educational communities in which educators use different forms of scientific results for educational purposes (Q4).

The structural design of the courses could be based on the VSO project [21]. The structure of the VSO project provides a basis for a course's structure. The VSO's components and related workflows could be used to design VLLs in accordance with the learning models offered in [30]. Assessment packages may be required for the measurement of the level of automation of knowledge and skills. The text resources for learning courses can be rapidly formed using the scientific documentation and papers offered in the [31] approach. Therefore, we have the ability to rapidly implement the scientific results mentioned in this paper in educational processes, including cloud services. We showed in the paper [21] that changes in the SWFs and packages are automatically reflected in related VLLs and courses. The same paper shows, that one VSO-object (or VSO-model) maybe used in different educational purposes.

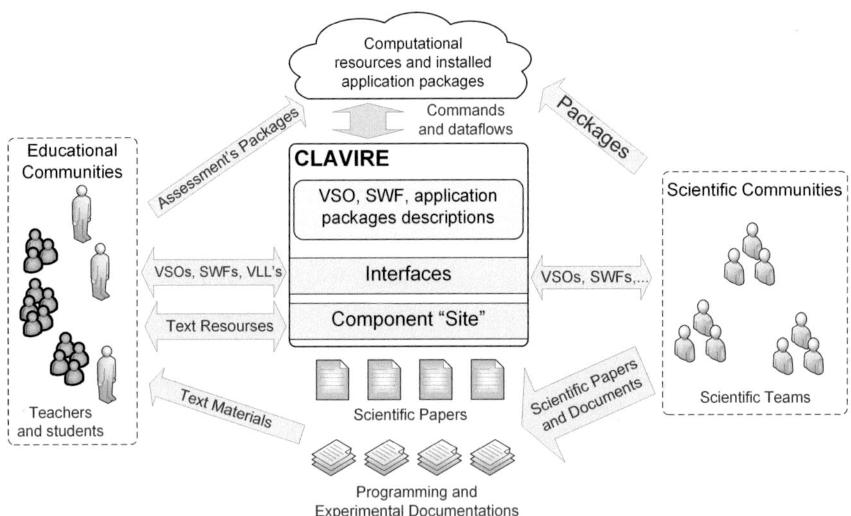

Fig. 5 Possible collaboration between scientific and educational communities

8 Conclusion and Future Work

Modern scientific society is characterized by many rapid changes in methods, approaches and technologies. Modern researchers should respond to these changes quickly, and should offer rapid, novel scientific results in a vivid and interactive manner.

In this paper, we developed an approach to extend the CLAVIRE platform in order to share traditional scientific documents, data sources, executable services, and 3D-visualization for collaborations and peer review (Q1). We developed a web-based intellectual editor, the "Package Manager," for the rapid and easy integration of scientific packages into CLAVIRE (Q2). We created a C++-based graphical library for developing visualization tools with motion-capture sensor (Microsoft Kinect) support. The IWF method is offered to develop a composite application with the interaction between the scientific package and the visualization tool (Q1). As shown above, the approach, the CLAVURE's interfaces, and the developed tools, fall within the principles of the five RO ontologies. Only the C++-based graphical library requires skills in programming, but this library significantly facilitates the development of visualization tools (Q2).

The approach mentioned above was examined by 20 young, evolving scientists from more than 10 cities located across the Russian Federation. All of the participants successfully integrated their scientific packages into CLAVIRE through PM (we created a prototype of the scientific environment (Q3), which contains 20 new cloud services). Seven of the ten participants developed visualization tools that presented a good quality 3D picture, and a relevant composite application. On the basis of the results mentioned above, we offer a possible approach to scientific and educational communities' collaboration for the rapid design of courses (Q4). This approach allows us to sustain courses that are up to date.

In the future, we are planning to develop a graphical platform for making a virtual environment based on the existing graphical model and the highly realistic graphical engine, "Fusion" [32]. This platform should allow scientists to create 3D scenes without programming. In addition, we will perform experiments by evolving scientific teams, which work on inter- and multidisciplinary problems, via the use of VSO interfaces. Furthermore, we are planning to apply the developed methods in the implementation of double-degree international educational programs [33]. Moreover, we intend to carry out social research on the subject of users' opinions regarding the approach offered.

Acknowledgment This paper is supported by the Russian Scientific Foundation, grant #14-21-00137 "Supercomputer simulation of critical phenomena in complex social systems"

References

1. Hardaway, D.: Sharing research in the 21st century: borrowing a page from open source software. Commun. ACM **48**, 125–128 (2005)
2. Meyer, D.: Academic social network ResearchGate aids debunking of stem cell study, (2014). In: GIGAOM. http://gigaom.com/2014/03/14/academic-social-network-researchgate-aids-debunking-of-stem-cell-study/. Accessed 31 Dec 2014
3. IJsbrand, A., Atzeni, S., Koers, H., Elena, Z.-S.: Bringing digital science deep inside the scientific article: the elsevier article of the future project. Lib. Q. **24** (2014)
4. Mell, P., Grance, T.: The NIST definition of cloud computing recommendations of the National Institute of Standards and Technology, Gaithersburg (2011)
5. Barseghian, D., Altintas, I., Jones, M.B., et al.: Workflows and extensions to the Kepler scientific workflow system to support environmental sensor data access and analysis. Ecol. Inform. **5**, 42–50 (2010). doi:10.1016/j.ecoinf.2009.08.008
6. Elmroth, E., Hernández, F., Tordsson, J.: Three fundamental dimensions of scientific workflow interoperability: Model of computation, language, and execution environment. Future Gener. Comput. Syst. **26**, 245–256 (2010). doi:10.1016/j.future.2009.08.011
7. Manuel, S., Batista, V., Dávila, A.M.R., et al.: OrthoSearch: a scientific workflow approach to detect distant homologies on protozoans, pp. 1282–1286. In: Proceedings of the ACM Symposium on Applied Computing, New York, New York, USA (2008)
8. Deelman, E., Gannon, D., Shields, M., Taylor, I.: Workflows and e-Science: An overview of workflow system features and capabilities. Futur. Gener. Comput. Syst. **25**, 528–540 (2009). doi:10.1016/j.future.2008.06.012
9. Oinn, T., Addis, M., Ferris, J., et al.: Taverna: a tool for the composition and enactment of bioinformatics workflows. Bioinformatics **20**, 3045–3054 (2004). doi:10.1093/bioinformatics/bth361
10. Bochenina, K.: A comparative study of scheduling algorithms for the multiple deadline-constrained workflows in heterogeneous computing systems with time windows. Proc. Comput. Sci. **29**, 509–522 (2014). doi:10.1016/j.procs.2014.05.046
11. Lu, S., Zhang, J.:Collaborative scientific workflows supporting collaborative science. Int. J. Bus. Process. Integr. **15**, 39–47 (2011). doi:10.1109/MIC.2011.87
12. Förstner, K., Hagedorn, G., Koltzenburg, C., et al.: Collaborative platforms for streamlining workflows in open science. Nat. Preced. (2011). doi:10.1038/npre.2011.6066.1
13. Bechhofer, S., De Roure, D., Gamble, M., et al.: Research objects: towards exchange and reuse of digital knowledge. Nat. Preced. (2010). doi:10.1038/npre.2010.4626
14. Belhajjame K., Zhao J., Garijo D.: The research object suite of ontologies: Sharing and exchanging research data and methods on the open web. (2014)
15. Dukhanov, A., Bolgova, E., Bezgodov, A., Boukhanovsky, A.:The Approach to Extension of the CLAVIRE Cloud Platform for the Researchers' Collaboration. In: Proceedings of the 9th International Conference on Knowledge, Information and Creativity Support Systems, pp. 370–383 (2014)
16. Knyazkov, K.V., Kovalchuk, S.V., Tchurov, T.N., et al.: CLAVIRE: e-Science infrastructure for data-driven computing. J. Comput. Sci. **3**, 504–510 (2012). doi:10.1016/j.jocs.2012.08.006
17. Knyazkov, K.V., Nasonov, D., Tchurov, T.N., Boukhanovsky, A.V.: Interactive workflow-based infrastructure for urgent computing. Proc. Comput. Sci. **18**, 2223–2232 (2013). doi:10.1016/j.procs.2013.05.393
18. Smirnov, P.A., Kovalchuk, S.V., Dukhanov, A.V.: Domain ontologies integration for virtual modelling and simulation environments. Procedia Comput Sci **29**, 2507–2514 (2014). doi:10.1016/j.procs.2014.05.234
19. Shank, J.D.: The emergence of learning objects: the reference librarian's role. Res Strateg **19**, 193–203 (2003). doi:10.1016/j.resstr.2005.01.002
20. Grunwald, S., Reddy, K.R.: In: Concept Guide on Reusable Learning Objects with Application to Soil, Water and Environmental Sciences, pp. 1–12 (2007)

21. Dukhanov, A., Smirnov, P., Karpova, M., Kovalchuk, S.: e-Learning Course Design Based on the Virtual Simulation Objects Concept. In: Proceeding of the 2014 IEEE 8th International Conference on Application of Information and Communication Technologies, pp. 508–513 (2014)
22. Karsakov, A., Bilyatdinova, A., Hoekstra, A.: 3D virtual environment for project-based learning. In: Proceedings of the 2014 IEEE 8th International Conference on Application of Information and Communication Technologies, pp. 468–472 (2014)
23. 7 Things You Should Know About Learning Tools Interoperability (2013)
24. Bilyatdinova, A., Karsakov, A., Bezgodov, A., Dukhanov, A.: Virtual environment for creative and collaborative learning. Proc. 9th Int. Conf. Knowledge, Inf. Creat. Support Syst. 313–320 (2014)
25. Sloot, P.M.A., Boukhanovsky, A.V.: Young Russian researchers take up challenges in the computational sciences. J. Comput. Sci. **3**, 439–440 (2012). doi:10.1016/j.jocs.2012.08.009
26. Tyutlyaeva, E., Kurin, E., Moskovsky, A., Konuhov, S.: Abstract: Using Active Storage Concept for Seismic Data Processing. 2012 SC Companion: High Performance Computing, Networking Storage and Analysis, IEEE, pp 1389–1390. (2012)
27. Suvorov, E., Akhmedzhanov, R., Fadeev, D., et al.: On the peculiarities of THz radiation generation in a laser induced plasmas. J. Infrared Millimeter Terahertz Waves **32**, 1243–1252 (2011)
28. Biktimirov, M.R., Biryaltsev, E.V., Demidov, D.E., et al.: Information infrastructure of tatarstan: from «SENet-Tatarstan» to «SEGrig-Tatarstan». Russ. Natl. Supercomput. Forum (2012)
29. Butyugin, D.S., Il'in, V.P., Petukhov, A.V.: Comparative Analysis of Approaches for High Frequency Electromagnetic Simulation. In: Proceedings of the Progress in Electromagnetics Research Symposium (PIERS), Moscow, pp. 1483–1487 (2009)
30. Dukhanov, A., Karpova, M., Bochenina, K.: Design Virtual Learning Labs for Courses in Computational Science with Use of Cloud Computing Technologies. Proc. Comput. Sci. **29**, 2472–2482 (2014). doi:10.1016/j.procs.2014.05.231
31. Bochenina, K., Dukhanov, A.: An approach to a rapid development of the problem-oriented educational services based on the results of scientific researches. WIT Trans. Eng. Sci. **93**, 877–884 (2014)
32. Bezgodov, A., Esin, D., Karsakov, A., et al.: Graphic toolkit for virtual testbed creation: application for marine research and design. Dyn. Complex. Syst. Century **7**, 34 (2013)
33. Dukhanov, A.V., Krzhizhanovskaya, V.V., Bilyatdinova, A., et al.: Double-degree master's program in computational science: experiences of ITMO University and University of Amsterdam. Proc. Comput. Sci. **29**, 1433–1445 (2014). doi:10.1016/j.procs.2014.05.130

iDAF-drum: Supporting Practice of Drumstick Control by Exploiting Insignificantly Delayed Auditory Feedback

Kazushi Nishimoto, Akari Ikenoue and Masashi Unoki

Abstract To achieve excellent drum performances, sufficient use of the extensor muscles of the wrists is important. However, it is actually very difficult and there have been no efficient methods and tools to train them. This paper proposes iDAF-drum, which is a novel training system of the extensor muscles in everyday drum practice. "iDAF" is an acronym of "insignificantly delayed auditory feedback" and usual people cannot perceive such a very slight delay. We found an interesting phenomenon that drummers raise the drumsticks higher than usual by inserting the unperceivable delay between impact and sound. By exploiting this phenomenon, iDAF-drum can efficiently train the drummers' extensor muscles without giving them any unusual feeling. We demonstrate the efficiency of iDAF-drum based on user studies.

Keywords Unperceivable factors · Drum practice · Delayed auditory feedback · Illusory feelings

1 Introduction

Good control of drumsticks is very important in drum performance. A drummer must not only drum at accurate tempo with adequate strength but also control the tone of the drum through the motion of the sticks. Wrist motion is the key to stick control. The flexor muscle, which contracts to cock the wrist, and the extensor muscle, which

Akari Ikenoue: Currently with Next Co., Ltd.

K. Nishimoto (✉) · A. Ikenoue · M. Unoki
Japan Advanced Institute of Science and Technology, Nomi, Japan
e-mail: knishi@jaist.ac.jp
URL: http://www.jaist.ac.jp/ks/labs/knishi/

A. Ikenoue
e-mail: identity0811@yahoo.co.jp

M. Unoki
e-mail: unoki@jaist.ac.jp

S. Kunifuji et al. (eds.), *Knowledge, Information and Creativity Support Systems*, Advances in Intelligent Systems and Computing 416,
DOI 10.1007/978-3-319-27478-2_36

contracts to extend the wrist, govern the wrists' motion. Balanced usage of both muscles allows the drummer to drum fast for a long period as well as to improve the tone of the drum [6–8, 18].

However, it is generally difficult to master the technique of stick control where the extensor muscle is sufficiently used. Special heavy sticks have often been used for practicing the intentional use of the extensor muscle in traditional drum training methods. However, using such heavy sticks causes damage to wrists due to overload. A new method for training the dominant use of the extensor muscle has recently been developed [14]. In this method, a special drumming form that forces the drummer to use the extensor muscle is proposed. However, this form is quite different from normal ones. The drummer has to master the special form only for training the extensor muscle. Thus, this method is not so efficient because it requires extra training time.

This paper proposes a novel method exploiting an "insignificantly delayed auditory feedback (iDAF)" for training drum stick control and a support system for drumming practice named iDAF-drum [11, 17]. Here, insignificantly delayed auditory feedback means a very short time delay between the impact of the stick with the drumhead and the emission of sound. Typically, humans cannot recognize the existence of iDAF: iDAF is unperceivable. Different from the conventional methods, drummers will not find any difference between drumming under our method and the conventional one. Nonetheless, it allows drummers to efficiently train in a way where the extensor muscle is sufficiently used.

2 Related Works

Several support and augmenting systems for drum performance have been studied so far. Jam-O-Drum [3] is a collaborative multimedia percussion system for performing interactive improvisations. Voice Drummer [15] is a system for inputting a percussion score by so-called "voice percussion." Many such systems focusing on drumming performance have been proposed, created and studied. However, they did not support practice while training how to use the extensor muscle of the wrists.

In contrast to performance training, there have not been so many attempts to support practice of the drum. Iwami and Miura [12] studied a computer system to help drummers practice loop patterns of the drum. It visualizes situations of drumming such as fluctuation of timing and impact strength to allow drummers to self-check their performances. This system shows where mistakes are made, but it does not tell the drummer how to practice to correct them. Beatback [9] is a system for supporting rhythm practice. However, "practice" here means "exploration" or "creation" of novel rhythmic patterns. The system encourages such generating processes of rhythm by working as a virtual partner of musical performance. Thus, this system does not truly focus on correction of the wrong drumming form.

Tsuji and Nishitaka [22] developed a system to improve drumming form. By showing rhythm lapses, drum-form lapses, hand-stroke amplitude, and striking strength, it leads the drummer to correcting his/her wrong form. This objective of that study is quite similar to ours. However, the way to correct the wrong form is indirect: No concrete instructions are provided. In contrast, we attempt to directly correct wrong usage of the muscles.

Several patents to improve drumming form have been applied for. "Practice aid device for percussionists" [1] proposed a small spacer for correcting the way of holding the drumsticks. "Muscle control development system and kit therefor" [4] proposed drum pads made of elastic materials. By preparing multiple pads whose degree of rebounding are different and by drumming on those pads, users can train in ways of stick control while hitting drums having different rebounding features. They aimed at improving drumming form, but they did not focus on training the extensor muscle.

"Drummer stick control up-stroke practice method and device" [2] proposed a special attachment that is mounted over a drumhead: A horizontal bar is set tens of centimeters above the drumhead. By intentionally hitting the bar by the up-stroke of the sticks just after hitting the drumhead, users can learn ways to intentionally raise the sticks after impact. The objective of this patent is similar to ours: It focuses on the up-stroke that requires the drummer to use the extensor muscle. However, it also imposes special and unusual ways of performance on the drummer. There is the risk of he/she adopting the bad habit of excessively raising the sticks.

Consequently, although various systems for supporting the practice of drumming performance have been proposed, created and studied, no system has focused on training the extensor muscle. Furthermore, there has been no attempt to exploit the effects of unperceivable factors like iDAF.

3 Proposed Method and System

3.1 Definition of iDAF and Method

Delayed auditory feedback (DAF) usually means a feedback of voice to its speaker with a 100–200 ms delay. It is well known that such a DAF prevents the speaker from smoothly speaking, since it leads to the phenomena of repeating syllables and stuttering [13].

If a DAF is applied to the performance of a musical instrument, behaviors in the performance change. In the case where a person repeatedly taps using his/her forefinger, the raising height of the forefinger tends to increase if the tapping sounds are delayed [20]. Therefore, by applying this result to drumming, we can expect that the raising height of the drumsticks will increase and that this will in turn provoke a motion that makes much greater use of the extensor muscle. However, such a large delay as 100–200 ms makes it difficult for people to play musical instruments, and they became unable to keep accurate rhythm [19]. As a result, it becomes practically impossible to practice drumming.

Insignificantly delayed auditory feedback (iDAF) is an auditory feedback with a very short delay that people normally cannot perceive. It is known that people can usually perceive a time lag between an event and its sound if it is longer than 20–30 ms [5, 16]. Therefore, we define iDAF, in this paper, as an auditory feedback with a delay not exceeding 30 ms.

We propose a novel method for drumming practice that exploits iDAF. It provides an unperceivable delay between the impact of the stick and the sound emission from the impact. If it can change a drummer's motion similar to the behavior change of the forefinger tapping with a long delay [20], the extensor muscle would come to be used much more without obstructing the drumming.

3.2 System Setup

Figure 1 illustrates the system setup of iDAF-drum. iDAF-drum consists of an electrical drum pad (YAMAHA TPS80S), a trigger module (YAMAHA DTXPRESS), a MIDI sound module (Roland SD-50), a USB-MIDI interface (YAMAHA UX-16), and a Windows PC (Windows VISTA). An impact signal from the electrical drum pad is input to the trigger module and then converted to a MIDI signal. The MIDI signal is input to the Windows PC. After a given time (shorter than 30 ms) passes, the PC inputs the signal into the MIDI sound module. Finally, a hitting sound is emitted with an insignificant delay.

For convenience of data analysis, when the Windows PC receives the MIDI signal from the trigger module, the PC outputs a pulse signal from its serial port that is used for synchronizing with the myoelectric potential data of the muscles (we call this pulse signal "synchronization signal" hereafter). We implemented the software runs on the PC for adding the delay while using C♯. In order to achieve 1 ms-order resolution for adding the delay, we used the Windows API functions. All of the sounds of the performance and a metronome are output from a headphone. We assigned a snare drum tone as the performance sound.

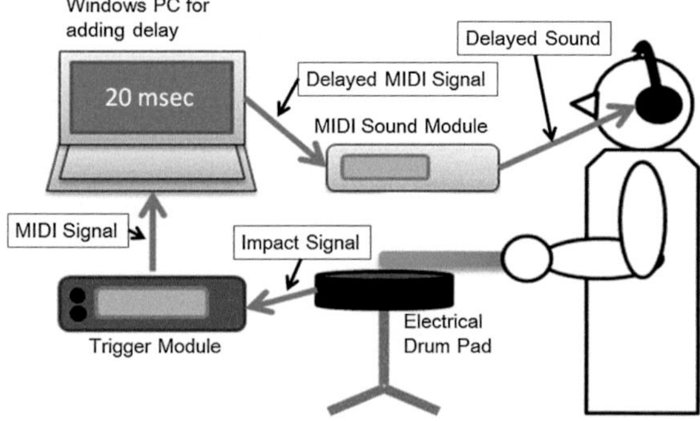

Fig. 1 System setup of iDAF drum

4 Estimating Effects of iDAF on Drumming

To the best of the authors' knowledge, there has been no study on estimating the effects of such a very slightly delayed auditory feedback as iDAF. Therefore, this section investigates the effects of iDAF on drumming performance. First, we confirm that iDAF produces no negative effects, and then we investigate whether iDAF changes the drumming behavior so that the drummer begins to use the extensor muscle to a much greater extent.

4.1 Experimental Procedure

We employed 12 subjects (Sex: 7 males and 5 females; Age: Average 20.67 y/o, STDV 3.94; Drum experience: Average 5.78 years, STDV 3.67, Max 13.0 years, Min 0.5 years). We asked them to hit the electrical drum pad using sticks held by the right and left hands alternatively along with a metronome sound that ticks every 250 ms. We instructed them to hold the sticks in a matched-grip manner.

Before starting experimental drum performances, we measured the maximum voluntary contraction (MVC) of the ulnar flexor muscle of the wrist and the radial extensor muscle of the wrist three times for each of the subjects. In the experiment, we asked each subject to conduct four performance sessions whose delay times changed in the order of 0, 20, 10 and 30 ms. The subjects were asked to perform for 90 s in each session. How they synchronized with the metronome was individually different, e.g., some subjects attempted to set the timing of sound emission to the metronome sound while others attempted to set the timing of impact to the metronome. Therefore, to standardize the way of synchronization, we asked the subjects to wear an eye mask and to attempt to set the timing of the sound emission to the metronome as much as possible. No warm-up was permitted. During each session, the myoelectric potential data of the extensor muscle and the flexor muscle were measured using electromyography (TEAC Polymate AP1532). In addition, we recorded the performances using a high-speed video camera. After finishing each session, we asked each subject whether he/she felt or found anything unusual. Finally, after all four sessions were finished, we asked about his/her musical experiences so far.

To increase the signal-to-noise ratio of the myoelectric potential data between the synchronization signal and 250 ms before that, we applied a signal-averaging method to the obtained data, and the root mean square (RMS) of the cleaned data was calculated. Then, the data were normalized using the MVC of each muscle so that the MVC value was set to 100 %. On the other hand, for each muscle and for each delay time, we calculated the accumulated electromyogram and the average of the gross amount of muscle activities of all subjects. In addition, we calculated the average and variance of inter-onset interval (IOI) values of the synchronization signals within the last 60 s data of all subjects for each delay time. Based on the literature [10], we calculated a relative difference signal (RDS) as an index of co-contraction of the flexor and extensor muscles of the wrist by the following equation:

$$RDS = \frac{F - E}{F + E} \quad (1)$$

where F and E correspond to RMS values of the flexor muscle and the extensor muscle, respectively. $RDS \cong 0$ means that both muscles simultaneously contract (co-contract), while $RDS \cong 1$ means that they reciprocally contract.

4.2 Results

The ANOVA for average IOI values of all subjects showed no significant main effect with or without delay ($F(3, 44) = 0.4, p < 0.754$). In the previous studies on DAF with 100–200 ms delay, an expansive speaking phenomenon was observed. However, such an expansive phenomenon does not arise under the iDAF condition. We also calculated the coefficient of variation (CV), which is usually used as an indicator of confusion caused by DAF. As a result, we could not find any significant difference between with and without insignificant delay ($F(3, 44) = 0.436, p < 0.728$). From these results, we can conclude that iDAF does not at all cause the confusion that arises in DAF. From the inquiry results obtained after each session, only one subject reported that he felt a slightly strange feeling when delay was given. However, he did not become unable to perform drumming and he could play as usual. None of the other subjects felt any difference or strangeness.

Figure 2 shows snapshots taken with the high-speed camera at 1 min from the beginning of each session. The small red circle in each picture shows the head of the

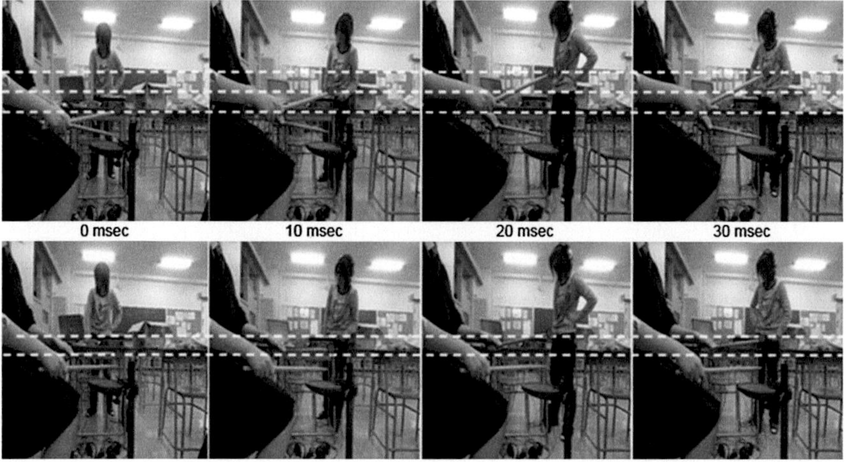

Fig. 2 Top reach points of the stick for 4 delay times. *Upper row* right hand, *Lower row* left hand

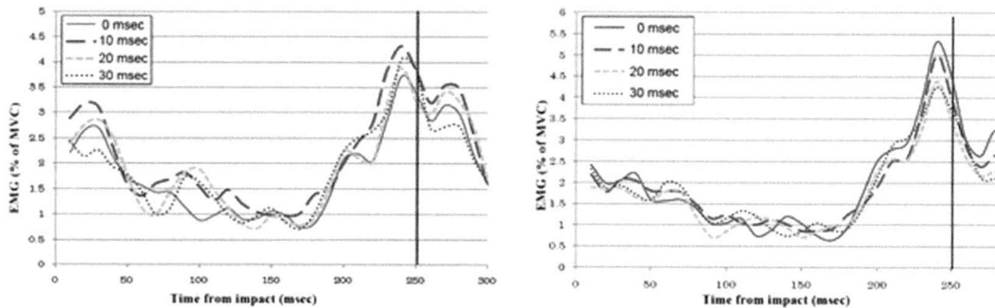

Fig. 3 Electromyograms of extensor muscle of left hand (*left*) and right hand (*right*) for each delay time

stick. The upper row shows the top reach point of the stick held by the right hand and the lower row shows that of the left hand. The delay times were 0, 10, 20 and 30 ms from the leftmost to the rightmost pictures. It can be seen that the more the delay time increases, the higher the stick is raised. This tendency was observed for most of the subjects.

Figure 3 shows average electromyograms (EMGs) of the extensor muscle of the left and right wrists for each delay time. The vertical line drawn at 250 ms shows the impact timing. The left hand data of Fig. 3 shows that the peak value of EMG with no delay (blue line) is smaller than the other peak values with delays. In particular, the peak value with 10 ms delay (broken red line) is about 0.6 % larger than that with no delay. In contrast, The right hand data of Fig. 3 shows that the peak value of EMG with no delay is larger than those with delays. We calculated gross value of EMG for each delay and compared the gross value with no delay to the gross value with each delay by the t-test. As a result, the difference between the gross value with 10 ms delay and that with no delay for the left hand is marginally significant ($t(11) = 1.892, p < 0.085$). However, no significant difference could be found in all other combinations. As shown in Fig. 2, the subjects raised the sticks higher under the iDAF conditions. Although this suggests that they used the extensor muscle much more, it was not supported by the electromyograms.

Figure 4 shows RDS values for a condition using heavy sticks, a condition with 20 ms delay, and a condition with no delay. The RDS values when using the heavy sticks are the smallest for both hands. From the t-test results, the difference between the value with no delay and that when using the heavy sticks is significant for the right hand ($t(4) = 3.004, p < 0.004$). In contrast, there was no significant difference between the values with no delay and those with 20 ms delay.

4.3 Discussion

From the results shown in Fig. 2 and responses to inquiries after each session, we obtained very interesting findings: The motions of drumming performance changed due to iDAF and the subjects tended to raise the sticks much higher than usual

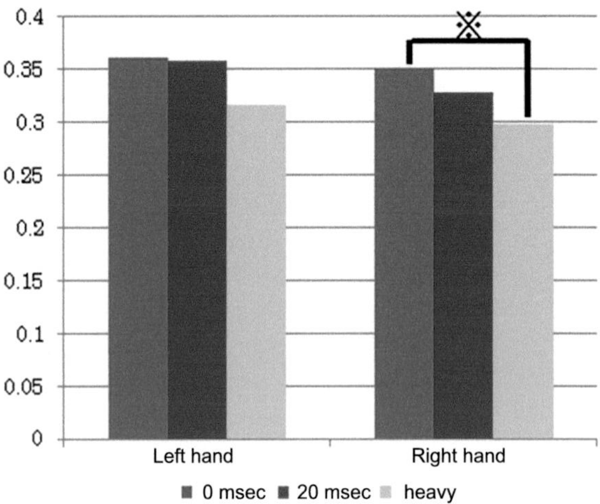

Fig. 4 RDS values with no delay, with 20 ms delay, and using heavy sticks for each hand

(with no delay), although they were not aware of the delays. This suggests that iDAF can successfully induce the performer to sufficiently use the extensor muscle in drumming performance without perceiving differences. It remains unrevealed why humans react to such unperceivable auditory delays. This is a very interesting research issue, but it is out of the scope of this paper; the objective of this paper is to create a useful support system for practicing drum performance. We only apply this phenomenon to the support system in this paper; elucidating its mechanism remains a future work.

Although iDAF may be able to provoke sufficient use of the extensor muscle, the results shown in Fig. 3 does not support the idea that the subjects came to use the extensor muscle more than usual when iDAF was given. Rather, although no significance was obtained, the longer the delays were, the less the extensor muscle was used for the right hand. For the left hand, the extensor muscle was mostly used with 10 ms delay, and then the longer the delays became, the less the extensor muscle was used.

It can be assumed that the reason why the sticks were raised higher when the delays were given is that the subjects unconsciously control the impact timing. However, to do so, they probably use not only their wrists but also their elbows.

It becomes impossible to effectively make subjects use their wrist extensor muscle if their elbows move. A possible method of making their elbows not move is to immobilize their upper arms. However, such a method imposes too great a load on the extensor muscle, and it can damage the extensor muscle in a similar way to the method using heavy sticks. Therefore, we think a better method is to make the impact-impact interval so short (namely, to make play faster) that the drummers need not move their elbow to raise the sticks higher for adjusting the impact-impact interval. As a result, we can expect iDAF to effectively make them use the extensor muscle. Conversely, it becomes difficult for them to greatly exercise the extensor muscle. However, people usually cannot immediately master such ways of using the extensor muscle: Long-term training is required. Therefore, it is preferable, from a safety viewpoint, to make them use this muscle little by little.

Figure 4 shows that the RDS value using heavy sticks is smaller than that of the other methods. This means that the extensor muscle and the flexor muscle tend to simultaneously contract (co-contraction) when the heavy sticks are used. Co-contraction of these muscles fixes the wrists. As a result, it becomes difficult to absorb reaction force by the impact, which causes damage of the wrist joints. In addition, co-contraction cancels the torque generated by the muscle that should work with the torque generated by the other muscle that should not work. Thus, it became evident that the traditional practice method using heavy sticks is not good because it may not only cause injuries but also accustom the drummer to bad drumming habits that disturb smooth control of the sticks. In contrast, there was no significant difference in RDS values between no delay and a 20 ms delay. Accordingly, we can expect our proposed method using iDAF to ensure compatibility between improvement of drumming technique so that the extensor muscle is sufficiently used while avoiding co-contraction of the extensor and flexor muscles.

5 Estimating Efficiency of iDAF-Drum

In the previous section, we showed the possibility that iDAF-drum leads the drummer to sufficiently use the extensor muscle. This section investigates whether the drummer can eventually master the correct drumming way of sufficiently using the extensor muscle by continually using iDAF-drum. If this can be achieved, the myoelectric potential of the extensor muscle without delay will increase along with the progress of training and finally become as strong as that with delay.

5.1 Experimental Procedure

In this experiment, we employed five subjects who are members of a brass band of a high school and are included in the 12 subjects of the experiment shown in the previous section. We asked them to perform a 10 min practice session every day, which includes 2 min single-stroke practices for the left and right hands, a 3 min change-up practice, and a 3 min drumroll of 16th notes (Fig. 5). Before starting this session, each subject was allowed a warm-up performance.

We set the metronome to a 500 ms interval. In the previous section, we pointed out that it is preferable to set the performance speed relatively fast. However, some subjects did not have so much experience. If we set the performance speed too fast, they might not have been able to perform the test pieces. Therefore, we examined maximum speed for each subject before the experiment and set the speed as fast as all of the subjects could perform the test pieces. Under this metronome interval, the speed when 8th notes were performed was the same as in the experiment conducted

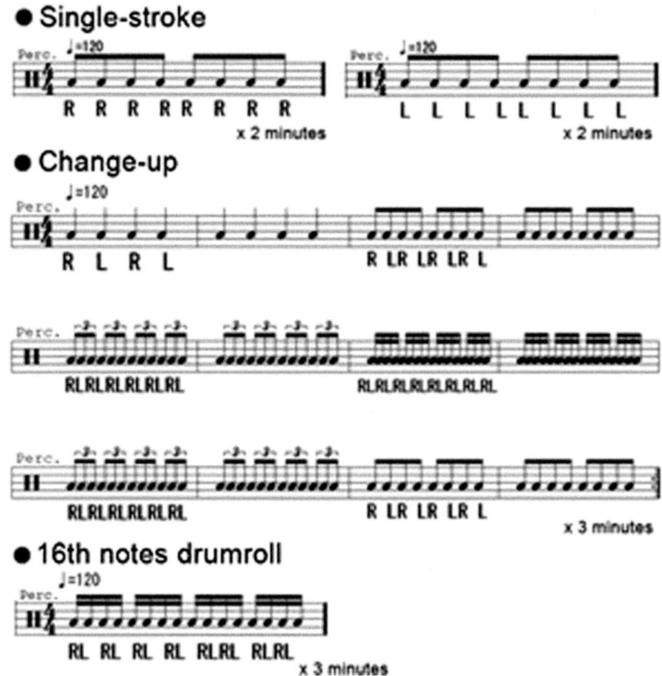

Fig. 5 Test pieces

in the previous section, and it became faster when shorter notes like 16th notes were performed. Therefore, the overall performance speeds of this experiment are faster than those in the experiment of the previous section.

The system setup of iDAF-drum is the same as that used in the previous section. However, in this experiment, we used two sets of iDAF-drum.

Each subject practiced 10 min every day using the iDAF-drum with a 20 ms delay for twelve days. On the first day, the sixth day, and the last day, we measured the MVC of the extensor muscle and the flexor muscle of each subject three times, and then we asked each subject to continue drumming along with the metronome at 250 ms intervals for 1.5 min using iDAF-drum with no delay and with 20 ms delay. During these performances, we measured electromyogram of the extensor muscle and the flexor muscle of both arms together with synchronization signals.

We applied the signal-averaging method to the obtained data to reduce noise and calculated the root mean square (RMS) of the cleaned myoelectric potential data between the synchronization signal and 250 ms before that. Then the data were normalized using the MVC of each muscle so that the MVC value was set to 100 %. In addition, for each muscle and for each delay time, we calculated the accumulated electromyogram data between 30 and 90 s from the beginning and the average of the gross amount of muscle activities of all subjects.

5.2 Results

Figure 6 shows the averaged electromyograms of the extensor muscles of both hands of all subjects with no delay on the three measuring days. In the figure, vertical black lines drawn at 250 ms show the impact timing. From this figure, we can see that the peak values became higher day by day. We calculated the averages of the myoelectric potential data between the synchronization signal and 250 ms before that. The ANOVA for the averaged myoelectric potential data showed a significant main effect of the measuring day for the left hand ($F(2, 12) = 4.815, p < 0.029$). However, no such significance was shown for the right hand ($F(2, 12) = 2.062, p < 0.170$).

Figure 7 shows the averaged electromyograms of the flexor muscles of both hands of all subjects with no delay on the three measuring days. In the figure, vertical black lines drawn at 250 ms show the impact timing. From this figure, we can see that the peak values of the flexor muscles of both hands substantially decreased from the first day to the sixth day, and then it recovered to almost the same level as the first day's level from the sixth day to the last day. However, the ANOVA for the averaged myoelectric potential data showed no significant main effect of the measuring day for both hands.

A supervisor of the brass band of the high school pointed out that one of the subject's drumming sound changed during the experiment: The sound became sharp and clear.

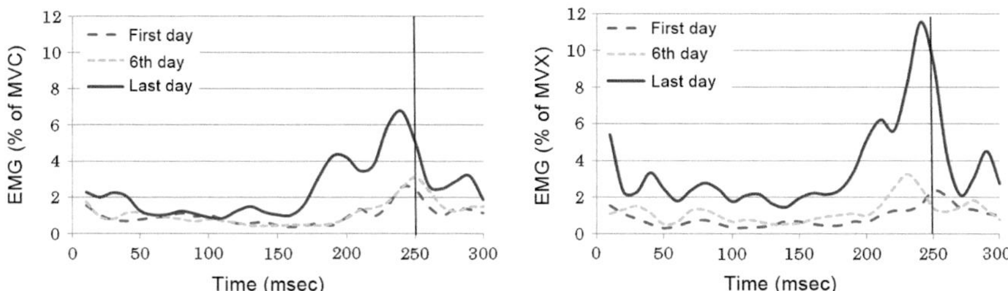

Fig. 6 Electromyograms of extensor muscle of left hand (*left*) and right hand (*right*) without delay

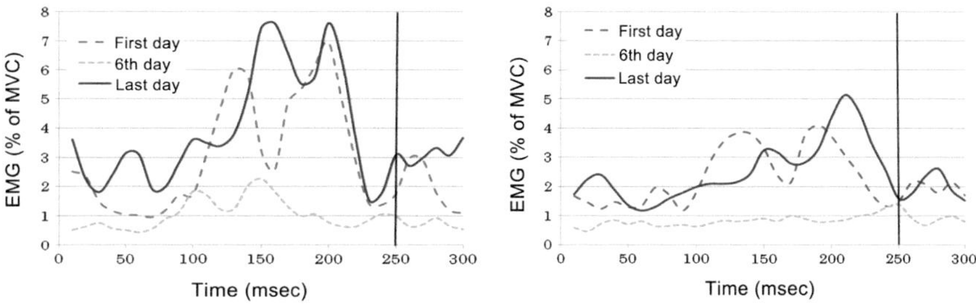

Fig. 7 Electromyograms of flexor muscle of left hand (*left*) and right hand (*right*) without delay

5.3 Discussions

If the drummer learns how to use the extensor muscle while being forced to use it by iDAF-drum, we can expect that the extensor muscle will come to be used even without delay day by day. From the results shown in Fig. 6, although no significance could be obtained for the right hand, a day-by-day tendency for the extensor muscle to be used more without delay was observed.

We should conduct a control experiment by preparing a control group whose subjects practice using only a normal drum without delay. This time, unfortunately, we could not do so due to the lack of experimental equipment. However, most of the subjects we employed have practiced the drum for a long time (Average: 4.71 years, STDV: 3.29). If such experienced drummers can immediately master techniques for using the extensor muscle in only 12 days' practice, we can assume that they have already mastered the techniques. Therefore, it is unlikely that a control group would show changes within these 12 days. In contrast, we obtained some evident changes in this experiment. This fact supports the idea that iDAF-drum is effective for improving the use of the extensor muscle.

All of the subjects were right-handed, and they could use the right hand at will. From this, we can infer that they already had some skill in using the extensor muscle of the right hand. This is likely a reason why no significance was obtained for the right hand's myoelectric potential data (Fig. 6). On the contrary, they could not use the left hand at will in the same manner as the right hand, and they did not have enough skill to use the extensor muscle of the left hand. As a result, the effect of training by using iDAF-drum clearly appeared. This implication was supported by the result that significance was obtained for the left hand's myoelectric potential data (Fig. 6).

In Fig. 7, the peak values of the myoelectric potential data of the flexor muscle substantially decreased from the first day to the sixth day, and then they recovered toward the last day for both hands. We believe that these results reflect a learning process related to usage of the extensor (not flexor) muscle.

According to Suwa's study [21], in the learning process of bowling skills, the body parts that players are conscious of change day by day. In the initial stage, they are conscious of specific parts like fingers. As they become proficient and able to earn high scores consistently, it was found that their consciousness spreads to the entire body. In our experiment, the subjects were learning new skills. Therefore, analogous to Suwa's study, we can infer that the targets that the subjects are conscious of are changing.

At the beginning of the experiment, the subjects had skill in using the flexor muscle, while their skill in using the extensor muscle was still rough. As a result, on the first day, the myoelectric potential of the flexor muscle was high while that of the extensor muscle was low. Through the training using iDAF-drum, their consciousness moved to the extensor muscle and they became unable to be conscious of the flexor muscle, which caused a decrement of the myoelectric potential of the flexor muscle on the sixth day. Meanwhile, they had not yet mastered the use of the extensor muscle, and the myoelectric potential of the extensor muscle was still not so high. On the last day, they learned the use of the extensor muscle and its myoelectric potential became higher. At the same time, their consciousness, which had been forced to concentrate on the extensor muscle, returned to the flexor muscle again. Eventually, the myoelectric potential of the flexor muscle recovered to be as strong as that on the first day. We think such a learning process reflects the transition of the myoelectric potential shown in Figs. 6 and 7.

6 Conclusions

In this paper, we proposed iDAF-drum to exploit the effects of insignificantly delayed auditory feedback in human behavior. Our purpose was to support drummers in mastering the sufficient use of the extensor muscle in controlling the drumsticks. Using iDAF-drum, we conducted user studies and investigated its efficiency. As a result, we found that the subjects came to raise the sticks higher than usual and that iDAF did not confuse the drummers in their performance. The subjects could perform drumming using iDAF-drum even as well as using a normal drum without delay. In addition, we revealed that the proposed training method using iDAF-drum was physically safer than a traditional training method using special heavy sticks: iDAF-drum imposed fewer loads on the wrist muscles by avoiding co-contract of the flexor and extensor muscles. From the experiments in which the subjects continually used iDAF-drum for 12 days, the myoelectric potential of the extensor muscle became stronger day by day. Furthermore, by comparing the electromyogram of the extensor muscle with that of the flexor muscle, we could observe the process of mastering a new skill of drumming. Consequently, we could confirm the usefulness of the proposed training method of drumming performance as well as iDAF-drum.

In the case of iDAF-drum, the drummers' actions changed although they could not perceive the existence of the delay. We have applied iDAF to a keyboard such as a piano. Similar to the drum cases, performers could not perceive the delay. However, they reported that its keys were heavier when iDAF is inserted between key-touch and sound emission than those without delay. Thus, people involuntarily react to the unperceivable factors and they perceived them as a different feeling. These phenomena are very interesting. It suggest possibilities of the unperceivable factors that people's behaviors can be changed without being realized the existence of the factors by them and we can provide some different feelings without actually changing structure of an object (such as the keyboard) to provide the feelings. In future, we will further explore other cases that provide such illusory feelings and apply them for effectively bringing out and fostering buried creative abilities.

Acknowledgments We thank the members of the light-music circle of Miyazaki University and the brass band of the high school associated with Kumamoto Gakuen University for their kind cooperation to our experiments. We also thank Dr. Kohei Matsumura of Ristumeikan University for his valuable advices on measuring electromyogram. This work was supported by JSPS KAKENHI Grant Number 26280126.

References

1. Barke, S.J.: Practice aid device for percussionists, United States Patent, US 7,473,836 B2 (2009)
2. Bennett, H.: Drummer stick control up-stroke practice method and device, United States Patent, US 2004/0025664 A1 (2004)
3. Blaine, T., Perkis, T.: The Jam-O-Drum interactive music system: a study in interaction design. In: Proceedings of the 3rd conference on Designing Interactive Systems: Processes, Practices, Methods, and techniques, pp. 165–173. ACM Press (2000)
4. Carter, D.: Muscle control development system and kit therefor, United States Patent, US 5,744,737 (1998)
5. Cox, J.: The minimum detectable delay of speech and music. In: Proceedings of IEEE International Conference on Acoustics, Speech, and Signal Processing, ICASSP'84, pp. 136–139 (1984)
6. Fujii, S., Kudo, K., Shinya, M., Ohtsuki, T., Oda, S.: Wrist muscle activity during rapid unimanual tapping with a drumstick in drummers and nondrummers. Motor Control **13**, 237–250 (2009)
7. Fujii, S., Moritani, T.: Rise rate and timing variability of surface electromyographic activity during rhythmic drumming movements in the world's fastest drummer. J. Electromyogr. Kinesiol. **22**, 60–66 (2012)
8. Fujii, S., Moritani, T.: Spike shape analysis of surface electromyographic activity in wrist flexor and extensor muscles of the world's fastest drummer. Neurosci. Lett. **514**, 185–188 (2012)
9. Hawryshkewich, A., Pasquier, P., Eigenfeldt, A.: Beatback: A real-time interactive percussion system for rhythmic practise and exploration. In: Proceedings of 2010 Conference on New Interfaces for Musical Expression, pp. 100–105 (2010)
10. Heuer, H.: Control of the dominant and nondominant hand: exploitation and taming of nonmuscular forces. Exp. Brain Res. **178**(3), 363–373 (2007)
11. Ikenoue, A., Ogura, K., Unoki, M., Nishimoto, K.: A supporting system to improve drumming form by applying insignificantly-delayed auditory feedback. Trans. Human Interface Soc. **15**(1), 1–14 (2013). (in Japanese)
12. Iwami, N., Miura, M.: A support system for basic practice of playing the drums. In: Proceedings of International Computer Music Conference, pp. 364–367 (2007)
13. Lee, B.S.: Effects of delayed speech feedback. J. Acoust. Soc. Am. **22**(6), 824–826 (1950)
14. Nagano, Y.: Relax! Physical drumming—An American way of stroke enables clear, fast and loud performance. Ritto Music (2011) (in Japanese)
15. Nakano, T., Goto, M., Ogata, J., Hiraga, Y.: Voice drummer: a music notation interface of drum sounds using voice percussion input. In: Proceedings of the 18th Annual ACM Symposium on User Interface Software and Technology, pp. 49–50. ACM Press (2005)
16. Nishibori, Y., Tada, Y., Sone, T.: Study and Experiment of Recognition of the Delay in Musical Performance with Delay, IPSJ SIG Notes, vol. 2003-MUS-53, No.9, pp. 37–42 (2003) (in Japanese)
17. Nishimoto, K., Ikenoue, A., Unoki, M.: iDAF-drum: supporting everyday practice of drum by adding an unperceivable factor. In: Proceedings of 9th International Conference on Knowledge, Information and Creativity Support Systems, pp. 384–395 (2014)

18. Osu, R., Franklin, D.W., Kato, H., Gomi, H., Domen, K., Yoshioka, T., Kawato, M.: Short- and long-term changes in joint co-contraction associated with motor learning as revealed from surface EMG. J. Neurophysiol. **88**(2), 991–1004 (2002)
19. Pfordresher, P.Q., Palmer, C.: Effects of delayed auditory feedback on timing of music performance. Psychol. Res. **16**, 71–79 (2002)
20. Pfordresher, P.Q., Bella, S.D.: Delayed auditory feedback and movement. J. Exp. Psychol.: Human Percept. Perform. **37**(2), 566–579 (2011)
21. Suwa, M.: A cognitive model of acquiring embodied expertise through meta-cognitive verbalization. Trans. Japan. Soc. Artif. Intell. **23**(3), 141–150 (2008)
22. Tsuji, Y., Nishitaka, A.: Development and evaluation of drum learning support system based on rhythm and drumming form, Electronics and Communications in Japan (Part III: Fundamental Electronic Science), vol. 89, Issue 9, pp. 11–21 2006. Anderson, R.E. Social impacts of computing: codes of professional ethics. Soc. Sci. Comput. Rev. **10**(2), 453–469 (1992)

Novel Methods for Analyzing Random Effects on ANOVA and Regression Techniques

Gergely Bencsik and Laszlo Bacsardi

Abstract The constantly increasing data volume can help to execute different analyses using different analyzing methods. Since there are many outgoing research on different fields, analysis can be performed on big data sets and can be interpreted from different points of view. The entire process is controlled by the research methodology precisely. However, there are increasing numbers of contradictory results which follow the same methodology but interpret their results differently. Our research focuses on how is possible to get different inconsistent results according to a given question. The results are proofed by mathematical methods and accepted by the experts, but the decisions are not valid since the correlations originated from a random nature of the measured data. This random characteristics —named as random correlation—could be unknown to the experts as well. But this phenomenon needs to be handled to make correct decisions.

Keywords Decision support • Random correlations • Big data

1 Introduction

Decisions must be made at continues time. The decision making processes are support by data analyses to predict the possible future effects. The standard research methodology is defined by many state-of-art publications [1, 2]. Specialized research methodologies also appear corresponding to the given research fields. Following these methodologies, different results were born. Our research started with analyzing data originating from different fields using decision support systems and we also experienced that contradictory results could be made based on the same

G. Bencsik (✉) · L. Bacsardi
Institute of Informatics and Economics, University of West Hungary, Sopron, Hungary
e-mail: bencsikg@inf.nyme.hu

L. Bacsardi
e-mail: bacsardi@inf.nyme.hu

S. Kunifuji et al. (eds.), *Knowledge, Information and Creativity Support Systems*,
Advances in Intelligent Systems and Computing 416,
DOI 10.1007/978-3-319-27478-2_37

data rows. Our main question was the following: is it possible that the result is not correct despite of the precise analyze process? Therefore, we focused on circumstances from which the result could be originated unexpectedly.

The literature contains many inconsistent statements. In biology, the salt consumption is always generates opposite publications. There are papers supporting it and do not disclose any connection between consumption and high blood pressure [3]. Another research group states that the high salt consumption causes not only high blood pressure but kidney failure as well [4]. In forestry, two publications with contradictory title were published related to Amazon rainforest green-up [5, 6]. In medicine, one publication states that risk of malaria related to climate change can be high, another tells it can be low [7, 8]. In Earth science, many researches were performed about the debris flows in Swiss Alps. One research group states that debris flows may increase [9]. Another group's result was that it may decrease [10]. This another research group published that it may decrease, then increase [11]. In sociology, there are arguments related to data based analysis and because these results do not produce the real predictions, a new methodology was proposed [12]. Another example is based on questionnaires and scores. Seeking the answer for internet addiction, Lawrence et al. showed that the increasing Internet using among young people increases the chance of depression [13]. But Shen et al. showed that Internet is critical for daily satisfaction of the children [14].

We assumed that these contradictory results could be originated from phenomena named random correlations [RC]. This paper is based on the [15] but it contains further analysis process on RC classes and more precise results are showed on Analysis of Variance (ANOVA) and regression techniques.

The paper is organized as follows. Section 2 introduces the definition of random correlations and its different classes. Section 3 deals with our RC analyzing framework, while RC analyzing processes are introduced as examples in Sect. 4. Calculation processes are detailed in Sect. 5. Our results in the field of ANOVA and regressions are discussed in Sect. 6, while Sect. 7 concludes our paper.

2 Classes of Random Correlations

Having inconsistent results in the same research could be useful, since it enlightens a question from different points of view. But having contradictory results could lead to wrong decision. The RC theory states that there can be connection between data rows randomly which could be misidentified as a real. The main idea is that data rows as variables present the revealed, methodologically correct results, however, these variables are not connected really, and this unconnected property is hidden from the experts as well. For an easier understanding, we defined different classes of RC as the randomness can be originated from different causes.

Class 1 The first class is related to the number of used methods. Different methods can be applied for a given problem set, and if we cannot find good results with one, then we choose another. This is a classical view for finding some kind of results. The number of chosen methods can raise with different input parameters range and seeking and removing outliers. It is not defined when the data are not related to each other. When more and more methods are used with different circumstances, i.e., different parameters and error rates, then one could be uncertain whether a real correlation was found or just a random one. Incrementing the number of usable methods leads with higher probability to such a case when a correlation was identified. The "non-correlated" statement is a statement from the view of methods which cannot produced "correlation" statement. From the inverse, the "correlated" statement is just a statement from the view of the method which we succeed with.

Class 2 The second class focuses on inconsistency and includes cases when two or more methods produce opposite results. In the process of data analyses, we generally stop at the first method which we used to obtain the satisfying result and no further correlations are seek with further methods. It is rather typify finding more precise parameters based on the "correlation found" method. However, we are continuing with other method and it is possible that one or more methods present different results and the decision could be different from the original one. The given methods could present the inconsistent results occasionally or always, based on given parameters and/or data characteristics. The term "one or more methods" is used since even one method can be inconsistent with itself. For example, when it produces different types of results near given circumstances, i.e., sample size.

Class 3 The third class is based on number of data items. Data volume is growing fast nowadays. The classical approach is that having more data equals with a more precise results. However, it could lead to problem when a part of the data rows produce different results then the larger amount of that same data rows. For example, if we measure a data row from the start to a time t, and this part of the data row leads inconsistent result from another part which was measured from the start to time $t + k$. This is a critical problem, because the time interval for the measurement is unknown. For this problem, the cross validation can be a solution. If all subsets of the data rows for a given time period t are not perform the same result, only a random model could be found. If they fulfill the "same result" condition, we find a true model likely.

3 Random Correlation Analyzing Framework

Based on RC definition and classes, a framework is constructed to analyze random correlations environments related to specific cases. Our framework use the following entities:

Parameters Every measured data has its own structure. Data items with various but pre-defined form are inputs for the given analyzing method. To describe all of data structure, matrix form was chosen. Therefore parameter k, which is the number of data rows [also the columns of the matrix], and n, which is the number of data items contained in the given data row [also the rows of the matrix], are the first two

RC parameters. The third parameter, range r means the possible values, which the measured items can have. To store these possibilities, only the lower (a) and upper (b) bounds need be stored. For example, $r(1, 5)$ means the lower limit is 1, the upper limit is 5 and the possible values are 1, 2, 3, 4 and 5. In this paper, integers are used, however, it is possible to extend this notation for real numbers. The continuous form can be approximate with discrete values. In this case, the desired precision related to r can be reached with the defined number of decimals. The sign $r(1, 5, \cdot\cdot)$ means the borders are the same as before, but this range contains all possible values between 1 and 5 by two decimals. Parameter t is the number of the performed methods. This parameter is strongly related to Class 1. For example, $t = 3$ means 3 different methods are performed after each other to find a correlation.

Models and Methods In the context of random correlations, there are two main models: (1) to calculate the total possibility space [Ω]; (2) to determine the chance when we get collision, e.g., find a correlation.

In the case of (1), all possible measurable combinations are produced. In other words, all possible n-tuples related to $r(a, b)$ are calculated. Because of parameter r, we have a finite part of the number line, therefore this calculation can be performed. That is why r is necessary in our framework. All possible combinations must be produced which the researchers can measure during the data collection. After produced all tuples, the analyzing method is performed for each tuple. If "correlated" judgment occurs for the given setup, then we increase the count of this "correlated" set S_1 by 1. After performing all possible iterations, the rate R can be calculated by dividing S_1 with $|\Omega|$. R can be considered as a measurement of the "random occur" possibility related to RC parameters. In other words, if R is high, then the possibility of finding a correlation is high with the given method and with related k, n, r and t. For example, if R is 0.99, "non-correlated" judgment can be observed only by 1 % of the possible combinations. Therefore, finding a correlation has a very high possibility. Contrarily, if R is low, e.g., 0.1, then the possibility of finding a connection between variables is low. This can be good, if we accept the rule of thumb, that correlation possibility should be lower than the "non-correlated" case. However, there is a third option as well. If the correlated and non-correlated judgments can be meaningful in the view of the final result. For example, in the case of Analysis of variance (ANOVA) both H_0 and H_1 can be meaningful, therefore R should be around of 0.5 (see the Sect. 4.2). Related to the whole possibility space, this RC model is named *Ω-model.*

In the case of (2), rate C is calculated. This shows how much another data rows are needed to find a correlation with high possibility. Researchers are usually have a hypothesis and then they are trying to proof their theory based on data. If one hypothesis falls, another comes up. In practice, we have a data row A and if this data row does not correlate with another, then more data rows are used to get some kind of connection related to A. The question is how many data rows are needed to find a certain correlation. We seek that number of data rows, after which correlation will be found surly. This method is named *Θ-model.* There is a rule of thumb stating that two variables (as data rows) are correlated from 10, but we cannot find any proof, it rather is a statement based on experiences. (We would like to further analyze this rule.) The calculation process can be various and depends on the given analyzing method and RC parameters.

4 Applying RC Analyzing Processes in Practice

4.1 RC Analyzing Session

From the viewpoint of RC, the main question is whether the observed data could provide another results or just the given one. In other words, could we measure data, which can provide another (e.g., "non-correlated") result as well? If we would like to perform a research starting from data management ending with results, then the given process should be analyzed from the viewpoint of random correlations as well.

To perform an RC analyzing session, six steps were defined named as steps of the RC framework:

1. Introduce the analyzed method's basic mathematical background;
2. Determine which random correlations' class contains the given case;
3. Define the role of random correlations;
4. Select the relevant parameters of random correlations;
5. Perform calculations;
6. Validate the results with simulations.

4.2 RC Analyzing Related to ANOVA and Regression Techniques

RC Session 1*: ANOVA with Ω-model* There is a questionnaire with 4 questions and each question can be answered with a number from 1 to 3 and 5 persons were asked. We would like to analyze this questionnaire with ANOVA. ANOVA is used to determine whether the groups' average are different or not and it is applied widely in different scientific fields [16]. The null hypothesis H_0 states that the averages are equal statistically and the alternative hypothesis H_1 declines the equality statistically. Let us apply the six steps of our method. ANOVA basic mathematical background can be read in [15] [*Step* 1]. As for the random correlations, ANOVA belongs into *Class* 2 and it has a specific place: both H_0 and H_1 can be meaningful [*Step* 2]. The "non-correlated" can be defined as the means are statistically similar [H_0], therefore the influencing variable has no effect on the subject. The "correlated" means that it has influence [H_1]. In this case, the random correlations mean that the H_0 or the H_1 can took priority over against the other according to parameters defined later in Step 4. The seeking rate should be around

0.5 in each case [*Step* 3]. It is allowed to get a little difference from that rate, but huge distortion is dangerous. The following entities are used in this case [*Step* 4]: $k = 4$ is the number of data rows [questions], and $n = 5$ is the number of data items [asked people]. The possible answer values are between 1 and 3, therefore $r(1, 3)$. ANOVA is one method, therefore $t = 1$. To calculate all combination and perform ANOVA in each case [*Step* 5], a program was developed in .NET framework and written in C# language. Finally, the rates were calculated [*Step* 6].

RC session 2: regression techniques with Ω-model The same questionnaire is analyzed as earlier, but regression techniques are used to show correlations. There are many different regression methods, so we focused on linear, quadratic, exponential and logarithmic techniques. In each case, we seek the best fitting entity. The mathematical background of the used regression techniques is detailed in [15] [*Step* 1]. Regressions are in *Class* 1 of random correlations [*Step* 2]. If we have a set of data items and we perform more and more regression techniques, then the chance of finding a correlation will be increased [*Step* 3]. Four regression types are used, therefore $t = 4$. There are always two columns [x and y coordinates], therefore $k = 2$ is constant so it can be skipped from the calculations. There are two different parameters for r: r_1 determines the range of x values [$r_1(a_1, b_1)$], while r_2 stands for range of the y values [$r_2(a_2, b_2)$]. The calculation of Ω is based on k, n, r_1 and r_2 [*Step* 5]. We need to perform all regression types on each different combination. The count of S_1 set will be increased by 1, if a correlation related to the given combinations (x and y data rows) and the acceptance level, is found. The acceptance level can be changed, we defined as $r^2 > 0.7$. The seeking R is calculated by dividing $|S_1|$ and $|\Omega|$. The different rates can be compared, while just one, two or all regression techniques are used. The contrasting and the whole simulation process is performed with a self-developed computer program [*Step* 6].

Producing of each case is not simple task since Ω increases exponentially with increasing parameters, this is why a computer program was developed.

5 Total Possibility Space Reducing Techniques

5.1 *Space Reducing Techniques and the Finding Unique Sequences Algorithm*

By increasing k, n, r, the Ω cannot be calculated in real time even with a fast computer. To make this calculations possible, Space Reducing Techniques (SRT) must be applied. SRT depends heavily on the given method. Therefore in each case of method, the own space reducing algorithm (SRA) must be developed. The SRT is a set, and an SRA is an item in this SRT set. A self-developed SRA named Finding Unique Sequences algorithm (FUS) is applied for ANOVA.

The outer variance is the variance of group's means, but one group's mean is determined by data values, which are related to the inner variance. Therefore, we proceed from the inner variance: we have to calculate only one column total possibility space and repeat it k times. Note that is not enough to store the *Sum of Squares Within* (*SSW*) parameter only, because one *SSW* can belong to one m mean, but one mean can belong to several *SSWs*. If we have (i) [1, 2, 2] and (ii) [1, 1, 3], the $SSW_{(i)} = 0.6666$ and $SSW_{(ii)} = 2.6666$, while the means are the same in both cases (1.6666). This leads to different F values. Therefore, we must store *mean–-SSW* tuples. This is the first level of decreasing.

At the second level, only combinations with different *SSW* needs to be calculated. For this, repeated permutation technique is used. Besides this, the frequency for each *mean–SSW* needs to be stored. The frequency of one tuple can be calculated with the repeated permutation

$$\frac{n!}{s_1! * s_2! \ldots s_i!}, \tag{1}$$

where n is the number of elements, s_i is the number of repetitions. We produce all repeated combinations for one group, then we calculate each combination's *mean*, *SSW* and *frequency*. Based on these triples, we can calculate *SSB* and *F*. The calculated F frequency can be derived as

$$F_{i,\, k,\, n,\, a,\, b} = \left(\prod_i^k C(SSW_i) \right) * C(m_i), \tag{2}$$

where $C(SSW_i)$ is the count of the *SSW* frequency and $C(m_j)$ is the count of the given means combination. A given F is compared with the $F_{critical}$, and since the frequency of that comparison is known, we can define how many times this judgment occurs. Finally, R can be calculated.

5.2 *Handling Methods Assumptions*

The analyzing methods have different conditions which must be satisfied. Generally, the first assumption is that sampling must be done randomly. In the view of RC, we can assume it is passed. Usually, another basic assumption is normality. Therefore all produced candidates must be checked in the view of normality. The combinations, which do not follow the normal distribution, must be deleted in the set of candidates. We used the D'Agostino-Pearson test to check normality [17]. Another assumption can be that the variances must be statistically equaled. This was checked with Bartlett test [18]. ANOVA has these three assumptions and they were implemented in our program, and the assumptions of regression techniques are also handled. If further methods are analyzed in the view of RC, then assumptions must be handled during the process.

5.3 Simulation Levels

It is possible that SRTs cannot grant enough reducing in the case of huge RC parameter values. Therefore simulation techniques must be applied to approximate the seeking of R.

Level 1 The trivial way to generate data rows randomly according to given k, n and r. We perform the analyzing method and notify the number of "correlated" and the iteration number i. Based on these numbers, an R can be calculated. Based on the definition of possibility, R is approximated by R in the case of large number of i. This is the fastest way to get an estimation of R, however, calculating R will be precise only if i is large enough. After a certain level, performing i iterations cannot also be possible in real time.

Level 2 In the case of FUS, the SRT first phase can be done quickly because of the square function. The problem related to k is that all k subsets must be produced from the result of the FUS's first phase. If we produce all first phase candidates, i.e., use repeated permutation, and then we use simulation technique, i.e., randomly chosen k subsets in each iteration, then the second phase has an input which contains only the accepted normal candidates. With this method, more precise R can be determined.

Level 3 Thc first phase candidates and the related frequency F can be combined. At *Level* 2, we can pick up k data rows, but its weight is 1, i.e., one judgment is calculated. If a data row was chosen in an iteration, we can define a weight based on F since it is known. For example, in $k = 3$ case, there is only one judgment at *Level* 2, however, $F_1 * F_2 * F_3$ judgments are produced at *Level* 3. In other words, when three given data rows are selected, then all theirs permutations are chosen as well, because in the first phase, one row represents a combination with its own all permutations, i.e., frequency F. We know that all F_k have the same result as in *Level* 2. Therefore, we get more than one information in one iteration. In the next iteration, these 3 data rows [neither theirs permutations] cannot be selected. This level produces more precise R^* with i iteration.

6 Results

6.1 RC Session I: ANOVA

FUS is performed on all combinations according to RC parameters. This provides the exactly Rs. However, the R^* is also noted since the average error rate of the approximation can be determined by comparing R and R^*. The significance level $\alpha = 0.05$.

Table 1 has two main parts. In the first part, R and R^* can be compared. The results show that approximation R^* is appropriate. To calculate R^*, 1000 iterations

Table 1 ANOVA results using FUS and simulation

r(*a*, *b*)	*k*	*n*	*R*	*R**	*r*(*a*, *b*)	*k*	*n*	*R**
(1, 3)	3	30	**0.9523**	**0.9344**	(1, 3)	4	100	**0.9151**
(1, 3)	3	50	**0.9544**	**0.9737**	(1, 3)	7	100	**1.09E-9**
(1, 3)	5	10	**0.9722**	**0.9629**	(1, 3)	10	100	**0**
(1, 3)	5	15	**0.9604**	**0.9899**	(1, 3)	10	500	**0**
(1, 5)	3	10	**0.9774**	**0.9241**	(1, 5)	4	100	**0.5889**
(1, 5)	4	5	**0.9580**	**0.9782**	(1, 5)	5	100	**0.0040**
(1, 5)	4	9	**0.9598**	**0.9537**	(1, 5)	7	100	**7.19E-19**
(1, 10)	3	5	**0.9577**	**0.9437**	(1, 10)	4	10	**0.9717**
(1, 10)	4	5	**0.9565**	**0.9671**	(1, 10)	4	19	**0.9601**

were performed. The second part deals with cases which *R*'s cannot be calculated with FUS either. In these cased, only R^* can be calculated in real time.

First, the rates are high in the favor of H_0. But in the case of large enough *k* and *n*, the rates are heavily turn to H_1. If the same experiment is performed with relatively small RC values, getting the result H_0 and the "non-correlated" judgment is very high. Contrarily, the chance of H_1 is increased with large enough values and "correlated" decision will be accepted at high possibility. However, this is a paradox in the view of big data. In general, if we have a conclusion with smaller number of data items, i.e., sample, then more data should enhance the conclusion. However, our results show that we can get contradictory results comparing cases with few data and big data. By increasing *k*, the chance to find statistically equaled data rows after each other in *k*-times can be "difficult". In other words, increasing *k*, the chance is high to choose one data row (the *k*th), which is not equaled statistically with the other already chosen data rows $(1, \ldots, k-1)$. The answer can be proven by *Θ-method*. However, this answer do not affect the conclusion about the ANOVA contradictory property.

If wide range is chosen, e.g., *r*(1, 10), and *n* high enough (more than 30), the candidates cannot be stored in memory because of theirs count. However, the *r*(1, 10) results which are illustrated in Table 1 suggest that the distortion is independent from parameter *r*.

6.2 RC Session II: Regression Techniques

In regression case, FUS can be used only partly: the first level of reducing cannot be applied since the order of the coordinates is important. For example, the $x' = \{2, 1, 2\}$ and $y' = \{1, 3, 1\}$ do not provide the same r^2 as $x = \{1, 2, 2\}$ and $y = \{1, 1, 3\}$. Therefore, all possibilities need directly produced in the first phase. The second level of reducing can be used without any modification. As we mentioned before, 4

Table 2 Results in case of regressions

$t = 4$	$r_1(1, 5); r_2(1,3)$	$r_1(1, 10); r_2(1, 3)$	$r_1(1, 3); r_2(1, 5)$	$r_1(1, 3); r_2(1, 10)$
$n = 5$	0.2873	0.3071	0.3122	0.3288
$n = 6$	0.2092	0.2161	0.2530	0.3239
$n = 7$	0.1387	0.1379	0.2204	0.3102
$n = 8$	0.1142	0.1027	0.1947	0.3029
$n = 9$	0.1057	0.0796	0.1894	0.2927

regression techniques were used and the cases with $r^2 > 0.7$ are only accepted as "correlated". It is easy to see, if we use less regression techniques, the chance to find a correlation is also less. For example, the $t = 4$ rates must be higher than or at least equal to the case of $t = 2$ (Table 2).

We can conclude that the $r_1(1, 3);r_2(1,10)$ case is very stable around 0.3. This is a significant results, since the "correlated" and "non-correlated" judgments have the same chance in each case. On the other side, it is rightful assumption, that the theoretically rate cannot be around 0.5 in regression case, because 0.5 would mean that the "correlated" judgment is not more, like a simple coin fifty-fifty rate. To consider it "correlated", the rate must be stricter. Therefore, the parameters related to rate 0.3 could also be suitable. However, the number of used regression methods can be increased with other regression techniques, it seems regression is not so sensible to RC.

7 Conclusion

A new random correlations theory was presented in this paper. Getting mathematically proven correlations between data rows as variables can be manipulated by RC parameters values. This manipulation is hidden from the researchers as well. Two RC analyzing sessions were performed which showed that ANOVA is very sensible, regression techniques are less sensible to random correlations.

We have to state that this work is not against the real existing correlations since we do not assume that real correlations are not exist. We just would like to avoid making false decisions based on false results because of this random characteristics.

Acknowledgments Gergely Bencsik's work was partially supported by the TAMOP-4.2.2. C–11/1/KONV-2012-0015 (Earth-system) project sponsored by the European Union and European Social Found. Laszlo Bacsardi's research was supported by the European Union and the State of Hungary, co-financed by the European Social Fund in the framework of TÁMOP 4.2.4. A/2-11-1-2012-0001 "National Excellence Program".

References

1. Khan, J.A.: Research Methodology. APH Publishing Corporation, New Delhi (2008)
2. Kuada, J.: Research Methodology: A Project Guide for University Students. Samfundslitteratur, Frederiksberg (2012)
3. Hooper, L., Bartlett, C., Smith, G.D., Ebrahim, S.: Systematic review of long term effects of advice to reduce dietary salt in adults. Br. Med. J. **325**, 628–632 (2002)
4. Pljesa, S.: The impact of hypertension in progression of chronic renal failure. Bantao J. **1**, 71–75 (2003)
5. Huete, A.R., Didan, K., Shimabukuro, Y.E., Ratana, P., Saleska, S.R., Hutyra, L.R., Yang, W., Nemani, R.R., Myneni, R.: Amazon rainforests green-up with sunlight in dry season. Geophys. Res. Lett., **33**(L06405) (2006)
6. Samanta, A., Ganguly, S., Hashimoto, H., Devadiga, S., Vermote, E., Knyazikhin, Y., Nemani, R.R., Myneni, R.B.: Amazon forests did not green-up during the 2005 drought. Geophys. Res. Lett. **37**(L05401) (2010)
7. Gething, P.W., Smith, D.L., Patil, A.P., Tatem, A.J., Snow, R.W., Hay, S.I.: Climate change and the global malaria recession. Nature **465**, 342–345 (2010)
8. Martens, P., Kovats, R.S., Nijhof, S., de Vries, P., Livermore, M.T.J., Bradley, D.J., Cox, J., McMichael, A.J.: Climate change and future populations at risk of malaria. Glob. Environ. Change **9**, 89–107 (1999)
9. Rebetez, M., Lugon, R., Beariswyl, P.A.: Climatic change and debris flows in high mountain region: the case study of the ritigraben torrent (Swiss Alps). Clim. Change **36**, 371–389 (1997)
10. Stoffel, M., Beniston, M.: On the incidence of debris flows from the early little ice age to a future greenhouse climate: acase study from the Swiss Alps. Geophys. Res. Lett. **33** (2006)
11. Stoffel, M., Bollschweiler, M., Beniston, M.: Rainfall characteristics for periglacial debris flows in the Swiss Alps: past incidences–potential future evolutions. Clim. Change **105**, 263–280 (2011)
12. Savage, M., Burrows, R.: The coming crisis of empirical sociology. Sociology **41**, 885–899 (2007)
13. Lam, L.T., Peng, Z.W.: Effect of pathological use of the internet on adolescent mental health. Arch. Pediatr. Adolesc. Med. **164**, 164–174 (2010)
14. Shen, C.X., Liu, R.D., Wang, D.: Why are children attracted to the internet? The role of need satisfaction perceived online and perceived in daily real life. Comput. Hum. Behav. **29**, 185–192 (2013)
15. Bencsik, G., Bacsárdi, L.: Effects of random correlation on ANOVA and regression. In: Proceedings of the 9th International Conference on Knowledge, Information and Creativity Support Systems, pp. 396–402 (2014)
16. Sahai, H., Ageel, M.I.: Analysis of variance: fixed, random and mixed models. Springer Birkhäuser Science, Berlin (2000)
17. D'agostino, R.B., Belanger, A., D'agostino, R.B.: A suggestion for using powerful and informative tests of normality. Am. Stat. **44**, 316–321 (1990)
18. Abell, M.L., Braselton, J.P., Rafter, J.A.: Statistics with Mathematica. Academic Press, Boston (1999)

nVidia CUDA Platform in Graph Visualization

Ondrej Klapka and Antonin Slaby

Abstract Many today's practical problems, e.g. bioinformatics, data mining or social networks can be visualized and better examined and understood in the form of a graph. Elaborating big graphs, however, requires high computing power. The performance of CPUs is not sufficient for this purpose but graphics processing unit (GPU) may serve as a suitable high performance, well optimized and low cost platform for calculations of this kind. The article deals with the Fruchterman-Reingold graph and brings solution to this problem; how its layout algorithm can be parallelized for the GPU using nVidia CUDA computing model. This article is continuation and extension of (Klapka and Slaby, The 9th international conference on knowledge, information and creativity support systems, 2014) [8] and gives some other facts and details.

Keywords CUDA · GPU computing · Graph layout · Parallelization

1 Introduction

In recent decades, importance of information and consequently the amount of information that is collected and subsequently processed has been dramatically increasing. One possibility to represent the information in a structured form, suitable for better understanding and further processing is expressing them in the form of a graph. Graphs can be used in many application fields such as data mining, bioinformatics, social networks, etc. [4].

O. Klapka (✉) · A. Slaby
Department of Informatics and Quantitative Methods, Faculty of Informatics and Management, University of Hradec Kralove, Hradec Kralove, Czech Republic
e-mail: Ondrej.Klapka@uhk.cz

A. Slaby
e-mail: Antotnin.Slaby@uhk.cz

S. Kunifuji et al. (eds.), *Knowledge, Information and Creativity Support Systems*, Advances in Intelligent Systems and Computing 416,
DOI 10.1007/978-3-319-27478-2_38

The amount of information in some applications raises the need to handle very large graphs, sometimes having hundreds of thousands to millions of nodes. Increasing size of graphs then requires more and more computing power for their processing. Computing power of current processors becomes insufficient for very large graphs and processing them by supercomputers is often not possible or is very expensive.

As a cost-effective solution to the above mentioned problem seems parallelization of calculations and using the graphics processing unit (GPU) for calculations. GPU can process a large number of parallel tasks (called threads) simultaneously [5].

The following text provides a brief description of the computing platform CUDA based on nVidia graphics chips and also presents a comparison of processing speed tested on randomly generated graphs of various sizes.

2 The Problem of Graph Visualization

Graphs are usually represented for processing by computers in some way enabling to execute this process easily. The most widely used ways include incidence matrix, adjacency matrix, edge list, a list of neighboring nodes, etc. [12]. These forms of representation provide a structured and precise description of the graph, but at the same time these representations are confusing and therefore inappropriate to human perception. On the other hand the best and the most intuitive way to express the graph for human is a graphic representation of nodes and edges. The graphical representation however is useful only if it meets certain basic aesthetic and cognitive criteria. There are not generally formulated and generally accepted criteria specifying how a properly represented and displayed graph should look like. It is clear, however, that a graph with plenty of mutual edge crossings is not easily understandable/readable for people. Here are some basic recommendations that should be followed [3, 6]:

- Evenly spaced peaks in the workspace.
- Minimization of edge crossings.
- Uniform (not very different) length of edges.

A key challenge of the problem of solving visualization of the graph is therefore to achieve a deployment of peaks in an appropriate compliance with the above rules.

There can be found a number of ideas and linked algorithms to distribute graph nodes. Generally, the largest and most widely used group of algorithms for visualization/rendering graphs is called force-directed algorithms group. Force-directed algorithms are based on the idea of conception of the graph as a mechanical system in which nodes and edges are subjected to two kinds of forces; the nodes mutually repel each other and at the same time edges cause attraction of nodes connected to each other [9].

Force-directed algorithms have quadratic computational complexity which arises from the necessity to calculate the repulsive forces interacting between each pair of nodes [9]. Quadratic complexity can be reduced to the logarithmic complexity by different simplification heuristics (usually branch and bound based). As a good example of this approach may serve Barnes-Hut algorithm which divides the graph into smaller parts and repulsive forces are calculated only between nodes in a given part of it [6, 9, 10]. This way complexity of the algorithm can be reduced, but this alteration results in lower global precision of rendering algorithms at the same time [6].

Quadratic complexity of algorithms is generally acceptable for small graphs, comprising hundreds to thousands of nodes. For very large graphs it causes noticeable slowdown of the algorithm. Consequently even today's most powerful processors cannot draw very large graphs in real time. CUDA platform is capable to provide high performance GPU computing and it could enable real-time draw of graphs having hundreds to thousands times more nodes in comparison with a number of nodes current processors are able to elaborate.

3 Description of the Computing Platform

Implementation of the non-graphic computation on the GPU enabled the arrival of Shader Model 4.0 and the birth of a unified architecture of separate GPU units in 2006. Since that time a very specific hardware, designed and optimized for applications and calculations in graphics, enables running not only graphics algorithms. GPU has become in such a way cost-effective, computationally very efficient architecture [4]. Modern GPU is a high-performance, parallel and programmable architecture. However, GPU programming, originally intended solely for graphic calculations, is unlike the programming for the processor very limited [2, 4]. Consequently effective use of this specific architecture requires at least a basic understanding of its principles. In the next part of the text the architecture of the GPU GeForce 9300M G from nVidia will be described. Non-graphic computing using CUDA computing model can be run via this GPU. All experiments were conducted in this environment.

Architecture GeForce 9300M G is based on Shader Model 4.0. It has 16 graphics processors, called stream multiprocessors (SM). All SM are programmable units with unified architecture. Every SM can thus be used as a powerful multi-core computing architecture accessible through programming interface of CUDA architecture. Supported interface CUDA for the GeForce 9300M G is interface of version 1.1 [1, 4].

CUDA architecture is a set of libraries that allow running the GPU codes in the modified C language. The modified C compiler creates the executable code for a device supporting CUDA computing model.

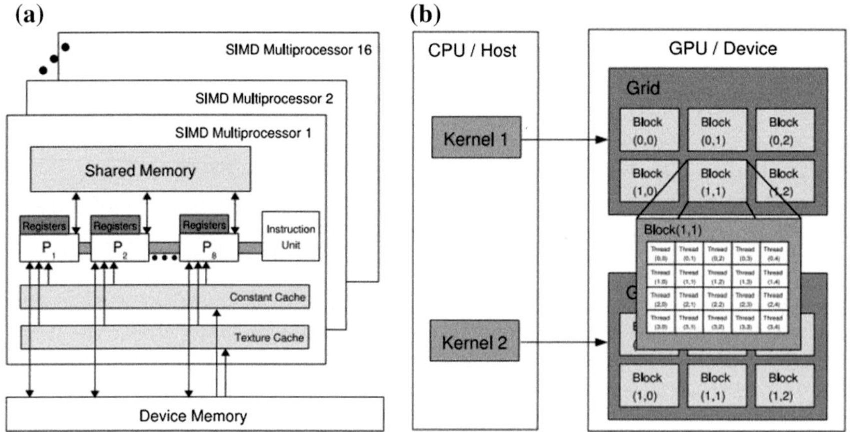

Fig. 1 **a** CUDA hardware model. **b** CUDA computing model [4]

3.1 CUDA Hardware Model

As to hardware, the GPU is a set of stream multiprocessors and each multiprocessor consists of several processors (in case of GeForce 9300M G there are 8 processors for each SM). Each SM has a shared memory that is available for all processors in SM. Each processor executes during one cycle the same instruction but this instruction is provided over specific data that may differ. Communication between multiprocessors is organized through global device memory that is shared for all multiprocessors of the GPU [4]. Diagram of CUDA hardware model is shown in Fig. 1a.

3.2 CUDA Computational Model

CUDA computing model is organized as a large number of parallel threads. The code that each thread executes is called a kernel function, and this code is the same for all threads. The threads are triggered off in groups, called blocks and the blocks are grouped in the grid. Each block of threads currently runs on a single multiprocessor. In case that more blocks are running than is the number of GPU multiprocessors, several blocks share one multiprocessor. The thread and block scheduling on multiprocessors are supervised by GPU planner and so the programmer is completely shielded from the issue of planning [4, 11].

From this follows that, depending on the total number of running threads and on capabilities enabled by the GPU, all threads don't necessarily have to run actually in parallel. The threads which actually run in parallel in the same block are called warp. Typical size of warp is 32 threads.

In terms of memory, each thread has a 32-bit register, and all the threads in the block have access to shared memory [11]. For GeForce 9300M G the size of the shared memory for each block is 16 kB [1].

Identification of individual threads in the program code of kernel function is enabled through the built-in variables. Individual threads within a block have their own numerical identifier that denotes built-in variable *threadIdx*, which may take integer values from the interval <0, the block size −1>). Similarly blocks in the grid are uniquely identified by the value of the built-in variable *blockIdx*, which takes integer value from the interval <0, the grid size −1>) [11]. The unique identification of each threads within the grid is then given by the formula (expression):

$$blockIdx \times number\ of\ blocks + threadIdx$$

Thanks to the easy identification of individual threads the programmer can easily ensure that each thread elaborates different set of data even though all the threads execute the same code.

4 Fruchterman-Reingold Algorithm Using NVidia CUDA

Fruchterman-Reingold algorithm from the family of force-directed algorithms was chosen in this article for demonstration of work with nVidia CUDA for visualization of graphs. The main advantage of Fruchterman-Reingold algorithm is in its speed, achieved by using a very simplistic model of interacting forces in the graph [3]. The interacting forces in the graph are modeled as follows [3, 7]:

- Attractive force (between each pair of nodes connected edges) is given by the equation:

$$f_p(d) = d^2/k$$

- Repulsive force (between each pair of nodes) is defined like this:

$$f_o(d) = -k^2/d$$

Function parameter d is the actual distance between the nodes and the variable k is the ideal distance between nodes in the area on which the graph is drawn. The k value is defined by the following equation:

$$k = C\sqrt{\frac{drawing\_area\_size}{num\_of\_nodes}}$$

Variable *C* is a suitably chosen constant, allowing adjusting the appropriate distance between the nodes of the graph [3, 7]. The more detailed model of the forces acting in the graph is given in the following part of the article.

4.1 Parallelization of the Code

Transformation of the code which is traditionally executed sequentially on the CPU is necessary so that it could enable parallel execution and make full use of the advantage of great potential of GPU computing.

The traditional code one iteration of Fruchterman-Reingold algorithm for the CPU is as follows [3, 7]:

```
For each pair of nodes v1 and v2 of the graph repeat:
  {d is a vector of distance between nodes}
  d := v1.position - v2.position;
  f_rep := - k * k / d; {repulsive force}

  {each node has two vectors: .position and .displacement
   Vector .displacement defines the distance by which the
   node will be transformed in given iteration}
  v1.displacement := v1.displacement + d/|d| * f_rep;

  If nodes v1 and v2 are connected by the edge do:
    f_attr := d * d / k; {attractive force}
    v1.displacement := v1.displacement - d/|d| * f_attr;
    v2.displacement := v2.displacement - d/|d| * f_attr;
  end
end
{limitation of displacement for the node v}
For each node v of the graph do:
  limitation := min(max_displacement, |v.displacement|);
  v.position := v.position + v.displacement /
|v.displacement| * limitation;
end
```

It is noticeable that the code that is in that version of the algorithm is repeated for each pair of nodes. Consequently it must necessarily consist of two cycles (quadratic complexity of the algorithm mentioned in the previous text in the chapter Problems visualization graphs). In case of a parallel version of the code, it is possible to remove one cycle as it may be replaced by a large amount of threads. The modified algorithm will calculate for each thread forces acting between the nodes denoted by the index of the current thread and leading all the other nodes.

Consequently the parallelized code of one iteration Fruchterman-Reingold algorithm is as follows:

```
{unique identification of the thread}
thread_id = (blockIdx * blockDim) + threadIdx;
{each thread will read one node according to it's ID in
such a way one cycle in the algorithm is removed}
v1 = nodes_list[thread_id];

For all other nodes in the thread v2 from the array
nodes_list do
  d := v1.position - v2.position;
  f_rep := - k * k / d; {repulsive force}
  v1.displacement := v1. displacement + d/|d| * f_rep;

  If nodes v1 and v2 are connected by the edge do:
    f_attr := d * d / k; {attractive force}
    v1. displacement:= v1. displacement - d/|d| * f_attr;
    v2. displacement:= v2. displacement - d/|d| * f_attr;
  end

limitation:= min(max_displacement, |v.displacement/);
v1.position := v1.position + v. displacement / |v. dis-
placement / * limitation;
```

5 Graph Representation

Processed graph can reach sizes of tens of thousands of nodes. With regard to the size of the graph it is necessary to use memory saving graph representation. As a suitable representation with reasonable memory consumption was chosen a graph representation by the list of neighbors that has complexity:

$$o(n+m)$$

where n is the number of edges and m is the number of nodes. The disadvantage of this approach is the time complexity of the testing operation whether the nodes are connected by an edge (which is a very frequent operation in force-directed algorithms). This complexity in that is:

$$o(n)$$

where n is the number of edges [12].

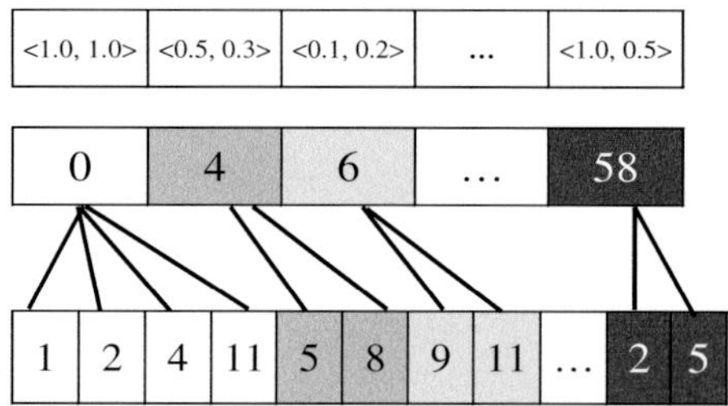

Fig. 2 Graph representation using three arrays: array of coordinates of the nodes, array of indexes of edges, and array of indices of nodes. If the node has degree 0, then in the array of index of edges is value −1

The selected graph representation for processing on GPU is as follows: The nodes are represented by two-component vectors, recording the x and y coordinates of the rendering canvas. The full list of nodes is then one-dimensional array of the elements of the float data type, which contains the individual coordinates of nodes. Edges are represented as two one-dimensional arrays of elements of data type int. They are the array of indices of edges and the array of indices of nodes. The array of indices of edges has the same length as is the length of the list of nodes in the graph. It contains a list of indices defining edges of each node in the second field, in the way shown in Fig. 2. The array of node indexes then has the length equal to the number of edges of the graph [12].

6 Results

Layout of position of graph nodes was set by Fruchterman-Reingold algorithm. The algorithm was tested on the CPU and the GPU for different sizes of graphs and different density of edges in the graph. The comparison was carried out on the CPU Intel Core 2 Duo T8100, running at 2100 MHz and GPU nVidia GeForce 9300M G with a G86 M core processor running at 400 MHz. Time needed for execution of 20 iterations for a random graph with an average node degree 5 and maximum node degree 20 was measured. Selected results of testing are shown and analyzed in details in our article [8]. Since prior to the actual calculation on GPU, firstly the processed data had to be copied to the GPU memory, this copying time was also included in the resulting calculation time (usually around 20–25 ms). Another investigation that was established to determine how the speed of the algorithm, running on the CPU and the GPU, reflects the density of edges in the graph follows. Calculation and measurements were carried out for a graph with 5,000 nodes, where in previous measurements were found very similar processing times. The values measured are shown in Table 1.

The following Fig. 3 shows the graph which was created on the basis of data from Table 1. It can be deduced from Fig. 3 that increasing density of edges in the

Table 1 Measured values for the 20 iterations of the Fruchterman-Reingold algorithm for a randomly generated graph with different densities of edges

Number of edges	Max degree	Execution time on CPU (ms)	Execution time on GPU (ms)
1,000	10	12,422	10,405
5,000	50	10,655	10,419
10,000	100	10,602	10,751
20,000	200	10,698	10,582
50,000	500	10,735	10,712
100,000	1,000	10,580	10,898
200,000	2,000	10,815	10,888
500,000	5,000	11,099	16,997
1,000 000	10,000	10,895	16,926

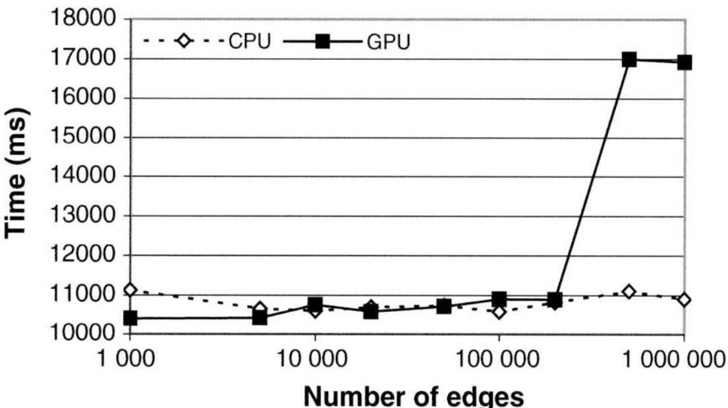

Fig. 3 Graph showing the dependence of the execution time for various densities of edges in the graph

graph results into the slowing of execution on the GPU, while the CPU execution time still fluctuates around 10,000 ms.

Increasing density of edges in the graph then leads to more frequent blocking of the threads and subsequently to the fall of the GPU performance. The GPU is therefore more powerful than the CPU for sparse graphs only. The CPU is but faster for the graphs with a large number of edges, which is caused mainly due to the fact that the logic for verifying whether the two nodes are connected by the edge may cause mutual blocking of some threads on the GPU.

7 Conclusion

The article is the contribution to the problem of visualization of very large graphs. At the same time the method of parallelization of Fruchterman-Reingold scheduling algorithm code for automatic graph layout was largely tested. Testing was focused on comparing the differences in processing performance (execution time) of selected graph by the CPU and the GPU.

Testing has shown that the small size graphs having up to several hundreds of nodes can be processed faster by the CPU than by the GPU. Furthermore, it was verified that the GPU is due to the huge amount of running threads and high optimization for simple arithmetic operations capable for layout nodes of very large graphs faster than the CPU. For graphs having hundreds of thousands or more nodes can then be processing of the graph even several times faster. It was also found that the difference between the CPU and the GPU in performance is especially noticeable for sparse graphs, with a large number of nodes and a reasonably small number of edges.

The article is continuation of Klapka and Slaby [8] where other facts and details can be found.

References

1. CUDA GPUs—nVidia Developer Zone. https://developer.nvidia.com/cuda-gpus
2. Frank, D., Kumanan, Y.: Exploring the Limits of GPUs With Parallel Graph Algorithms. School of Computer Science. Carleton University, Ottawa (2010)
3. Fruchterman Thomas, M.J., Reingold Edward, M.: Graph Drawing by Force-Directed Placement. University of Illinois, Department of Computer Science (1991). http://pdf.aminer.org/001/074/051/graph_drawing_by_force_directed_placement.pdf
4. Harish, P., Narayanan, P.J.: Accelerating large graph algorithms on the GPU using CUDA. In: Center for Visual Information Technology, International Institute of Information Technology Hyderabad, India. http://citeseerx.ist.psu.edu/viewdoc/download?doi=10.1.1.102.4206&rep=rep1&type=pdf
5. Hennessy, J.L., Patterson, D.A.: Computer Architecture: A Quantitative Approach. Morgan Kaufmann Publishers, Los Altos (2011)
6. Hu, Y.: Efficient and High Quality Force-Directed Graph Drawing. Wolfram Research Inc, USA. http://yifanhu.net/PUB/graph_draw_small.pdf
7. Klapka, O.: Vizualni analyza dat: Vizualni analyza vlastnosti a vztahu dat. Hradec Králove: Univerzita Hradec Králove, Fakulta informatiky a managementu, Katedra informatiky a kvantitativnich metod (2013)
8. Klapka, O., Slaby, A.: Graph visualization performed by nVidia CUDA Platform. In: The 9th International Conference on Knowledge, Information and Creativity Support Systems, pp. 408–414. KICSS'2014 Proceedings, Nicosia (2014)
9. Kobourov, S.G.: Force-Directed Drawing Algorithms. University of Arizona. http://cs.brown.edu/~rt/gdhandbook/chapters/force-directed.pdf
10. van der Maaten, L.: Barnes-Hut-SNE. Pattern Recognition and Bioinformatics Group, Delft University of Technology, The Netherlands (2013). http://arxiv.org/pdf/1301.3342v2.pdf
11. Rafia, I.: An Introduction to GPGPU Programming CUDA Architecture. Mälardalen Real-Time Research Centre. http://www.diva-portal.org/smash/get/diva2:447977/FULLTEXT01.pdf
12. Vajdik, R.: Reprezentace Grafu. Technicka Univerzita Ostrava, Fakulta elektrotechniky a informatiky, Katedra informatiky, Ostrava (2009). http://homel.vsb.cz/~vaj049/AlgoritmyII/reprezentace_grafu.pdf

A Method for Opinion Mining of Coffee Service Quality and Customer Value by Mining Twitter

Shu Takahashi, Ayumu Sugiyama and Youji Kohda

Abstract In this work, we focus on customer value of Japanese coffee service by analyzing the huge tweet data set. We suggest the quantitative evaluation method of customer value by using the morphological analysis of the tweets and generated value dictionary. Our suggested methodology can be applied to any service and product to improve the commoditization problem from the viewpoint of service science.

Keywords Customer value · Natural language processing · Service value · Opinion mining

1 Introduction

Recently, big data utilization in social network data (e.g. Facebook, Twitter, Flicker) has attracted much attention from researchers and research divisions of companies in order to analyze the market with high "objectivity" and high "completeness" [1, 3, 4]. In this study, we focus on the Twitter data from the viewpoint of customers value of products and services in the commoditization of service. Specifically, we examine coffee service in Japan as an example of commoditization.

Demand for coffee beans has been increasing, and coffee provisioning method becoming more diversify in the world coffee markets and service. In the decade from the year 2000, coffee bean consumption in the importing countries has increased to

S. Takahashi · Y. Kohda
School of Knowledge Science, Japan Advanced Institute of Science and Technology, Ishikawa, Japan
e-mail: takahashi.shu@jaist.ac.jp

Y. Kohda
e-mail: kohda@jaist.ac.jp

A. Sugiyama (✉)
Faculty of Humanities, Yamanashi Eiwa College, Yamanashi, Japan
e-mail: ayumu@yamanashi-eiwa.ac.jp

S. Kunifuji et al. (eds.), *Knowledge, Information and Creativity Support Systems*, Advances in Intelligent Systems and Computing 416,
DOI 10.1007/978-3-319-27478-2_39

more than 8 %. In addition, coffee beans consumption in the exporting countries has increased by 48 % or more.[1]

With the advent of a convenience store coffee, a competition in the coffee market has intensified in Japan. The total number of sales of convenience store coffee has increased to 500 million cups in less than one year from the start of sales in 2012. Coffee consumption in Japan was 440 thousand tons in 2013 [2]. This number increased by 4.3 % from the previous year because of the hit of a convenience store coffee. Therefore, commoditization of Seattle-type coffee and convenience stores coffee has progressed intensify other result of price competition. In addition, a new type of coffee shop called Third-wave appeared around the west coast of North America, and is spreading all over the world.

Controlling commoditization in coffee service is important for coffee market not only Japan, but also within the emerging countries. In this study, we explore the possibility to control the commoditization in coffee service. Therefore, we analyze the values of the customers within among the Seattle-type coffee and convenience store coffee, to clarify the needs of the customers for coffee service.

2 Problem Statement

The type of coffee service is classified into three categories. The first one is a convenient-store-style coffee that is provided by convenience stores (CVS) and fast food shops (e.g. Mac cafe by McDonalds). This kind of coffee has the authentic flavor and is inexpensive price. The second type is cafe service providing relaxing place. Many number of Seattle-type coffee present such kind of service (e.g. Tullys Coffee, Starbucks Coffee). Among these, by providing the third place [5] nor at work or at home, the large chain store Starbucks have succeeded in to satisfy the needs of customers. The third type is a Third-wave coffee. Third-wave coffee is only uses single origin coffee beans that were harvested from seedlings of a single species, and is made by drip.

In Japan, intensification of price competitiveness has proceeding between Seattle-type coffee and convenience stores coffee. In this study, we analyze by three points "taste", "price" and "place", the values of the customers to the CVS and Starbucks. Using the tweets of customers obtained from the Twitter Streaming API data, to be analyzed by the following procedure.

(i) Extract the Japanese tweets that are related to coffee service using the Twitter Streaming API.
(ii) Generate dictionary about "taste", "price" and "place" by the Japanese tweets extracted.
(iii) Map a feature space of "taste", "price" and "place" about CVS and Starbucks using the generated dictionary.
(iv) Fit of the statistical methods to analyze tweet vectors in the feature space.

[1] International Coffee Organization—Historical data.

3 Data Collection and Feature Extraction

In order to extract the sense of values from the tweets of customers, we created a dictionary about sense of values (value dictionary) related to "taste", "price" and "place" [8]. In the dictionary about "taste", "price" and "place" have been produced, the top five of the words are shown in Table 1.

Extract the tweets about CVS and Starbucks among 118,855 target tweets. Among the tweets, we defined as CVS that contains [, convenience], and Starbucks that contains [, , Starbucks]. Table 2 shows the counts of tweets related to CVS and Starbucks extracted. The target data is 5,141 tweets out of 118,855 Japanese tweets about the coffee.

We represent the tweets of the target data by a Vector Space Model [6] based on the value dictionary. As the N words presented in the value dictionary such as $V_1, V_2, V_3, \ldots, V_N$ tweet T is expressed as tweet vector t such as the following.

$$t = \{(V_1, T), (V_2, T), (V_3, T), \ldots, (V_N, T)\}$$

Table 1 The top 5 frequent words to define the of dictionary from the tweet data related to coffee tweet

Word (JP)	Part	Meaning
(a) Taste		
美味しい	n.	Delicious
濃い	adj.	Strong
熱い	n.	Hot
甘い	n.	Sweet
薄い	adj.	Weak
(b) Price		
円	n.	Yen
高い	adj.	Expensive
得	n.	Economical
価格	n.	Price
安い	adj.	Inexpensive
(c) Place		
雰囲気	n.	Mood
いい	n.	Niceness
好き	adj.	Like
店内	n.	Interior
豊富	n.	Wealth

Table 2 Counts of tweets

Class	Count
Starbucks	3612
CVS	1529
Total	5141

In the subsequent analysis, we analyze from tweet vectors, which are mapped to the feature space.

4 Analysis

If customers are focused on value related to "taste", it can be expected that the tweet vectors include value in the "taste" dictionary. We compare value on "taste" of customers about CVS and Starbucks through the target tweets. Table 3 shows the number of tweet vectors containing the feature space that is generated from the dictionary.

We test with difference of population proportion using the Pearson's χ^2 tests on the Table 3a. From the results of the Pearson's χ^2 tests, there was no difference in the population proportion at level of significance 0.05. Because there is no difference in the population proportion of tweet related to taste, it can be said that consumers are focused in the same way as CVS and Starbucks on "taste".

On the other hand, we find the difference in Pearson's χ^2 tests on "price" and "place" at level of significance 0.05 shown in Table 3b, c. These results indicate that CVS customers are focused more strongly on "price" than Starbucks customers.

Table 3 The number of tweets related to three categories

	F	T	Total
(a) Price			
CVS	1280	249	1529
Starbucks	3017	595	3612
Total	4297	844	5141
(b) Place			
CVS	1344	185	1529
Starbucks	3327	285	3612
Total	4671	470	5141
(c) Taste			
CVS	1216	313	1529
Starbucks	2722	890	3612
Total	3938	1204	5141

Table 4 Difference in the population proportion

Word	Meaning	CVS	Starbucks
嬉しい	Happy	2	55
高級	High-quality	3	48
大人	Adult	2	41
若者	Youth	2	25
オシャレ	Fashionable	1	20

On the other hand, Starbucks customers are focused more strongly on “place” than CVS customers.

We analyze that the customers are focused on what kind of value on “place” when they buy CVS or Starbucks. Table 4 shows five words with a great difference at level of significance 0.05 in the population proportion, which corresponds to the value dictionary about place.

There are many positive words for Starbucks coffee, such as “happy”, “high-quality” and “fashionable”. It is evaluated that the place offering Starbucks is a high quality compared to the CVS, and customers are satisfied. In addition, many words about Generation like “adult” and “youth” have appeared.

Above all, it is seen from generation words appeared from Starbucks tweet vectors that third place has been evaluated for not only equipment but also generations and people who gather there. In contrast, the place of CVS has not been evaluated, it is found that customers are not interested in the place for drinking CVS.

Recently, positive-negative judgement of SNS text by machine learning technices suggest some useful prediction about market and political trends [10]. Takumura et al. provide the Japanese semantic orientations of words by constract the lexical network [9]. In their study, they the emotion was treated as the spin of electrons and succeeded in extracting semantic orientations with high accuracy, even when only a small number of seed words are available. In our study, we estimate the sentimental coefficient of custumers by Japanese semantic orientations of tweet words.

Sentimental coefficient $S(T)$ of tweet T is written by berow.

$$S(T) = \frac{\sum_{i=1}^{t}(L(a_i) * F(a_i))}{\sum_{i=1}^{t} F(a_i)}$$

Here, score $L(a_i)$ and frequency $F(a_i)$ are calculated by score of a verb, a noun and an adjective phrase $a_1, a_2, a_3, \ldots, a_i, \ldots, a_t$ in semantic orientation map. $L(a_i)$ value is from -1(negative) to 1(positive). Distribution map of sentimental coefficient about “price” and “place” in dictionary are shown in Fig. 1.

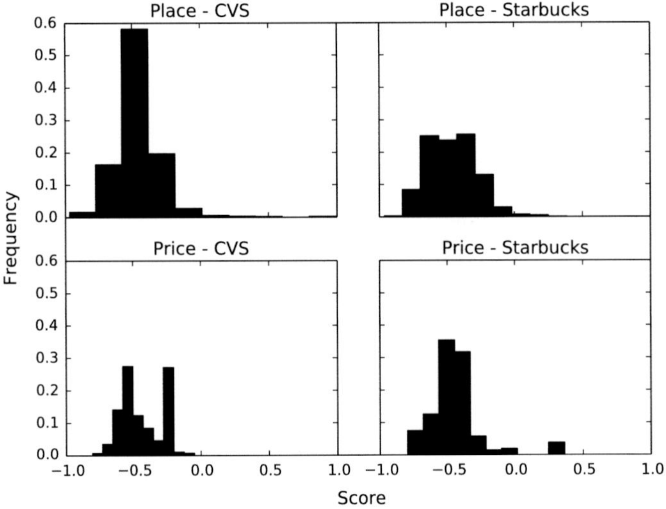

Fig. 1 Distribution map of sentimental coefficient

Table 5 Sentimental coefficient of "price" and "place"

Class	μ	σ	Frequency
Place—CVS	−0.4495	0.2529	860
Place—Starbucks	−0.4604	0.2224	1079
Price—CVS	−0.4964	0.1445	284
Price—Starbucks	−0.4024	0.2485	628

Moreover, significant difference are not observed in the results of t-test about sentimental coefficient of "price" and "place" shown in Table 5. These results indicate the castumer value evaluate by not only sentimental coefficient of SNS text data, but also the quanitative analysis of tweet is important for consider the service value.

5 Discussion

In this study, we discuss the customers value of coffee service by analyzing the huge tweet data set. We suggest the quantitative evaluation method of customer's value by using the morphological analysis of the tweets and generated value dictionary.

The characterizes of all service and product can be classified into three qualities: search qualities, experience qualities and credence qualities [11]. With commoditization, the value of product and service fall into the search qualities [7]. Originally, as Japan is represented by "hospitality", so it is good at imparting credence qualities and experience qualities. It is important to think how to scientifically handle experi-

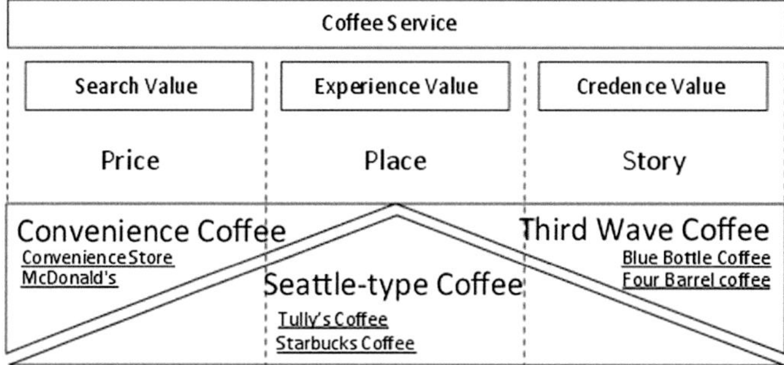

Fig. 2 Service value in coffee service

ence qualities and credence qualities and how to incorporate product and service in order to avoid commoditization.

The value dictionary used in this analysis, it is assumed that the value for "price" is the search qualities, and the value for "place" is experience qualities. Therefore, the conceptual diagram of each field in the coffee service is shown in Fig. 2. For the CVS, customers have a consumption with a focus on value related to "price" compared to the Starbucks. Accordingly, commoditization in the service area as "Convenience coffee" and "Seattle-type coffee" is because the values of the two regions are depressed in the search qualities. Therefore, it is considered that by making the service deployment focused on the searched value "place", "Seattle-type coffee" will become possible to control commoditization.

Thus, SNS data have a useful value to analyze the service value of any market service and customer satisfaction in search qualities, experience qualities and credence qualities. Our suggested methodology can be applied to any service and product to improve the commoditization problem from the viewpoint of service science.

References

1. Anderson, C.: The end of theory. Wired magazine **16**(7) (2008)
2. Eric Pfanner, H.K.: 7-eleven gets sales boost from coffee drinkers in Japan. Wall Str. J. (2014)
3. Godes, D., Mayzlin, D.: Using online conversations to study word-of-mouth communication. Mark. Sci. **23**(4), 545–560 (2004)
4. Liu, Y.: Word of mouth for movies: its dynamics and impact on box office revenue. J. Mark. **70**(3), 74–89 (2006)
5. Oldenburg, R.: The Great Good Place: Cafes, Coffee Shops, Bookstores, Bars, Hair Salons, and other Hangouts at the Heart of a Community. Marlowe, New York (1999)
6. Salton, G., McGill, M.J.: Introduction to Modern Information Retrieval. New York, McGraw-Hill (1983)
7. Schmitt, B.: Experiential marketing. J. Mark. Manag. **15**(1–3), 53–67 (1999)
8. Takahashi, S., Sugiyama, A., Kohda Y.: Quantitative evaluation of the coffee service quality and customer value by mining twitter. In: Proceedings of The 9th International Conference on Knowledge, Information and Creativity Support Systems. pp. 415–420 (2014)

9. Takamura, H., Inui, T., Okumura, M.: Extracting semantic orientations of words using spin model. In: Proceedings of the 43rd Annual Meeting on Association for Computational Linguistics. pp. 133–140. ACL2005 (2005)
10. Xu, K., Li, J., Song, Y.: Identifying valuable customers on social networking sites for profit maximization. Expert Syst. Appl. **39**(17), 13009–13018 (2012)
11. Zeithaml, V.A.: How consumer evaluation processes differ between goods and services. Mark. Serv. **9**(1), 25–32 (1981)

Using Wiki as a Collaboration Platform for Software Requirements and Design

Irit Hadar, Meira Levy, Yochai Ben-Chaim and Eitan Farchi

Abstract Wiki is a collaboration platform allowing building a corpus of knowledge in interlinked web pages created and edited by different users. Wiki has been applied in different domains and usage contexts in business and education. This paper reports on an exploratory study conducted as a part of an ongoing research regarding the use of wiki in software development projects. The objective of this study was to capture the perceptions of users regarding wiki as a collaboration platform. Specifically, we aimed at understanding what hinders and what motivates users' contribution to documents constructed within the wiki environment. Based on data obtained via interviews with wiki users we found, among other things, that they do not perceive wiki as a stand-alone communication and collaboration tool, and tend to conduct off-line discussions prior to changes made in the wiki. We also identified several wiki features, which, if improved, may enhance wiki usage.

Keywords Wiki • Collaboration • Software development • Design • Requirements • Knowledge creation • Knowledge capture

---

I. Hadar (✉) · Y. Ben-Chaim
Information Systems Department, University of Haifa, Haifa, Israel
e-mail: hadari@is.haifa.ac.il

Y. Ben-Chaim
e-mail: yochai.benchaim@gmail.com

M. Levy
Industrial Engineering and Management Department,
Shenkar College of Engineering and Design, Ramat-Gan, Israel
e-mail: lmeira@shenkar.ac.il

E. Farchi
IBM Haifa Research Lab, Haifa, Israel
e-mail: farchi@il.ibm.com

S. Kunifuji et al. (eds.), *Knowledge, Information and Creativity Support Systems*,
Advances in Intelligent Systems and Computing 416,
DOI 10.1007/978-3-319-27478-2_40

1 Introduction

Wiki is a platform that allows different users to collaboratively build and share body of knowledge [2, 9, 13, 20]. It has been applied in different domains and usage contexts including the most commonly known and used Wikipedia, educational wiki applications, and various business-related wiki-based collaboration systems.

Wikis are Web2.0 applications used in various settings for supporting collaborative processes. One such setting is in schools and colleges (e.g., [6]), where their effectiveness is measured according to qualitative post experience attitude surveys, showing pitfalls of integrating wiki in educational context. Wikis are also used to enable business interactions. For example, Graupner et al. [11] report on a wiki for consultants, using specific document template-based pages for projects, which make it difficult to implement in other wiki settings. The use of wikis in a corporate context (e.g., [11]) incorporates the two concepts of technology and management, without which, the process cannot fully succeed.

In our ongoing research [4], we focus on wiki as a collaboration platform in software development, and specifically for developing requirements and design documents. Using wikis for improving software documentation [7], managing documentation of a software project [1], supporting collaborative requirements engineering [8] and as a basis for development environments [15] have been suggested in the past. When using wiki for design and requirements, overcoming cultural issues should be considered [5], otherwise these wikis become inactive after a short period of time [3].

In a previous research [4], we analyzed users' actual behavior as recorded in the wiki log, such as instances of accessing the wiki per user, pages visited, changes made in each visit, and more, and identified accordingly changes in behavior correlated with changes in different parameters. While finding differences in actual users' behavior in different circumstances, a general impression was formed that project stakeholders tend to avoid changing wiki pages, even when they are engaged and collaborative in other aspects, e.g. reading wiki documents, actively participating in team meetings, contribute to discussions via email, etc.

In this study we explore this phenomenon by investigating the perceptions of wiki users in software projects using wiki as a collaboration platform. Furthermore, we aim to understand what motivates users to be more engaged in projects via wiki usage and what hinders their engagement, and to contribute to the corpus of knowledge built in the wiki environment. An earlier report on this study can be found in [12].

The next section briefly describes the wiki platform and its existing and potential benefits as a collaboration tool. Next we present the settings of the exploratory study, followed by its preliminary results. We discuss and analyze our findings and, finally, we conclude and discuss future work.

2 Wiki Collaboration Platform

A wiki is a set of linked web pages created through incremental development by a group of collaborating users [13], The first wiki was developed by Ward Cunningham in 1995, as the Portland Pattern Repository, to communicate specifications for software design. The term wiki (a Hawaiian term meaning "fast") gives reference to the speed with which content can be created.

Wiki is defined as "a system that allows one or more people to build up a corpus of knowledge in a set of interlinked web pages, using a process of creating and editing pages" [9]. The wiki system is designed to increase collaboration and poll of ideas, as a centralized system designed to increase efficiency and reduce uncertainty [20]. Since the first wiki in 1995, wiki has been used in many contexts: in the public domain, most notably, the online encyclopedia Wikipedia project; in education applications [18] and in the business context [16]. Wiki use was found successful and increasing in the software development industry because it is an excellent means to collect asynchronous contributions from a group of distributed people in a centralized repository of textual artifacts [14]. Wikis are particularly valuable in distributed projects as global teams may use them to organize, track, and publish their work.

Gonzalez-Reinhart [10] describes the three generations of wiki support tools. The first generation wikis were open-source wiki pages used for conversational knowledge management solutions, based on group participation and voluntary social connections. These wiki platforms became favorable due to their reasonable economical and technological demand. However, since it was an open-source solution, the companies were neither supported by a specific provider nor had any content propriety guarantee. In order to solve these problems, the second generation wikis were developed by companies that offered the product as well as support services and guaranteed the firms' content propriety. Next, the third generation wikis were developed as application wikis, intended mainly for corporate use. Application wikis provide functionality beyond features provided in the first and second generations of wikis, such as improved editing, improved files' attachment management, and improved collaboration, allowing anyone who can type the ability to edit pages.

Using wiki as an intranet and content management system in a company holds benefits and challenges [21]. For example, a previous study of the use of wiki at IBM showed that it was simple for the creation of information artifacts and for access and use of shared information, resulting in better overview of a project's status [17].

3 The Empirical Study

3.1 Settings and Method

The objective of the exploratory study was to identify the perceptions of wiki users in software development projects regarding the wiki as a collaborative platform. The participants in the study were software developers from four different software development projects in IBM, exhibiting multiple case studies [22]. In these projects, the wiki platform was used mainly for the requirements analysis and design stages. The software development team members in these projects were globally distributed, spread over three different continents, and from a wide range of managerial roles and seniority. Eleven local team-members participated in the current study.

The use of wiki for maintaining all project requirements and design documents was enforced in all projects participating in the study. The wiki was also used to monitor the project status and highlight the main requirements and design issues agreed upon. The wiki hosted informative documents such as requirements and design documents, and presentations. All the wiki pages were available to all the users for viewing and changing, according to the "wiki way" [13].

The use of the wiki incorporates automatic wiki-based characteristics:

- Data is visible to all, and always online and available.
- Changes are automatically visible to all as soon as they happen.
- Notifications are sent to stakeholders when changes are made.
- Wiki usage is logged and monitored (this was common knowledge to the users).

Our research method was based on the grounded theory approach [19] employing semi-structured interviews as a data collection tool. The interviews included questions regarding the usage of the wiki in general, specific usage behaviors, and about opinions regarding exiting wiki features, for the purpose of probing for parameters that may explain user behaviors and attitudes toward the wiki. The collected data underwent an inductive analysis, in which the transcripts of the interviews were divided to segments, followed by their analysis including open, axial and selective coding [19], in which categories emerged from the data. These categories will be presented next.

3.2 Findings

Tables 1 and 2 summarize the emergent categories regarding the usage and perceptions of wiki as a collaboration platform, and proposed improvements.

Our analysis was aimed at understanding the aspects of wiki that may hinder wiki usage. Therefore, the categories presented in Table 1 and 2 are focused specifically on these aspects. Moreover, these categories are not independent, and

Table 1 Categories of usage and perceptions of wiki as a collaboration platform

Main category	Sub-category	Comments/Examples
General usage of wiki	Frequency: several times a month	This was also validated by monitoring the users' actual behavior (log)
	Action: mostly change existing material	Some users have never inserted new material
Communicating via wiki	Perceived as a waste of time	"Wiki is an un-needed overhead"
		"Wiki is used in addition to other communication means, thus doubles the efforts"
	Prefer F2F or email discussions	"We discuss issues in 1:1 meetings or emails"
		"Using comments blocks the page; so using an off-line communication tools is better"
	Only few respond to wiki email notifications	Several interviewees indicated that they never respond to wiki notifications
Attitude towards changing text within wiki	Only following an off-line discussion	"If a page is owned by someone, I would not change it without discussing with them first"
	Writing notes or highlight changes rather than erasing/changing	"I would add but not erase or overwrite [..] I would mark with a different font [to indicate suggestions for change]"

Table 2 Required improvements for the wiki platform

Main category	Sub-category	Examples
Wiki user interface features to be improved	Personalized view	"Automated user-specific view: see what is meant for you first
	Editor	"Word processor is much more convenient"
	Visual representations	"It's easier to draw UML diagrams in a UML tool"
	Attachment	"Using attachments is not comfortable"
Wiki communication features to be improved	Notes writing	"More comfortable commenting is needed"
	Reducing the overwhelming number of notification	"The notification mechanism should be personalized, so that notifications will be sent only to the specific users interested in the change made"
		"Filter pages I receive notifications on"
Accessibility	Should be included in single sign-on	"An additional login is not comfortable"

specifically, some categories in Table 1 (the current state) are directly linked to categories in Table 2 (possible improvements). For example, "Using comments blocks the page; so using an off-line communication tools is better" explains both—the current reluctance and the proposed improvement to enhance wiki as a communication tool by adding notes.

3.3 Discussion

The most salient finding from the interviews is that the interviewees *do not perceive wiki as a stand-alone collaboration tool*, in which they are able to communicate with their peers, discuss the issues at hand and accordingly update the knowledge constructed in the wiki. This drives them to communicate in other means prior to making changes in the wiki environment, which in turn facilitates their perception of the wiki as an inconvenient and redundant overhead rather than an enabler of collaboration.

One reason for this perception may stem from the lack of understanding and embracing the "wiki way" [13]: no individual owns a wiki page, but rather all involved users have a mutual responsibility to develop it and keep it up to date. An additional obstacle is the technical limitations inherent in the specific wiki environment used in the investigated projects. In this environment, communicating with peers in order to discuss required changes is perceived as inconvenient and restricting, making the users reluctant to change text inserted by their peers without preceding off-line discussions. Once the discussion has been conducted and decisions have been made, documenting it in the wiki seems to be documentation-related overhead rather than a collaborative construction of knowledge. This, again, is not in line with the wiki way where knowledge is constructed in the wiki environment rather than merely documented.

Many advantages of the wiki are relevant in the settings of this research. For example, wiki is known to be beneficial in the workplace for groups requiring a collaborative medium, with relatively small number of participants who are geographically distributed, as is the case of software development projects [14] such as those participating in this research. Nevertheless, we found the above documented obstacles that prevent realizing the full potential of the benefits of wiki.

The findings of the current research may explain previously reported results regarding the lack of long-term continuity in wiki usage (e.g., [3]). Furthermore, our findings lay the foundations for possible future enhancements of the wiki platform in order to motivate its usage and augment its benefits. These enhancements can be implemented and then evaluated utilizing social network analysis for analyzing users' actual behavior, combined with qualitative perception analysis.

While this paper focuses on qualitative, interview-based data, our on-going research includes data collected via additional qualitative and quantitative methods, providing further findings and validation. Nevertheless, some limitations are to be taken into consideration. The study was conducted with the participation of team-members in four distributed software development projects in IBM. The specific wiki platform configuration used in these projects was selected, and its use was guided, by IBM personnel. As in any case study, the findings of this study reflect the setting of the specific cases investigated, and generalization of the conclusions should be done cautiously, taking the factors reported here into consideration. Additional case studies in various contexts are needed for further validation and generalization of the results.

4 Conclusion

Wiki is a collaboration platform that allows different users to collaboratively build and share body of knowledge. Wikis are used in various settings and contexts for supporting collaborative processes. This research examined the use of wiki in distributed software development projects, and specifically in the requirements analysis and design phases. Our aim was to explore the perceptions and attitudes of software developers using wiki in this context towards the collaborative nature of the wiki environment, and specifically to identify what hinders users from exhausting the full potential of the wiki as a collaborative platform.

Our findings suggest that the wiki environment as implemented in the IBM software development projects participating in this study is not perceived as a stand-alone communication and collaboration tool, and that its users tend to conduct off-line discussions prior to changes made within the wiki. We also identified several wiki features as well as cultural aspects, which, if improved, may enhance wiki usage and its benefits. Future research will examine the long-term usage of the wiki in the firm of the investigated case studies. In the future we plan to develop and implement enhancements for the wiki environment according to the findings of this study, and evaluate their contribution.

This research is based on multiple case studies from a single organization. Generalization of the findings is limited, however possible, in cases where similar implementation of wiki is observed. Further research may examine additional cases of wiki usage in software development, to further validate and generalize the identified factors and stemming perceptions hindering the usage of wiki as a collaboration platform.

References

1. Alonso, J.M. G., Olmeda, J.J.B, Rodriguez, J.M.M.: Documentation center—simplifying the documentation of software projects. In: The Fourth Workshop on Wikis for Software Engineering, Porto, Portugal (2008)
2. Andersen, E.: Using wikis in a corporate context. In: Hohenstein, A., Wilbers K. (eds.) Handbuch E-Learning, Deutscher Wirtschaftsdienst, vol. 11, pp. 1–15 (2005)
3. Arazy, O., Croitoru, A., Jang, S.: The life cycle of corporate wikis: an analysis of activity patterns. In: The 19th Workshop on Information Technologies and Systems (WITS), Phoenix, AZ, USA, pp. 163–168, Dec 14–15 (2009)
4. Ben-Chaim, Y., Levy, M., Hadar, I., Farchi, E., Bronshteine, A.: Achieving the successful engagement of stakeholders in a globally distributed process. In: Proceedings of the 5th Mediterranean Conference on Information Systems (MCIS), Tel-Aviv-Yaffo, Israel, 12–14 Sept 2010
5. Biesack, D.: The culture of collaboration. In: The Third Workshop on Wikis for Software Engineering, WikiSym2007, Montreal, Quebec, Canada, 21 Oct 2007
6. Cole, M.: Using wiki technology to support student engagement: lessons from the trenches. Comput. Educ. **52**(1), 141–146 (2009)

7. Correia, F.F.: Extending and integrating wikis to improve software documentation. In: The Fourth Workshop on Wikis for Software Engineering, Porto, Portugal (2008)
8. Ferreira, D., Alberto, R.S.: Wiki supported collaborative requirements engineering. In: The Fourth Workshop on Wikis for Software Engineering, Porto, Portugal (2008)
9. Franklin, T., Van Harmelen, M.: Web 2.0 for content for learning and teaching in higher education, Technical Report, Bristol: JISC (2007)
10. Gonzalez-Reinhart, J.: Wiki and the wiki way: beyond a knowledge management solution, Information Systems Research Center, pp. 1–22 (2005)
11. Graupner, S., Singhal, S., Basu, S., Motahari, H.: Enabling a semantic wiki to drive business interactions, HP Labs Technical Report: HPL-2009-193 (2009)
12. Hadar, I., Levy, M., Ben-Chaim, Y., Farchi, E.: Using Wiki as a collaboration platform for software requirements and design. In: Proceedings of the 9th International Conference on Knowledge, Information and Creativity Support Systems KICSS'14, Limassol, Cyprus, 6–8 Nov 2014
13. Leuf, B., Cunningham, W.: The wiki way: quick collaboration on the web. Addison-Wesley, Boston, MA (2001)
14. Louridas, P.: Using wikis in software development. IEEE Softw. **23**(2), 88–91 (2006)
15. Luer, C.: A component-based, agile software development wiki. In: The Third Workshop on Wikis for Software Engineering, WikiSym2007, Montreal, Quebec, Canada (2007)
16. Majchrzak, A., Wagner, C., Yates, D.: Corporate wiki users: results of a survey. In: The International Symposium on Wikis '06, Odense, Denmark, pp. 99–104, 21–23 Aug 2006
17. McCarty, E.: How IBM uses an intranet to connect a global audience. Knowl. Manage. Rev. **11**(2), 28–33 (2008)
18. Ravid G.: Open large shared knowledge construction systems dominance. In: The Wikipedia Social Structure, Academy of Management Annual Meeting (2007)
19. Strauss, A.L., Corbin, J.: Basics of Qualitative Research: Techniques and Procedures for Developing Grounded Theory. Sage Publications, Newbury Park (1998)
20. Topper, C.M., Carley, K.: M: A Structural perspective on the emergence of 32 network organizations. J. Math. Sociol. **24**(1), 67–96 (1999)
21. Trkman, M., Trkman, P.: A wiki as intranet: a critical analysis using the Delone and McLean model. Online Inf. Rev. **33**(6), 1087–1102 (2009)
22. Yin, R.K.: Case Study Research, Design and Methods, 3rd edn. Sage Publications, Newbury Park (2003)

Enhancing Software Architecture via a Knowledge Management and Collaboration Tool

Sofia Sherman, Irit Hadar, Meira Levy and Naomi Unkelos-Shpigel

Abstract Software architecture is an important part of software development, aiming at ensuring a high-quality product. Recent research has shown that collaboration and knowledge management are important parts of the architecture process, and have significant role in architecture design and review. In this paper we present a prototype for a tool we developed, as part of our ongoing research on the software architecture process, for supporting collaboration, communication and knowledge sharing during all steps of the architecture development process. This tool was developed based on the findings of a case-study research in a global, large software organization.

Keywords Software architecture · Knowledge management · Collaboration

1 Introduction

Over the last two decades, there has been increased focus on architecture within soft-ware development. Software architecture design is an essential stage in the software development process and has a significant impact on software quality.

S. Sherman (✉) · I. Hadar · N. Unkelos-Shpigel
Information Systems Department, University of Haifa, Haifa, Israel
e-mail: shermans@is.haifa.ac.il

I. Hadar
e-mail: hadari@is.haifa.ac.il

N. Unkelos-Shpigel
e-mail: naomiu@is.haifa.ac.il

M. Levy
Industrial Engineering and Management Department,
Shenkar College of Engineering and Design, Ramat-Gan, Israel
e-mail: lmeira@shenkar.ac.il

S. Kunifuji et al. (eds.), *Knowledge, Information and Creativity Support Systems*,
Advances in Intelligent Systems and Computing 416,
DOI 10.1007/978-3-319-27478-2_41

Literature in this field usually refers to technological aspects of architecture such as design methods, architectural notations, patterns, analysis, etc. Only few empirical studies focused on human and collaborative aspects of software architecture [e.g. 1, 8, 11].

During the software architecture process many obstacles arise, most of which stem from collaboration-related issues and lack of specific knowledge resources [12]. One of the main problems caused by insufficient collaboration between stakeholders is the misalignment between the architecture review and the architecture design processes, and between different architecture review process instances and their outcomes. Knowledge preservation, accessibility, and reuse are vital for good and efficient software architecture design and review. Using already created and proven knowledge reduces time and resources invested, and typically results in higher-quality outcomes.

This paper presents a prototype for a tool we developed, as part of our ongoing re-search on the software architecture process, for supporting different aspects of the architecture process, including collaboration, communication, knowledge sharing, and review. This tool was developed based on the findings of a case-study research in a global, large software organization. A subset of these findings can be found in [12].

In the next section we briefly present the motivation for the proposed solution and relevant background. In Sect. 3 we present the conceptual framework of the solution followed by a description of the tool. In Sect. 4 we present and discuss a preliminary evolution of the solution and finally, in Sect. 5 we conclude.

2 Motivation and Background

Several studies have shown the importance of collaboration and knowledge management during software development in general and during architecture design and review in particular [1, 11]. In this section we provide a brief overview of the topics we combine in our research.

Software architecture design consists of two steps: (1) choosing an overall strategy for the architecture; and, (2) specifying the individual components that make up the application [7]. These result in an architecture document that includes architecture views and design rationale. Architecture design includes extracting and understanding architecturally significant requirements, making choices, synthesizing a solution, exploring alternatives, validating them, and more [8]. While architects make many architectural decisions, they neglect documenting them [10]. While architects do wish for efficient support to retrieve useful—previously stored—knowledge, they refrain from codifying it in the first place [10]. Most definitions of software architecture in literature typically relate to technological aspect of architecture design, and do not emphasize the required resources, including tools and human stakeholders. Architects need to work and collaborate with other stakeholders in order to produce high-quality architecture. Insufficient collaboration and knowledge sharing between the stakeholders involved in the process leads to difficulties and obstacles towards successful design of a software architecture solution [12].

Software architecture review is a vital phase in the software architecture process, aiming at ensuring and validating the quality of a proposed architecture. The reviewers focus on identifying project problems before they become costly to fix; providing information to various stakeholders for better decision making; and improving product quality [10]. Literature review reveals a wide range of evaluation methods used in review process [e.g. 1], however, little attention has been given to the review process in general and we found no evidence for tool that supports this process [2].

Knowledge management (KM) KM plays a central role in architecture process, embodies knowledge and serves as a vehicle for communication among stakeholders [6]. In particular, software architecture captures early design decisions and it is a transfer-able abstraction of the system, which enables further reuse in other software systems [3]. Thus, managing architecture knowledge is an important and inherent part of the software development processes. According to Maranzano et al. [10], KM should be embodied in architecture review for capturing best practices and socialize such practices across the organization. However, while their research fosters cross-organizational learning and transfer of lessons learned and best practices, and highlights the need to identify and share corporate assets and reusable components, it does not provide a framework for embedding the KM practices within the review process itself for facilitating collaboration and engagement among the review participants.

Collaboration between software architects for sharing architecture knowledge has been the focus of several research works. Farenhorst et al. [5] proposed an architectural knowledge-sharing tool based on best practices of KM and characteristics of architecture knowledge. Liang et al. [9] proposed a collaborative architecting process and an accompanying tool suite that integrate the architecting activities through architecture knowledge sharing. However, we did not find any research focusing on the architecture review process, the knowledge produced from its outcomes, and the collaboration between the stakeholders involved. The solution presented in this paper aims to bridge this gap by providing an environment and guidelines for embedding collaboration and KM practices throughout the architecture process.

3 Proposed Solution

3.1 Framework

The main concepts this solution is based on are collaboration, knowledge management and sharing, between and among different stakeholders involved in the architecture process. Here we address only the stakeholders who are directly

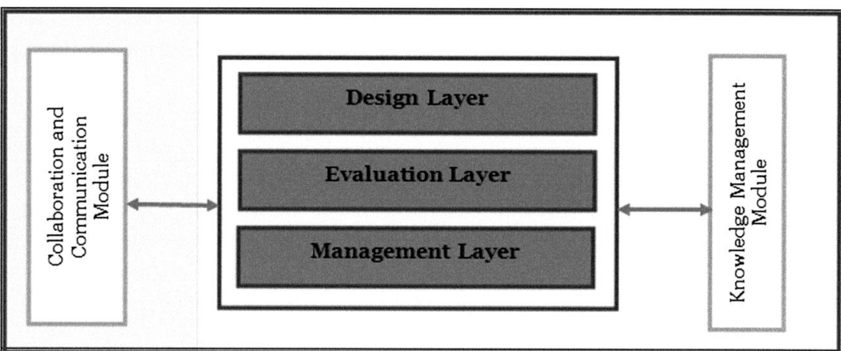

Fig. 1 A conceptual framework for architecture KM and collaboration

involved in the process of developing architecture and approving it for the next stages. The framework for the conceptual solution model (Fig. 1) presents a general view of the architecture process.

We divided these stakeholders to three main layers: (1) Management—stakeholders who manage the architecture process including: project allocation, review allocation and architecture knowledge. (2) Design—software architects who are responsible for designing and documenting the architecture solution, preparing it for review, and improving the architecture solution according to the reviewers' comments and instructions, if provided. (3) Evaluation—architects, who serve as architecture reviewers and operate in different levels of the architecture process, for example: peer review, manager review, or board review.

The proposed solution supports online collaboration and KM. Collaboration is enabled via the Collaboration and Communication module, which includes several types of collaboration means: forum, private messaging, and collaboration on documents under review. KM is enabled via the Knowledge Management module; each outcome produced during the architecture process via the collaboration module will be stored for future use in the KM module. These outcomes include discussions between reviewers and architects and among reviewers during the processes of the architecture design, review preparation, and review discussions and decisions. This information will be made accessible for architects and reviewers: architects will be able to reuse this knowledge for future architecture development and learn about reviewers' requirements and comments; reviewers will be able to access previous reviews' outcomes in order to see, for example, previously raised issues regarding the same product, customer or architect. Additional information included in the KM module is theoretical architecture knowledge extracted from textbooks, and checklists formulated by managers and reviewers, based both on theory and practice. These checklists evolve within the tool over projects and time.

3.2 Tool

Based on the framework presented above, we developed a prototype for a tool that manages and supports the architecture process. The architecture design and review processes are supported by the proposed tool as described below.

Architecture knowledge management and collaboration: There are two types of knowledge available in the tool. *Formal knowledge* produced by the management and evaluation layers based on different sources: company requirements, architecture educational sources (e.g., textbooks), and previous review outcomes. This knowledge is presented to the architecture design layer as guidelines (Fig. 2a) and is accessible directly in the guidelines or via the search engine. In the current version of the tool, we used [4] as the source for the quality attributes' description and as a basis for the first version of the checklist. *Informal knowledge* created while working with the tool. It allows relevant stakeholders to access documents from different projects, public conversations regarding projects, and relevant architecture artifacts. This knowledge is accessible in the tool via projects or the search engine (Fig. 2b).

The collaboration within the system provides two important capabilities. First, it enables communication and joint decision making during architecture design and architecture review. Second, since all comments and discussions are documented in the system, this knowledge becomes available, not only to stakeholders of the current projects, but also to architects and reviewers in other projects, who may benefit from the lessons learned and knowledge accumulated over previous projects.

The architecture design process: The main usage scenarios include: (1) Architects are allocated to projects by the management layer and design the architecture. During their work, architects can access formal knowledge in the guidelines section (Fig. 2a) or informal knowledge in forums or comments that reside in other projects (Fig. 2b). Architects can communicate with management and evaluation layer users, for questions and guidance. (2) During the architecture design, architects fill a checklist report (Fig. 3) regarding all aspects of the architecture under development.

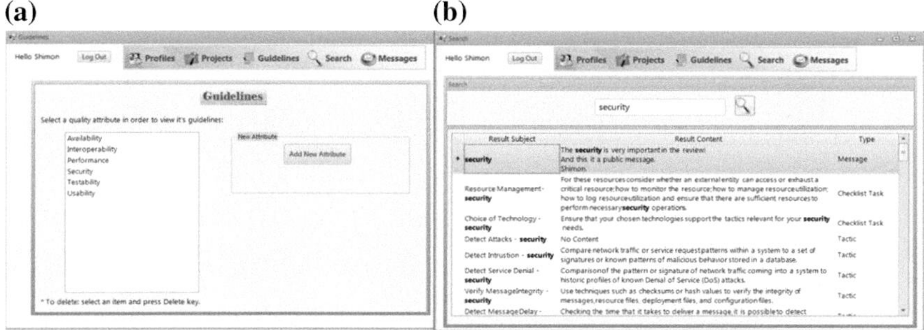

Fig. 2 Screen Shots. **a** Formal knowledge content. **b** Knowledge search engine

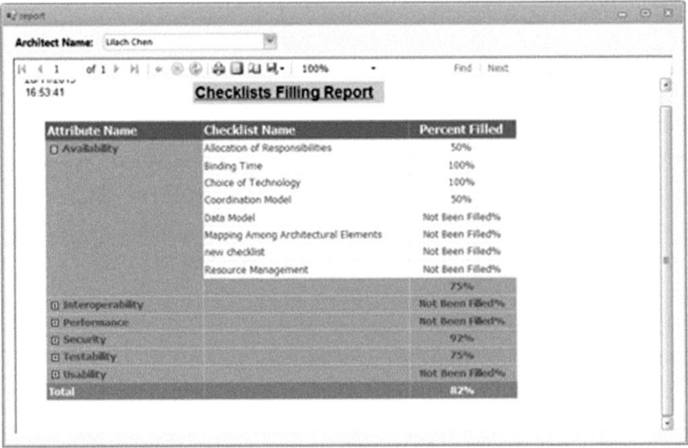

Fig. 3 *Screen Shot* Checklist report

(3) When ready, architects submit the architecture document and the checklist report.

The architecture review process: The main usage scenarios include: (1) Upon submission, the management layer representative allocates reviewers listed in the evaluation layer, according to their profiles. (2) While performing review, reviewers can to add comments within the architecture document and automatically follow comments in the document that were left by other reviewers (Fig. 4), as well as

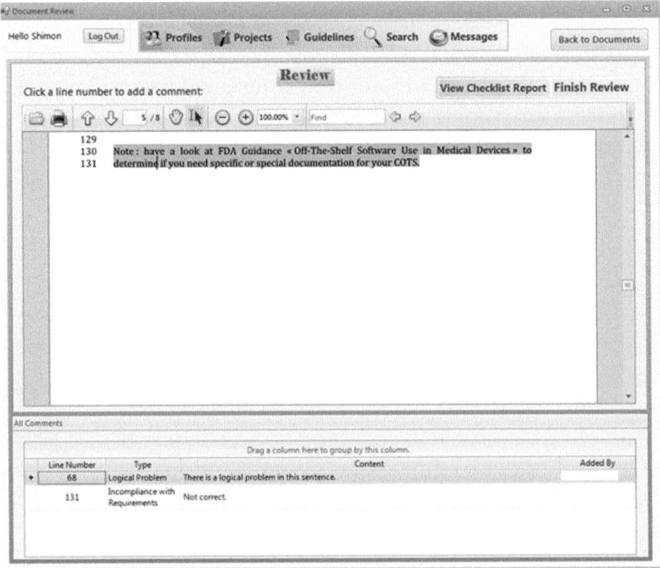

Fig. 4 *Screen Shot* Review screen

follow-up on changes made by the architect based on the review comments. During the review, reviewers can communicate with each other using the Collaboration and Communication module, look for history of previous reviews of the same or other products, read comments, and participate in discussions among reviewers regarding other relevant software reviews. Using the system for discussions among reviewers, contributes to managing knowledge history by documenting and storing it in the database and making it accessible for future needs.

4 Evaluation and Discussion

A demo of the solution proposed in this paper was presented in [13]. Based on the discussion following the demo presentation and additional practitioners' evaluations conducted afterwards, we bring some feedbacks and discuss their implications. We performed interviews with three potential users of this tool: one software architect, and two software architects who serve also as reviewers.

Functionality of the tool. Two features, additional to the exiting functionality, were proposed by the interviewees. The first referred to the structure of the review comments that are currently formed and accumulated in the system as free text. A template-oriented comment structure would make it easier for reviewers to form the comments and, more importantly, would make it easier for other stakeholders to retrieve relevant comments when searching for knowledge within the system. The second functionality-related set of comments referred to additional reports this tool should be able to produce, as a managerial means. It is important to note, however, that this tool was designated to support collaboration and knowledge management; its managerial features are a secondary aspect of this tool, and therefore their extension needs to be carefully considered.

Implementation concerns related to the characteristics of the organization. One of the challenges in implementing this tool, as in any solution implementation, is the required change in business processes. Specifically, our interviewees referred to these two extreme cases: (1) When there is no established structured architecture review process, implementing a tool that enforces such structured process requires a significant change in culture and procedures; (2) When a heavily structured architecture review process does exist, odds are that it would require some adaptation of the process (technological and procedural) in order to use the tool.

Social concerns stemming from sharing review comments. The tool allows writing public review comments attached to architecture documents. These comments become part of the accessible knowledge in the system. Two of the three interviewees were concerned that this would violate architects' privacy, as these comments can be perceived as critic on the architects' work. The third interviewee (interestingly, it was the architect who does not serve as a reviewer) found this as an excellent knowledge sharing opportunity and did not think that architects would objects. We believe that this may highly depend on organizational culture and should be investigated prior to implementation.

In addition to the feedbacks above, proposing opportunities for improvements, the interviewees also referred to points of strengths and contribution of the tool, thus supporting the conjecture that the proposed tool meets its objectives [12, 13]. For example, all three interviewees believed that the tool would be highly effective for managing and controlling the review process and its outcomes, especially in organizations where many projects are reviewed by multiple reviewers, and that software architects would be able to benefit from the knowledge managed in the system when constructing their architecture solution and preparing it for review.

5 Conclusion and Future Work

Previous research has shown that insufficient collaboration and knowledge management hinders the quality of software architecture. The solution we propose supports these two aspects of the architecture process and provides a platform for managing architecture design and review with emphasis on collaboration between stakeholders and knowledge creation and reuse. This solution was preliminary evaluated by relevant practitioners and will be further refined.

In the next step of this research we will evaluate the proposed solution in collaboration with a global, large software organization in order to learn about its strengths and weaknesses from practitioners' point of view, and possibly conduct a feasibility check via a controlled usage of the tool in real industrial settings.

References

1. Babar, M., Kitchenham, B., Zhu, L., Gorton, I., Jeffery, R.: An empirical study of groupware support for distributed software architecture evaluation process. J. Syst. Softw. **79**, 912–925 (2005)
2. Babar, M.A., Zhu, L., Jeffery, R.: A framework for classifying and comparing software architecture evaluation methods. In: Australian Software Engineering Conference (2004), pp. 309–318
3. Babar, M.A., Dingsøyr, T., Lago, P., van Viliet, H.: Software Architecture Knowledge Management—Theory and Practice. Springer (2009)
4. Bass, L., Clements, P., Kazman, R.: Software Architecture in Practice, 3rd edn. Addison-Wesley Professional (2013)
5. Farenhorst, R., Lago, P., van Vliet, H.: Effective tool support for sharing architectural knowledge. Int. J. Cooper. Inf. Syst. **16**(03n04), 413–437 (2007)
6. Farenhorst, R., Hoorn, J.F., Lago, P., van Vliet, H.: The lonesome architect. In: Software Architecture (2009) and European Conference on Software Architecture WICSA/ECSA (2009)
7. Gorton, I.: Essential Software Architecture. Springer, Berlin, Heidelberg, NY (2006)
8. Kruchten, P.: What do software architects really do? J. Syst. Softw. **81**, 2413–2416 (2008)
9. Liang, P., Jansen, A., Avgeriou, P.: Collaborative software architecting through knowledge sharing. In: Collaborative Software Engineering, pp. 343–367. Springer, Berlin, Heidelberg (2010)

10. Maranzano, J.F., Sandra, A., Rozsypal, S.A., Zimmerman, G.H., Warnken, G.W., Wirth, P.E., Weiss, D.M.: Architecture reviews: practice and experience. IEEE Softw. **22**(2) (2005)
11. Sherman, S., Hadar, I., Hadar, E., Harrison, J.: The overall value of architecture review in a large scale software organization. In: Proceedings of the IWSSA Co-located with CAISE (2011)
12. Sherman, S., Hadar, I., Levy, M.: Enhancing software architecture review process via knowledge management. In: Proceedings of the AMCIS (2010)
13. Sherman, S., Hadar, I., Levy, M., Unkelos-Shpigel, N.: Enhancing software architecture via a knowledge management and collaboration tool. In: Proceedings of the 9th International Conference on Knowledge, Information and Creativity Support Systems KICSS'14, 6–8 Nov 2014, Limassol, Cyprus (2014)

Collaborative Requirement Prioritization for an E-Recruitment Platform for Qualified but Disadvantaged Individuals

Ahmet Suerdem and Basar Oztaysi

Abstract Collaboration of system designers and social researchers to search for solutions to social problems is becoming increasingly important. In this context, G@together project funded by JPI-Urban Europe is an interdisciplinary project aiming to build an e-platform facilitating the employment of qualified but disadvantaged people. The use of qualitative social data and the participation of end-users to the system design process provide a significant potential for co-creation studies. In this study we used different social research and system engineering techniques to elicit the requirements of different stakeholders concerning an e-recruitment platform. These methods helped the collaboration of system designers, social researchers and user groups to prioritize the system requirements. Analytic Hierarchy Process, a commonly used multi-criteria decision making technique is used to integrate the different elements of this process.

Keywords Disadvantaged groups • E-recruitment • Agile software design • Co-creation • Analytic hierarchy process

1 Introduction

Determining the requirements of members of disadvantaged groups for developing systems and products designed for their specific use is now recognized as an important issue for social inclusion. Inclusive accessibility to the citizenship rights and services requires the participation of vulnerable groups and persons in the

A. Suerdem (✉)
Department of Business Administration, Istanbul Bilgi University, Istanbul, Turkey
e-mail: ahmet.suerdem@bilgi.edu.tr

B. Oztaysi
Department of Industrial Engineering, Istanbul Technical University, Istanbul, Turkey
e-mail: oztaysib@itu.edu.tr

S. Kunifuji et al. (eds.), *Knowledge, Information and Creativity Support Systems*, Advances in Intelligent Systems and Computing 416,
DOI 10.1007/978-3-319-27478-2_42

decision making process during the development of products that affect their lives [21].

In this paper, we share our experiences during the conceptualization of an online recruitment platform for qualified, yet socially disadvantaged individuals.

When online recruitment systems first emerged in the mid-1990s, popular management press heralded them as a ‘recruiting revolution’ [3] Despite this hype, some critics [11, 13, 19] have noted that these systems are still not an efficient alternative to traditional methods due to their clumsiness. These systems have some major shortcomings such as lack of talent pool diversity, being rigidly standard and lack of customization. Software architecture design plays an essential role in the system development process and determines the quality of the end product. In this respect, the major element underlying these shortcomings is the top-down design process from an expert and managerial perspective without consulting other stakeholders [16]. Such a design process essentially addresses the standardized skilled talent requirements of big corporates, while ignoring the talent pool diversity and the requirements of SMEs or NGOs [14]. An open, bottom-up and collaborative design process is essential for making e-recruitment systems to satisfy sometimes incompatible requirements of various user groups and stakeholders. Involving all stakeholders from the very beginning to different stages of design process will prevent a wrong conceptualization that may turn into an unattractive product for some of the end users [2].

Architecture design studies habitually address the technical aspects of software systems such as patterns, frameworks and notations. There are only few empirical studies highlighting the collaborative elements in the design process [18]. One exception is the agile software development approach which criticizes the top-down philosophy of the main-stream up front architecture design and highlights the importance of a collaborative process [6]. This approach emphasizes the significance of a bottom-up architecture design iterating between requirement analyses and solutions evolving through collaboration between self-organizing, cross-functional, interdisciplinary teams and stakeholders.

A collaborative conceptualization process where all stakeholders generate ideas and turn these into concepts before determining the properties of the system is highly important for an adaptive, evolutionary and agile architecture design [22]. Unsolved problems between stakeholders during the conceptualization phase can be costly at the later stages of the system development process [9]. Particularly, linear designs reducing complex system architectures involving fuzzy interactions between software components, agents, data structures and interfaces may lead to future system failures [23]. Hence, a collaborative and process distributed requirements analysis at the conceptualization stage is essential for designing agile, integrated and intelligent systems. This should involve a deliberative design process between stakeholders rather than just aggregating the preferences of an abstract average end user [7].

A collaborative architecture design can be considered as a multi-criteria decision problem requiring optimization among many requirements. The decision-maker has to trade-off certain requirements at the expense of the others. This is usually done by the aggregating the preferences by optimizing weighted sums over a criterion. Aggregation, however, usually tends to information loss at the expense of the minority groups whose requirements may not correspond with the majority. This creates an ethical issue which can be handled through social life cycle analysis. In general, life cycle analysis (LCA) is a methodological framework to ensure multi-criteria sustainability. Particularly, social LCA aims to maintain socially responsible product development by determining the human relationships that could be impacted by the life cycle of a product and identifying the ways in which social conditions could be improved [10]. Within this framework, Climaco and Valle propose an open exchange multi-criteria decision framework that involves interactive procedures avoiding the aggregation of preferences of different agents [5]. These procedures aim to combine optimization protocols with the experience and intuition of decision agents. The idea is to engage stakeholders from diverse backgrounds and viewpoints to the solution of a complex problem involving much diversified uncertainty issues.

In this vein, in this study we share our experiments regarding a collaborative process involving social scientists, system designers and different stakeholders for prioritizing the features of an e-recruitment platform for disadvantaged people. Within this framework, we suggest a method integrating social sciences and system engineering to conduct requirement analyses for handling the complexity in the system design. This paper gives a more detailed explanation of the qualitative procedures we have mentioned in KICSS proceedings [20]. The rest of the paper is structured as follows: First, we summarize the main steps combining qualitative and quantitative methods for determining and prioritizing the features of an e-recruitment system through a collaborative process. Then, we explained Analytic Hierarchy Process method and how we have integrated the steps mentioned in the previous part to the prioritization of system requirements. The following part demonstrates an application and findings of the method for prioritizing the system requirements of an e-recruitment platform designed for women who have interrupted their careers. Finally, we discuss results and implications.

2 Method

2.1 Main Steps of the Study

Our method involves six steps, combining qualitative and quantitative methods for determining and prioritizing the features of an e-recruitment system through a collaborative process:

1. *Identifying target groups*: To determine the scope of our case we identified the boundaries of the concept "qualified but disadvantaged" through a desk research of the statistics, reports, official documents and interviews with official employment agencies. Official definition of disadvantaged in Turkey is very

limited, covering just disabled people, ex-convicts and veterans. We extended this definition to cover socially disadvantaged groups defined according to the EU standards: individuals facing difficulties of accessing employment and other social and cultural resources because of their social positions and identities [8]. Hence we identified our target case as "women who have interrupted their carriers for family reasons and want to go back to the job market".

2. *Determining employment problems of the target group*: Face-to-face interviews are conducted with the target group members and employers to elicit information about the process of job seeking experience. Interview transcripts are coded from a grounded theory perspective [4]. We followed an inductive approach starting with generation of open codes from the transcripts and ending with organization of these codes into major categories: descriptive features of the interviewees; organisation of the job search; learning to search for a job; experience with discrimination and prejudices during job search; experiences with and expectations towards e-recruitment platforms. After the analysis of the transcripts with a CAQDAS (Computer Assisted Qualitative Data Analysis Software), we determined the relationships between existing issues.
3. *Determining the Good Practices:* We reviewed literature about existing systems and examples of good practices to determine the base features of the architecture design.
4. *Determining the potential system requirements through brainstorming method*: Brainstorming method is a widely used idea generation and problem formulation technique applied in creative product design and development [24]. We made several brainstorming sessions with the members of "Yeniden Biz Platformu", an NGO supporting highly qualified women who have interrupted their careers. During these sessions we have formulated the problems encountered during job search and generated ideas to solve these problems by means of an e-recruitment system. The agenda of the brainstorming session was set by the themes obtained from interviews and literature review. We organized the ideas into an affinity diagram to list all possible requirements of the system design.
5. *Filtering the requirements*: System designers and end users negotiated together to filter out the requirement list to exclude the basic (i.e. basic elements required in all systems) and trivial features (too sophisticated to be effective). We ended up with a simplified feature list.
6. *Prioritization of the system needs*: The remaining system features are prioritized by means of Analytic Hierarchy Process. The degree of importance of each system requirement is determined.

2.2 Analytic Hierarchy Process

Analytic Hierarchy Process (AHP) is a multi-criteria decision making method developed firstly by Thomas Saaty (1980) [17]. AHP method disintegrates the

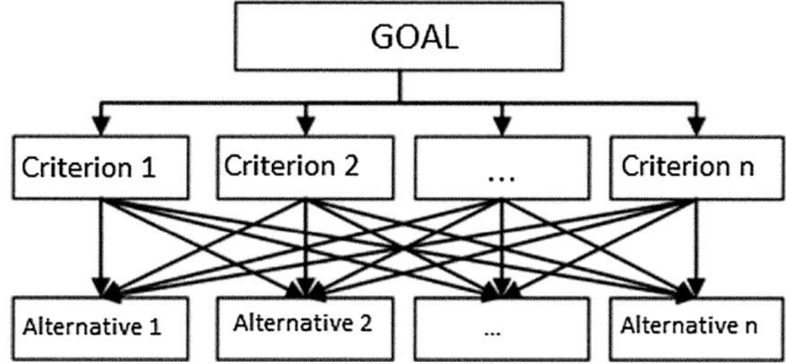

Fig. 1 Example of AHP hierarchy

complex decision problems according to different perspectives and deals with them in a hierarchical manner. Hence, complex problems are simplified into easier sub-problems. AHP method has an appropriate structure to make a decision by considering both the quantitative and qualitative factors.

The structure of AHP consists of a problem objective, the criteria to be used to decide for this objective and alternatives among which the preferences will be made (Fig. 1).

The steps and processes contained within AHP method are listed as follows [15, 17]:

Step 1: Identification of the Problem: Existing problems concerning the system are identified during the first stage of AHP. Subsequently, the objective to be achieved by the decision-maker(s) is determined. In our case, this step is handled during the abovementioned qualitative methods (interviews, desk research). The objective is to prioritize the features minimizing discrimination during online job matching process.

Step 2: Identification of the Criteria: System designers and end users come together to determine the criteria to be included in the system to solve the problems. This step was handled during brainstorming and requirement filtering stages. Stakeholders negotiated to agree on a simplified list of requirements and features that would minimize discrimination.

Step 3: Identification of the Alternatives: At this stage, system designers determine all the alternative options to achieve the decision-making objective.

Step 4: Establishing the Hierarchic Structure: The identified main objective is divided into several sub-goals and a hierarchy as shown in Fig. 1. When establishing the hierarchy, the items at the same level are assumed as independent of each other.

Step 5: Performing Pairwise Comparisons: Based on the established hierarchy, the items at the same level are exposed to item by item paired comparisons. All paired criteria are compared in terms of their significance on the objective (Fig. 1). Then, the decision-maker(s) make a pairwise comparison of the alternatives within each criterion. The evaluations are performed by using the relative significance levels (Table 1) developed by Saaty [17].

Table 1 Relative significance level to be used in the paired comparisons

1–9 Scale	Definition
1	Equal Importance (E)
3	Modarate Importance (M)
5	Strong Importance (S)
7	Very Strong Importance (VS)
9	Extreme Importance (Ex)
1.	2.

Step 6: Measuring the weight of the criteria: The weight of each criterion is calculated by means of pairwise comparison matrices. At this stage, pairwise comparison matrix A has been obtained provided that the significance level between the alternative a_{ij} and criterion c_{ij} is shown.

$$A = \begin{bmatrix} a_{11} & a_{12} & \dots & a_{1n} \\ a_{21} & a_{22} & \dots & a_{2n} \\ \vdots & \vdots & \dots & \vdots \\ a_{n1} & a_{n2} & \dots & a_{nn} \end{bmatrix} C = \begin{bmatrix} c_{11} & c_{12} & \dots & c_{1n} \\ c_{21} & c_{22} & \dots & c_{2n} \\ \vdots & \vdots & \dots & \vdots \\ c_{n1} & c_{n2} & \dots & c_{nn} \end{bmatrix} \quad (1)$$

For this reason, each paired comparison matrix is primarily normalized. The most commonly used method of normalization in practice is dividing each element of the column into the sum of the relevant column.
Then, each element of matrix A is divided into the sum of the relevant column.

$$b_1 = \sum_{i=1}^{n} a_{i1} \quad c_{ij} = \frac{a_{ij}}{b_i} \quad (2)$$

As a result, matrix C with normalized values is obtained. The significance level of the criteria with each other is calculated with the arithmetic mean method by using the values in the matrix C.

Step 7: Consistency Analysis and Revision: After the paired comparison matrices are calculated, the consistency of the evaluation is tested. The decision-makers are asked to revise their opinions for the paired comparison matrices considering consistency problems.

3 Application and Findings

Within the scope of the project, qualified—disadvantaged group is defined as "women who have interrupted their carriers for maternity reasons and want to go back to the job market". In accordance with this definition, we conducted a collaborative design process with "Yeniden Biz Platformu" (yenidenbiz.com). An NGO working supporting women who have interrupted in their careers.

We also conducted a literature research to investigate the examples of worldwide good e-recruitment practices. Subsequently, we have performed interviews with the HR departments and woman job seekers. Finally, through brainstorming and interviews we elicited a final list of end user requirements for the e-recruitment system to be designed.

Elicited system criteria may be grouped under six titles: (i) *Branding and information*. This part contains the main theme and message about the organization and useful information and links related to the self-improvement of the job seekers. Consistent branding and graphics, intuitive navigation and a privacy statement are the basic elements that need to be complemented by intuitive features such as information about training and/or coaching opportunities or forums for experience sharing. (ii) *Tools and utilities for entering the job and job-seeker profile data:* This part includes interface features enabling data entrance and checking such as building, storing and editing online resumes and candidate and company profiles and online reference checks. (iii) *Database queries, selection, sorting and ranking:* This part includes search and advanced search features enabling to filter according to multiple profile criteria such as industry or location. (iv) *Assessment and online testing:* The system features to evaluate the competences, personalities and suitability of the job seekers for the job. (v) *User friendliness:* Includes features such as integration to corporate websites and informing about available job opportunities through instant messaging. (vi) *Communication before the interview:* Includes features to ensure communication between the candidates and the companies such as chat or Skype conversations.

As a result of a second round of interviews, sixth group is excluded from the model. Pair-wise comparison matrices were established for the remaining system criterion groups and system features and the evaluations of the decision-makers regarding the priorities were summed up.

The example of paired comparison matrix for AHP is given in Table 2. The decision-makers were asked to compare the five criterion groups. The diagonal of the matrix was selected to be equally-weighted. Each evaluation involved in the matrix should be read as the evaluation of the criterion group in the relevant row over the group in the respective column. For example; the *Tools for data entrance* requirements have been found "Moderately Important" with respect to the *Branding* requirements.

Table 2 Paired comparison of the requirement groups as per the platform

→	C1	C2	C3	C4	C5
C1	Eq	1/M	VS	S	M
C2	→	Eq	EX	VS	S
C3	→	→	Eq	S	M
C4	→	→	→	Eq	M
C5	→	→	→	→	Eq

Table 3 Numeric values of the paired comparisons and the weights

→	C1	C2	C3	C4	C5	Priorities
C1	1	0.33	7	5	3	0.260
C2	3	1	9	7	5	0.503
C3	0.33	0.2	1	5	3	0.134
C4	0.2	0.14	0.33	1	3	0.068
C5	0.14	0.11	0.2	0.33	1	0.035

The verbal evaluations stated in Table 2 are converted into numeric values in the next step. The numeric values are given in Table 3. Unlike the previous table, the empty cells are completed as the inverse of the value symmetrical to the diagonal.

The processes defined in the previous section are performed on this matrix and their significance levels are calculated. The significance weights are given in the last column of Table 3. Accordingly, "Tools for data entrance" is the most relevant criterion group with 0.503 which is followed by "branding" and "matching" criteria.

The above-mentioned pair-wise comparison matrices were prepared for each criterion groups, and thus the significance of each feature in the relevant group was found.

The significance weight of each feature in the relevant group is stated in the local significance column in Table 4. The significance level of the feature in the general platform is obtained by multiplying the significance level of the relevant group by the local significance. The significance level of the features relative to the total is listed under the general significance.

Table 4 Local and general significance level of the characteristics

	Local significance		General significance		Local significance		General significance
Branding	0.26			*Matching*	0.13		
R1.1		0.277	0.072	R3.1		0.197	0.026
R1.2		0.248	0.065	R3.2		0.153	0.021
R1.3		0.298	0.078	R3.4		0.309	0.041
R1.4		0.176	0.046	R3.5		0.341	0.046
Data enter	0.50			*Tests*	0.07		
R2.1		0.036	0.018	R4.1		0.264	0.018
R2.2		0.080	0.040	R4.2		0.076	0.005
R2.3		0.341	0.171	R4.3		0.263	0.018
R2.4		0.130	0.066	R4.4		0.397	0.027
R2.5		0.413	0.208	*User friend*	0.04		
				R5.1		0.250	0.009
				R5.2		0.750	0.026

R2.5 (structured data entry), R2.3 (depersonalized application procedures) and R1.3 (online training and courses) are the most important requirements among the system features.

4 Implications

When the agencies involved in the employment field in Turkey are reviewed, it can be seen that İŞKUR (Government employment agency) and municipality employment centers generally address unskilled, under-qualified job seekers. Kariyer.net and Linkedin address people seeking mid-level managerial jobs, and private HR and "head hunter" companies are utilized for seeking high level managerial jobs. These systems either do not have any specific procedures for disadvantaged people or have very general, bureaucratic procedures. Customized systems appropriate for the specific requirements of specific social groups will bring more effective solutions into the employment process. This study may provide an example for a collaborative process to design such a system. As a result of a multi-stage negotiation level between stakeholders, we have found that women who want to go back to their careers prioritize structured data entry, depersonalized application procedures and links to online training and courses for the design of an e-recruitment platform that minimizes discrimination during the employment process. Besides these potentials, our method may have some limitations. Criticisms of AHP emphasize that its over-reliance on quantitative methods and its hierarchical structure makes it a weak instrument for handling complex system designs [1]. Future research should consider methods more convenient for the fuzzy character of the social life cycle analysis [5, 12].

Acknowledgments This work is a part of Urban Europe Project entitled "Gettogether without Barriers" and is supported by the Scientific and Technological Research Council of Turkey (TÜBİTAK), Grant No: 113K027.

References

1. Barzilai, J.: Measurement and preference function modeling. Int. Trans. Oper. Res. **12**, 173–183 (2005)
2. Bhattacharya, S., Krishnan, V., Mahajan, V.: Capturing the effect of technology. Sequence design methodology. IIE Trans. **30**, 933–945 (1998)
3. Boydell, M.: Internet recruitment helps HR careers. Canadian HR Reporter, Feb 11 2002, Canada (2002)
4. Charmaz, K.: Grounded Theory. The SAGE Encyclopedia of Social Science Research Methods. SAGE Publications (2003)
5. Climacol, J., Valle, R.: MCDA and LCSA—a note on the aggregation of preferences. In: The 9th International Conference on Knowledge, Information and Creativity Support Systems KICSS'14, Limassol (2014)

6. Dingsøyr, T., Nerur, S., Balijepally, V., Moe, N.B.: A decade of agile methodologies: towards explaining agile software development. J. Syst. Softw. **85**, 1213–1221 (2012)
7. Driessen, P.H., Hillebrand, B.: Integrating multiple stakeholder issues in new product development: an exploration. J. Prod. Innov. Manage **30**, 364–379 (2013)
8. Council of the European Union, Joint report by the Commission and the Council on Social Inclusion (2004)
9. Florén, H., Frishammar, J.: From preliminary ideas to corroborated product definitions: managing the front end of new product development. Calif. Manag. Rev. **54**, 20–43 (2012)
10. Guidelines for Social Life Cycle Assessment of Products, http://www.unep.org/publications/search/pub_details_s.asp?ID=4102
11. Harris, M.M., Van Hoye, G., Lievens, F.: Privacy and attitudes toward Internet-based selection systems: a cross-cultural comparison. Int. J. Sel. Assess. **11**, 230–236 (2003)
12. Jørgensen, A., Le Bocq, A., Nazarkina, l., Hauschild, M.: Methodologies for social life cycle assessment. Int. J. Life Cycle Assess. **8**(Online First), 1–8 (2008)
13. Kehoe, J.F., Dickter, D.N., Russell, D.P., Sacco, J.M.: E-selection. In: Gueutal, H.G., Stone, D.L. (eds.) The Brave New World of e-HR: Human Resource Management in the Digital Age, pp. 54–103 (2005)
14. Maurer, S.D., Liu, Y.: Developing effective e-recruiting Websites: insights for managers from marketers. Bus. Horiz. **50**(4), 305–314 (2007)
15. Özden, H.Ü.: Primary School selection using analytic hierarchy method. J. Marmara Univ. FEAS **24**(1) (2008)
16. Palmer D., Kaplan, A.: Framework for Strategic Innovation. Blending Strategy and Creative Exploration to Discover Future Business Opportunities, http://www.1000ventures.com/business_guide/innovation_strategic_byip.html (2007)
17. Saaty, T.L.: The Analytic Hierarchy Process. McGraw-Hill, New York (1980)
18. Sherman S., Hadar, I., Levy, M., Unkelos-Shpigel, N.: Enhancing software architecture via a knowledge management and collaboration tool. In: The 9th International Conference on Knowledge, Information and Creativity Support Systems KICSS'14, Limassol (2014)
19. Stone, D.L., Dulebohn, J.H: Emerging issues in theory and research on Electronic Human resource management (eHRM). Human Resour. Manage. Rev. 1–5 (2013)
20. Suerdem, A., Oztaysi, B.: Prioritization of the requirements for the platform to be used for the employment of qualified—disadvantaged individuals. In: The 9th International Conference on Knowledge, Information and Creativity Support Systems KICSS'14, Limassol (2014)
21. The European Social Fund and Social Inclusion, http://ec.europa.eu/employment_social/esf/docs/sf_social_inclusion_en.pdf
22. Vijayasarathy, L., Turk, D.: Drivers of agile software development use: dialectic interplay between benefits and hindrances. Inf. Softw. Technol. **54**, 137–148 (2012)
23. Wood, D.J., Jones, R.E.: Stakeholder mismatching: a theoretical problem in empirical research on corporate social performance. Int. J. Organ. Anal. **3**, 229–267 (1993)
24. Yuizono T., Munemori J.: Evaluation indexes to understand the creative problem solving process in the distributed and cooperative KJ Method. In: The 9th International Conference on Knowledge, Information and Creativity Support Systems KICSS'14, Limassol (2014)

Text Comparison Visualization from Two Information Sources by Network Merging and Integration

Ryosuke Saga

Abstract This paper describes a comparison visualization method based on network integration. A topic may be discussed by many people, resulting in various information sources with different viewpoints and perspectives. To comprehend a topic, network visualization has been proposed. However, this method cannot compare claims and opinions. Thus, the current study investigates network visualization methods that individually compare texts from multiple information sources.

Keywords Knowledge visualization • Comparison analysis • Co-occurrence graph • Graph integration

1 Introduction

A person's impressions of things change depending on viewpoint. For instance, we often use the term "curious" to express a personal characteristic, but this characteristic often corresponds to "easy to get tired". Viewpoints are also called positions, perspectives, and fields of view. For consistency, the term "viewpoint" is used in this paper.

Traditionally, numerous arguments originate from a difference in viewpoints between two individuals. For example, competitors (plaintiff and defendant) in a court of law discuss a claim. We can also consider the review system of scholarly papers to be a viewpoint-based discussion of originality and feasibility between authors and reviewers. With the progress of the Internet, two or more persons can argue extensively. On review sites such as Amazon and TripAdvisor, consumers evaluate items according to their opinions, and service providers such as hotels and

R. Saga (✉)
Graduate School of Engineering, Osaka Prefecture University,
1-1 Gakuen-Cho, Nakaku, Sakai 559-8531, Japan
e-mail: saga@cs.osakafu-u.ac.jp

S. Kunifuji et al. (eds.), *Knowledge, Information and Creativity Support Systems*,
Advances in Intelligent Systems and Computing 416,
DOI 10.1007/978-3-319-27478-2_43

producers reply to their customers. In anonymous bulletin board systems, many people submit opinions on a topic based on their viewpoints. Therefore, arguments and discussions are derived from the viewpoints of two or more persons.

Thus, outputs related to an object depend on the viewpoint although related facts are known. In other words, data are generated according to viewpoints (in this case, the viewpoint is known as a model or distribution). Viewpoints are extracted using several methods; for quantitative data, the methods of multivariable analysis and data mining are mainly used (principal component analysis, Data Development Analysis, and so on) [1, 4, 5].

Texts (bags of words) are mined through keyword and topic extraction to comprehend textual viewpoints. Methods of clustering, Principle Component Analysis (PCA), probabilistic PCA, and latent Dirichlet allocation are used [2, 3] to extract important keywords and topics. Using these keywords and topics, we can infer and understand their backgrounds and the viewpoints associated with them. However, these methods cannot directly determine the differences and gaps among the viewpoints of n persons. For example, we can analyze a topic in one document set (e.g., a newspaper) using PCA, but we cannot comprehend the variation in the results of two document sets.

Given this background, my project aims to determine the difference between the viewpoints of n persons through network visualization [9]. The previous paper has proposed the method to visualize the difference between the viewpoints of two individuals [7]. This paper summarizes the theory as well as confirms its usability and feasibility.

2 Network Visualization for Text

Information visualization techniques are highly important in result comprehension. Thus, my research aims to apply network visualization in the clarification of viewpoints.

In this study, network visualization is based on graphical representation in mathematics. A network consists of vertices and edges that possess attributes. A vertex set is V, an edge set is E, and a network G_i is presented as $G_i = \mathrm{G}(V_i, E_i)$. V and E are also known as network elements, whereas V_i and E_i are elements of network Gi. Each vertex $v_i \in V_i$ and each edge $e_i \in E_i$ has n and m attributes, that is, $v_i = (v_{i1}, v_{i2}, \ldots, v_{in})$, $e_i = (e_{i1}, e_{i2}, \ldots, e_{im})$. The edge between two vertices A and B is expressed as $e_{A \to B}$ ($e_{A \to B}$ is equal to $e_{B \to A}$ in an undirected network).

This network visualization has been applied in areas such as human relationships and social network analysis. In the network visualization of text data, a network is generated from a document set, which is derived from an information source. Each vertex is labeled with a keyword or a key phrase, and each edge often displays a co-occurrence coefficient. This study focuses on network visualization of text data within such domains.

The network visualization of text data involves themes such as network generation from data, graph layout, and graph representation. My research emphasizes the comparison of graphs obtained from several text datasets that correspond to viewpoints. If we analyze the differences in viewpoints, our network output must depict the viewpoint of each information source and verify the relationships of corresponding network elements though pairwise comparison.

However, network visualization contains many significant vertices for text data; thus, the analysis cost of comparing and recognizing the differences among networks is high. For instance, the number of vertices and edges of two networks may be similar in number, but they may have unique characteristics, and two networks may also possess some similar vertices and edges. Thus, each vertex and edge must be verified between two networks, and the comparison cost increases when the network is large.

3 Text Comparison on Network Visualization

3.1 Principal Strategy

The principal strategy for the visualization of text comparison is detailed in the steps shown in Fig. 1. First, the data is pre-processed from the information source to determine the attributes of network elements. These attributes are used for final representation. Subsequently, network elements and construct networks are extracted based on each information source. To compare and integrate these networks, we merge the network elements from two networks. Finally, the integrated and compared network is outputted.

3.2 Network Merging and Integration

We assume that the network derived from datasets D1 and D2 are $G_{D1} = \mathrm{G}(V_{D1}, E_{D1})$ and $G_{D2} = \mathrm{G}(V_{D2}, E_{D2})$ and that the integrated network between G_{D1} and G_{D2} is G_I. G_I is given by the integration function "Integrate" as follows:

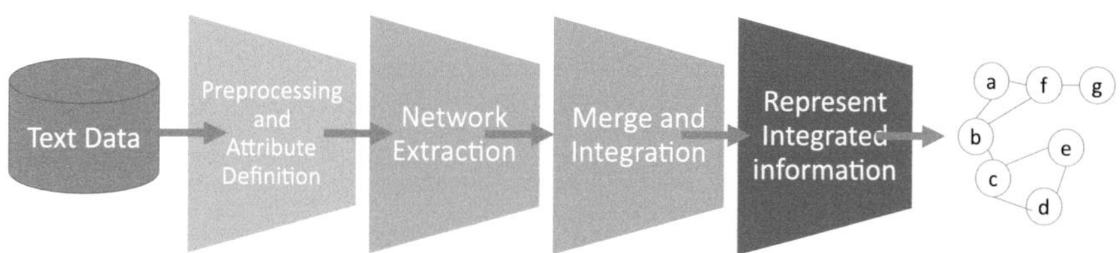

Fig. 1 Principal strategy for integrated graph

$$G_I = \mathrm{G}(V_I, V_E) = \mathrm{Integrate}(G_{D1}, G_{D2}) \tag{1}$$

In this study, the integration function matches the elements of two graphs and sets the method of integration. In general, we can add, subtract, multiply, and divide. When the integration function obtains the maximum elements of a certain attribute, each vertex and edge is applied as follows:

$$\mathrm{Integrate}(G_{D1}, G_{D2}) = \begin{cases} \max(v_{D1}, v_{D2}) \\ max(e_{D1}, e_{D2}) \end{cases} \tag{2}$$

An element may also be observed in one graph but not in another. For example, one graph has a vertex "a," but another does not. Thus, the Eq. (2) does not aggregate elements according to the given function. Therefore, *Set* function compensate for limited elements and can be compared with the integration function; for instance, the Set function is union operation and the compensated graphs of G_{DI} and G_{D2} are denoted by G'_{DI} and G'_{D2}. The vertices and edges are then merged using the following formula:

$$\left(G'_{D1}, G'_{D2}\right) = \mathrm{Set}(G_{D1}, G_{D2}) = \begin{cases} V_{D1} \cup V_{D2} \\ E_{D1} \cup E_{D2} \end{cases} \tag{3}$$

Formula (1) is completely converted into

$$G_I = \mathrm{Integrate}(\mathrm{Set}(G_{D1}, G_{D2})) \tag{1'}$$

We suppose that a graph G_A has three vertices labeled "A," "B," and "C," whereas graph G_B has "B," "C," and "D". Furthermore, G_A has two edges $e_{A \to B}$ and $e_{B \to C}$, and G_B has three edges $e_{B \to C}$, $e_{B \to D}$, and $e_{C \to D}$. Then, the integrated network has four vertices and four edges shown in Fig. 2.

Using the integration function, we can determine a significant amount of information. Assuming that multiple companies enter a single market, we can visualize their documents. Moreover, frequent and attractive keywords regarding this market can be extracted and summed up. We can also extract the gaps between companies by the differences in word frequency. Furthermore, we can determine emergent patterns based on the calculation of edge coefficients by quotient computation.

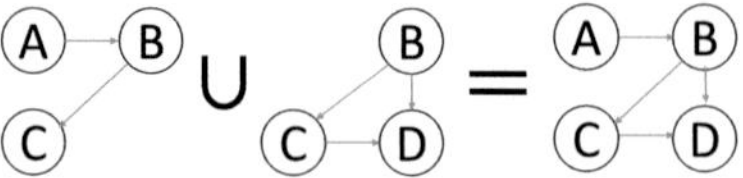

Fig. 2 Example of two graphs by union operation

3.3 Integration Representation

In the final step, the integrated information is presented. Such information can be represented by variables such as color and figure. Color, in particular, is more useful.

4 Case Study

To confirm the feasibility and usability of additions to the previous work [7], a case study is performed on the dataset of amazon.co.jp called "amazon dataset". Data for 365 healthcare goods are collected including 1,746 comments from reviewers (that is, *goods consumers*) with each good having more than two comments. The explanation for each good can be regarded as the use in value, characteristics, and claims from goods providers. Meanwhile, the visualization result can show the gap in evaluation between goods providers and goods consumers.

An improved FACT-Graph [6] is used to visualize these datasets. The FACT-Graph has been developed as a trend visualization method. In this case, the improved FACT-Graph is used for text visualization based on a previous work [8] by implementing integration and set functions. The overview is presented in Fig. 3. For visualizations, the integration and set functions are the maximum function and union operation shown in Eqs. (2) and (3), which apply to the attributes of term frequency, document frequency, and TF-IDF index.

Each vertex and edge is assigned a specific color to differentiate the goods provider from the goods consumer. The color thickens as the value in use between the two groups becomes more different. Figure 4 shows the color map. Blue is assigned to the goods explanation, and red is assigned to the goods consumer (the reviewer comments). The vertex size reflects the term frequency of a keyword, and

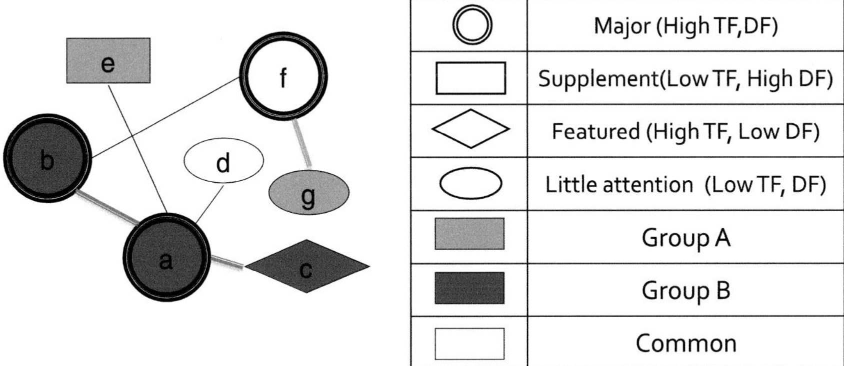

Fig. 3 Interpretation of FACT-Graph for comparison visualization

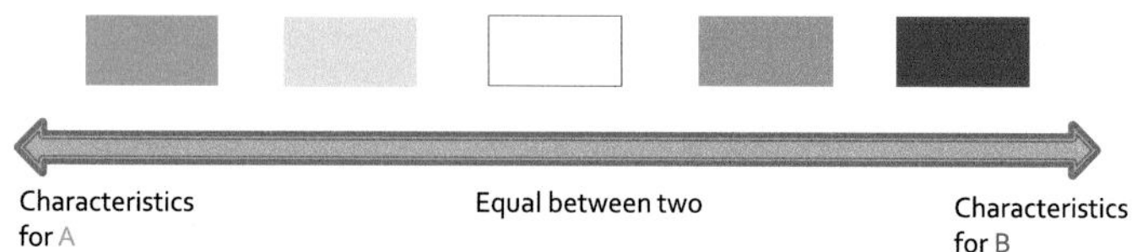

Fig. 4 Color map in visualization

the shape of a vertex reflects the keyword status. The four keyword classes according to TF and DF (in Fig. 3) are used to identify the keyword status. Major keyword class is shown in the graph as a double circle, supplement keyword class as a square, featured class as a diamond, and little attention class as a circle.

Figure 5 shows a part of text comparison visualization of amazon datasets. This figure shows the text of a protein drink. In this figure, "protein" is a white vertex and "soybean" is a blue double circle. However, the soybean vertex connects red edges. Therefore, both the goods provider and the goods consumer place the same value on "protein," whereas the goods provider uses the term "soybean" more frequently than the goods consumer does. The goods consumer also focuses on "soybean" but usually discusses and refers to such a term by using other keywords.

Similar to a figure, the proposed method clearly shows the differences between the goods provider and the goods consumer.

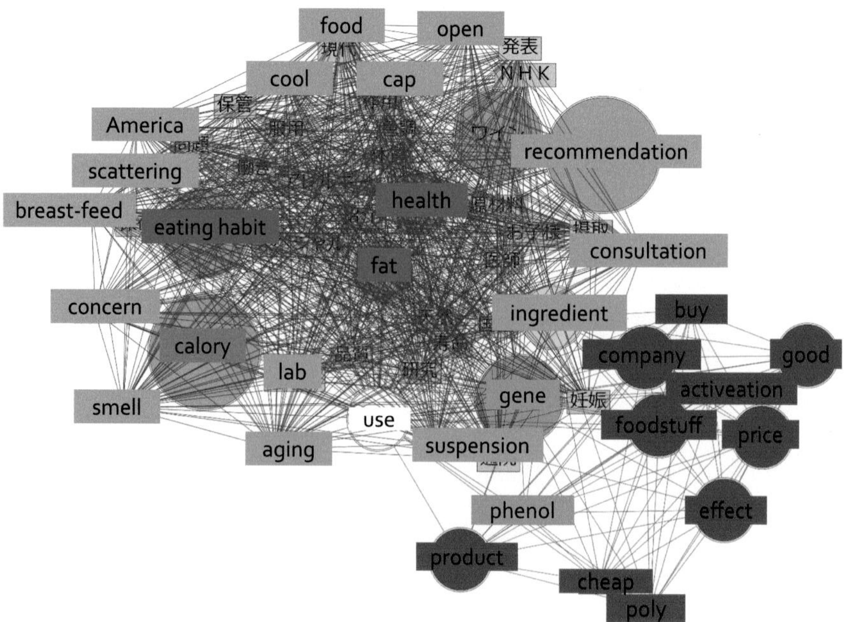

Fig. 5 Text comparison visualization of the amazon dataset between goods explanation and reviewer comments

5 Conclusion

This paper described the comparison of text using network visualization. It also listed the principal strategy applied to compare text datasets as well as discussed the network merging and integration process. A case study using the "amazon dataset" could show the comparison between goods explanation and reviewer comments.

In the future, the proposed method can be extended to multiple text sources, that is, $n = 3$ or more. The analysis environment for integrated networks should also be developed.

References

1. Aoki, S., Toyozumi, K., Tsuji, H.: Visualizing method for data envelopment analysis. SMC **2007**, 474–479 (2007)
2. Bishop, C.M.: Pattern Recognition and Machine Learning. Springer (2010)
3. Blei, D.M., Ng, A.Y., Jordan, M.: Latent Dirichlet allocation. J. Mach. Learn. Res. **3**, 993–1022 (2003)
4. Han, J., Kamber, M., Pei, J.: Data Mining: Concepts and Techniques. Morgan Kaufmann (2011)
5. Manly, B.F.J.: Multivariate Statistical Methods A Primer, pp. 93–106. Chapman & Hall/CRC (1986)
6. Saga, R., Terachi, M., Tsuji, H.: FACT-Graph: trend visualization by frequency and co-occurrence. Electron. Commun. Jpn. **95**(2), 50–58 (2012)
7. Saga, R.: Visualization of comparison of texts from two information sources by network integration. In: 9th International Conference on Knowledge, Information and Creativity Support Systems, pp. 441–445 (2014)
8. Saga, R.: Comparison analysis for text data by integrating two FACT-Graphs. Intell. Interact. Multimedia: Syst. Serv. Smart Innov. Syst. Technol. **14**, 143–151 (2012)
9. Ward, M., Grinstein, G., Keim, D.: Interactive Data Visualization: Foundations, Techniques, and Applications. A K Peters/CRC Press (2010)

Collective Knowledge and Creativity: The Future of Citizen Science in the Humanities

Milena Dobreva

Abstract Citizen science is a contemporary reinvention of some research practices of the past when 'unprofessional' researchers contributed to scientific projects led by academics; a worth-noting peak of research undertaken in this paradigm had been observed in the 19th century. In the 21st century, citizen science mostly resides in digital environments and depends upon eInfrastructures which not only provide citizens with access to research data management, but also play the role of novel scientific communication tools aiming to engage and support citizens in their research contributions. This paper's main purpose is to introduce the concept focusing on citizen science within the Humanities where its use is still limited compared to other research domains, as well as frequently confused with crowdsourcing. We also present some initial outcomes of the user studies undertaken within the EC-funded *Civic Epistemologies* project featuring a set of three international focus groups and a web questionnaire; these help to understand better the current attitudes and challenges in this area. Finally the paper delves into some possible reasons for the slower uptake of citizen science in both the humanities domain and digital cultural heritage and explores to what extent such projects contribute to 'collective knowledge' as well as to creativity.

Keywords Citizen science models · Crowdsourcing · eInfrastructures · Motivation · Activities

1 Introduction

The Green Paper on Citizen Science commissioned by the EC defines citizen science as "general public engagement in scientific research activities when citizens actively contribute to science either with their intellectual effort or surrounding

M. Dobreva (✉)
Library Information and Archive Sciences Department,
Media and Knowledge Science Faculty, University of Malta, Msida 2280, Malta
e-mail: milena.dobreva@um.edu.mt

S. Kunifuji et al. (eds.), *Knowledge, Information and Creativity Support Systems*,
Advances in Intelligent Systems and Computing 416,
DOI 10.1007/978-3-319-27478-2_44

knowledge or with their tools and resources" [12]. Although this term gained popularity recently to reflect on the engagement of 'unprofessional' researchers in scientific inquiry and currently is associated with big groups of such contributors, the practice of involving citizens in research in domains such as astronomy, lexicography and biology was well established in the 19th century; the phenomenon is currently studied in depth within the AHRC-funded project 'Constructing Scientific Communities: Citizen Science in the 19th and 21st Centuries' based in the Universities of Oxford and Leicester [4].

One example of a long-running study integrating citizen science is the Christmas Bird Count [3] which started in 1900 and is still continuing; this effort aims to gather data on amounts and types of birds across different geographic areas and involves volunteer birdwatchers. Yet another wide-ranging effort is the creation of a dictionary of Mediaeval Latin which took 101 years to complete. This project produced seventeen lexicographic volumes the first of which was published in 1975 and the last one in 2013; however the contribution on them launched as early as in 1913 [5].

The advancement of ICT, Internet and mobile technologies opens new prospects for bringing together different communities unified by their interest to contribute to research. This resulted in a rapid growth of the citizen science initiatives around the globe, and subsequently in an increased body of academic publications discussing various aspects of it [8].

The current technological infrastructures facilitate two dimensions of citizen science: *scale* and *substance of tasks performed*. The current social media culture makes it easy to bring together big groups of people but also the modern technology offers mobile devices and a wide range of tools which could engage citizens in a variety of research-related tasks. Thus it is not coming as a surprise that the number of projects experimenting with citizen involvement across various sciences constantly grows. The most typical scenario is the one of citizens directed by professional researchers in studies which revolve mostly around observation of natural phenomena and notation in multiple locations or across longer time spans.

The interest to such projects grew to the extent that specialised platforms which allow to define research tasks and involve users had been created; e.g. Zooniverse [20], Curio [15], and CrowdCrafting [6] developed in a collaboration between the Citizen Cyberscience Centre and the Open Knowledge Foundation. These platforms are used for research in different domains, but mostly in the Sciences with few implementations in the Humanities.

There was also a substantial interest to the potential of citizen science in funding agencies that currently are the main source of funding of such projects. Wiggins and Crowston [21] invited some 840 projects to respond to a survey about citizen science and received 128 responses. They summarised the most popular funding sources of the projects as follows: federal and other grants—68 projects; in-kind contributions—31 projects, private donations—23 projects, participant fees: 11 projects.

The potential of citizen science had been as well addressed in a number of EC-funded projects, e.g. **socientize** (http://www.socientize.eu/) which is working on a white paper on citizen science, and **Civic Epistemologies** (http://www.civic-epistemologies.eu/) that aims to develop a roadmap for citizen science use in the cultural heritage domain.

Since research is one of the most creative human activities, a still under-explored area is to what extent unprofessional researchers are involved in trivial repetitive tasks as opposed to creative activities related to research, and how citizen science could provide a creativity outlet for the members of the wide public. It is also not completely clear what different aspects of collective knowledge are most prominent in citizen science. As an initial study related to this question we decided to explore how citizen science is used in the research in the Humanities. The paper is organised as follows. In Sect. 2 we provide some examples of citizen science and crowdsourcing initiatives in the Humanities. While crowdsourcing is not necessarily aimed at research outcomes, it is a form of mass public involvement which is well established in the cultural heritage sector and could be used as to illustrate the logistics for citizen science initiatives. Section 3 presents some of the findings of initial user studies implemented within the Civic Epistemologies project. Finally, Sect. 4 discusses aspects of creativity, collective knowledge creation and areas for further research.

2 Citizen Science in the Humanities

2.1 Crowdsourcing Use in the Humanities

The application of Citizen Science in the field of Humanities has been less common than in the sciences, but some examples of projects of the British Library [9] and the Welsh National Library [7] for example illustrate the potential of this approach.

Johan Oomen and Lora Aroyo [18] list several examples where amateur researchers and labourers contributed to process or gather data. They highlight six different types of crowdsourcing related to digital cultural heritage.

The first is **correction and transcription** where the citizen is granted access to a database of texts, generally scanned manuscripts, and asked to transcribe or make correction to text which was already transcribed electronically via a computer programme. **Contextualization** happens when citizens submit data such as letters, photographs, stories or other materials in order to gather a meaningful context. Submitting data in databases with the aim of completing them or making them more sound is instead known as **complimenting collection**. **Classification** is the practice of tagging the data, or labelling it, in order to easily group similar data or locate relevant information in a short period of time. **Co-curation** seems to occur mostly with projects involving the aesthetic arts and allows the citizens to interact with institutions regarding selection activities for publication. Lastly there is **crowd-funding** where the citizens gather together money and resources in order to support efforts initiated by others. Recently, Noordegraaf et al. suggested a different model for crowdsourcing in the cultural heritage context which explores six pillars: institution, collection, goal, crowd, infrastructure, and evaluation [17].

Crowdsourcing is not necessarily aimed at research activities but is familiar to many cultural heritage institutions and can be used to explain how citizen science projects can be organised in real life practice.

2.2 Some Project Examples

An example of a recent humanities centred project based on data gathered via citizen science is a project aimed at bringing together personal correspondence "Letters of 1916" [16]. It uses a website interface to gather and provide access to letters to/from Irishmen submitted from people all around the world. People can also opt to translate submitted letters. This project has helped unveil various details surrounding the lifestyle of people back in those days, thus creating a new intimate perspective of the early twentieth century. One interesting feature of this project is that it develops a hybrid collection featuring letters belonging to the collections of cultural heritage institutions and letters belonging to personal archives.

The British Library also experiments in the crowdsourcing domain with its project "Georeferencing: help us place our digitized maps" [11] where citizens are encouraged to help the British library identify their historic maps and their modern day location. Helpful users are cited and thanked. These examples as well as the aforementioned forms that citizen science can take show that these are mainly data-focused projects and that citizens are not really taking part in other stages of the research process, such as formulating research questions, choosing methodology and discussing the results.

2.3 Creativity, Citizens and Crowdsourcing in the Humanities

The range of activities to which unprofessional researchers contribute in citizen-science projects as suggested in [21] includes the following:

1. Define question
2. Gather information
3. Develop hypothesis
4. Design study
5. Data collection
6. Analyse sample
7. Analyse data
8. Interpret data
9. Draw conclusions
10. Disseminate results
11. Discuss results and ask new questions

Those activities assume different levels of creativity. The tasks of transcribing historical letters or providing geolocations would normally be considered to be quite trivial and are from the contributive type of citizen involvement as defined in [1]. Thus one research question for the future is how citizens involved in Humanities research could contribute to creative rather than trivial tasks? Furthermore, it is essential to understand what is the motivation of citizens to contribute to such projects. Some initial research on the motivation in citizen science projects in biodiversity had been done in [19] but studies in the Humanities-related citizen science initiatives are still lacking.

3 Some Initial Findings from the Civic Epistemologies Project

The Civic Epistemologies project started in August 2014 and applied a mixed methods approach to understand the different demands and expectations of citizens and stakeholders from the citizen science domain (cultural institutions, academic institutions, activist organisations, infrastructure providers):

- Existing body of knowledge (the project team studied existing publications and examples of projects from the domains of Humanities and Arts—which could later be adopted as best practice examples);
- User studies conducted within the project. The methodology adopted was mixed methods combining expert consultations within the project consortium with focus groups aiming to capture the opinions of different stakeholders/users (policy makers with a focus group held in Malta; citizen activist organisations with a focus group held in Sweden, and citizen scholars with a focus group held in Spain). The project also conducted a web questionnaire study which gathered 85 responses mostly from European countries.

Some of the findings from these studies indicate that:

- There is still a noticeable confusion of citizen science and crowdsourcing within the cultural heritage institutions.
- There is a difference in the major motivation of researchers for using citizen science in the Humanities and in the Sciences. In the Humanities, according to the Civic Epistemologies online survey both citizen science and crowdsourcing are seen as useful tools mostly to expand the institutional knowledge on a certain topic and to facilitate the progress of existing research (see Fig. 1). The potential to save time are not that clearly stated in the Humanities while in the sciences there is evidence on substantial contribution to saving researcher's time:

  > For example, in the Planet Hunters project originated in December 2010, participants look for transits in data from NASA's Kepler satellite. In May 2011, five months after launch, Planet Hunters was still attracting activity that amounts to the equivalent of 51 Full Time Equivalent staff. [2]

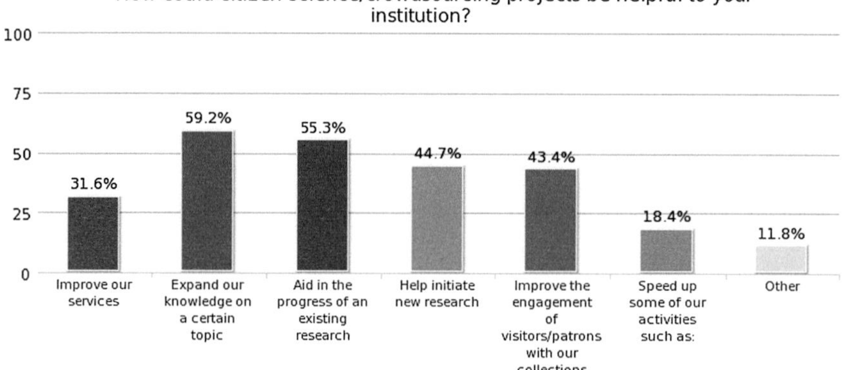

Fig. 1 Ranking of benefits from citizen science/crowdsourcing according to the online questionnaire of Civic Epistemologies

- The practical experience in implementing citizen science projects shows that about 25 % of the respondents in the web survey and 40 % of the focus group participants had contributed to citizen science and/or crowdsourcing initiatives. These should not be taken as indicative across the sector because the decision to respond to the survey or to take part in focus groups was depending on the personal interest of contributors to the theme.

The detailed outcomes of the user studies will be made available for consultation on the project website, http://www.civic-epistemologies.eu.

4 Discussion and Future Work

4.1 Citizen Science: Between Collective Knowledge and Creativity

The notion of 'collective knowledge' is one which can be analysed in the context of citizen science initiatives. By their very nature, having multiple contributors to research tasks, but not bound to a formal organization, makes of citizen science an interesting case. Recently, Achim Hecker suggested three conceptualisations of collective knowledge [13]:

- Collective knowledge as shared knowledge, where knowledge is "held by each member of a collective" [13, p. 430]
- Collective knowledge as complementary knowledge, where knowledge is "constituted by the complementary interaction of distributed individual knowledge within a collective" and

- Collective knowledge as knowledge embedded in collective artefacts, where knowledge is "incorporated in collective artifacts".

The nature of citizen science might make one expect that they would tap into the third type of conceptualization due to the outcome of such initiatives which could be seen as artifacts. However the contribution of multiple citizens happens in serendipitous manner would mean that collective knowledge in this case is a mixture of the three types of conceptualisations.

Understanding better the specific features of collective knowledge within the citizen science context would allow for better organization of such initiatives and constitutes and interesting area for further research. Further aspect of it is to understand how such contributions of multiple actors could form a solid basis not only to codify collective knowledge, but also to boost creativity.

4.2 *The Lower Use of Citizen Science in the Humanities*

While citizen science grows in popularity in general, the majority of citizen science oriented projects take place in scientific areas. For example only one out of 47 projects surveyed by Franzoni and Sauermann [10] was in the area of the Humanities (Archeology) and another single project was in the area of arts (Music). Similarly, only 4 % of some 270 projects within the CrowdCrafting platform are in the Humanities (see Fig. 2).

Why citizen science is underutilised in the Humanities when it is a profitable and inexpensive means of expanding research? The currently ongoing project Civic Epistemologies is striving to develop a roadmap for the use of e-Infrastructures to aid in the inclusion of citizens in research related to cultural heritage and digital humanities.

While citizen science has not been taken into account when determining the value and impact of digital resources even in best practice projects [14] we can expect that in the future it will play such a role.

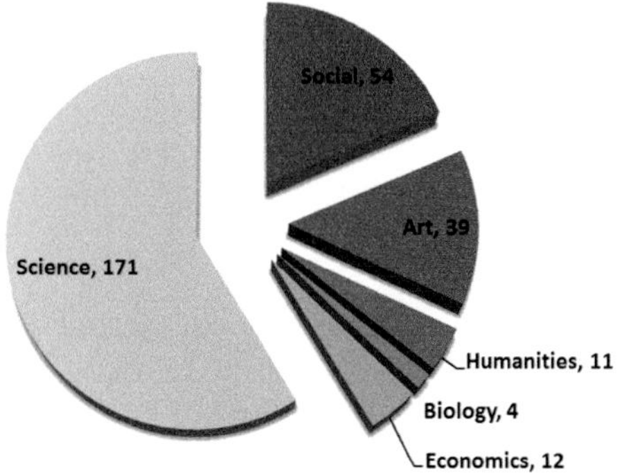

Fig. 2 Distribution of citizen science projects which are making use of the CroudCrafting platform as of December 2014

Another domain which can be explored further is related to the use of citizen science as an approach to use more intensively eInfrastructures offering access to cultural heritage content for educational purposes and enhancing skills of the citizens. This is an area addressed by some of the research activities of the Advanced Computing for Innovation (ACoMin) project.

We could expect that in the near future we will have better understanding how Humanities research could benefit more from citizens' contribution, and to what extent creative tasks constitute part of the motivation of citizen researchers.

Acknowledgments This research is supported by the EC-funded projects Civic Epistemologies funded under FP7 grant agreement 632694 and AComIn "Advanced Computing for Innovation", Grant 316087, funded by the FP7 (Research Potential of Convergence Regions).

References

1. Bonne, R., et al.: Public participation in scientific research: defining the field and assessing its potential for informal science education. A CAISE Inquiry Group Report. http://files.eric.ed.gov/fulltext/ED519688.pdf (2009)
2. Christian, C., Lintott, C., Smith, A., Fortson, L., Bamford, S.: Citizen Science: Contributions to Astronomy Research. arXiv:1202/1202.2577 (2012)
3. Christmas Bird Count. http://birds.audubon.org/christmas-bird-count
4. Constructing Scientific Communities: Citizen Science in the 19th and 21st Centuries. http://conscicom.org/
5. Coughlan, S.: Dictionary reaches final definition after century. http://www.bbc.co.uk/news/education-28952646 (2014)
6. CrowdCrafting. http://blog.okfn.org/2013/09/17/crowdcrafting-putting-citizens-in-control-of-citizen-science (2013)
7. Dafis, L.L., Hughes, L.: What's the Welsh for "Crowdsourcing"? Citizen science and community engagement at the National Library of Wales. In: Ridge, M. (ed.) Crowdsourcing our Cultural Heritage. Ashgate (2014)
8. Dobreva, M., Azzopardi, D.: Citizen science in the humanities: a promise for creativity. In: Papadopoulos, G. (ed.) Proceedings of the 9th International Conference KICSS, Limassol, Cyprus, 6–8 Nov 2014, pp. 446–451 (2014)
9. Ellis, S.: A history of collaboration, a future in crowdsourcing: positive impacts of cooperation on British librarianship. Libri **64**, 1–10 (2014)
10. Franzoni, C., Sauermann, H.: Crowd science: the organization of scientific research in open collaborative projects. Res. Policy **43**(1), 1–20 (2014)
11. Georeferencer Project. http://www.bl.uk/maps/
12. Green Paper on Citizen Science: European Commission. http://www.socientize.eu/sites/default/files/Green%20Paper%20on%20Citizen%20Science%202013.pdf (2013)
13. Hecker, A.: Knowledge beyond the individual? Making sense of a notion of collective knowledge in organization theory. Organ. Stud. **33**, 423–445 (2012)
14. Hughes, L., Ell, P., Dobreva, M., Knight, G.: Assessing and measuring impact of a digital collection in the humanities: LLC. J. Digit. Scholarsh. Humanities (2013)
15. Law, E., Dalton, C., Merrill, N., Young, A., Gajos, K.: Curio: A platform for supporting mixed-expertise crowdsourcing. Human computation and crowdsourcing. In: Works in Progress and Demonstration Abstracts. AAAI Technical Report CR-13-01. An Adjunct to the Proceedings of the First AAAI Conference on Human Computation and Crowdsourcing, pp. 99–100 (2013)

16. Letters of 1916. Creating history. http://dh.tcd.ie/letters1916/
17. Noordegraaf, J., Bartholomew A., Eveleigh, A.: Modeling crowdsourcing for cultural heritage. *Museums and the Web 2014*, In: Proctor, N., Cherry, R. (eds.) Museums and the Web. http://mw2014.museumsandtheweb.com/paper/modeling-crowdsourcing-for-cultural-heritage/ (2014)
18. Oomen, J., Aroyo, L.: Crowdsourcing in the cultural heritage domain: opportunities and challenges. In: Proceedings of the 5th International Conference on Communities and Technologies (C&T'11), pp. 138–149. ACM, New York, NY, USA (2011)
19. Rotman, D., et al.: Dynamic changes in motivation in collaborative citizen-science projects. In: Proceedings of the ACM 2012 conference on Computer Supported Cooperative Work (CSCW'12), pp. 217–226. ACM, New York, NY, USA (2012)
20. Smith, A.M., Lynn, S., Lintott, C.J.: An introduction to the Zooniverse. Crowdsourcing: Works in Progress and Demonstration Abstracts. AAAI Technical Report CR-13-01. (2013)
21. Wiggins, A., Crowston, K.: Describing Public Participation in Scientific Research, iConference Toronto. http://crowston.syr.edu/system/files/iConference2012.pdf (2012)

Aligning Performance Assessments with Standards: A Practical Framework for Improving Student Achievement in Vocational Education

Metwaly Mabed and Thomas Köhler

Abstract The academic achievement of students is a top concern of educators and teachers given the evidence for its links to future career opportunities, especially in vocational education field. Therefore, the aim of this study was twofold. The first purpose was to identify the learning performance that reflects students' assimilation of the electrical engineering content. The second purpose was to describe the construction of a reliable and valid academic achievement test designed to measure students' learning performance. To achieve these goals, the researchers developed the academic achievement test in several phases by following the systematic approach.

Keywords Academic achievement · Learning assessment · Learning outcome · Knowledge measurement · Tests methodology · Vocational education

1 Introduction

There is clear link between learning performance and future career opportunities for students, especially in vocational education fields. Since learning outcome may reflect human abilities, competencies, and many other psychological attributes such as interest and attitude, it is the most common concept that has received more attention from educational organizations from primary schools to high schools. Learning assessment aims to provide objective evidence that represents the truth about learning level through an achievement test as a learning performance indicator. Academic achievement is often used to refer to the knowledge that is

M. Mabed
Faculty of Education, Suez University, Suez, Egypt
e-mail: metwalys@gmail.com

T. Köhler (✉)
Faculty of Education, Dresden University of Technology, Dresden, Germany
e-mail: thomas.koehler@tu-dresden.de

S. Kunifuji et al. (eds.), *Knowledge, Information and Creativity Support Systems*,
Advances in Intelligent Systems and Computing 416,
DOI 10.1007/978-3-319-27478-2_45

obtained by a student through a school program or curriculum. According to many researchers, academic achievement is defined as "the competence of a person in relation to a domain of knowledge" or the proficiency of students' performance in a certain course [2].

Given the importance of this topic in an educational context, the process of exploring, assessing, and evaluating academic achievement by a particular test reveals itself as being critical. This is one of the reasons why achievement tests are a top concern in education. The American Educational Research Association, American Psychological Association, and National Council on Measurement in Education pointed out that academic achievement tests "are measures of academic knowledge and skills that a person acquired in formal and informal learning opportunities" (p. 124) [3]. Another definition of academic achievement testing can be found in the Dictionary of Education, where it is defined as "a test that measures the extent to which a person has acquired certain information or mastered certain skills, usually as a result of planned instruction or training" (p. 3) [21]. Consequently, academic achievement tests tend to measure recent learning performance and are closely tied to particular subjects or courses [4]. However, the test of academic achievement may serve three major functions [23]:

- measuring the effectiveness of student performance at a particular point of time and educational program;
- predicting the behavior of a student across different situations; and
- assessing various psychological traits and characteristics.

Moreover, the information collected by the achievement test supports decisions concerning both the placement or the diagnostic issues during the instruction process. Furthermore, there is a wide range of reasons for using academic achievement tests, among others [22] to: (a) identify where the student falls along continuum of knowledge acquisition; (b) classify students in groups according to their scores at the end of the course; (c) determine eligibility for particular educational programs; and (d) measure the effectiveness of the instruction process. Whether the achievement test is a norm- or criterion-referenced measure, the core issue is that its results should represent accurately the evidence concerning student performance in order to enhance the learning process as a central goal in any educational organization.

Indeed, in vocational education the precise specification of what to measure is poorly understood. The development of achievement tests to assess student learning performance is more beneficial for research and practice. Although many researchers are interested in measuring achievement [5, 7, 8, 12, 18–20], no existing measure was found to be adequate to deal with the electrical engineering content in this study. The necessity for the construction of such a test is related to the influence of the achievement factor in many aspects of the students' professional skills in vocational secondary school. Therefore, the construction and development of an appropriate test to measure students' achievement will enhance the quality of learning outcomes.

2 Development Phases of the Achievement Test

2.1 Clarifying the Purpose of the Test

The starting point for creating any instrument such as an achievement test is to identify its psychometric purpose. The test development should be grounded on the purpose for which the test scores will be used [9]. In this study, the main function of the academic achievement test is to assess student performance in electrical engineering. Additionally, the student score provided more information about a student's performance concerning different instructional methods.

2.2 Identifying the Educational Objectives

During the initial preparation of the achievement test, the authors were guided by a fundamental issue to construct an adequacy academic achievement test. This issue focused on the idea that the test items should be formulated to represent accurately the knowledge delivered by the electrical engineering course. The best elements which describe the electrical engineering topics clearly are the educational objectives, which are defined as something that "specifies an observable behavior that students should be able to exhibit after completion of a course instruction" (p. 68) [9]. Accordingly, the course outline, the textbook, and the curriculum guide were reviewed to identify the educational objectives in the course. As a result, a checklist of the potential educational objectives in electrical engineering was the output of this phase. The objectives in the checklist were involved in the four categories of Bloom's taxonomy [6] i.e., knowledge, comprehension, application, and analysis.

2.3 Performing Panel of Experts

The first draft of the educational objectives checklist was distributed to some experts to judge the face validity of the electrical engineering objectives. Subsequently, three subject matter experts were asked to complete a questionnaire in order to check the educational objectives that should be covered by the electrical engineering course and the academic achievement test as well. Based on the previous identification process, 76 objectives were revealed in five main topics in the electrical engineering course as the following: principles of the alternative current (AC) generation, single circuits of AC, series circuits of AC, parallel circuits of AC, and resonance circuits.

2.4 Developing Test Blueprint

Implementing the test accurately should also be based on the electrical engineering topics which form a table of item specifications. Therefore, the next phase was to develop a table of specifications or a blueprint for the test, which was considered to be one useful indicator that takes into account both the educational content and process [9, 16]. In the same line, standard 3.3 [3] provided a comprehensive description of this element: "the test specifications should define the content of the test, the proposed number of items, the item formats, the desired psychometric properties of the items, and the item and the section arrangement" (p. 43).

Consequently, the numbers of the educational objectives that were identified in the earlier step were assigned in a table based on the corresponding content areas and taxonomy levels. Afterward, the relative weights for each topic concerning the content area and taxonomy level were determined. The percentage values are based upon the importance of each topic, the number of pages for each topic in the textbook, and the percentage of the educational objectives.

Sixty items were suggested to be appropriate to involve in the academic achievement test. Subsequently, the number of items for each content area, as viewed in Table 1, was calculated from the results of the relative weights as well as the number of total items (60 items) in the academic achievement test.

2.5 Determining and Generating Test Questions

Since the essential aspect in developing the academic achievement test is determining the item format, the multiple-choice format is employed widely to assess student performances. This format consists of two parts. The first part is the stem sentence which presents a problem or a question. The second part is formed of four or more alternative responses, one of which is more suitable to answer the stem part

Table 1 Test blueprint for electrical engineering course

Content	Taxonomy level				Total
	KN	CO	AP	AN	
Principles of AC generation	5	4	4	1	14
Single AC circuits	3	3	4	3	13
Series AC circuits	2	3	5	2	12
Parallel AC circuits	2	2	6	3	13
Resonance circuits	3	3	2	–	8
Total	15	15	21	9	60

[a] *KN* Knowledge; *CO* Comprehension; *AP* Application; *AN* Analysis

than the others. In addition, multiple-choice format is the most frequently used in tests for its advantages; for example, this format [16]:

- offers more flexibility for assessing diversity of content processes than other item formats;
- allows for the evaluating of complex mental skills, ones that would be high up on the list of cognitive abilities, such as higher order thinking skills;
- allows for precise interpretation, which can lead to important evidence for content-related test validity; and
- can be simply used and objectively scored, especially when compared to other item formats such as essays.

Although these benefits are gained from using multiple-choice (MC) items, the preparation and writing the MC questions require professional skills. Furthermore, this format cannot assess student ability to organize, generate, and formulate ideas. [17]. Therefore, 76 MC items were generated for the trial run. The researchers took into account that the items should be congruent with the instructional objectives as a paramount concern during question production.

2.6 *Preparing Test Instructions*

The next step of the creation of the achievement test was focused on producing the test directions or instructions. One of the standards for creating tests [3] stated that "the instructions presented to test takers should contain sufficient detail so that test takers can respond to a task in the manner that the test developer intended" (p. 47). Therefore, to ensure that the students understand what is required from them, the directions of the academic achievement test appeared on the first page of the test. These directions clarified the test purpose, the test content, the instructions concerning responses, and an example of the correct way to record the answers.

2.7 *Performing Panel of Experts*

Important parts of test preparation which should be checked are the items' accuracy, appropriateness of relevance to educational objective, and level of readability [9]. Thus, a print copy from the initial test items and educational objectives was distributed to a panel of experts consisting of four experts. Each expert was asked to evaluate each item according to its clarity for the reader and the relevance to the course objectives, as well as rate of them from (1) to (5) to mark the quality. The average rating of the four specialists concerning each item in the test was computed. The researchers made the decision to remove any item from the test when it had an average score of (1) or (2). Whereas the average score of (3) indicated that the item should be accepted with minor changes, the average score of (4) or (5) suggested that the item should be accepted without revision. By applying these criteria, two items were changed and six items were removed from the test. As a result, the number of the items in the achievement test decreased to 70 items.

2.8 Conducting the Pilot Study

The aim of this step was to gather data to determine the validity and reliability of the test. For that reason, the pilot study of academic achievement test was administered at the end of the second semester in the academic year 2009/2010. The achievement test was given in a paper and pencil format to a total number of 87 students in Tanta Electrical Secondary School in Egypt. After eliminating the incomplete responses, the usable subject population was 69 students with an overall response rate of 79 %.

3 Findings

3.1 Test Validity

The researchers were keen to utilize a valid achievement test. Therefore, it was central to check the validity of academic achievement test. However, test validity is defined as "the process of collecting evidence to establish that the inferences, which are based on the test results, are appropriate" (p. 21) [1]. From this meaning, the validity term is related to the scores which are obtained from applying the test more than the test items themselves.

The process of assessing validity includes many actions, such as asking test takers and experts about their subjective opinions on the content of the test items, as well as using factor analysis to evaluate the internal item structure [14]. In this study, the principle factor analysis with a varimax rotation procedure was employed to calculate the factor loadings and eigenvalues by using the Predictive Analytics SoftWare (PASW), version 18.0. Two criteria were determined to achieve satisfactory validity: (a) the factor loadings should exceeded the minimum threshold of 0.5 as recommended by [13]; and (b) the items with multiple loadings should record higher load on their related level than the load value on other levels [11]. The results revealed that the factor loadings of five items were less than the cut-off of 0.5. Moreover, four items were loaded greater onto another level than their corresponding level. In summary, nine items out of the 70 items were removed from the achievement test. Then, the factor analysis with varimax rotation was run again over the dataset of 61 questions.

The results showed that the factor loading value of each item was at 0.5 or above and there was no cross loadings in between the test items. Accordingly, all

Table 2 Results of total variance explained by academic achievement test

	Components			
	1	2	3	4
Initial Eigenvalues				
Total	20.89	9.86	5.00	3.22
% of variance	34.26	16.17	8.19	5.28
Cumulative %	34.26	50.42	58.62	63.89
Extraction sums of squared loadings				
Total	20.89	9.86	5.00	3.22
% of variance	34.26	16.17	8.19	5.28
Cumulative %	34.26	50.42	58.62	63.89
Rotation sums of squared loadings				
Total	13.17	11.45	7.26	7.09
% of variance	21.59	18.78	11.90	11.62
Cumulative %	21.59	40.37	52.28	63.89

eigenvalues were more that the cut-off value of one. Table 2 presents the total variance explained by each level of the academic achievement test. Moreover, it was yielded that the four levels accounted for 63.98 % of the total variance.

3.2 *Test Reliability*

The reliability is another indicator calculated during the construction of the achievement test. While the test is defined as a scale describing student behavior in a specified domain, the test reliability refers to the stability of this scale when the testing procedure is repeated on a population of individuals or groups [3]. The value of the reliability ranges from zero to one. Whereas the value of one indicates the test has perfect reliability, zero value refers to no reliability. Cronbach's alpha was higher than the restrictive criterion of 0.7 [10], with 0.966.

Since the academic achievement test consisted of four levels (knowledge, comprehension, application, and analysis), it is not sufficient to compute the reliability only for the entire test when the score of a sub-level will be used. According to the standard 2.1 [3], "For each total score, sub-score, or combination of scores that is to be interpreted, estimates of relevant reliabilities and standard errors of measurement or test information functions should be reported". Thus, the reliability for each level in the test was also reported, as shown in Table 3. As a result, the items of the four levels in the achievement test presented satisfactory reliability criteria.

Table 3 Results of knowledge level reliability on the academic achievement test

Test level	No. of items	Alpha	Mean	Variance	SD
Knowledge	15	0.967	11.42	28.22	5.31
Comprehension	15	0.939	9.71	26.71	5.17
Application	22	0.958	14.68	56.43	7.51
Analysis	9	0.936	4.54	13.14	3.62

3.3 *Item Analysis*

The level of item difficulty can be used to determine if the item is useful or not. The difficulty level of an item refers to the proportion of students answering an item correctly. The item difficulty index is calculated by dividing the number of students who answered the question correctly by the total number of students who answered the question [24]. Furthermore, the difficulty index can range from zero to one. While a zero value indicates that no student answered the item correctly, a one value indicates that all students answered the item correctly. In the present study, the item was considered difficult when the difficulty value was less than 0.25. Also, the item was considered to be easy when the difficulty value was greater than 0.80. The results illustrated that the test items provided an acceptable difficulty value range from 0.30 to 0.78.

The item discrimination provides a suggestion of the degree to which an item correctly differentiates among the test takers on a certain domain [24]. The steps of calculating the discrimination index were described by [14]. First, students who have the highest and lowest overall test scores are grouped into two groups: the upper group, which is made up of the 25–33 % who have the highest overall test scores, and the lower group, which is made up of the bottom 25–33 % who have the lowest overall test scores. Afterward, the value of each item for the upper and lower groups is computed. Finally, the item discrimination index is provided simply by calculating the difference between the p-values of the two groups, the upper group and lower group.

The overall test scores were arranged in descending order. Subsequently, the highest 19 students (27 % of the total number 69 of students) were included in the upper group, while the lowest 19 student (27 % of the total number 69 of students) were selected in the lower group. The item discrimination indices were calculated. The items were categorized concerning their discrimination indices according to recommended criteria [15]. Any item with an index of discrimination of 0.40 and up was considered excellent, and when the value ranged from 0.3 to 0.39 as a good item. Moreover, any item with a discrimination value of 0.2–0.29 was considered to be acceptable, whereas the ratio from zero to 0.19 refers to a test item that should be revised. An item with negative discrimination index value should be removed from the academic achievement test. Fortunately, the findings reported that all discrimination indices for the test items were positive values. Moreover, the results showed that the test items provided sufficient discrimination value ranging from good to excellent items.

Based on the results of item analysis, 61 items were found to be valid to evaluate student achievement in electrical engineering. On the other hand, the total number of test items in the table of specification was 60 items. Subsequently, the number of items for each content area was calculated from the results of the relative weights as well as the number of the total items (60 items) in the achievement test. A complete copy of the achievement test would exceed the length of this paper but can be received in either Arab or English language from the authors.

4 Conclusion and Limitation

The present study addressed the construction and validation of an achievement test to measure a student's learning performance in electrical engineering. The development of the achievement test included a rigorous process of planning, creating, and evaluating. However, careful attention must be paid to the purpose of the test, because the usefulness of test interpretations depend on the agreement with which the purpose and the domain represented by the test have been explicated [3]. Therefore, the purpose, the educational content, and the target population of the academic achievement test were identified. This test aimed to assess a student's learning performance in electrical engineering at a vocational secondary school and was implemented in Egypt.

Ideally, one or more actions may be applied during development of an instrument, such as content analysis, review of the literature, critical incidents, direct observations, expert judgment, and instruction objectives [9]. The researchers took into account several activities to overcome the problems which might be inherent in a test. For that reason, it is recommended that the test items reflect the knowledge directly involved in the electrical engineering course. Therefore, the content of the electrical engineering course was analyzed to identify the educational objectives. Expert review was employed to test face and content validity. The output of this stage was 76 objectives reflecting five main topics in electrical engineering. To increase the content validity of the test, the importance of each topic and the number of pages were determined while preparing the blueprint of the academic achievement test.

A 76 item list was generated to assess student learning performance in electrical engineering. The academic achievement test items covered four taxonomy levels, which are knowledge, comprehension, application, and analysis. Presenting a clear, simple, and concise direction are an integral part of well-constructed test items [16]. Therefore, the first draft of the academic achievement test and the directions for the test takers were given to a panel of experts to judge the appropriateness of them. The pilot study was conducted to collect evidence which determine the validity and reliability. The characteristics of the pilot study sample were similar to the group which the final test was intended for. The results from the pilot study suggested that the academic achievement test demonstrated acceptable reliability across the test. The alpha coefficient for the whole test and all the sub-levels was reasonable.

The academic achievement test showed evidence of discriminant and convergent validity. The principle component's factor analysis as the extraction technique and a varimax rotation as the orthogonal rotation method procedure were executed by using the statistical program PASW. The results of factor analysis revealed that the application level explained the higher proportion of the variance, while the comprehension level accounted for the lower percent of the variance. Moreover, the academic achievement test had a good item difficulty, which fell within the range of 0.30–0.78. Also, the discriminative index of the academic achievement test items ranged from 0.41 to 0.79. Since the academic achievement test items were designed to reflect predetermined objectives, this variation in value for the items would be expected. In sum, the findings indicated that the 60 item test holds promise as a valid and reliable academic achievement test to measure students' learning performance in electrical engineering. A complete copy of the achievement test would exceed the length of this paper but can be received in either Arab or English language from the authors.

The construction and development of the academic achievement test in the present study should be considered in the light of a few limitations. The academic achievement test is relative, i.e. limited to electrical engineering contents in the vocational education domain. Moreover, although the research sample of the pilot study was appropriate for the correlation matrix and factor analysis, it was relatively small. Future research utilizing a larger sample size is necessary to sufficiently provide evidence regarding the validity of the test.

Acknowledgments This research is funded by the Ministry of Higher Education in Egypt and the Free State of Saxony/Germany.

References

1. Adams, J.: The language of assessment, 2nd edn. In: McDonald, M. (ed.) The Nurse Educator's Guide to Assessing Learning Outcomes, pp. 9–26. Jones and Bartlett, Sudbury, MA (2007)
2. Algarabel, S., Dasi, C.: The definition of achievement and the construction of tests for its measurement: a review of the main trends. Psicologica **22**(1), 43–66 (2001)
3. American Educational Research Association (AERA), American Psychological Association (APA), & National Council on Measurement in Education (NCME): The standards for educational and psychological testing. AERA, Washington, DC (1999)
4. Ariyo, A.O.: Construction and validation of a general science aptitude test (GSAT) for Nigerian junior secondary school graduates. Ilorin J. Educ. **27**, 20–29 (2007)
5. Bayrak, B., Bayram, H.: The effect of computer aided teaching method on the students' academic achievement in the science and technology course. Procedia – Soc. Behav. Sci. **9**, 235–238 (2010)
6. Bloom, B.S. (ed.). Englehard, M.D., Furst, E.J., Hill, W.H., Krathwohl, D.R.: Taxonomy of Educational Objectives: The Classification of Educational Goals, Handbook I, Cognitive Domain. David McKay, New York (1956)

7. Carle, A., Jaffee, D., Miller, D.: Engaging college science students and changing academic achievement with technology: a quasi-experimental preliminary investigation. Comput. Educ. **52**(2), 376–380 (2009)
8. Choy, J.L., O'Grady, G., Rotgans, J.I.: Is the study process questionnaire (SPQ) a good predictor of academic achievement? Examining the mediating role of achievement-related classroom behaviours. Instr. Sci. **40**(1), 159–172 (2012)
9. Crocker, L., Algina, J.: Introduction to classical and modern test theory. Holt, Rinehart and Winston, New York (1986)
10. Fornell, C., Larcker, D.F.: Evaluating structural equation models with unobservable variables and measurement error. J. Mark. Res. **18**(1), 39–50 (1981)
11. Gefen, D., Straub, D.W.: A Practical guide to factorial validity using PLS-Graph: Tutorial and annotated example. Commun. Assoc. Inf. Syst. **16**(5), 91–109 (2005)
12. Haislett, J., Hafer, A.A.: Predicting success of engineering students during the freshman year. Career Dev. Q. **39**(1), 86–95 (1990)
13. Hulland, J.: Use of partial least squares (PLS) in strategic management research: A review of four recent studies. Strateg. Manage. J. **20**(2), 195–204 (1999)
14. Kline, T.: Psychological testing: A practical approach to design and evaluation. Sag Publications, Inc., Thousand Oaks, California (2005)
15. Mitra, N.K., Nagaraja, H.S., Ponnudurai, G., Judson, J.P.: The levels of difficulty and discrimination indices in type a multiple choice questions of pre-clinical semester 1 multidisciplinary summative tests. Int. E-J. Sci. Med. Educ. **3**(1), 2–7 (2009)
16. Osterlind, S.J.: Constructing test items: multiple-choice, constructed-response, performance, and other formats, 2nd edn. Kluwer Academic Publishers, New York (1998)
17. Quellmalz, E., Hoskyn, J.: Classroom assessment of reasoning strategies. In: Phye, G.D. (ed.) Handbook of Classroom Assessment, pp. 103–130. Academic Press, San Diego, CA (1995)
18. Rivkin, S.G., Hanushek, E.A., Kain, J.F.: Teachers, Schools, and Academic achievement. Econometrica **73**, 417–458 (2005)
19. Romano, J., Wallace, T., Helmick, I., Carey, L., Adkins, L.: Study procrastination, achievement, and academic motivation in web-based and blended distance learning. Internet High. Educ. **8**(4), 299–305 (2005)
20. Selçuk, G., Sahin, M., Açıkgöz, K.: The effects of learning strategy instruction on achievement, attitude, and achievement motivation in a physics course. Res. Sci. Educ. **41** (1), 39–62 (2011)
21. Shukla, R.: Dictionary of Education. APH Publishing, New Delhi (2005)
22. Smith, D.R.: Wechsler individual achievement test. In: Andrews, J.J., Saklofske, D.H., Janzen, H.L. (eds.) Handbook of Psychoeducational Assessment: Ability, Achievement, and Behavior in Children, pp. 169–193. Academic Press, San Diego (2001)
23. Srivastav, G.N.P.: Management of teacher education: a handbook. Concept Publishing Company, New Delhi (2000)
24. Whiston, S.C.: Principles and applications of assessment in counseling, 2nd edn. Thompson, Brooks Cole, Belmont, CA (2005)

Author Index

A
Abari, Kálmán, 351
Abelha, Vasco, 273
Amaral, Luís, 117
Ayoola, Anthony, 439

B
Bacsardi, Laszlo, 499
Ben-Chaim, Yochai, 529
Bencsik, Gergely, 499
Benford, Steve, 227
Bezborodov, L.A., 467
Bezgodov, A.A., 467
Bezgodov, Alexey, 371
Bilyatdinova, Anna, 371
Biró, Piroska, 351
Bolgova, E.V., 467
Boukhanovsky, A.V., 467
Braun, Richard, 317

C
Chujyou, Shinobu, 177
Clímaco, João Carlos Namorado, 105
Csenoch, Mária, 351

D
Demertzis, Konstantinos, 289
Didaskalou, Alexandros, 145
Dobreva, Milena, 565
Dukhanov, A.V., 97, 467
Dukhanov, Alexey, 371

E
Esswein, Werner, 317
Etzioni, Yuval, 431

F
Fadda, Gianfranco, 305
Farchi, Eitan, 529
Fenu, Gianni, 305
Forsgren, Robert, 81

G
Gabryel, Marcin, 259
Gervás, Pablo, 211
Glover, Tony, 227
Gonzalez, Jose Miguel Garrido, 145
Greffier, Françoise, 383

H
Hadar, Irit, 529, 537
Hammar, Peter, 81
Hanson, Paul, 421
Harter, Paul, 227
Higuchi, Takeo, 49
Hoffmann, Oliver, 37

I
Ikenoue, Akari, 483
Iliadis, Lazaros, 289
Ivanov, S.V., 97

J
Jändel, Magnus, 81

K
Kantorovitch, Julia, 145
Karampiperis, Pythagoras, 447
Karsakov, Andrey, 371
Kato, Hiroshi, 177
Ketui, Nongnuch, 193
Klapka, Ondrej, 511

S. Kunifuji et al. (eds.), *Knowledge, Information and Creativity Support Systems*,
Advances in Intelligent Systems and Computing 416,
DOI 10.1007/978-3-319-27478-2

Knyazkov, K.V., 97
Kohda, Youji, 521
Köhler, Thomas, 575
Koukourikos, Antonis, 447

L
Lasica, Ilona-Elefteryja, 359
Leimeister, Jan Marco, 335
León, Carlos, 211
Lev-On, Azi, 131
Levy, Meira, 431, 529, 537
Liapis, Aggelos, 145

M
Mabed, Metwaly, 575
Machado, José, 273
Marshall, Joe, 227
Máth, János, 351
Méndez, Gonzalo, 211
Mettouris, Christos, 65
Moffat, David C., 421
Moolwat, Onuma, 161
Motta, Enrico, 145

N
Neves, João, 273
Neves, José, 273
Nishimoto, Kazushi, 483
Niskanen, Ilkka, 145
Nowak, Bartosz A., 243, 259
Nowicki, Robert K., 243, 259

O
Oztaysi, Basar, 547

P
Panagopoulos, George, 447
Papadopoulos, George A., 65, 457
Pechsiri, Chaveevan, 161
Pinto, Agostinho Sousa, 117
Piriyakul, Rapepun, 161
Pitsikalis, Stavros, 359

S
Saga, Ryosuke, 557
Schmitt, Ulrich, 391, 409
Sherman, Sofia, 537
Shlomi, Yaron, 431
Shmelev, V.A., 97
Sielis, George A., 457
Skulimowski, Andrzej M.J., 17
Slaby, Antonin, 511
Spano, Lucio Davide, 305
Steinfeld, Nili, 131
Stilman, Boris, 1
Suerdem, Ahmet, 547
Sugiyama, Ayumu, 521

T
Tajariol, Federico, 383
Takahashi, Shu, 521
Tennent, Paul, 227
Theeramunkong, Thanaruk, 193
Thillainathan, Niroshan, 335
Tzanavari, Aimilia, 457

U
Unkelos-Shpigel, Naomi, 537
Unoki, Masashi, 483

V
Valle, Rogerio, 105
Vicente, Henrique, 273

W
Walker, Brendan, 227
Woźniak, Marcin, 243, 259

Y
Yoshitaka, Atsuo, 177

Z
Zafeiropoulos, Anastasios, 145

Printed in the United States
By Bookmasters